36th Center for Chemical Process Safety International Conference (CCPS 2021)

Topical Conference at the 2021 AIChE Spring Meeting and 17th Global Congress on Process Safety

Online
18-23 April 2021

ISBN: 978-1-7138-5727-3

36th Center for Chemical Process Safety International Conference (CCPS 2021)

Topical Conference at the 2021 AIChE Spring Meeting and 17th Global Congress on Process Safety

Online
18-23 April 2021

AIChE
The Global Home of Chemical Engineers

Printed from e-media with permission by:

Curran Associates, Inc.
57 Morehouse Lane
Red Hook, NY 12571

CURRAN ASSOCIATES INC.
proceedings
.com

Some format issues inherent in the e-media version may also appear in this print version.

Printed with permission by Curran Associates, Inc. (2022)

For permission requests, please contact AIChE
at the address below.

AIChE
120 Wall Street, FL 23
New York, NY 10005-4020

Phone: (800) 242-4363
Fax: (203) 775-5177

www.aiche.org

Additional copies of this publication are available from:

Curran Associates, Inc.
57 Morehouse Lane
Red Hook, NY 12571 USA
Phone: 845-758-0400
Fax: 845-758-2633
Email: curran@proceedings.com
Web: www.proceedings.com

TABLE OF CONTENTS

621490 The latent triggers of disasters are alive - deep inside your company

Abstract:

Never in our history have our workplaces been so safe and our workers so well protected. Yet in recent years, something strange started to happen. While many companies have 'hit a wall' in safety performance, some have seen a slow but visible increase in serious accidents, and most alarmingly, a few 'best-in-class' companies suffered catastrophic events. It seems as if the closer we get to 'zero accidents', the more vulnerable we become for serious events to happen... Something is wrong. Making the systems safer, puts the organization at higher risk? This presentation will demonstrate how the few seconds before a disaster were years in the making – by well-intended, but dangerous ideas, beliefs and delusions that become destructive forces in the culture. It will outline the new roles of leaders and practitioners in a different world awaiting us.

<u>Case studies - Unrevealing the Common "Hidden Hazards" Missed during Risk Assessment - The Known Unknown!</u>

Only if the Chemicals could speak human language! These complex entities are our friends as well as foe based on how we take good care in terms of understanding them at their core.

All the methodologies available in terms of a good Process Hazard Assessment can be as good as the detailing and competence level of the team involved.

The abstract on the subject is culmination of **speaker's real life experience based on 500+ HAZOP** assessments carried out on chemical processes. It talks about the Known hazards in the industry where incidents have occurred and could have been averted by means of systematic hazard evaluation. The topic focuses on the importance of gathering right process safety information and questioning the basics during process hazard analysis.

The case studies that will be covered are related to known hazards which are controlled with highest level of controls but are often missed for other routine as well as non routine activities such as :

- Vent Header mix up
- Reaction hazards related to scrubber operations
- Waste transfer and handling of unstable organic residues
- Sampling of hazardous chemicals such as shock sensitive or pyrophoric
- Handling of waste in drums and at waste management facility
- Chemical Overpressurisation during chemical processing due to Chemical Reaction Hazards
- Impurity profiling during chemical reaction hazard study
- Temperature as an Ineffective control in thermally decomposing powders
- Product changeover mishaps
- Goof ups with BPCS controls

The author details out further on a methodology that has helped in terms of listing down such hazards and has helped in reducing the significant process safety incidents by 70% within 2 years.

Few of the important imperatives followed by the Author's organisations for a good Process Hazard Assessment are:

- Strong Leadership Commitment to provide adequate resources
- Competent and Right structure of Team
- Process Safety Information
- PHA Assessor
- Process Safety driven by Operations team and not the Safety Managers
- Continual assessment even post successful commissioning
- Ensuring risk assessments remain as LIVE status with the status of Plant.
- Encouraging reporting of deviations and Plugging in learnings

616549 Opportunities for Continual Risk Reduction By Applying Inherently Safer Design Strategies throughout the Process Lifecycle

Abstract:

Process safety management addresses the risk treatment of process hazards to tolerable levels using a variety of methodologies and technologies. Application of inherently safer design (ISD) strategies allows for leveraged risk reduction and cost reduction (loss avoidance) for process designs by focusing on treating the hazards directly rather than merely relying upon add-ons to control associated risks.

Why?

The drivers behind an initiative to seek out and capitalize on the opportunities for applying ISD strategies include the same drivers for operational excellence: effectiveness (e.g. risk reduction, operating performance) and efficiency (e.g. reduced cycle time, waste and cost).

What?

Risk treatment includes ISD strategies and traditional risks controls (engineering controls, administrative controls). More specifically, ISD strategies include elimination, minimization, substitution, simplification aimed at inherent hazardous properties and operating conditions.

When?

Each stage of the process lifecycle offers timely opportunities for examining potential applications of ISD strategies. These include design of the product, process concept, detailed engineering, construction and start-up, operation and maintenance, and abandonment. For example, even the safeguards themselves intended for risk control can have aspects of inherent safety "designed in" to them.

Where?

You can assess the benefits of ISD strategies wherever the design is being reviewed, such as design reviews, process hazard analyses (PHAs), audits or incident investigations; for example, during Concept Hazard Reviews or Risk Assessment, Initial/Revalidation PHAs, MOC Hazard Reviews.

How?

Successful application of ISD strategies requires a systematic approach for managing opportunities. The organization must first establish a policy articulating the value of achieving inherent safety, communicate that policy to instill a culture to seek out opportunities, set goals against which progress can be measured, and implement the policy through documented plans. Management system processes with stated requirements for resources and activities should be supported by formalized procedures that organize and allocate sufficient resources, define and assign roles and responsibilities, describe the steps and methods to perform the ISD review, and specify the means of measuring results. A well-established concept for management systems is the continual improvement lifecycle involving the steps: Plan-Do-Check-Act (PDCA). Models for effective process safety management (PSM) systems also utilize the PDCA cycle. The PDCA concept is useful for managing the whole ISD initiative towards achieving its goals, and also the process for integrating ISD strategies to reduce risks associated with the process design (or re-design).

Attendees will benefit from practical takeaways of how ISD strategies can be applied to treat process safety risks at three key lifecycle stages: design, operation and maintenance. These examples will be presented as three design case histories to demonstrate how ISD strategies have been utilized to reduce risks at the product level where chemical properties were changed, at the process level where process chemistry was changed, and at the process equipment level where process risk controls were changed; all towards the goal of benefiting from incorporating ISD strategies.

This paper will draw upon the authors' experiences in designing, developing, implementing and auditing PSM systems, as well as facilitating PHAs/LOPAs, building siting evaluations, audits and incident investigations.

SPRING21
+17TH GCPS
A Joint AIChE and CCPS Meeting

Protection Layer Integrity Management, an Illusion?

Rajender Dahiya, CSP, MIChemE
Professional Process Safety Engineer
Senior Technical Services Manager – Loss Control
AIG PC Global Services, Inc.
A member company of American International Group, Inc. (AIG)
Houston, Texas USA
rajender.dahiya@aig.com

Prepared for Presentation at
American Institute of Chemical Engineers
2021 Spring Meeting and 17th Global Congress on Process Safety
Virtual
April 18 - 22, 2021

Protection Layers Integrity Management, an Illusion?

Rajender Dahiya, CSP, MIChemE
Professional Process Safety Engineer
Senior Technical Services Manager – Loss Control
AIG PC Global Services, Inc.
A member company of American International Group, Inc. (AIG)
Houston, Texas USA
rajender.dahiya@aig.com

Keywords: protection layers integrity, safety critical, incident prevention

Abstract

Illusions about protection layer integrity can leave a facility exposed to loss. It doesn't have to be that way.

Trending scenarios where employees experienced an illusion of process safety were identified as the author conducted risk assessments and interacted with site mangers at complex high-hazard oil and chemical processing plants. In each of these scenarios, robust safety systems with multiple layers of protection were in place, yet because of gaps in operating management systems and process safety knowledge of the workforce, the facility remained exposed to potential loss. Similarly, the author observed that some organizations believe that a process hazard analysis (PHA), once completed, provided a backdrop of safety. This perception is yet another illusion. PHA alone is not intended to guarantee safe operations. A robust integrated management system and a strong process safety culture are essential for safe operations.

These findings raise the question, "How do we know if the systems in place are effective and the process safety culture is sufficient?" This question can be answered in one of two ways: 1) proactively, where gaps or opportunities for improvement are identified before a loss or 2) through a reactive approach in which gaps in the safety systems are realized after a loss as a lesson learned, often a costly one.

In this paper, the author identifies real-life examples of "illusion" and explains how a proactive approach to identifying gaps in management programs, employee competency and safety culture can help to ensure layer of protection integrity. Emphasis is given to the integrity of maintenance programs, operating discipline, training, and risk-modeled exercises. The author also shares several surprises that were revealed during the fact-checking discussions along-side relevant industry best practices.

The paper concludes by explaining how a well-written and implemented management program in general, and the inspection, testing and preventive maintenance program in particular plays a vital role in maintaining the integrity of the protection layers that can help to prevent potentially serious and major process incidents.

1 Introduction

The purpose of the protection layers (PL) is to either prevent or mitigate a potential incident. These protection layers are typically independent of the basic process control systems (BPCS) and act when the BPCS are unable to prevent the escalation. These safety systems can be either automatic or manual or in combination. A combination of management programs, tools, techniques, and human actions are required to maintain the integrity and availability of these critical systems.

Failure to maintain the integrity of safety systems is one of the primary root causes of major incidents.

In this paper, the author shares numerous industry best practices that may be helpful.

For the purpose of this paper, world class operators (companies) are those – small or large – that have implemented very strong management programs, a competent (trained and experienced) workforce and go beyond the minimum requirements of local codes, regulations and manufacturer's recommendations. The practices that world class companies put into place beyond what is minimally required are often then identified as "best industry practices" and shared with peers in that industry.

In this paper, it is presumed that the hazards identification and risk management steps are completed, and adequate safeguards are installed. The scope is limited to the integrity, availability, and operability of these safety critical systems during the operation phase of the facility.

Note: the terms "protection layers", "safety systems", "safeguards", and "safety barriers" are used interchangeably across various industries and also in this paper.

2 Protection Layers – Life Cycle

The Hazard identification and risk assessment (HIRA) process starts early in the project design and engineering stages and continues throughout the lifecycle of the project employing various techniques and tools. These may include hazard identification studies (HAZID), PHA, layer of protection analysis (LOPA), quantitative risk assessment (QRA) etc. Protection layers or safeguards designed to prevent an incident or mitigate the consequences of an incident are identified during this process.

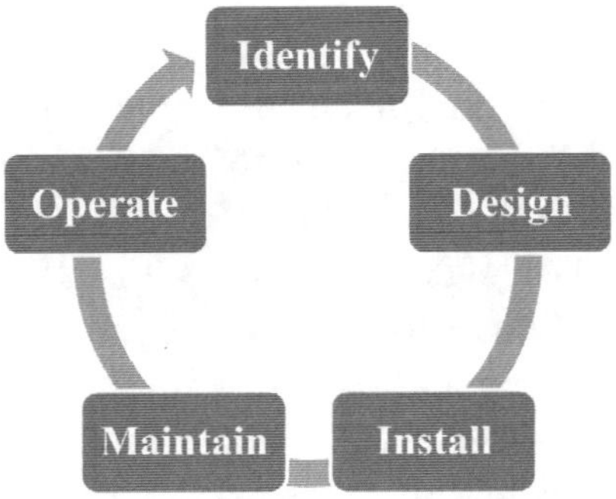

Figure 1: Protection Layers Lifecycle

The HIRA process is mandatory and critical to demonstrate that all potential hazards are identified and risks managed using a hierarchy of controls and the facility is safe to operate. These identified safeguards are then designed, installed, and tested for the proper functioning and the project is handed over to Operations.

Responsibility then becomes that of the plant management to operate and maintain the equipment and systems as described in design specifications, and as per licensor, vendor, codes, and industry best practices. Figure 1: Protection Layers Lifecycle.

The following scenarios demonstrate how knowingly or unknowingly, the integrity of the safeguards can deteriorate with time. This integrity deterioration can lead to a facility operating at a higher and potentially unacceptable level of risk.

2.1 Design Stage – Acceptable Risk

Figure 2. Design Stage - Risk Management illustrates how risks from identified potential hazards are managed through adequate safeguards to an acceptable level as per corporate risk criteria. This is normally the case when the site is built and handed over to Operations. The expectation is to then keep the protection layers intact during the operating stage. For companies who maintain the facility as it was designed, the protection layers can stay in place "as is" and the plant operates within an acceptable level of risk.

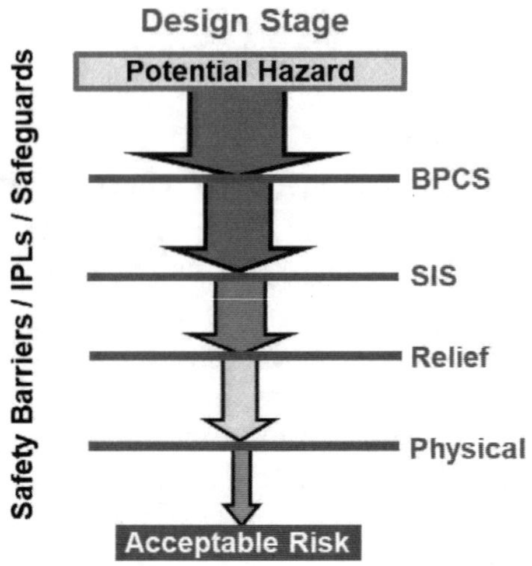

Figure 2: Design Stage – Risk Management

2.2 Operation Stage – An Illusion!

A well-designed and protected facility does not typically stay the same after the design stage. Safety systems can become disabled, degraded, or bypassed. See Figure 3. Over time, gaps in the protection layers develop. In this case, knowingly or unknowingly, the facility is operating at a higher, potentially unacceptable risk. And management may be unaware this is occurring.

It is commonly heard from the management personnel that "My plant is built to codes and we are following the codes." No doubt, meeting code requirements is a must; however, codes are a minimum requirement to operate a facility. Industry Best Practices and Recognized and Generally Accepted Good Engineering Practices (RAGAGEP) can further help in minimizing incidents and losses.

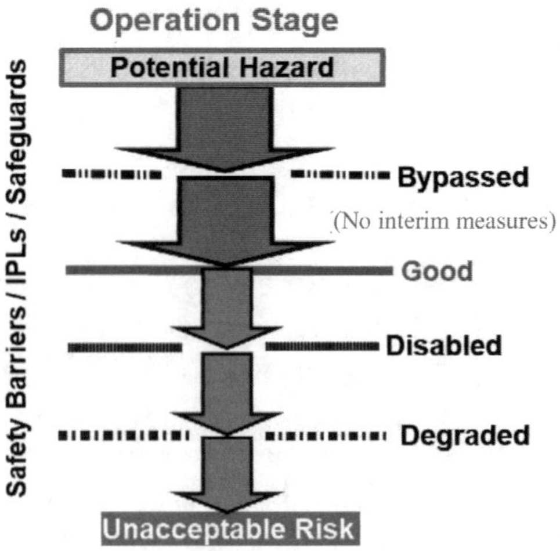

Figure 3: Operation Stage – Risk Management

For the scenario depicted in Figure 3, it can be seen that one "Good Safeguard" can prevent an incident, but only if it is an independent protection layer (IPL). However, to reach the point where 75% of the safeguards were not maintained, many warning signs must have been neglected. There are several proactive actions that can help minimize the potential for this type of scenario: identifying process weaknesses by investigating process safety near miss (PSNM) incidents, auditing, training, management reviews, key performance indicators (KPI), emergency operating drills, and more. Plant management should know the current state of the risk through documentary proof.

The root causes of these weaknesses are discussed in the following sections.

2.3 Reactive Approach – An Incident

Management failure to maintain the integrity of safety systems is one of the primary root causes of major incidents, and management failure to learn from the incidents is the root cause of repeat incidents. Figure 4 - Swiss Cheese Model demonstrates a scenario where all preventive and mitigative safeguards failed, and an incident occurred. There were opportunities to identify these gaps in protection layer integrity through warning signs and weak signals. This is the reactive approach of risk management and it does not stop with one incident. Following an incident, complacency continues on, the incident reoccurs and a vicious cycle begins. This scenario explains the results of weaknesses in the management programs to maintain the integrity of safeguards - one of the leading causes of major incidents and their recurrences due to reluctancy after a reactive approach. The holes in the safeguards represent a failure mode and indicate weaknesses in the management program and / or human performance. Thus, through robust ITPM practices, the size of the holes can be minimized and, sometimes, certain failure modes (holes themselves) can be eliminated through engineering.

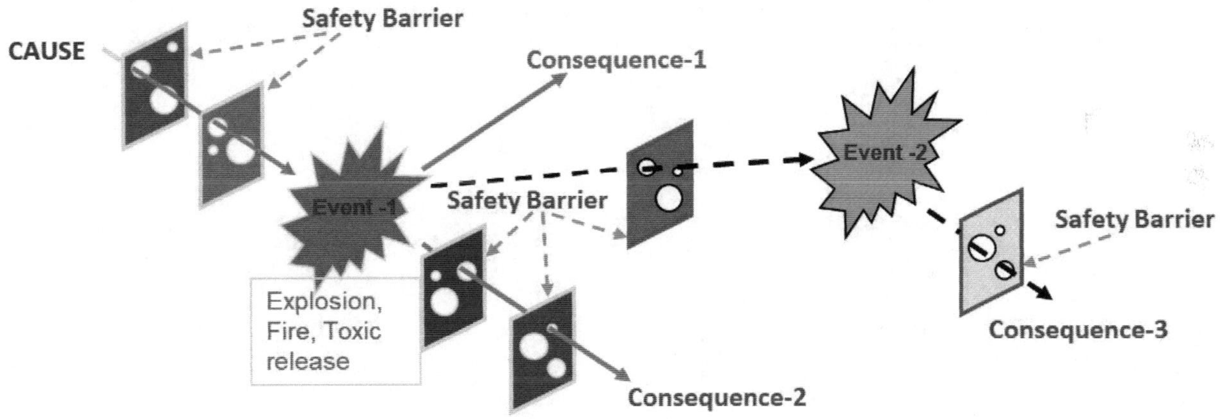

Figure 4: Incident Scenario

The management system weaknesses and their causes are explained in the following sections.

3 Walkthrough Process

This section is based on author's personal experience derived from loss control discussions with site management and employees of several large and complex high hazard industries including oil refineries, chemicals, petrochemicals and gas processing plants, oil terminals, medium and small size chemical batch process and midstream operations. Some of the sites visited were world class leading industries with best practices in loss prevention and risk management. However, those also have improvement opportunities that are also highlighted in this paper. The improvement opportunities identified in this paper are from all kind of industries mentioned above. The best practices shared are learned from the leading industries, insurance brokers and insurance risk engineers, US Chemical Safety Board (CSB), industry organizations such as American Institute of Chemical Engineers (AIChE), Center for Chemical Process Safety (CCPS), Institution of Chemical Engineers (IChemE), American Petroleum Institute (API) and other web resources. The following sections will cover high level questions and deep dive to reveal the gaps in the management programs and human performance required to maintain the integrity and availability of protection layers.

3.1 High Level Screening Questions (Conference Room)

To start, the facility manager and/or discipline managers are asked high-level questions found in Table 1. All "YES" answers were verified in detail. The findings often surprised the site management.

Table 1: High Level Questions

Site Discipline Manager's Answer	Process Safety Management (PSM) Covered Sites			Non-PSM
Questions	**World Class**	**Large and Complex**	**Small-Medium**	**Low Hazards**
Do you know and understand the facility's hazards?	YES 100%	YES 100%	YES 100%	Yes 100%
The answer is a big "YES" by all site managers with full confidence.				
Are those hazards identified and risks managed adequately?	YES 100%	YES 100%	YES 100%	Yes >90%
The answer is a big "YES" by PSM covered facilities and >90% by others. In most cases PHA is the answer. However, poor quality PHAs are not uncommon in which either the risks are not assessed or managed adequately.				
Are there management systems in place to maintain the safeguards?	YES 100%	YES 100%	YES >80%	Yes >50%
The confidence starts to diminish as the questioning moves towards documentation. The common answer is the PM program however, there are gaps.				
Is there documented evidence to demonstrate that all safeguards are maintained and available – fully	YES 100%	YES >90%	YES <60%	Yes <50%

functional and the current level of your facility's risk is tolerable?				
More than 60% of the time, documentary proof could not be made available due to inadequate and inconsistent management programs.				
Does the management system in place adhere to the RAGAGEP as a minimum requirement?	YES 100%	YES >70%	YES <30%	Yes <5%
Mid and small size PSM covered facilities think they adhere but in reality, NO, they do not. Most programs of the large site's adhere but not all.				
Is documented evidence externally verified and benchmarked with peers?	YES >90%	YES <30%	YES <5%	No
Most PSM covered sites employ a third party for audits; very few companies benchmark against their peers.				
Note: The values reported in Table 1 are based on author's estimates based on experience.				

All the YES answers are reviewed further and verified in more detail by reviewing the management programs, records, interviewing personnel, and physical check in the field. Most common findings are discussed in the following sections.

3.2 Process Safety and Process Engineering

Process engineering and process safety personnel normally own the hazard identification and risk assessment process. They are accountable for communicating the outcomes of the hazard studies to the affected personnel and monitoring the integrity, availability, and operability of the safeguards. Process Safety is also responsible for educating the management and workforce about process risks and the management of those risks. The answer to the following question was verified in detail.

Question 3. Are all hazards identified and the risks acceptably managed?
Answer: The response of most of the process safety / engineering manager(s) is YES.

The following gaps were identified during discussions, document reviews, site visit, and interviews.

3.2.1 Most common findings

- Lack of process safety competency throughout – management to staff level.
- Poorly done hazard studies, particularly PHA.
- No formal process to communicate PHA outcomes to affected personnel.
- PHA report is not made available to affected personnel for reference.

- Current process safety information (PSI) such as C&ED, safety interlock list, piping and instrumentation drawings (P&IDs), interlock loop drawings are not readily available and / or not referenced.
- Lack of understanding of the purpose of a safety bypass.
- Production and personnel safety are still a priority due to poor process safety culture.

Illusion: Management assumes that relevant staff are trained in process safety management. However, in more than 50% of the cases, only a 45-60-minute computer-based (CBT) PSM training that covers all 14 OSHA PSM elements is provided to management and employees. Therefore, a process safety training and competency program is implemented however, not effective due to the inadequacy of the training. In a few instances board operators have never seen a PHA report.

3.3 Integrity of Safeguards – Inspection, Testing and Preventive Maintenance (ITPM)

Inspection and maintenance are responsible for maintaining the integrity of all safety systems. The answer to the following question was verified in detail.

Question 1. Are all safety critical systems (safeguards) maintained as per manufacturer's recommendations, local codes and RAGAGEP?

Answer: The response of most of the inspection and maintenance manager(s) is YES. The most common answer is that all safety systems are inspected, tested, and maintained but there are no test records.

The following gaps were identified during discussions, document reviews, site visit, and interviews.

3.3.1 Most Common Findings

a. No list of safety critical equipment and devices.
b. Inadequate maintenance procedures.
c. Inspection and testing methods are not written.
d. No formal maintenance training program.
e. Incomplete or no records of inspection and testing.
f. Required PSI not available – C&ED, P&IDs and interlock loop drawings.
g. Inadequate or indecisive KPIs.
h. No or poorly audited program(s).

3.3.2 Most Common Findings – World Class Companies

a. Lacking current and consistent PSI.
b. Incomplete testing records.
c. Lacking data analysis, learning and continuous improvement.
d. Shallow internal audit. Checkbox approach still prevails.
e. Deficient benchmarking with other sites or peer companies.
f. Instrumented safety systems test failures are not investigated.

Illusion: Management assumes that all safety systems are maintained. Yet in the majority of the cases where formal inspection, testing and a preventive maintenance program were implemented and audited, no or only basic findings were reported. In more than 50% of the cases, there was no list of safeguards and test records were either incomplete or not documented at all.

3.4 Availability and Operability – Operations and Process Control

Operations, process engineering and process control personnel are responsible for ensuring the availability, accessibility, and operability of safety barriers. The answer to the following question was verified in detail.

Question 2. Are all safety critical systems (safeguards) available and operable?
Answer: The response of most of the operations/process manager(s) is YES.

The following gaps were identified during discussions, document reviews, site visit, and interviews.

3.4.1 Most common findings

- Availability – Safety Bypass
  - Inadequate formal safety bypass program – procedure, permit, log, audit, KPIs.
  - In most cases, consequences of safety bypass, and interim measures sections are either missing from the bypass permit or poorly completed.
  - Incomplete bypass permits signed off by the approvers.
  - Missing duration of bypass.
  - Permits active for several month after reinstating the safeguard.

- Operability – Training and Accessibility
  - Users are not trained when to use the ESD and what happens when it is activated. This requirement is not documented in operating procedures.
  - Missing C&ED or shutdown keys.
  - Inconsistency between the critical interlocks in the field and operating procedure.
  - The ESD button is not installed at a safe location or is not accessible.
  - ESD button is not visible, no, or poor labeling.

3.4.2 Most Common Findings – World Class Companies

- Bypassed without appropriate equivalent interim measures. In most cases, critical safeguard (IPL) is bypassed without assessing the risk of the bypass and providing equivalent temporary measures.
- More than 50% of the closed bypass permits reviewed were found incomplete and signed off by multiple level of authorities.
- PHA/LOPA and C&ED are not referred to understand the consequences of bypass.

Illusion: Management assumes that there is a formal safety bypass program implemented and audited periodically with minor findings. However, in more than 70% cases, a bypass permit is used just to get management approval (signature) to bypass a safeguard. The consequences of the bypass and interim measures are either not on the permit or poorly articulated. Interestingly, due to lack of process safety competency and poor process safety culture, management signs these permits without ensuring that the process meets the intent. In this case, bypass simply means safeguard is disabled.

4 Example – Disabled Safeguard (IPL)

The following example describes how a critical safeguard was disabled for 7 years and was not be captured by any of the various programs and activities due to ineffective and inconsistent management programs and poor culture. If all the required activities were performed per corporate policies, then how is it that such a big gap in front of everyone's eyes could not be recognized and fixed in 7 years??? Many new sets of eyes have come to this place in those years.

During the plant control room visit, a visual alarm (RED Window) was observed on the distributed control system (DCS) screen. After discussions with the board operator and site management, the following observation and facts were identified.

- A process with a runaway reaction hazard with an auto trip safety system, which was set at 355^0F, was disabled; the board operator controlled the process which reached 372 ^{0}F, 17^0F above its trip set point.
- The trip system was in disabled mode for more than 7 years, so the unit did not trip automatically. This event was documented in the logbook like other process data / parameters. It was not discussed in any meeting and no concerns were raised. No incident was submitted.
- This was the third incident in one year with that particular board operator. There may be more such incidents on other shifts.
- The board operator and operation supervisors don't understand what went wrong in this case and why a controlled temperature runaway event was considered an incident when there is no injury, no realized loss, and the unit was not shutdown. On the other hand, they think this demonstrates that the operators are competent to manage the emergency operations. In other words, "Job well done!" However, they do not understand the potential risk associated with this incident. This is also a normalization of deviation and the company is very fortunate not to face an accident so far.

Considering the high potential consequences, this incident (PSNM) should be a candidate for a detailed root cause analysis (RCA). The investigation will not only point at the weaknesses in the barriers and human interface but also identify missing barriers, unidentified hazards or underestimated / unmitigated risks.

The team performing the investigation would be expected to identify the following system causes, at a minimum:

- **Management failure to maintain the integrity of critical safeguards.**
  Considering the potential consequences, this risk should be one of the highest risks at the site and be taken seriously.
    a. The trip system was disabled for more than 7 years. There was no documentation for disabling this critical barrier such as safety bypass or management of change.

b. Operation, maintenance, and process safety staff were not aware of this change.
c. No one questions the "trip window" RED on the DCS for 7 years.
d. Annual emergency operating procedures recertification and operator training could have captured it. The operating procedure review process and training was not effective.
e. Maintenance could not capture it during the preventive maintenance and testing of this critical safety system in 7 years. The maintenance program was not effective. This critical barrier was not in the list. In realty, there was no formal list of safety critical equipment (SCE).

- **Management failure to identify the hazards and manage the risks.**
  a. This was not captured during the last PHA revalidations. Since there was no MOC or bypass, credit for this barrier was taken and risk was considered acceptable which was not true. A senior operator must have raised the alarm in PHA. This is an example of a poor quality PHA.
  b. This is one of the highest consequence scenarios and must be evaluated well. This may be a candidate for a more detailed risk study. Any change or bypass must be escalated to the plant manager for approval.
  c. There were no process safety key performance indicators (KPI). KPIs could have raised the alarm to the repeat incidents.
  d. Several audits were performed during this time. No one questioned this disabled trip system in DCS or asked for the SCE list and testing records. The auditing process was ineffective.
  e. Process safety near miss were not defined and included in the investigation procedure. Therefore, they were never reported and investigated.
  f. Ineffective training and competency of site management and employees. In this case, training was more important for management personnel than shop floor staff.

All operation, maintenance and engineering activities mentioned above are performed as planned. There were several opportunities to detect this deficiency. The management programs, procedures and practices were ineffective individually as well as inconsistent and lacked integration.

- PHA, P&ID, C&ED, MOC were inconsistent as well as ineffective individually. Competency of PHA team and process safety culture was deprived.
- DCS, EOP, Training, EOP Recertification, shift handover, operations upset reporting, were not consistent as well as not effective individually and in combination.
- Inspection and maintenance programs were inadequate.

Examples of good process safety culture and practices are:

a. Management should enquire about PSM activities along with production and personnel safety.
b. Process safety is given the priority by management and transfer downward.
c. Process safety near miss incidents are reviewed in daily management meetings.
d. Plant supervision to review the critical alarms in the control room.

The root causes of this failure are lack of process safety competency and poor process safety culture misalignment between site management and shop floor employees.

The example shows a fragmented and ineffective management system; however, an integrated management system could have identified and the problem fixed sooner. A cohesive management system, competent site management and a strong process safety culture is must to identify and eliminate such gaps.

Illusion: What can be a bigger illusion than this as explained in the example? These weaknesses are not uncommon in industries. They exist from high hazard industries to small operations.

5 What is Next

As discussed earlier, management programs and human performance is required to maintain the integrity and availability of safeguards. As a minimum the facility should develop and implement the following processes and programs and document the results.

5.1 Process Safety Engineering

Process safety team is normally responsible for educating the site personnel about the hazards associated with the plant and ensuring that the integrity and availability of safety systems is maintained, and facility is operated at acceptable risk.

a. Provide adequate initial PSM training to management and employees. Then continue need-based and refresher training for continuous improvement.
b. Educate management about the importance of process safety systems in improving the productivity, safety, and integrity of the plant by preventing untimely shutdown and losses.
c. Define what is critical safety equipment, safeguard, or protection layer. Provide the list to maintenance and operations.
d. Ensure management and employees are trained for maintaining and using the safeguards.
e. Communicate the outcome of hazard studies to management and affected employees.
f. Keep PHA report a living document and educate employees how and when to use it. Make it available to them.
g. Make process safety information available to relevant employees which include at least P&IDs, C&E drawing, shut down keys, safety interlock list, loop drawings.
h. The integrity and availability of safeguards should be audited and any opportunity for improvement should be implemented. Provide need-based training and improve the overall process.
i. The status of all safeguards should be tracked using formal key performance indicators, compare against the goal and review by management. Keep continuous improvement cycle on.

5.2 Maintenance Management

If one or more of the following programs do not exist, those should be developed and implemented. If all required programs exist, then review them for the effectiveness and improve them per RAGAGEP and industry best practices. Continuous improvement should be an ongoing process. Equally important is that the process be effective.

It doesn't always matter what you do…what matters is how good you do it. To help ensure effectiveness:

Develop/write:

a. A list of all the safeguards. PHA/LOPA reports are the best sources.
b. Performance standards.
c. Preventive maintenance procedures.
d. Inspection and testing methods/procedure.
e. Inspection and testing frequency per manufacturer or risk-based guidelines.
f. Inspection and testing checklist/form. Use a current cause and effect diagram to develop the checklist.
g. Formal ITPM deferral program. Ensure it meets the industry practices, not just a paper copy for signature and records.
h. Formal training program for maintenance employees and contractors. It should include ongoing and refresher training.

Quality assurance:

i. Process to verify each equipment function test results documented and complete. Final sign off by instrument and control, process control and operations.
j. Process to ensure the inspection and test records are reviewed by a competent person.

Review and improve:

k. Process to report test failures as PSNM.
l. Formal audit program consistent with industry best practices.
m. KPIs, track and analyze. Communicate to all affected and review by management.
n. Action tracking program and review by management. Ensure actions are prioritized with due dates. Implement a formal approval process for changing the due date if needed.
o. Formal learning process. Ensure the process is effective for sharing and learning from others. Incorporate lessons learned into the procedures and training program in addition to the toolbox talk.

5.3 Operations Management

Safety barrier failures to prevent or mitigate an incident do not depend merely on its readiness but also on the correct use and upkeep by operations or other users.

Operations or the other user should ensure that the safeguards are available, accessible and operable. This can be achieved by implementing the following programs.

5.3.1 Availability

Safety barriers must be ever ready to perform their function. However, there are circumstances when a safeguard is required to be bypassed. The purpose of safety bypass process is to ensure that the risk of bypassing a safeguard is managed by equivalent interim measures and is restored back to its original operating condition. The intent of the program is not for a worker, to just get approval signatures and file the form. The following steps should be followed to ensure that the risk is managed, and the safeguard is restored in its working condition.

The majority of the facilities have a written safety bypass program and some use temporary MOC for this purpose. There are improvement opportunities in the procedure as well as its implementation. The program should include at least the following.

a. Written procedure, bypass permit and a bypass log. Some companies have a partly or fully electronic bypass management program.
b. Active bypasses are readily available in the DCS in a paper or electronic log.
c. In addition to other requirements, there should be an approval escalation process based on the criticality (for e.g., bypassing an alarm versus safety interlock) and duration of bypass (e.g., SIS bypass allowed for a maximum of 24 hours).
d. Ensure that consequences of bypass, and interim measures are included in the procedure and permit. Refer PHA/LOPA to know the consequences of bypass.
e. Communicating safety bypass should be a mandatory item during shift handover process.
f. Tracking process of transition from bypass to MOC to closure.
g. Before signing, approving authority should ensure that all sections of the permit are correct and complete especially consequences of bypass, and interim measures. These are the most critical steps of bypass and are poorly done.
h. Ensure duration of bypass comply with the procedure.

5.3.2 Operability and Accessibility

Most of the instrumented safeguards are interlocked and function automatically. The most common manual safety system is an emergency shut down system (ESD) which is normally hardwired and can be activated from the field (in some cases more than one location) as well as DCS. Following improvements are identified:

a. As mentioned earlier, more than 50% operators do not know when to push the ESD button and what happens after the button is pushed. Even if they know, it is not documented. ESD system criteria should be clearly documented in the operating procedures (emergency procedures) and communicated to the operators (end users) including authority to shut down, when to shut down and what happens afterwards. A Cause and Effects diagram is the most relevant document to help a person understand the logic of the ESD. This diagram should be made available in the DCS. Figure 5 depicts the steps for ESD in a typical high risk process unit with multiple IPLs.

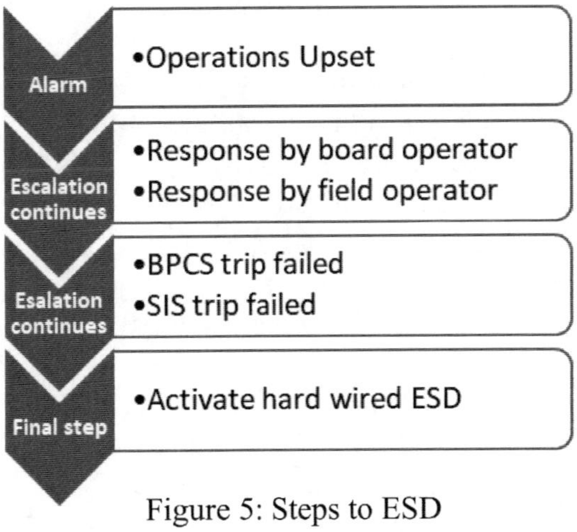

Figure 5: Steps to ESD

b. ESD buttons should be labeled appropriately for identification and visibility. A safety cover should be provided to prevent accidental activation and weather protection if the button is outdoors.
c. The location of ESD button should be at a remote location away from the hazard along the evacuation route. For example, an ESD button could be located on either side of a truck loading rack instead of in the center of the rack next to the loading pump.
d. Operator should sign the testing form for each ESD tested and field verified.

5.4 Process Safety Competency and Culture

5.4.1 Process Safety Competency

Process safety training for management / supervisory level is equally important as the shop floor employees. They are the people who verify and approve all the risk management decisions. Competent management and employees will ensure the quality and completeness of each activity and will prevent the check box mentality. Everyone should be trained as per their roles in PS activities instead of just 45-60 minutes CBT PSM overview. In the earlier example, there were many opportunities to catch the disabled safeguard but lack of PSM competency was one of the root causes. A training matrix should be developed for all affected management personnel and employees.

5.4.2 Process safety culture

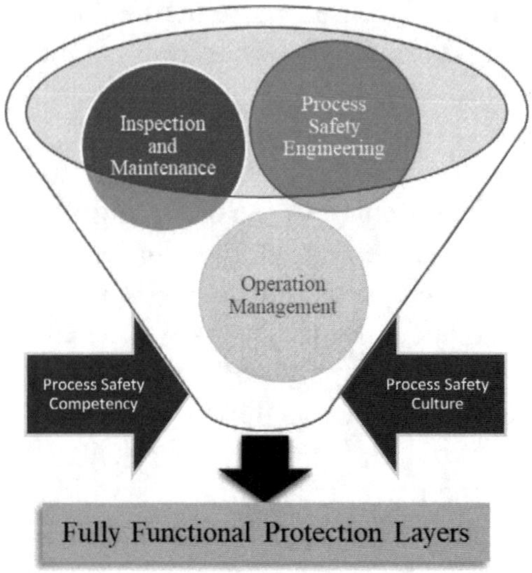

Figure 6: A Safer Workplace

One of the leading causes of overlooking the process risks and their management is poor process safety culture. Production is a must but safety first should be the one of the key slogans of the company, posted widely and discussed in meetings. More than that, safety first should be practiced by the management as a role model to help influence shop floor employees. A strong positive top down process safety culture is must for developing and implementing required programs and continuous improvement. Figure 6 demonstrates that a robust management system, competent workforce and a strong process safety culture are the foundation of a safer workplace. In other words, a competent workforce in a good process safety culture will be able develop and implement a robust management system.

5.5 Industry Best Practices

Following are some of the industry best practices used in world class and leading industries.

Testing and Maintenance:

a. A formal procedure and testing records. An example of Safety System (SIS) Testing and Inspection Procedure is given in Table 2.

Table 2: Safety System (SIS) Testing and Inspection Procedure – Sample

<table>
<tr><th colspan="5">Scope</th></tr>
<tr><td>Test frequency</td><td colspan="4">▪ Sensor: XX Months at the maximum
▪ Logic Solver: XX Months at the maximum
▪ Final Element: XX Months at the maximum</td></tr>
<tr><td>Requirements</td><td colspan="4">Qualifications of persons doing testing and others</td></tr>
<tr><td>Required documents</td><td colspan="4">P&IDs, C&ED, data sheets, loop drawing, testing sheet, other</td></tr>
<tr><td>Tools and equipment</td><td colspan="4">As applicable</td></tr>
<tr><td>Safety requirements</td><td colspan="4">▪ JSA
▪ Gas monitoring
▪ PPE
▪ ……</td></tr>
<tr><td>Environmental impact</td><td colspan="4">Leak, spill or emission</td></tr>
<tr><td>Responsibilities</td><td colspan="4">▪ Process control engineer
▪ Instrument and electrical technician
▪ Operator (DCS and Field)</td></tr>
<tr><td>HIRA</td><td colspan="4">▪ Safety Bypass
▪ Procedure deviation
▪ Risk mitigation plans</td></tr>
<tr><td>Description of safety function</td><td colspan="4">▪ Description
▪ Critical action
▪ Loop drawing</td></tr>
<tr><td rowspan="2">Test procedure</td><td colspan="4">All required step including permit to work, safety and test outcomes recorded.</td></tr>
<tr><td>▪ Step 1
▪ Step 2
▪ ….. and so on</td><td>Y</td><td>N</td><td>NA</td></tr>
<tr><td>Findings</td><td colspan="4">All corrective actions documented and communicated to relevant parties.</td></tr>
<tr><td rowspan="2">Sign-off</td><td>Department</td><td>Name</td><td>Date</td><td>Signature</td></tr>
<tr><td>▪ Process Control
▪ Maintenance
▪ Operations</td><td></td><td></td><td></td></tr>
</table>

b. Testing failures and challenges to safety systems are investigated as a process safety near miss incident. The ITPM frequency is changed in case the SIS or PSV fail the test. (for e.g., 5 years to 2.5 years).

c. A formal ITPM deferral program and deferred equipment listed as overdue. ITPM completed at the earliest opportunity.

Safety Bypass Management

a. Some facility's policy is not to bypass a Safety Trip and ESD system, instead shutdown the unit and fix the problem. This is possible mostly in smaller units with batch processes or individual units not impacting the entire site.
b. Safety bypass approval escalation process based on the risk level and duration of the bypass. The duration of SIS bypass can be allowed up to maximum of one shift to 24 hours. Further evaluation and approval escalation are required to extend after 24 hours or use an MOC.
c. Electronic reminders for a bypass in every shift especially with newer control systems.
d. Live status of bypass. In modern DCS, active bypasses are visible on the DCS or dedicated SIS console. A list of active bypasses available.
e. An example of Safety System (SIS) Bypass Permit is given in Table 3.

Table 3: Safety System (SIS) Bypass Permit – Sample

Safety System (SIS) Bypass Permit/Approval Form			
Requested By		Unit:	MOC#: (if >24 hrs.)
Date and time start		Date and time end	(Max 24 hrs.)
Device being bypassed			
Bypass description			
Reason for bypass			
Are any other initiators in the system currently bypassed? If yes, list.			
Risk of operating in bypass (mode of failure)			
Mitigation measures	Backup Variable if any: ▪ Description and Tag#: ▪ Trip Point:		
Method of bypassing			
Approval	Name	Date / Time	Extension-Date/Time
Board Operator			
Shift Leader			
Operations Manager			
Installed by			
Removed by			
Verified by			
Note: safeguards other than SIS can be bypassed for longer time as per the procedure. Criteria for approval escalation based on time and criticality of bypass should be documented in the procedure, training and add to this permit accordingly.			

Audits

a. Process safety, ITPM and operations programs are benchmarked with peers and externally verified through auditing.

KPIs

a. Process safety, ITPM and operations KPI tracking and management review and lessons learned shared and incorporated accordingly.
b. Process safety competency and culture are given high priority - management walks the talk.

6 Proactive Approaches

The following are the proactive approaches most closely associated for improving the integrity and availability of protection layers.

6.1 Inspection, Testing and Preventive Maintenance: In case, ITPM frequency is not met, a formal deferral must be documented and perform missing ITPM at the earliest opportunity. Inspection and testing frequency can be adjusted based on the learning and best practices.

6.2 Incident Investigation and Reporting: PSNM incidents provide the warning signs about the weaknesses of the safety barriers and management programs. Safety system test failures and safety system challenges are investigated, and lessons learned implemented to prevent the failure on demand. PSNM incidents are investigated for their potential consequences.

6.3 Supervision and Verification: The verification of work completion is very effective for ensuring the effectiveness of programs and practices. Verification will ensure that the actions completed meet the intent.

6.4 Management Reviews: Any changes to safety systems which increased the risk to unacceptable level are escalated to the senior management for their review and approval using an appropriate process such as an MOC or bypass. This includes ITPM deferral and / or safety system bypass.

6.5 Drills and Simulations: Tabletop and Red Tag drills are conducted on all emergency operating procedures (EOP). The critique of the drill is very critical, and the findings must be communicated and incorporated in the relevant procedures and training programs. For example, if an operator action is not performed in the prescribed time and manner as mentioned in PHA then the time assumption should be updated in the next PHA accordingly.

6.6 Process Safety Audits: An effective audit will identify these gaps if done correctly in many ways such as deep dive to verify the list of SCE is current, ITPM records are up to date, C&E diagram is current, available, used and consistent with other PSI, DCS alarms and indications are reviewed and verified, and PHA quality.

6.7 Key Performance Indicators: If correct KPIs are developed, useful data is collected and analyzed, then such weaknesses could be captured sooner depending on the update frequency of KPIs, for example daily or weekly or monthly. These KPIs should be reviewed by the management and accessible to all relevant personnel. In the example above, several repeat operations upsets, repeat PSNM, no ITPM on SCE etc. could have captured in multiple KPIs.

7 Summary and Conclusion

Maintaining protection layer integrity requires 1) a well written and implemented management system 2) competent management and workforce and 3) a strong process safety culture. The first step is documentation of the safeguard. Then only, an inspection and testing and maintenance program can be developed, and employees trained accordingly. Because supervisors, engineers, and senior management make most of the risk management decisions, they need a more comprehensive process safety training than the workforce to implement the management program effectively and efficiently. Process safety near misses are the warning signs for weaknesses in the management program required to maintain the safeguards. This is one of the proactive approaches for identifying the weaknesses of the safeguards. It does not matter how detailed the processes and programs are, if the quality and integrity of these processes and programs are low. Keep in mind the very old saying that is always current: that what is not measured cannot be improved. Good audits can deep dive and verify the consistency in the relevant documents and check the completeness and quality of the activities.

8 References

- U.S. Chemical Safety Board https://www.csb.gov/
- American Petroleum Institute https://www.api.org/
- Occupational Safety and Health Administration – Process Safety Management https://www.osha.gov/process-safety-management
- IChemE Safety Centre https://www.icheme.org/knowledge/safety-centre/
- AIChE-CCPS. Guidelines for Defining Process Safety Competency Requirements. https://www.aiche.org/ccps/resources/publications/books/guidelines-defining-process-safety-competency-requirements

SPRING21
+17TH GCPS
A Joint AIChE and CCPS Meeting

Improving Integration of Reactive Chemistry into an Established Risk Assessment Culture

Karl Rufener
LyondellBasell, Principal Research Engineer
8280 Sheldon Road, Channelview TX 77530
Karl.Rufener@LYB.com

Jessica Albert
LyondellBasell, Manager, Process Safety Americas
1221 McKinney Ste 300, Houston, TX 77010
Jessica.Albert@LYB.com

Prepared for Presentation at
American Institute of Chemical Engineers
2021 Spring Meeting and 17th Global Congress on Process Safety
Virtual
April 18 - 22, 2021

Improving Integration of Reactive Chemistry into an Established Risk Assessment Culture

Karl Rufener
LyondellBasell, Principal Research Engineer
8280 Sheldon Road, Channelview TX 77530
Karl.Rufener@LYB.com

Jessica Albert
LyondellBasell, Manager, Process Safety Americas
1221 McKinney Ste 300, Houston, TX 77010
Jessica.Albert@LYB.com

Keywords: Hazard analysis; reactive chemical hazard; RCH; PHA

Abstract

Traditional Process Hazard Analysis (PHA) and Layers of Protection Analysis (LOPA) toolsets provide systemic approaches for assessing potential hazards associated with a process and identifying appropriate mitigation. However, events involving reactive chemistry may stem from infrequent, complex combinations of initiating events that can be more difficult to identify and address through traditional PHA or LOPA analyses. These reactive chemistry events may result in significant consequences[1] and mitigations may be desired even if the probability of occurrence of any one scenario is very low. This paper describes the current state of LyondellBasell's recently developed reactive chemical hazard (RCH) methodology which is being applied to help identify lower probability but potentially high consequence events contributed to or caused by an unintended or uncontrolled reaction. This work also describes key successes and challenges experienced with development and initial implementation of this program over the last two years, and outlines options for institutionalizing this change into an established management system with existing strong risk assessment programs.

1 Introduction

Hazard analysis techniques, including various process hazard analysis (PHA) and layers of protection analysis (LOPA) methodologies, are widely used within the chemical manufacturing and processing industries. However, infrequent and significant reactive chemistry related consequence events continue to occur. Examples of recent reactive chemistry related events include the organic peroxide decomposition after Hurricane Harvey in Crosby, Texas in 2017[2] and the runaway polymerization of styrene in Vishakhapatnam, India in 2020 which resulted in more than ten fatalities and hundreds of other individuals needing medical attention[3]. A new initiative was chartered at LyondellBasell, subsequently known as the Reactive Chemical Hazards Methodology (RCH Methodology), to enhance the process for evaluating reactive chemistry hazards (RCH) and to strengthen associated systems to support a strong awareness of reactive chemistry hazards.

2 Description of the LyondellBasell RCH methodology

A team consisting of experienced LyondellBasell process safety, engineering, research, and project management professionals was chartered to develop a methodology to improve identification, characterization, and management of high consequence reactive chemical hazards, with targeted application for processes with higher levels and concentration of RCH.

The RCH methodology consists of three major steps. These steps are outlined here and described in more detail below. The first step is to develop a single repository of potential reactive chemistry hazards in the specific technology being reviewed (RCH Inventory). The second step is for an expert review to challenge the RCH Inventory (RCH Challenge) to identify missed potential chemistries or scenarios of interest. The third step is to integrate the RCH Inventory knowledge into the company's existing management systems or technology-wide information systems to inform the PHA process, function as a knowledge retention tool and strengthen and improve RCH awareness. For LyondellBasell, we anticipate the third step will be a one-time effort that will involve modifying our existing Operational Excellence Management System (OEMS). The basic process flow is illustrated in Figure 1.

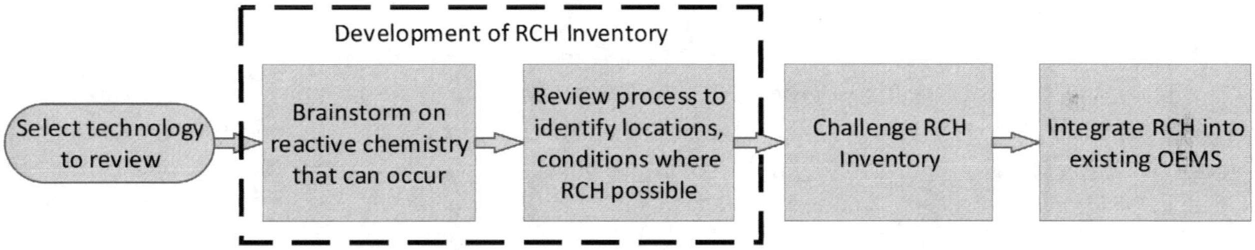

Figure 1. General Process flow for RCH Methodology

3 Development of the RCH Inventory

The RCH Inventory development is a critical step as all subsequent analysis depends on the quality of this work. The RCH Inventory session is a focused brainstorming that analyzes how the intersection of chemicals and process conditions can create a situation where a reaction becomes unwanted, unintended, or uncontrolled and could lead to an unacceptable consequence. The inventory process and tool were designed to trigger identification and discussion of scenarios that are not always identified in PHAs, such as multiple failure or other unlikely initiating events. While a PHA is more process deviation focused, the inventory is more focused on the possible potential chemistries and then identifying where these could occur.

3.1 Team Composition

The team composition is of high importance and includes the following: corporate technology experts, site technology experts (a representative from each site is desired) and at least one individual knowledgeable across broad areas of reactivity, such as someone with an R&D or chemistry background. The RCH inventory session is led by a facilitator, trained in the RCH Inventory process. The goal of the RCH facilitator is to encourage active brainstorming and avoid limiting the team's thoughts to known and well-understood reactions. Any proposed but previously unknown or unconsidered hazardous reaction mechanisms should be noted and considered in the RCH Inventory development until those reaction pathways are determined to be inviable or non-hazardous. If possible, a designated scribe can prove invaluable to ensuring all comments were captured. When putting the team together, the overall size team size is kept to a limited number to enhance brainstorming effectiveness.

3.2 RCH Inventory Preparation

A detailed listing of all potential chemicals should be well-researched and documented prior to conducting the RCH Inventory sessions. Existing chemical interaction matrices can be a good starting point but could potentially be limited in scope to the specific unit at the plant, thereby failing to include chemicals of surrounding processes. In practice, all on-site materials handled in drum quantities and larger were included in the RCH Inventory process. For smaller production facilities or for materials that can initiate particularly vigorous chemical reactions even in trace amounts, the volume threshold may need to be reduced. Grouping of potential contaminants into chemical families by chemical characteristics (acids, bases, nitrates, olefins, etc.) is useful to streamline the analysis and help focus on the most serious potential reactive hazards. There are a number of available resources that can be used to aid in this process as well[4–6].

Another powerful tool is a list of any known reactive chemistry events within the technology. Events are considered that have taken place both at the location being studied or that are known in other implementations of the technology. Incidents that have occurred within similar processes or technologies are also be considered.

Finally, the technology is divided up in advance into smaller nodes to facilitate the brainstorming. The nodes are larger than in a typical PHA and grouped by close-coupled processes. This helps to support brainstorming reactive chemistry hazards that can result from the interconnectivity of units. Boundaries for reactive hazard consideration included both inside and outside process battery limits, storage, utilities, transportation, and laboratory handling and analysis.

3.3 RCH Inventory Session

The first step in the RCH Inventory session is to identify the potential reactions that can occur within the node. During the following brainstorming session, creativity is encouraged to support identification of all possible ways for these chemicals to encounter each other, regardless of the probability or number of things that need to fail for it to occur. Mechanical integrity failures, contamination (e.g. from transport, transfer or cleaning activities), human failure and control failures are all considered. Similar to a PHA, all states of the process are considered, such as normal operating and transitory states related to shutdowns, maintenance, and upsets. The intent is that if it is possible within the realm of physics or chemistry for the chemicals and conditions to be combined such that a reactive chemistry event occurs then the scenario should be recorded as part of the RCH Inventory.

Brainstorming of this nature can be extensive, so it may be necessary to limit the brainstorming to RCH with the most serious potential consequences.

The brainstorming can be treated similarly to a cause-consequence analysis (CCA) by asking a series of probing questions to stimulate identification of RCH scenarios. Some example questions include but are not limited to:

1. What can cause these combinations of chemicals and subsequent reactions to occur in the process?

2. Where can these combinations occur?

3. What actions or events may happen to cause this combination?

4. How can I get chemical [X] into process [ABC] during a specific operation?

5. What internal or external factors will cause this reaction to run away?

Once the initial free-form brainstorming has concluded, the team works through a pre-defined list of initiating events to ensure the process was thorough and complete.

3.4 RCH Inventory Documentation

The RCH Inventory is documented in a series of matrices, organized by major process nodes. The matrices are designed to capture a summary of relevant chemistries in that node, indicate how the chemistry and a primary physical causal factor intersected, and then allow for documentation of the full scenario description and other notes and important factors. These collected matrices are the full RCH Inventory for a process. An example of a portion of a matrix is shown in Figure 2. A database could be developed to catalog and store this information but could be resource intensive to develop.

Figure 2. Example of an RCH Inventory matrix.

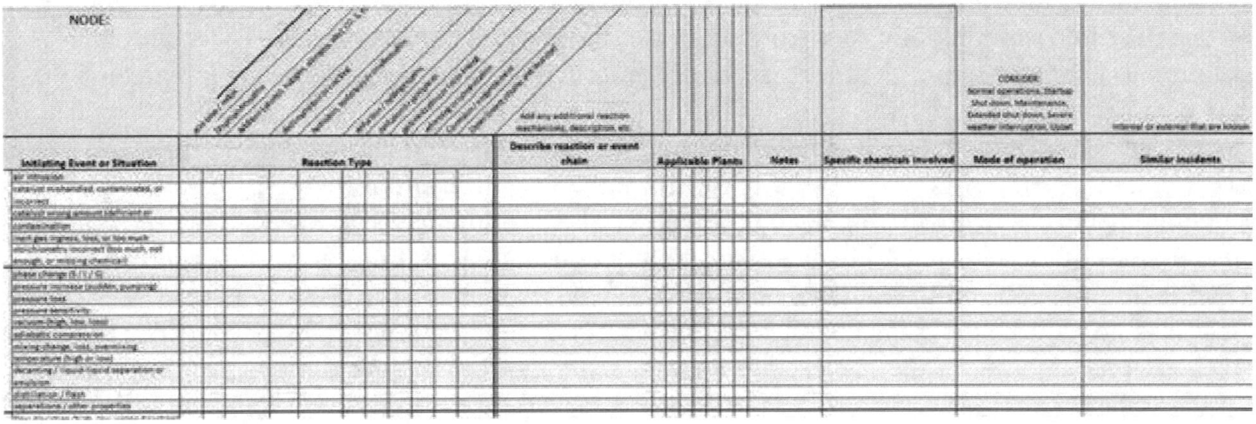

General reactive chemistry classes that were identified during the brainstorming and relevant to the specific process are listed across the top, a general listing of potential initiating events are listed at the left of the columns, and additional columns are shown for capturing the full event chain, additional notes, the specific chemicals involved, mode of operation, and a field for capturing any relevant information for known events that lend evidence to scenario credibility.

4 Challenging the RCH Inventory

The RCH Inventory development, although rigorous, benefits from a review by a person that is well-versed in general reactive chemistry hazards but is not an expert in the technology being analyzed. This ensures a "fresh eyes" review that can ask basic questions around assumptions and commonly held technology knowledge. This challenge to the inventory is intended to identify any potential chemistry or reaction pathways that may have been overlooked or discounted by the primary team and to review and flag reaction causal chain descriptions that need strengthening. The challenge can be performed concurrently with RCH Inventory development or afterwards.

A concurrent challenge to the RCH Inventory development involves embedding the challenger in the brainstorming sessions. A concurrent challenge is more efficient because the challenger can hear the brainstorming discussion, ask experts questions for clarity, and give immediate feedback on unclear scenarios. Potential disadvantages include the possibility that the challenger is influenced by the team, so it is critical that the challenger understands their role and continues to question the team.

Alternatively, a challenge can be conducted after the brainstorming sessions. Advantages include a reduced likelihood that the independent expert will be influenced by the viewpoint of the main RCH Inventory participants. In addition, the challenger can conduct a deeper review of existing documentation and how it is captured in the inventory process. Conversely, this is a less efficient process and may require significant support both from answering the questions of the external reviewer and in updating to the RCH Inventory final document with the results from any post-challenge report.

5 Integration of the RCH Inventory into the Company's Management Systems

Once the RCH inventory has been developed, the final step is to integrate the reactive chemistry knowledge into the company's existing processes and management systems. For example, it is important that the reactive chemistry scenarios in process hazard analyses are complete and the hazards mitigated, but it is also critical that the mechanical integrity department has clear processes and procedures for handling equipment in reactive chemistry service that is sensitive to contaminants. An assessment is used to determine how well reactive chemistry knowledge is integrated at each site practicing a technology (RCH Assessment). The RCH Assessment looks at key areas of interest including chemistry foundation, safe operating limits, management of change, training, RCH related incident history review, maintenance and equipment, laboratory practices, logistics, procurement and contracted services, and emergency response. The outcome of these assessments can then be used to identify potential future opportunities for improvement to existing management systems. Potential ways to integrate and institutionalize this new information include changes to company standards (e.g. Operational Excellence Management Systems) and technology specific information.

For technology specific information, the changes could focus on extracting the content of the RCH inventory and making it accessible to all, such as through updated training, as a new resource for conducting future process hazard analysis, or as a resource for conducting risk analysis as part of management of change.

For management systems, potential changes could include modifications to how reactive chemistry events are tracked and reviewed for incidents with no negative consequence. In addition, there may be opportunities to strengthen operating procedures and training with regards to reactive chemistry knowledge, in particular for processes and procedures completed by engineering/operations support personnel (e.g. maintenance, logistics and procurement, laboratory staff and contracted services) to help ensure that individuals responsible for handling, coordinating, working with or around reactive chemicals, or for the systems that directly interact or contact these materials are aware of specific reaction hazards and their potential impacts.

6 Observations on development and initial application of the RCH Methodology

Over the last two years of development and initial application of the RCH Methodology at LyondellBasell, the program has already had success. The RCH Methodology was applied to two technology applications that are relatively mature, in practice for over fifty years each with extensive reactive chemistry documentation and knowledgeable experts. Even so, new information was gleaned from the RCH Inventory and lessons learned from the site RCH Assessments are being integrated into LYB systems. As a result of these successes, this program will continue to be rolled out to other technologies with reactive chemistry hazards.

The biggest challenge with implementation of the RCH Methodology has been managing culture change associated with the different method of identifying potential hazards. The RCH Inventory session requires a different way of thinking to identify reactive chemical hazards as described here. This is a significant adjustment for staff that has been highly trained and accomplished at more traditional PHA tools. During RCH Inventory development a constant guard against slipping into PHA and LOPA analysis habits was initially required. However, once this new way of thinking about reactive chemical hazards is accepted and acknowledged on its own merit, personal ownership of the process and drive of all participants has been excellent.

7 Conclusion

Over the last two years since initial development and implementation, the RCH Methodology as described here and applied at LyondellBasell shows promise to reduce potential for reactive chemistry events. As a measure of success, the creation of RCH Inventories by using a CCA (cause-consequence analysis) style RCH Methodology has helped identify new information on potential reactive chemistry scenarios in two mature technologies.

Integrating the learnings from the RCH Methodology into a company's management systems has the potential to drive improvements across multiple areas including risk management, process hazard analysis practices, and safety culture.

8 Acknowledgements

The authors would like to thank and acknowledge their fellow current and past RCH team members Roberto Alvarez, Amy Caldwell, Martin de Zeeuw, Charles Lau, David Mannel, Michel Martin, Edward Metselaar, Raj Ramlal, and Rene Zamarripa for their work in developing and implementing the RCH Methodology.

9 References

[1] L. A. Nara, "The Importance of Considering the Highly Improbable (a.k.a. Black Swan Events)," Chemical Engineering Progress, pp. 20, February 2013.

[2] U. S. Chemical Safety and Hazard Investigation Board, "Organic Peroxide Decomposition, Release, and Fire at Arkema Crosby Following Hurricane Harvey Flooding," U.S. Chemical Safety Board; pp. 1-154, 2018.

[3] N. K. Prasad, et al., "The Report of The High-Power Committee on The Styrene Vapour Release Accident at M/s LG Polymers India Pvt. Ltd.," EFS&T Department GoAP, pp. 1-319, July 2020.

[4] M. J. Urben and P. G. Pitt, "Bretherick's Handbook of Reactive Chemical Hazards (8th Edition)," Elsevier, 2017.

[5] T. A. Davletshina and N. P. Cheremisinoff, "Fire and Explosion Hazards Handbook of Industrial Chemicals," William Andrew Publishing/Noyes, 1998.

[6] N. P. Cheremisinoff, "Handbook of Hazardous Chemical Properties," Elsevier; 2000.

Ecolab Process Safety Qualification Process

Nathan Thompson : Ecolab
Global Process Safety and Engineering Training Program Manager
St Paul, Minnesota, USA
Nathan.Thompson2@Ecolab.com

Prepared for Presentation at
American Institute of Chemical Engineers
2021 Spring Meeting and 17th Global Congress on Process Safety
Virtual
April 18 - 22, 2021

Keywords: Training, Learning, Total Productive Maintenance, TPM, Training and People Development Pillar, T&PD Pillar, Technical Network, PS Incidents, Competency, Upskill, Practice, Skill Application

Abstract

This paper focuses on the development of the Ecolab Process Safety Qualification Process (PSQ). The foundations of the PSQ are based in TPM Pillars and methodologies (Total Productive Maintenance), as well as key Process Safety metrics/data collected from all global plant facilities

PSQ Goals:

1. Ensure engineering employees have the knowledge and expertise to complete Process Safety tasks critical to reducing Process Safety events, and driving a consistent program across global plant sites.
2. Provide tools, support, and training for employees working through requirements in the Engineering Career Ladder
3. Drive deployment of the TPM Training and People Development Pillar (T&PD)

Outline of PSQ Process:

- 30% Training, Testing, Evaluation
- 70% Examples of experience and skill application

1.Employee completes competency self-assessment

2.Employee completes training and testing

a. Passes tests in all 12 core elements

4.Employee shows skill application with examples of their work and coaching

a. Submits summary of work to Review Panel to show skill application

6.Review Panel confirms qualification through

After test piloting, the PSQ was adjusted to include all plant employees with an interest in developing Process Safety competency, not just engineers. This adjustment significantly boosted participation, and influenced other training programs to follow a similar model. The PSQ was launched globally in a phased approach.

A second, higher level of PSQ is in development specifically for Principal/Chartered engineers. This evolution puts a stronger focus on complex technical concepts, and adds leadership competencies to the qualification requirements. Launch is planned for early 2021

This project was selected by the 5th CCPS Global Summit on Process Safety, Singapore, 2019. Nathan Thompson was also selected as vice-chair of the CCPS Early Career Professionals project.

1 Introduction

Ecolab, a global petro/chemical manufacturer, had recently seen their Process Safety performance begin to plateau, and not see the improvements of past years. They were in danger of not hitting their targets across their global plant sites for Process Safety Event Reduction.

They had also recently developed an Engineering Career Ladder within the Human Resources department. The in-house competency development approach was used for building long term career opportunities for employees.

With help from the Global Process Safety Team, and Global Engineering, the career ladder created consistent job titles and roles for all engineering roles with criteria that employees must achieve to move through the ladder. This created a path for engineers in the company to follow when developing their own careers.

With these two key data points in mind, the Process Safety Training Program, and senior leadership, recommended developing an internal Process Safety Qualification Process to drive performance improvement. There were three goals for developing/deploying this process:

1. Ensure engineering employees at all global plant sites have the knowledge and expertise to complete key Process Safety tasks at the plant level. These tasks are crucial to Process Safety Event Reduction, and driving a strong consistent program across all global plant sites.

2. Provide tools, support, and resources for engineering employees to develop their careers, and complete the requirements listed in the Engineering Career Ladder

3. Drive deployment of the Total Productive Maintenance (TPM) Training and People Development Pillar (T&PD)

2 PSQ Development Process

2.1 Competency Benchmark Inventory

The Project team members included Process Safety team employees (Subject Matter Experts - SMEs) from all global regions. Project was led by the Process Safety Training Program Manager

There were many data sources for the project team to begin developing a framework for the PSQ:

- *Process Safety Event Data reported by Ecolab global plant sites*
  - SMEs analyzed recent Process Safety Event data reported by global plant sites. This data was used to develop a core set of competencies to address priority risks. Additional competencies were also developed for future evolution of the program.

- *Risk Based Process Safety (RBPS)/ Process Safety Management (PSM) Criteria*
  - Industry standards such as CCPS, AiChe, OSHA PSM, and academic degree programs gave the framework for common core elements of Process Safety that are used to drive technical competency.
- *TPM T&PD Pillar Tools*
  - TPM criteria offered guidance on developing a Competency Profile for technical topics, as well as information about the desired audience/job role for the training, and required competency levels for each

Project team determined the following:

Table 1. RBPS Core Elements

Chemical Compatibility/ Red List	RBPS/PSM Program Knowledge SHE Management System	Managing/Using Process Safety Information	Regulatory Compliance/ RAGAGEP
HIRA/ PHA Tools	Management of Change	Compliance with Global Standards	Plant Design
Measurement and Metrics	Root Cause Analysis	Critical Equipment	Process Safety Culture

Table 2. Engineering Career Ladder Roles

ENGINEER I	CHARTERED ENGINEER I	ENGINEERING LEADER I
ENGINEER II	CHARTERED ENGINEER II	ENGINEERING LEADER II

Table 3. TPM Competency Levels

0. NK - No Knowledge	1. BA - basic awareness	2. WK - working knowledge	3. FO - fully operational	4. LE - leading edge
Has not heard of element or skill, and has no experience	Able to identfy and discuss primary elements of the skill with others	Able to apply the skill having a direct involvement in decisions and actions	Able to perform a majority of the activity of the skill area, including in non-standard situations	Significantly develop the skill and coach others (applies to minority of positions and individuals)

2.2 TPM T&PD Pillar: Mapping to Ecolab Needs

TPM (Total Productive Maintenance) is actively deployed at Ecolab facilities. TPM strives to improve safety, reduce costs, increase quality, and maximize reliability.

TPM is designed to systematically eliminate Safety, Production, and Monetary losses by collecting and using data. This data should identify losses, determine root causes, implement measures to eliminate them, and set a new standard to sustain the gains. TPM deployment is only successful with the

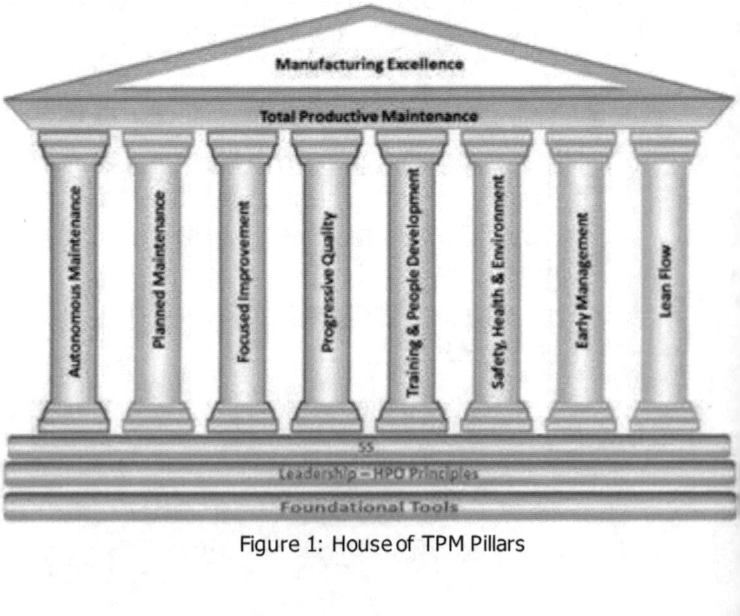

Figure 1: House of TPM Pillars

engagement and development of the people. Motivating and empowering them to reduce losses will lead to performance improvement of the site, and move toward world class operations. Leaders and teams on the production areas each play a key role in the cultural transformation of TPM methodologies and tools. Pillar structure helps to achieve this by owning a specific set of losses, develop systems to prevent loss, and support training and development of the organization. [1]

The Training and People Development Pillar (T&PD) is unique because its tools and methods are embedded across other TPM pillars to address training, competency, and definition of job roles on the team. Losses related to these tools and methods are identified both in the T&PD Pillar, as well as the other pillars at the facility.

2.3 Competency Development

Based on these benchmarks findings and Ecolab Safety data, individual competencies were developed for each of the 12 Core Elements, using the TPM Competency Levels, as well as Bloom's Taxonomy of Learning Objectives.[2] Examples of individual competencies are given in Table 4.

Table 4. Example Process Safety Competencies

Define common PHA terminology	Recognizes hazards and risks of site-specific chemistries	Identify on-site equipment in the critical category (Critical Equipment List)
Recognises different PHA methodologies	Initiates change management process	Supports open and effective communication about PS issues at the site

Competencies were developed into a profile for all plant job roles. These competencies set the level of learning required to address Process Safety issues at sites, and the foundational level of knowledge required for passing the PSQ. Approximately 150 competencies were developed for each of the core elements, across the TPM competency levels. Increasingly higher levels of Bloom Taxonomy action verbs were used to ensure progression of learning and competency.

The TPM Competency levels were mapped back to criteria for roles in the Engineering Career Ladder, and aligned to Risk Based Process Safety program elements. An example of this mapping is given in Table 5.

Table 5. Example Job Role Mapping to Competency Level

Engineering Career Ladder	HIRA/ PHA Tools	Plant Design
ENGINEER I	BA	BA
ENGINEER II	BA	WK
CHARTERED ENGINEER	WK	FO
CHARTERED ENGINEER II	FO	FO
ENGINEERING LEADER I	FO	FO
ENGINEERING LEADER II	FO	LE

To complete the PSQ process, candidates must meet or exceed the competency level of Engineer II in the Engineering Career Ladder

- This could include people in non-engineering roles (SHE Managers, Maintenance Team, Shift Supervisors), but who complete engineering work at the site.

2.4 Training and Evaluation Process Development

Once the competency profile was finalized, the PSQ was outlined in specific steps for completion. The steps included:

- 30% based on training/testing/evaluation
- 70% based on examples of Skill Application in their job
  - Daily work, project work, previous job roles

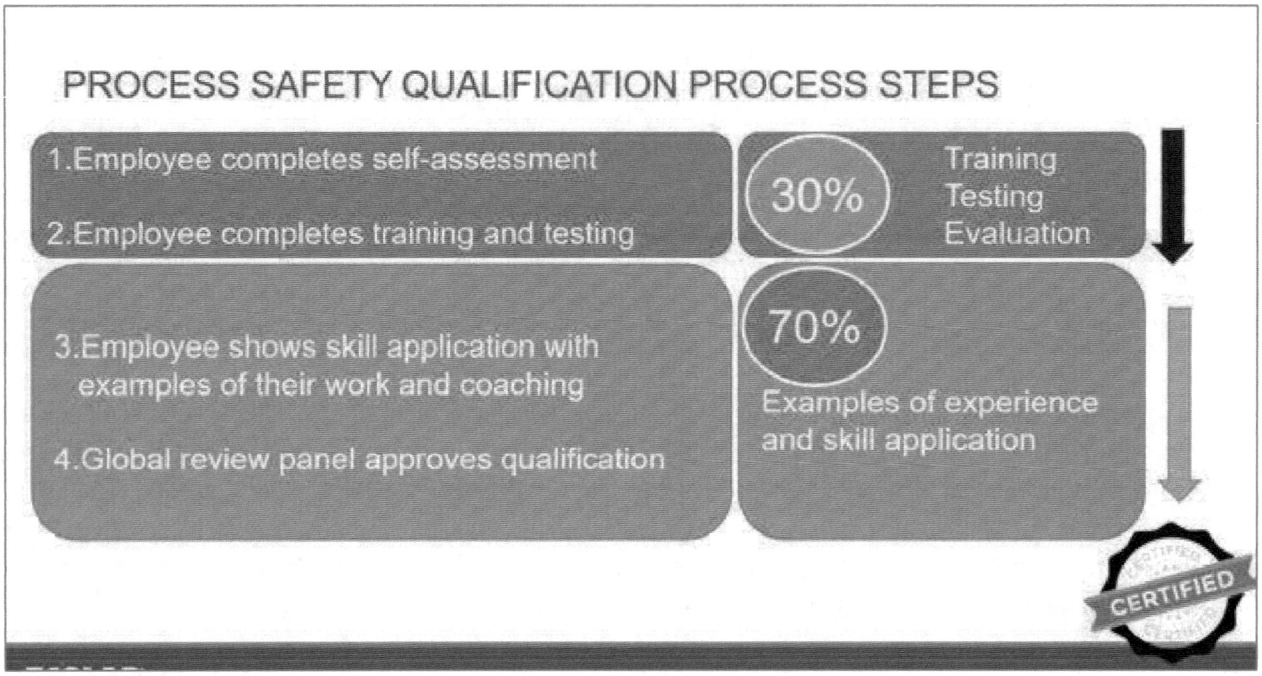

Figure 2: PSQ Completion Steps

A competency self-assessment tool was created for learners to identify where their biggest Process Safety competency gaps may be. Results from this assessment helps guide them towards projects and tasks where they can apply skills and gain more experience.

A series of online tests for each of the core competencies, as well as online/digital training materials were developed to support learners during the PSQ. This includes a series of online training sessions, as well as in-person, on the job training sessions.

A majority of the learning happens in the workplace, by completing different types of Process Safety activities and tasks. All 12 of the core Process Safety Elements must be applied in the workplace in order to increase competency.

A *Skill Application Summary* document is a brief, high level overview of work related to Process Safety core elements and is used to evaluate learner competency in the final step of PSQ. These are a 3-5 page document developed by the learner, and is similar to an executive summary. This is submitted to a panel of Process Safety SMEs for evaluation and approval. The document is evaluated against the competency profile for the Engineer II job role. One project can show competency in multiple elements, but it must also include daily work the learner has completed in their role. An example of the PSQ process completion is described below in Figure 3

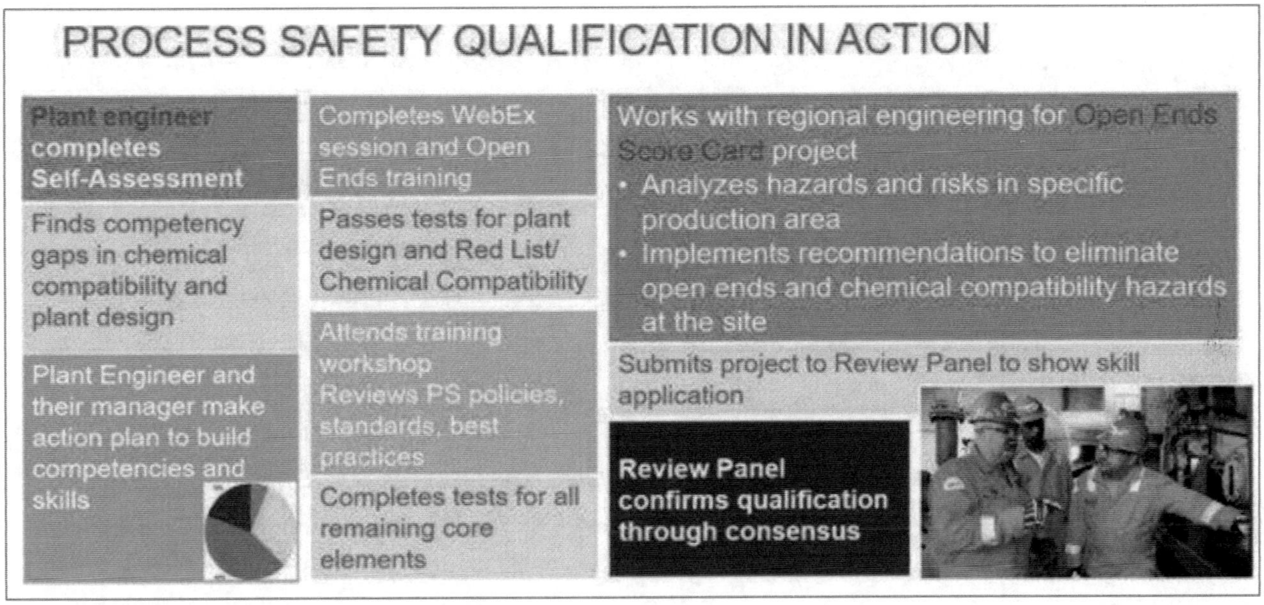

Figure 3: Typical path of PSQ Completion for a learner

3 PSQ Pilot and Deployment

3.1 Pilot Process and Feedback

A sub-group of global plant sites was selected for piloting the PSQ based on role in company, and the size of the plant. A deliberate blend of small, medium, and large plant sites was needed for an accurate pilot. Plant sites in all global regions had representation. Communications, assessments, test, training, and other PSQ materials were all created in English

There was already a strong training program in place for global operations teams, so training was updated to meet the needs of an Engineering/Technical audience. These training materials set the foundation of the Qualification Process

Upon launch, several training workshops were conducted in two regions (North America and Europe). Online learning, communication, and support were included in the training/piloting process.

A total of 24 learners participated in the pilot, with 9 of them completing the full final steps. Feedback was collected throughout the pilot to improve the overall PSQ process Based on feedback, several key updates were made before the full global deployment:

- Improve wording in the online tests to accommodate non-native English speaking learners
- Calibration of SME review panel for Skill Application Summary submissions
- Move to full online-digital learning environment to accommodate Global COVID-19 Pandemic travel restrictions

3.2 PSQ Deployment and Digital Cohorts

Full global deployment of the PSQ was planned for 2020, and included conducting a series of 3-day training workshops and coaching/mentoring session. These included regular training sessions, paired with on-the-job training sessions within the production areas of the plant. This method aligned with TPM T&PD methodologies.

The first goal for deployment was to have at least one person from half of all global plant sites to complete the PSQ process by the end of 2020. This would be about 65 people, and was considered a stretch goal.

3.3 Case Study: COVID- 19 Restrictions Evolve into PSQ+CPM

The Global COVID-19 pandemic forced a major change in the deployment strategy of PSQ. A quick change was made to Digital Cohorts style training sessions (Figure 4), and combined a series of other training topics related to the Engineering Career Ladder. Digital Cohorts is a training methodology designed to systemically move a group of people through a training curriculum at the same pace, and encourage interactions between learners during the training process.

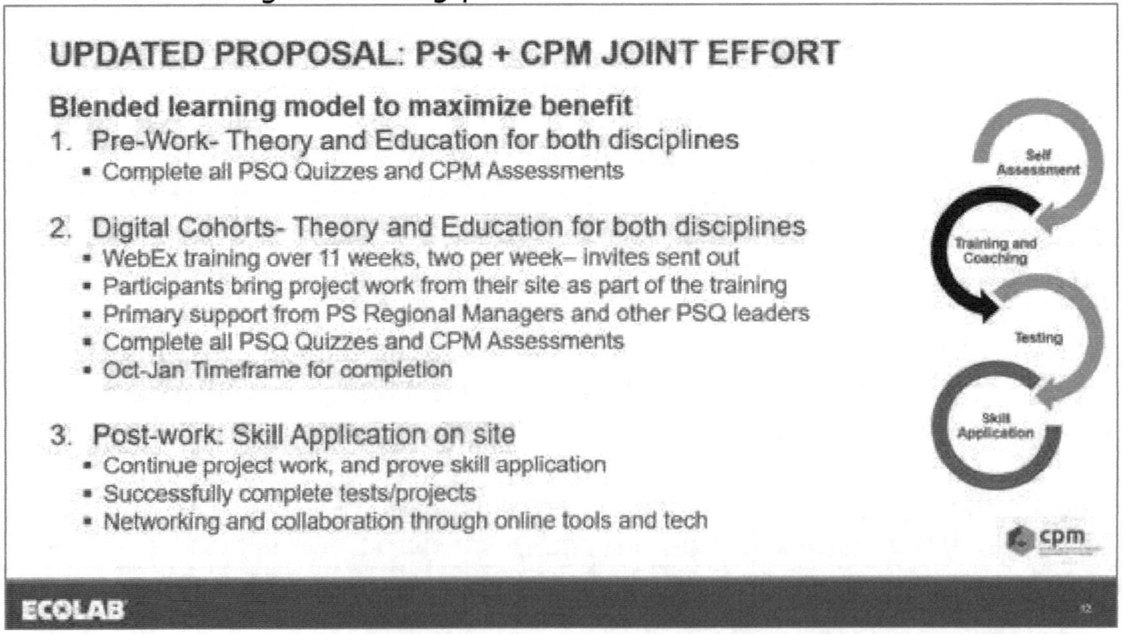

Figure 4: Example of Digital Cohorts Training Process

The adjustment to the digital cohorts allowed for momentum to continue, while meeting the needs of global travel restrictions. The Capital Project Management (CPM) training process was also added to the proposal. CPM aligns to the TPM Early Equipment Management (EEM) Pillar, applies to nearly the same audience as the PSQ.

The original 3 day Process Safety workshop was combined with the Capital Project Management training and was re-designed into a digital classroom environment. This included using videos, eLearning modules, and Microsoft Teams meeting software to drive learner engagement.
A 12-week training cohort was deployed for approximately 75 learners across all regions. The Digital Cohort consisted of two, one hour training sessions each week that learners were required to attend. Attendance was required for all training sessions. Attendance, test scores, completion averages, and overall program metrics were closely tracked using reporting functions within a Learning Management System (LMS). PowerBI software was used to analyze learning metrics, and visually display program progress fore leaders:

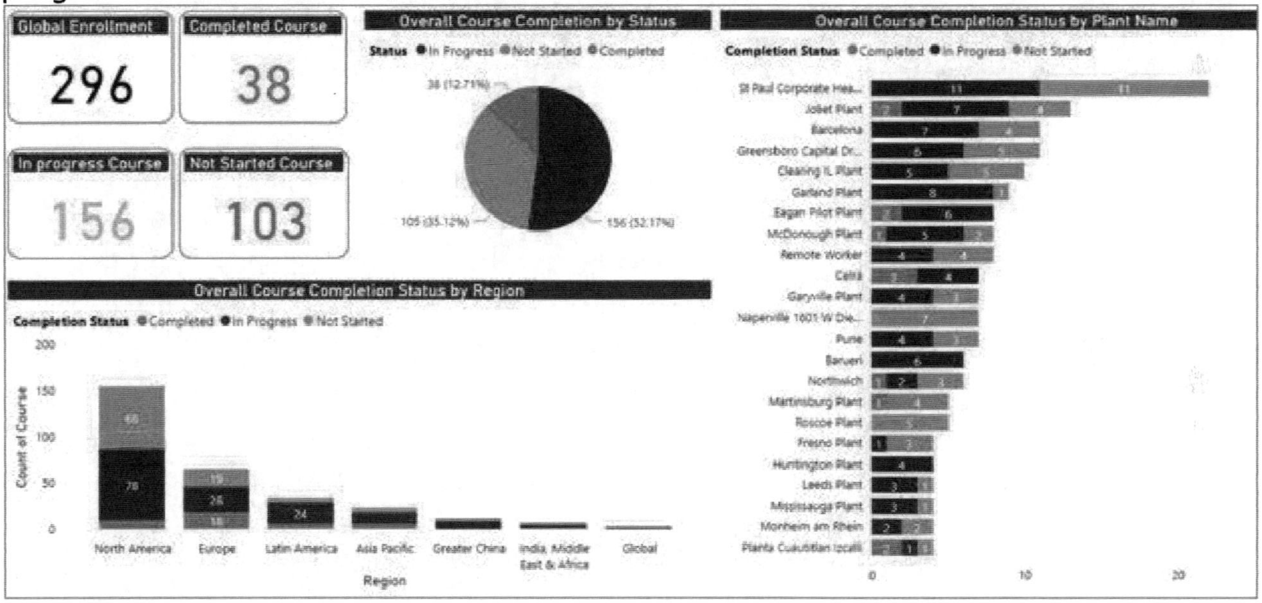

Figure 5: View of learning metrics data using the PowerBI visualizations dashboard

4 Results and Outcomes

4.1 Findings from the Digital Cohort

Of the 75 learners in the Digital Cohort, 41 were able to complete the PSQ within the first 12-week training series, including all training, tests, and the Skill Application Summary step. The remaining learners have been given more time to meet the Skill Application Summary requirements, and gain more experience in their own job role. A majority of the learners also completed the CPM steps in addition to PSQ.

A specific set of challenges was encountered with the change to digital cohorts. First, many learners were initially not comfortable with the Microsoft Teams platform, however, they were able to adjust over time. The instructors and SMEs leading the training sessions also had to adjust to this new platform.

The digital platform is only able to take learners to the BA-WK (Basic Awareness-Working Knowledge) competency level. The pilot workshops were an opportunity for hands-on/experiential learning that was no longer available. The project team realized additional coaching and mentoring would be necessary to push learners into the FO (Fully Operational) competency level that is desired for some of the core elements.

The adjustments made in the pilot seemed to have helped drive success. Test completion rates improved, as well as calibration activities for the global panel of SMEs Recognition included an email signature

Table 6. Learning Metrics for First Year of PSQ Deployment

Total Global Enrollment	296 as of February 2021
Digital Training Sessions	35 one hour sessions
Quizzes Completed in LMS	4.186
Total PSQ Training Hours	821 across lessons
Full Completion of PSQ	41 Learners

4.2 Continuing Momentum after initial deployment

Although face-to-face workshops are desired, digital cohorts will continue to be used for the time being. With this unique challenge in mind, additional guidance for coaching and mentoring learners through the PSQ has been added for managers in the plant sites (Figure 6). This supports the learners through the process and opens up communication between plant level teams, and regional/global teams.

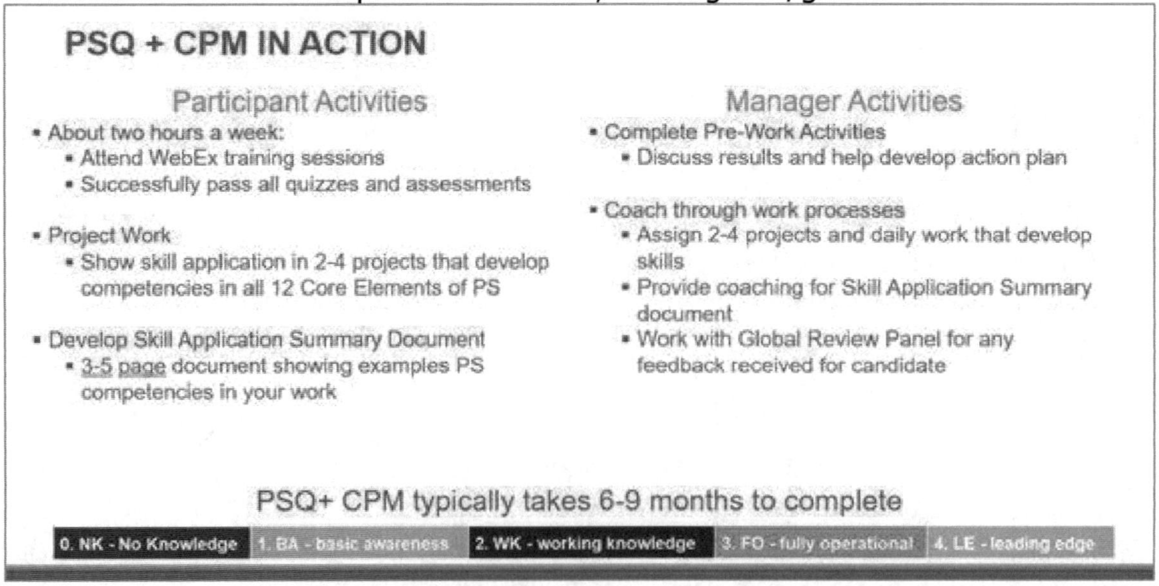

Figure 6: Outline of expectations of the learner and their manager for completing the PSQ

4.3 Scaling the PSQ Process into other disciplines in Ecolab

Other TPM Pillar teams have started using the competency based approach for training and upskilling people in the pillar teams. This is naturally built into the work that the pillars do for teams at the plant sites, however, some TPM Pillar Team members need onboarding or refresher training. The success of the PSQ development and deployment has been recognized as a best practice within the company, and a model for others to use as a template for success when building TPM Pillar teams.

The PSQ Process has also been applied to other engineering disciplines as a foundation for training and career development (Figure 7). Previously, there wasn't a strong framework in place for early career engineers to advance or learn new skills. The competency and skill application allows for continuous improvement for learners.

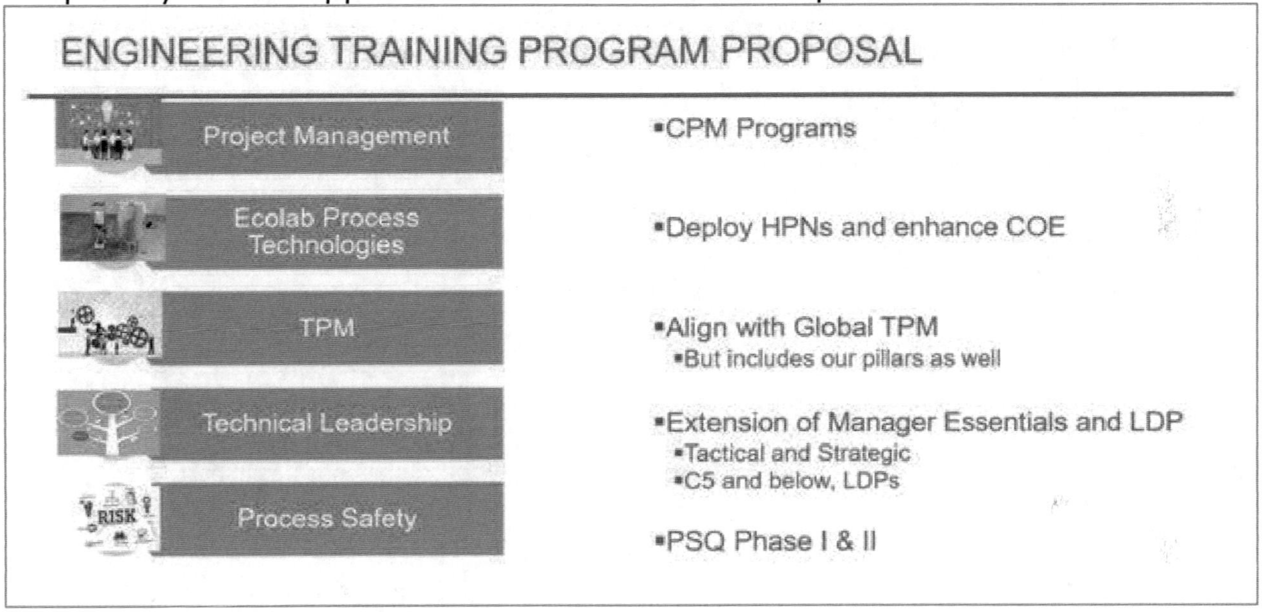

Figure 7: Proposed Framework for Global Engineering Training Program

5 PSQ Evolution: PSQ II

5.1 Continuing to upskill Engineers with Process Safety

The PSQ process was designed to continuously improve learner's skills and abilities through training and experience. The natural evolution of the PSQ is to continue supporting this process with additional tools and criteria mapped to a higher competency level (PSQII) and more advanced job roles (Chartered Engineer Level).

The PSQII continues the momentum of PSQ by creating development opportunities for more senior engineers (Figure 8). PSQII also supports the creation of networks of qualified engineers that leadership can pull from for knowledge and specific projects. Candidates for PSQII must have completed the first level of PSQ as a pre-requisite.

The cornerstone of development at the PSQII competency level (Fully Operational FO) is a priority for the engineering areas:

- PHA/HAZOP Leader
- Critical Equipment and TPM Preventative Maintenance Pillar Leader
- Plant Design Expert
- Process Chemistries Specialist
  - Learners chose one priority area as a pathway to full completion of PSQII
- Addition of Ecolab leadership competencies and membership with an externally facing professional organization also drives competency and skill application in new and challenging ways.

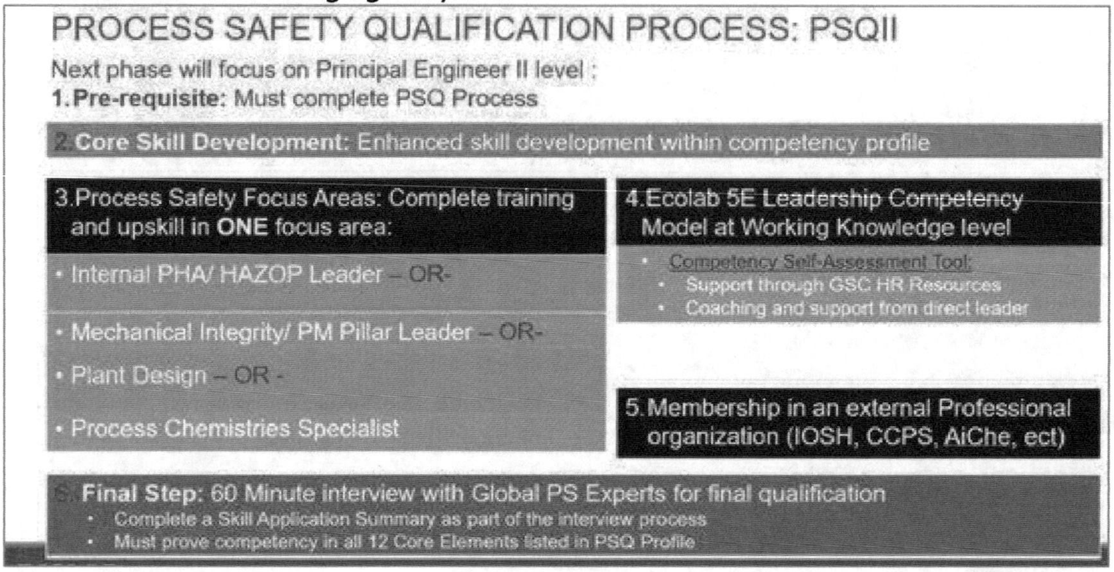

Figure 8: Steps in the PSQII Process

6 Conclusion

The development of the PSQ process has been challenging in many different ways, including alignment of job roles, correct competency level, and TPM Pillar methodologies. With those challenges addressed, the PSQ has delivered direct results, with the best Process Safety performance in the company's history for 2020! [3]

As momentum gains, PSQ continues to grow and become the benchmark for other disciplines within the organization. The Engineering Career Ladder has become a starting point for many in their career, especially with support from PSQ experiences.

Competency based training and evaluation tools continue to show results, and help Ecolab grow and develop their talent internally. This supports long term gains of the company, and the internal employees. This creates multiple career options for employees and drives a strong Process Safety Culture.

PROCESS SAFETY QUALIFIED

Figure 9: Email Signature logo for learners who have completed the PSQ Process

References

1. Ecolab Internal Standard, TPM T&PD Pillar Play book.
2. R. Heer, A Model of Learning Objectives–based on A Taxonomy for Learning, Teaching, and Assessing: A Revision of Bloom's Taxonomy of Educational Objectives; Available at https://www.celt.iastate.edu/teaching/effective-teaching-practices/revised-blooms-taxonomy/; Accessed: January 10, 2021.
3. Ecolab Internal Standard, PS-ST-05, Process Safety Metrics Standard, Global Supply Chain-Process Safety.

SPRING21
+17TH GCPS
A Joint AIChE and CCPS Meeting

Effective Pre-Startup Safety Review

Tips and Tools for Ensuring Operational Readiness

Jody E. Olsen P.E.
JE Olsen Consulting LLC
Anchorage, AK
jodyo@jeolsenconsulting.com

Prepared for Presentation at
American Institute of Chemical Engineers
2020 Spring Meeting and 17th Global Congress on Process Safety
Virtual
April 18 - 22, 2021

Effective Pre-Startup Safety Review

Tips and Tools for Ensuring Operational Readiness

Jody E. Olsen P.E.
JE Olsen Consulting LLC
Anchorage, AK
jodyo@jeolsenconsulting.com

Keywords: pre-startup safety review, PSSR, operational readiness, management of change, MoC, OSHA, process safety management, PSM, projects, safe startup

Abstract

The keys to effectiveness of pre-startup safety review (PSSR) and operational readiness involve establishing well-defined workable practices, providing effective tools, dedicating resources to complete the work, assuring management support, ensuring follow-through, and continuously improving the process through user and client feedback. Many of these concepts are described in general terms in various guidelines and reference materials. However, effectiveness of the implementation cannot be achieved through generalities. The devil is in the detail and the quality of the implementation.

This paper describes a number of specific practices and tools that have been proven effective in the field to drive better PSSR and project turnover. These practices may be scaled to apply to projects and modifications of any size as well as to procedure changes and turnaround restart applications. A number of specific experiences are shared that demonstrate the importance of selecting the correct tools and resources to support the work. The proof of success came by having more satisfied clients and users. Quality control (QC) personnel began applying the tools developed more widely than originally intended. Construction personnel were pleased to have clear deliverables and to be assessed by clearly defined standards. Operators who received the project turnover were engaged in the process and were satisfied with the improvements implemented.

Implementing successful PSSR is a combination of having strong work processes, tools, resources, commitment, and follow-through. Failure of any of these components may cripple the overall PSSR process and potentially compromise the success and safety of an otherwise well-implemented project.

1 Introduction

It is recognized within the process safety community that non-steady state operations, such as startup or shutdown of equipment and processes result in a large percentage of industrial incidents. As per a 2010 study by Ostrowski and Keim, 50+% of incidents occur during transient operations that account for less than 10% of operating time[1]. For these reasons, pre-startup safety review (PSSR) as required by the OSHA process safety management (PSM) regulation[2], or operational readiness, as described by CCPS Guidelines for Risk Based Process Safety[3], are essential elements of a process safety management program and are essential for preventing process safety-related incidents and losses.

So we know that PSSR must be completed before we operate new, modified, or reinstated equipment. But how do we make PSSR effective? How do we do it well?

The keys to effectiveness of PSSR and operational readiness involve establishing well-defined workable practices, providing effective tools, dedicating resources to complete the work, assuring management support, ensuring follow-through, and continuously improving the process through user and client feedback. Many references are available that describe the need for these PSSR and operational readiness activities. And many of these references describe the general process and may provide some tools and guidance. But experience in the field tells us what works and what does not.

This paper describes a number of specific practices and tools that have worked well. And this paper includes some cautions regarding practices, or pitfalls, to avoid. One size does not fit all and practices and tools must be fit-for-purpose and continuously improved. It is important to realize that there are no magic bullets. Value is achieved through the quality and completeness of the implementation. Thorough application requires rigor, commitment, and attention to detail. However, a successful practice is self-reinforcing. Successful practices are continued, expanded, and improved not primarily through the auditing of an outside investigator, but based on the compounding effects of successful results and pride from within the organization. The ideas and anecdotes that follow may offer some insights into some better practices that if applied may improve existing PSSR implementation.

2 Work flow process: who, what, when, where, how, plus why?

Having a well-defined, well-understood, and practicable work flow process is essential to implementing effective PSSR. The work flow process defines who, what, when, where, and how PSSR or operational readiness will be implemented. For the why, consider including a description of PSSR in company process safety overview training courses and exercises.

2.1 Why?

The Why? is best explained by highlighting industry experience and sharing major incidents that stemmed from inadequate review in advance of startup or re-start activities. These events are the drivers for the regulatory requirements. But the Why? can be reinforced by emphasizing the value to the operations customer of receiving a satisfactory handover of new, modified, or simply repaired equipment. PSSR is required by process safety regulation to assure that new or modified equipment may be operated safely. PSSR also functions as the "turnover" or "custody transfer" process from project owners and implementers to Operations. Operations is the ultimate customer and client and, therefore, must be satisfied with the project and the PSSR process.

Figure 1. Typical project management work flow showing PSSR point in the process

The Management of Change (MoC) and PSSR processes are closely intertwined. PSSR occurs at this later stage of the MoC and project process as shown in Figure 1. Some form of PSSR, or operational readiness review, should also occur on equipment returned to service following maintenance or replacement-in-kind (RIK) repair work, as well.

2.2 What?

The "What?" may be described as an essential hold point for inspection and validation before custody transfer occurs (Figure 1). The "What?" may also be explained by describing the core elements of PSSR. Custody transfer assurance and confidence in safe operation is accomplished by:

1. Inspecting installed equipment to assure that it matches the approved design:
    a. Addressing all discrepancies between approved design and installed equipment, assuring all follow-up issues are resolved, and assuring that all pre-startup requirements are complete.
    b. Documenting post-startup action items where closure may occur after the system is operational.
2. Testing installed equipment to assure that it functions as intended.
3. Assuring that accurate as-built drawings are available to operating, maintenance, and other key personnel.
4. Assuring that all other startup required process safety information (PSI) documentation has been gathered and submitted to the appropriate, accessible place.
5. Assuring that startup required procedures, operating and emergency procedures in particular, are complete and available.
6. Assuring that training is complete on the new process, equipment, and procedures.
7. Assuring that all action items from PHA, MoC technical reviews, MoC environmental, health, and safety (EHS) reviews, and PSSR inspection and testing are resolved and pre-startup requirements are complete.
8. Assure that the customer, the Operator, is satisfied with the condition of the equipment being turned over to Operations.

The PSSR process may be thought of as a combination of field inspection and testing along with administrative confirmation that documentation, data systems, procedures, training, and follow-up action items required for safe startup and operation are complete. See Figure 2.

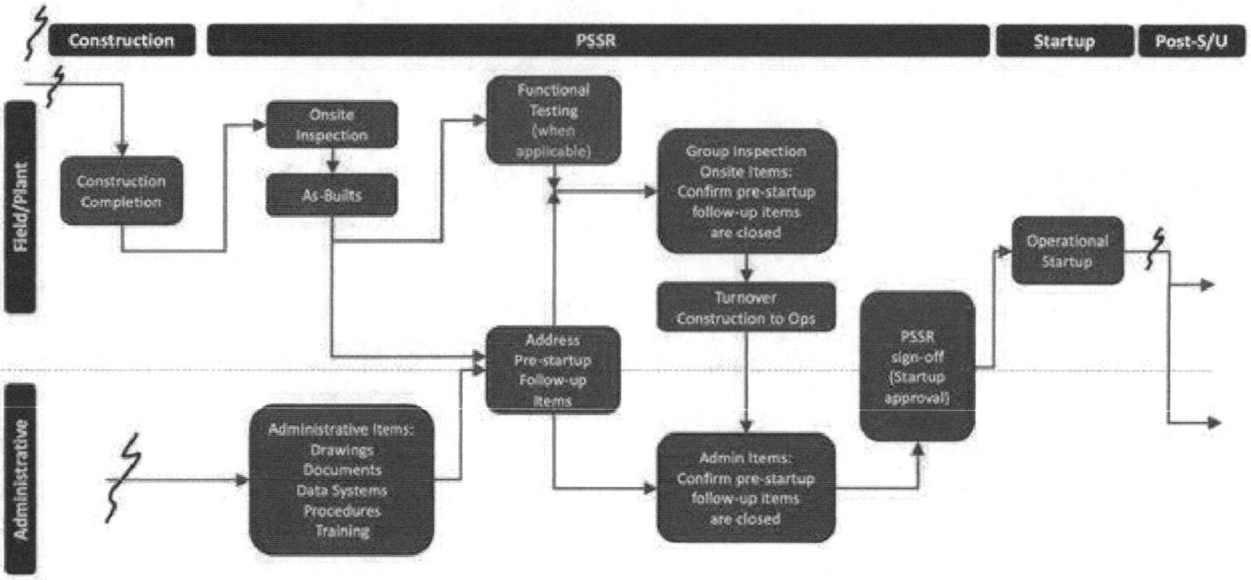

Figure 2. Simplified PSSR work flow showing field vs administrative elements

The work flow process diagrams (Figures 2, 3, 4, and 5) show the PSSR stage as well as stages leading up to PSSR and following after PSSR. This section, 2.2, touches on all aspects of the PSSR or operational readiness reviews. The latter sections of this paper are expanded discussion focusing on the field inspection and walkdown portion of the process, Section 2.2.1.

2.2.1 Onsite Inspection

The intent of the onsite inspection, also called walkdown, is to ensure that equipment has been completely and safely installed to the standards expected by the operating company. A primary objective of the onsite inspection process is to ensure that the equipment and the process configuration are installed as shown on approved for construction drawings. Any discrepancies must be annotated to show as-built condition and reconciled.

Red flag: When discrepancies are found between installed equipment and approved construction drawings, the process for addressing and reconciling those discrepancies must be defined and rigorous. These deviations are evidence that equipment was installed that may not have received process hazard analysis or other MoC reviews. Assumptions must not be made that deviations found were properly approved or will be approved.

Solutions: 1) Ensure that any and all post-PHA changes received or receive a belated but thorough MoC review including process hazard review, technical review, and other requirements in advance of sign-off and startup. 2) Reward rejection of unapproved changes and follow-on impacts in order to enable a fair, thorough review process. 3) After startup, address the causes for unapproved changes through post-mortem failure analysis, such as using cause mapping®[4], and implement corrective actions to prevent recurrence.

Figure 3 represents a typical work flow for projects that are smaller, involve less complex equipment, involve no changes or impacts to automation or safeguards, and may involve equipment types more frequently inspected. This work flow diagram includes a single group walkdown where no functional testing is required.

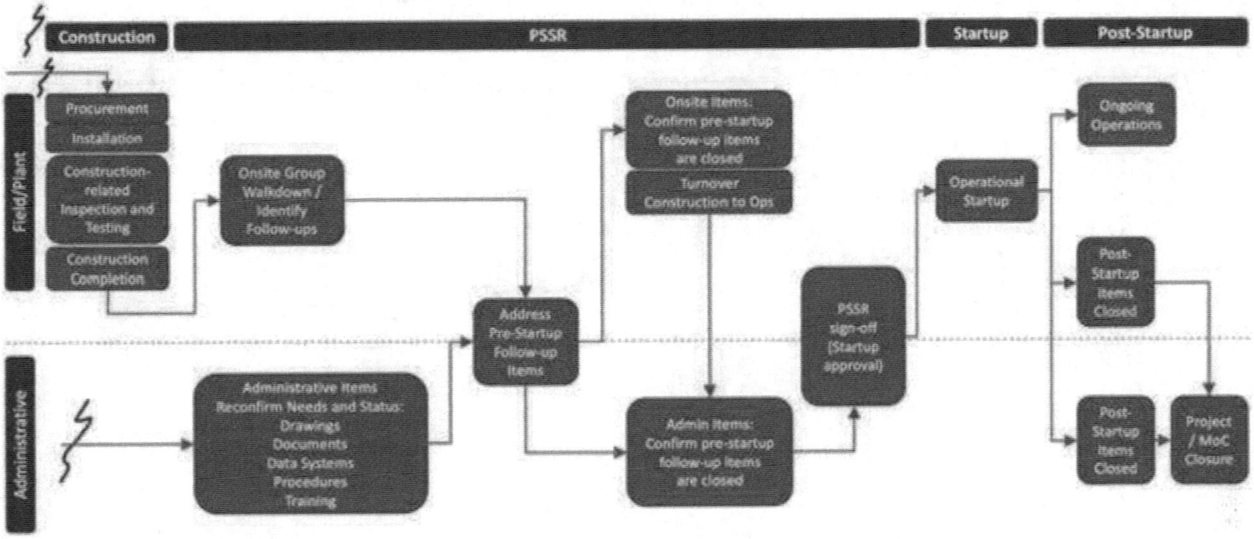

Figure 3. Sample PSSR work flow, smaller project, less complex equipment & components

Refer to Figure 4 and Figure 5, below, for an example work flow for projects requiring two onsite walkdowns, an individual and a group walkdown. See Section 2.3 and beyond for further discussion of the who, when, where, how, and tools used to support an effective inspection and field walkdown process.

2.2.2 Functional Testing

For projects involving automation, controls, and safety devices, function testing of the installed equipment and instrumentation typically occurs after confirmation of mechanical completeness and in advance of final system inspection. (See Figure 4.) Safety valves, process control valves, instruments, and communications are tested to ensure functionality of safety instrumented functions (SIFs), process control loops, and other alarm and indication instrumentation. Mechanical safety devices such as pressure safety valves (PSVs) and check valves should be function and leak tested. Functional testing is performed by technicians trained for this activity. Results are documented and retained. These technicians may be employees or contractors associated with commissioning teams. The commissioning team representatives will generally work in conjunction with company instrumentation and controls technicians.

Rotating equipment including drivers and associated controls and automation will require inspection, function testing, and commissioning that is specific to the equipment type. New, modified, or re-commissioned rotating equipment will have requirements such as alignment, checking for free rotation, and filling of lubrication and seal systems that occur during construction. Verification of rotation direction, i.e. bump testing, should occur in conjunction with functional testing. These activities will also be performed by trained commissioning technicians with the support of plant operating and maintenance personnel.

Figure 4 and Figure 5 (Section 2.3) represent a typical work flow for projects that are larger, involve more complex equipment, involve automation and safeguarding components, and may involve equipment types less frequently inspected. Figure 4 and Figure 5 include functional testing and two walkdowns, individual and group. For larger projects or turnarounds, these PSSR activities are conducted multiple times on sub-sets of the entire program.

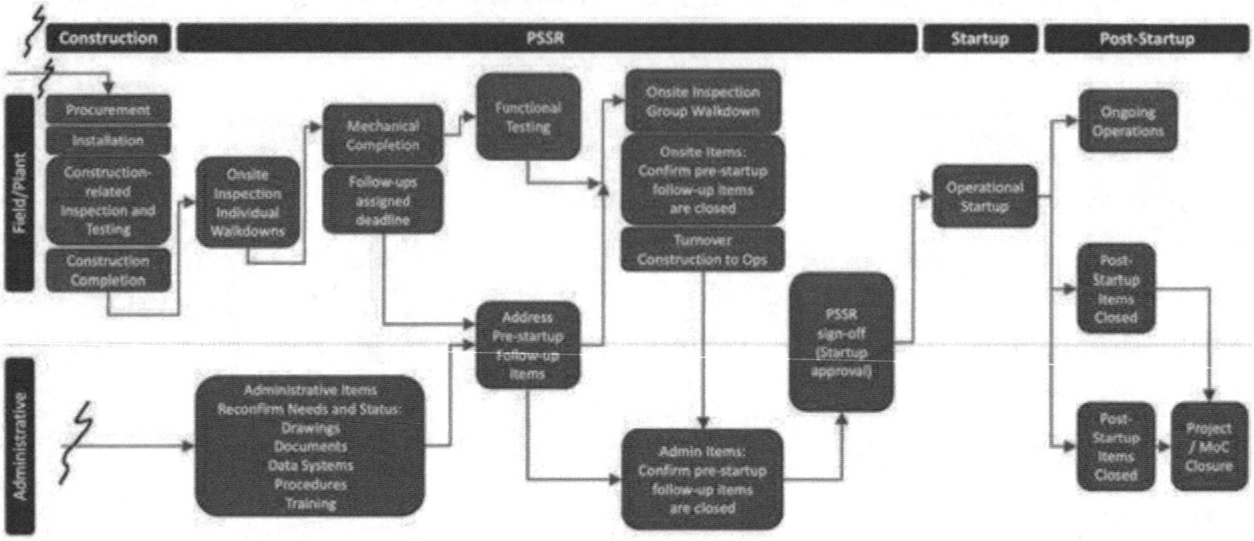

Figure 4. Sample PSSR work flow, larger project, more complex equipment & components

2.2.3 As-Built Drawings

Drawings such as piping and instrumentation diagrams (P&IDs) and/or process flow diagrams (PFDs) must be turned over and available to operations, maintenance, and other key personnel reflecting as-built condition, also called "as installed or redlined." These drawings may be in any format, hand mark-ups (in red), electronic mark-ups (in red), or issued digital drawings showing as-built condition as confirmed during field walkdown. Drawings must be accurate, properly signed and approved, and available before startup.

Note: Refer to "Red flag" above regarding discrepancies found between as-built condition and approved drawings and assure appropriate reconciliation of those discrepancies. Simply marking the change is not sufficient. A rigorous, though belated, MoC equivalent process must approve the change before startup may occur.

2.2.4 Additional PSI, Database, and Automation Updates

Additional process safety documentation including additional drawings, shutdown matrices, equipment information (via database or digital PDF), engineering calculations, relief and ventilation design documentation, various lists and logs (such as car seal lists, dead leg lists, spare parts lists, etc.) must be updated considering as-built condition and be included in project turnover documentation. Data management systems may include computerized maintenance management systems (CMMS), equipment data systems, document management systems, or enterprise systems that encompass all data elements. The process and timing for uploading into final electronic systems will vary from company to company and may occur post-startup for certain items.

Automated human machine interface (HMI) screens must also be complete and current. Any programmable logic controller (PLC) interface updates impacting operation, control, and safety responses must be complete before startup. Database updates must occur on the timing deemed appropriate for the particular application.

2.2.5 Procedures

Certain procedures, including operating procedures such as startup, shutdown, operating limits and responses, and troubleshooting, must be complete, available, and trained on by all on-shift personnel and any incoming personnel before he/she comes on-shift. Emergency response and other safety procedures must also be complete and available with training completed in order for safe startup to occur. Certain procedures that will not come due for months or years, such as some recurring inspection, testing, and maintenance procedures, could be completed safely after startup. However, regulatory requirements may dictate that procedures, including maintenance procedures, be complete or "in place" meaning that the procedure with accompanying implementation schedule has been loaded and is available in core maintenance management systems. Note that maintenance procedures may be considered to include preventative and predictive activities such as inspection and testing procedures, plans, and schedules. Repairs, or corrective maintenance, associated with either planned activities or unplanned events are performed, when needed. Be sure to consult the applicable regulation regarding compliance-related drivers.

Pitfall: An assumption that the initial version of startup procedures is final or "etched-in-stone."

Solution: After first use, startup procedures should be reviewed to determine if a better approach exists and then revised, as needed. All procedures require review and revalidation on defined intervals as well as continuous improvement, as needed; but immediately around startup is a good time for initial review, revalidation, and improvement of startup procedures.

2.2.6 Training

Training includes information on the new or modified equipment, process flow, process technology, chemicals, and/or procedures. Training typically falls into one of several categories, either pre-startup required, post-startup required, or awareness level training. Pre-startup training is required for on-shift personnel in critical roles requiring the training. Required training will likely need to be staged to address all critical personnel as they cycle into the critical role from being off-shift, off-rotation, time away from work, or for new incoming personnel. Mechanisms must be in place to track and maintain these training requirements and records. Tying new equipment or process training requirements into core operator competency training programs is equally important and frequently missed. Awareness level training should be prepared and distributed to other personnel. Records of awareness level training completion should also be tracked and maintained.

2.2.7 Follow-up Action Item Reconciliation and Closure

Follow-up action items are developed and added throughout the project process. The MoC design review process and the PHA, which occur before approval of construction and installation activities, will yield a list of follow-up recommendations and actions. Some of these action items must be completed before the final design is approved for construction. Other PHA or MoC review recommendations or action items may allow for completion during later stages of the project including either before or after startup. Likewise, the PSSR process itself through both the onsite inspection and testing process as well as the administrative reconciliation process will yield a list of additional follow-up action items.

For remaining discussion of follow-up action item management, refer to Section 4.

2.2.8 Turnover to Customer, Operations

The final step and the ultimate purpose of the PSSR process is to ensure that new or modified equipment is ready and safe to operate. The PSSR serves as the point of custody transfer from the project or maintenance team to Operations. Therefore, PSSR is a joint effort with multiple groups and individual positions having roles in the process. The ultimate customer is Operations.

2.3 Who? Define Participants and Roles

MoC and PSSR cross boundaries between groups and require the participation of cross-functional team members. Multi-disciplinary collaboration is essential to success.

Pitfall: In some instances, certain teams or team members may incorrectly believe that the MoC process itself or PSSR in particular are owned by one department or group.

Solutions: Ensure that all employees and contractors involved in PSSR understand the cross-functional nature of the PSSR work stage and recognize their role in the integrated process.

Regarding the onsite equipment walkdown, the following participants should be included. Also refer to Figure 5.

2.3.1 Operations

Ensuring that the customer, Operations, is satisfied with the condition of the equipment being turned over is a key element of a robust and self-sustaining PSSR process. The specific client that must be satisfied is the field or plant operator. Engaging the operator-owner of the equipment is essential to ensuring that the PSSR process is effective. An experienced field or plant operator must be involved in final walkdowns and sign-off on the equipment being turned over. His/her input will be a key component in continuously improving the process. The operator-owner will be either positively or negatively impacted during day-to-day operations depending on the quality of the design and turnover. Ultimately, the Operator is the most routinely at-risk position for being impacted by safety incidents. Therefore, the involvement of the Operator is a natural and clear connection not only from an operational perspective but from a safety perspective, as well.

Anecdote A: Company "A" PSSR process was started not due to PSM requirements, but due to Operator dissatisfaction regarding the projects that were being turned over to Operations by construction crews.

To address the problem, a more formal PSSR process was created. A project-specific and site-specific checklist was created for repetitive small pipeline replacement installations that were occurring as part of a broader gathering system replacement project. An agreement was reached by participants for an onsite walkdown process. A new PSSR checklist was created. Construction crews used the checklist during the course of their activity and in final preparation for the walkdown. All participants met on location to review the installation. All participants were signers of the PSSR checklist form while onsite. Follow-up punchlist items were agreed upon on the basis of the signed PSSR form.

Following the implementation of the PSSR process, the Operator was happier with the condition of the equipment being turned over to Operations. The construction crews were happy because they had a clear understanding of the requirements and they were satisfying their customer. This PSSR process was driven by customer need, not regulation, as the system under review was not under federal or state regulatory requirements at the time. The engineer that formulated the process later learned how closely this "home grown" internally derived process matched PSSR and operational readiness processes as described in literature and by regulation. And the project engineer was happy that all were satisfied and that PSSR had been achieved in a somewhat "organic" process. It was driven and adopted by an internal and inherent need rather than by being imposed from above and potentially being resisted or rejected making it ineffective.

Note: The Lead Operator or an Operations Foreman or Supervisor may be an additional participant and signer of the PSSR, but these positions should not replace the field or unit operator involvement nor represent the operator-owner position. The size, scope, and complexity of the project or collection of project PSSRs should govern the decision to involve the Lead Operator. For example, the Lead Operator should be involved in all PSSRs associated with a turnaround or large project to ensure continuity and consistency. The Lead Operator may not need to be involved in every repetitive installation walkdown in a wider program or campaign, but he/she should attend a sampling of those onsite group walkdowns to ensure that he/she has a good understanding of the process being used and to provide guidance, as needed. The Lead Operator can also participate to support less experienced site operators. Only the essential participants should be involved to avoid diluting the effort, shifting accountability, or confusing ownership. The decisions on participation should be project-specific and site-specific.

2.3.2 Inspection

The Quality Control (QC) representative or a third-party inspector having sufficient expertise regarding the equipment and process is an essential participant in the PSSR process. Having an outside set of eyes that specializes in this type of work is recognized as being an essential ingredient for final inspection and unbiased interpretation of the findings. The qualifications and redundancy of QC personnel are dependent on the size, scope, and complexity of the project or job being turned over.

2.3.3 Construction/Implementation

Construction or implementation personnel participation is required. The resource performing the construction activity is commonly a contract resource, but could be in-house personnel. At a minimum, the crew lead or foreman for the construction company or group that installed the equipment should participate. The preference would be for involvement from both a foreman level lead for ownership and job-specific knowledge, as well as a superintendent, for consistency across jobs. For maintenance work, in-house or contract, equivalent representation is required.

2.3.4 Engineering

Engineering involvement is essential. Engineering representation should be through the project lead engineer, a technical design engineer, or construction engineer. If the design deliverables were handed over to a construction group to manage implementation, the construction engineer should participate. It is important to note that intent may sometimes be lost in translation. Circling back to the design engineer and including more direct design engineer involvement in PSSR may be an important step if deviations begin to creep into the process. Knowing that those deviations exist in the first place may require design engineer involvement. Connection with the end product is also instructive to design engineers and builds more ownership and insight into implementation realities. For specific equipment inspections, the engineer must be sufficiently qualified in the discipline necessary to review the installed equipment type.

2.3.5 PSSR/Operational Readiness Lead

A PSSR Lead should be established for projects of any size. The PSSR Lead must understand the requirements for inspection and testing in the field as well as the requirements for administrative information gathering, review, and approval. The PSSR Lead must also have a complete understanding of the requirements for follow-up item prioritization, sign-off, and closure. The PSSR Lead should understand the relationship between PSSR and MoC. For larger projects, the PSSR Lead may be a member of the commissioning team. For smaller projects, the project lead engineer or another trained and qualified individual may fill the role. A third-party representative outside of the construction/implementation and operations teams, is needed.

Note: MoC Coordinator vs Turnover Coordinator vs PSSR Lead roles
MoC Coordinator roles, skill-sets, and training vary from company to company. Typically, the MoC Coordinator role is focused on the handling and processing of MoC documentation. The MoC Coordinator is frequently involved in managing the communication of MoC project status through various stages of approval. Depending on company structure, MoC Coordinator communications may trigger assembling some team members for certain project stages. Commonly, a different position, such as Project Turnover Coordinator, may be responsible for coordinating pre-walkdown communications and meetings, onsite walkdowns, action item reconciliation meetings, and/or post-walkdown turnover meetings particularly for large projects. The PSSR Lead oversees many of the responsibilities of these two roles during PSSR. Due to the wider responsibilities of the PSSR Lead role, typically neither the MoC Coordinator nor the Turnover Coordinator should be assigned as the PSSR Lead.

Pitfall: In instances where the MoC Coordinator is given authority to promote or signal progression of an MoC through various phases, misunderstanding of the process and/or roles may lead to progressing MoC stage approvals prematurely without proper field review and sign-

off. Beware of assuming that simply having the title of MoC Coordinator guarantees a complete understanding of all checks and balances within the MoC and PSSR progression process.

Solutions: In instances where the MoC Coordinator is granted or understood to have authority to progress or to signal progression of an MoC through various phases, the MoC Coordinator must understand the essential field reviews and technical approvals that are required before moving beyond the PSSR phase. Protocols and procedures must define requirements. And tools used must require sign-off by all essential team members.

2.3.6 Operations Management/Leadership:

Supervisor sign-off at some level should be included but for differing drivers. The signature of a mid-level or higher-level supervisor is required not for assurance of operational readiness, but to assure accountability and ownership by management for having a robust and fulfilled PSSR process. Pressure to move quickly to startup activities often originates from the top. Management signature on the PSSR form assures that the manager has been held personally accountable for the PSSR process. Whether the supervisor or manager is a mid- or higher- level supervisor or manager would typically depend on the size or scope of the project being prepared for startup.

To recap, essential and recommended equipment inspection participants should be considered as the following:

1. PSSR Lead (essential)
2. Operations
    a. Operator, operations-owner (essential)
    b. Lead Operator (essential or optional, decision is job-specific and site-specific)
3. Quality Control (QC) / Third-Party Inspector (essential)
4. Construction/Implementation company representative, either contract or in-house
    a. Crew foreman level (essential) and
    b. Superintendent level (recommended)
5. Engineer, project/job lead, design, or construction (essential)
    a. Project Lead Engineer may also serve as PSSR Lead, if qualified. However, the Construction Engineer should not serve as PSSR Lead.
6. Operations Manager or mid-level Operations Supervisor (sign-off required, but onsite review not necessary, See Section 2.5)

The participation and ownership of the steps within the PSSR process are depicted in Figure 5. Figure 5 reflects projects involving functional testing and having two walkdowns.

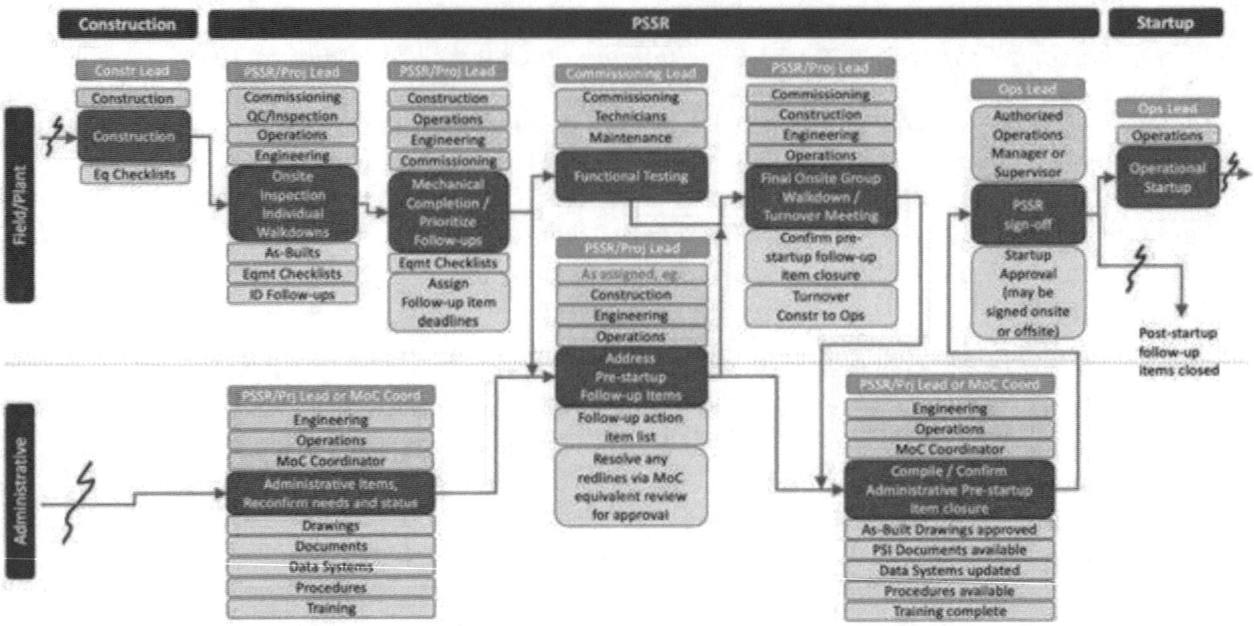

Figure 5. Sample PSSR work flow, larger project, more complex equipment and components including resources, personnel and tools

These positions and roles should be considered as guidance for PSSR participant requirements. Always consider the job-specific and site-specific factors when defining the PSSR team participants.

2.4 When? Define work sequence

When does PSSR occur? The simple answer is: before startup. (See Figure 1.) But defining the work flow process means more than simply saying pre-startup safety review occurs before startup. The work flow process must establish expectations that allow for the right level of inspection by the right people at the right time. Individual components requiring testing must be inspected for mechanical completeness in advance of function testing. Function testing will typically occur before final group PSSR walkdowns. Field walkdowns should be scheduled to allow time for individual, detailed inspections in advance of final group walkdowns that ensure concurrence on findings, approvals, and follow-up requirements. (See Figure 4 and Figure 5 representing larger projects or projects involving automation and safety devices.)

<u>Pitfalls</u>: Group walkdowns should generally be considered as a critical step for most PSSR processes, but be aware that group walkdowns may lead to group-think, time pressure, or peer pressure to wrap-up the process. Group walkdowns may not allow adequate time for detailed field review and inspection.

<u>Solutions</u>: A single group visit to the location may be appropriate for very simple or repetitive installations (See Figure 3.) But for more complex equipment or for installation types that are less frequently reviewed, inspectors, engineers, and/or operators should visit the location and perform their individual inspection in advance of the group walkdown.

Scheduling the individual onsite inspections in advance of the group walkdowns not only allocates time and schedule space for both activities, it also builds in time to address follow-up action items and functional testing simultaneously. Initiating work early to close action items supports an earlier startup and lessens pressure to downgrade critical follow-up items. (See Figure 4 and Figure 5.)

2.5 Where?

Part of defining the work flow process involves defining which portions of the process are conducted onsite in the field or plant. In other words, who must have "eyes on" the equipment to complete their inspection and sign-off? In general, all essential and recommended members of the approval team must complete a field walkdown to sign off for their position. The one exception would be the Operations management or supervisory representative. This exception is based on their unique role as a higher-level accountable representative. Their signature assures their support of the process itself, rather than ensuring the integrity and completeness of the installation. Likewise, administrative activities such as compiling and submitting documents, data, and training records occur offsite. Refer to Figures 2, 3, 4, and 5 for a "swim lanes" depiction of the "on location" vs administrative activities.

2.6 How?

Executing PSSR involves several parts. Implementing a work flow process which defines roles and activity sequencing is part of that process. But creating and utilizing effective tools is an equally important part of the process. Those tools include: broad PSSR process checklists, equipment installation PSSR checklists, equipment testing procedures, and appropriate forms and database tools for capturing and managing inspection, testing, and follow-up findings. These tools may be home grown or provided through vendors. But the content and format are key to the functionality of the tool. Be sure to modify the tool or checklist to reflect important reminders, triggers, language, or approval methods gleaned from references and/or experience.

<u>Pitfall</u>: The PSSR process occurs at the final stages of the entire project process. The PSSR stands in the way of a startup that everyone wants to achieve. Just as it is difficult to pay close attention to the final afternoon speaker before happy hour begins or the last class before the semester is over, it is natural to want to "check boxes" and "power through" the PSSR. Resist this inertia and the human desire to "get ur' done." This energy and impatience will lead to shoddy quality and completeness.

<u>Solutions</u>: Find internal triggers and practices to prevent the natural inclination to "get 'ur done" from impacting PSSR quality. Ensure that personnel, such as QC inspectors, are brought in for this stage where their sole responsibility and deliverable is this pre-startup inspection role. Hire detail-oriented people to perform inspections. Conduct internal audits that are quality focused, not task focused. Incentivize and reward findings of deviations and subsequent corrections of those deviations, even if, and particularly when, startup is delayed. This type of incentive mirrors stop-the-job rewards.

3 Effective Tools

3.1 Equipment Installation Checklists

Checklists are an important tool for effective PSSR. Checklists must be fit-for-purpose. The PSSR equipment installation checklist encompasses equipment requirements, project closeout requirements, and site-specific requirements.

3.1.1 Equipment category specific checklists

To be effective, checklists must be specific for a given equipment category. For example, a project including a pump should include a checklist specific to pumps and include line items and acceptance criteria for the pump and the driver. A project involving a generator should have a checklist and line items specific to generators. Most mid-sized to larger projects include piping and valves. A standard checklist involving piping and valve components is appropriate to include for all projects that involve more than the modification or replacement of a single piece of equipment or equipment component. As with all aspects of PSSR, it is important to find an effective and functional balance between too many checklists and too few.

Pitfall: Overly specific checklists may be overkill and add unnecessary complexity and paperwork to the PSSR process. Detailed conformance with manufacturer installation requirements and verification should be handled via appropriate installation and/or maintenance procedures.

Solutions: Installation checklists by broad equipment category should be used. The intent of the PSSR checklist is not to replicate very specific equipment and manufacturer requirements. The intent is to provide a hold point to inspect for items which may be site-specific or have a history of being commonly missed.

3.1.2 Standard sections for equipment checklists

A number of standard sections should be included in the PSSR checklist regardless of equipment category. These sections may be rolled into the standard template by equipment category or be captured on a standalone checklist. The preferred recommendation is to include these sections and line items on the standard checklist for each equipment category in order to allow for usage of a single checklist while visiting a specific location. Combining on a single form simplifies the walkdown process and gives better assurance that the item is reviewed sufficiently across the area for each equipment type.

Some of these standard sections for equipment installation checklists include:

1. Construction-related inspection and testing records (such as non-destructive examination (NDE), hydrotest), completed and records gathered
2. Function Testing, completed and records gathered
3. Bolt-up
4. Coatings
5. Insulation

6. Structural Supports/Bracing
7. Visual Systems, signs and labeling
8. Housekeeping
9. Drawings/Documents, as-builts
10. Follow-ups, action item work list initiated

Don't forget labeling and housekeeping! Labeling and housekeeping in the final inspection and walkdown process should not be underemphasized. The project or modification should be handed over as a completed deliverable. Labeling and housekeeping not only reflect pride in ownership but also directly impact safety. New projects should meet company guidelines and an equivalent or better standard with regard to the existing condition of equipment. If not, these projects become contributors to deteriorating expectations for the condition of equipment in the plant.

Pitfall: An assumption that labeling is not an important nor required deliverable from the project team.

Solution: Establish requirements for labeling as part of the delivered project. Include labeling requirements on equipment installation checklists. Provide a label maker and supplies to implementers. Ensure that a tag making resource is identified and available.

3.1.3 Acceptance criteria

Most checklists are important but all checklists are not created equal. The application and purpose will dictate the level of detail that is required for a given checklist. For PSSR, the checklist item description should include complete sentences that describe acceptance criteria. Several criteria may be grouped under a single inspection item, if related. The preference would be to limit to no more than 3 to 5 criteria, if needed. The criteria for a single line item should be closely related and brief while fully describing the condition which allows for acceptance. In applications other than PSSR, a single bullet item checklist may be appropriate. Examples that come to mind are: grocery lists, packing checklists, or camping checklists. Single word or phrase PSSR checklists are not appropriate.

Pitfall: Checklists having single word or phrase line items are insufficient as a tool for PSSR equipment walkdowns.

Solution: Acceptance criteria must be clearly communicated and easily available to all PSSR team members. Acceptance criteria should be included on the checklist.

Refer to Figure 6 for an excerpt of an example checklist that defines acceptance criteria for each line item as opposed to listing single word or phrase items.

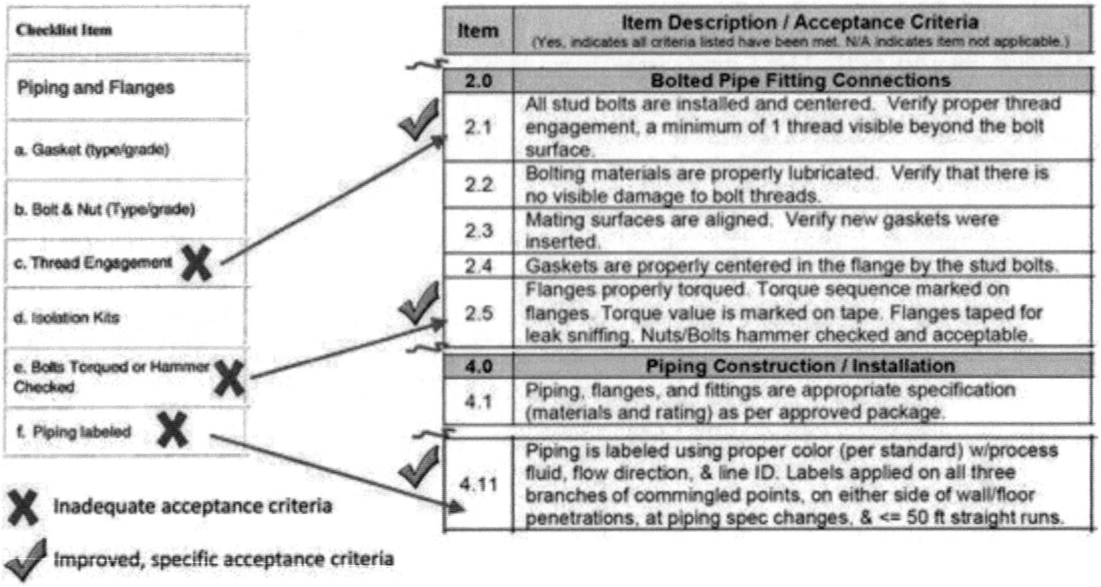

Figure 6. Comparison between PSSR checklists having inadequate acceptance criteria and better acceptance criteria

Combining criteria on a single line is done for brevity of the checklist and for streamlining of the walkdown and records submittal process. If any item of several listed does not meet the acceptance criteria, the line item is indicated as "Unacceptable." The follow-up action item will define the particular item that is out-of-conformance.

<u>Pitfall</u>: Having more than one criterion on a single line item creates some risk that the inspector or auditor checks the line item as acceptable when all criteria have not been met.

<u>Solution</u>: Checklist header instructions must be clear that all criteria must be met in order to check as "Acceptable." Emphasize during training. Audit the process occasionally.

A balance must be achieved that helps prevent the human tendency to move through the line item too quickly, i.e. "gloss over" or idly check boxes, without making the documentation and the process overly cumbersome and ineffective due to its length.

3.1.4 <u>Format considerations</u>

All line items on the list should be addressed and indicated as Applicable or Not Applicable. Requiring each line item to be addressed reduces the likelihood of skimming the list and missing the item altogether. For this reason, the checklist should be specific and applicable to the category of equipment being inspected in order to ensure the time invested is worthwhile. Applicability is implied by checking "Yes" (acceptable) or "No" (unacceptable) which saves space on field forms and lists. Phase the questions or statements so that "No" is always the unacceptable or undesirable response. All "No" responses require follow-up and closure.

Always allow space for comments. Not having a space to include any notes or comments is a lost opportunity. Even if only a couple items require comments, it is important to have a mechanism to capture these notes.

The signature and approval box on the equipment inspection form should include an indication of whether an onsite inspection was conducted by each signer. (See Figure 7.)

REVIEW / APPROVAL (indicate whether job was walked down in field or signed off based on walkdown by others)

	Field Y/N	Print Name	Sign Name	Date
Installer / Construction Foreman or Lead*				
Installer / Construction Superintindent				
QC Representative*				
Field/Plant Operator*				
Lead Operator				
Project/Design/Construction Engineer*				
PSSR Lead*				
Operations Manager*				

Figure 7. Sample equipment inspection signature box including "eyes on" indication

3.1.5 Ensure all checklist line items are appropriate to the PSSR phase of project

Occasionally, PSSR checklists may be found to include items that are design-related or items that must be completed as construction stage checks. These items should be removed from the PSSR checklist. All design considerations and reminders should be included on a design review checklist which is reviewed in advance of project design approval. These reviews are sometimes referred to as technical reviews or discipline reviews. Design approval precedes the construction stage. The exception to this rule would be where latitude is given to specific placement or orientation of installed equipment in the field. Where construction or installation discretion is allowed or required, those items should be field checked relative to acceptance criteria.

		Yes	No	N/A
	Relief Devices			
1	Relief devices and lines are adequately sized.			
2	Relief device calculations account for inlet and outlet piping.			
3	Relief devices relieve into a piping system or other containment.			
4	Relief devices relieving to atmosphere do not relieve in vicinity of personnel or equipment not designed to contain relief fluids.			
5	All installed relief lines drain as required, avoiding pockets or low points where fluids may collect in the line.			
6	Relief devices are secured and supported as per approved drawings.			

		Yes	No	N/A
	Piping/Valves			
1	Piping installed meets the approved for contruction design criteria.			
2	Flanges installed meet the approved for contruction design criteria.			
3	Gaskets installed meet the approved for construction design criteria.			
4	Gaskets installed were not reused or damaged.			

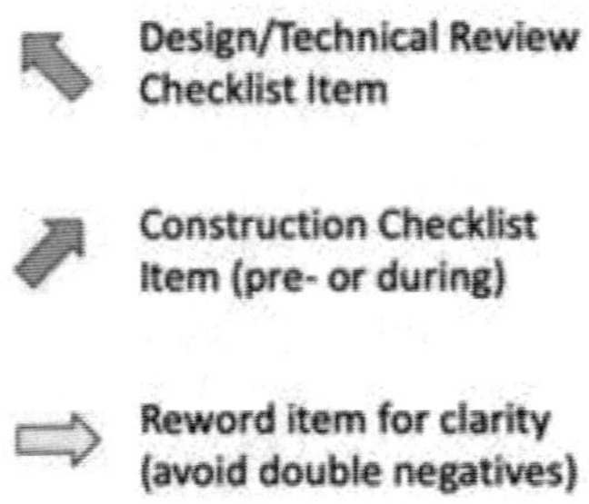

Figure 8. Checklist items requiring earlier project stage verification, not during PSSR

Certain items cannot be visually checked in the field following installation. These items should be included on pre-construction or during construction checklists. (See Figure 8.)

3.2 Paper vs e-Tools

Paper checklists or electronic tablet checklists may be considered for use. Choose whichever option is most functional, safe, and efficient for the particular application. The end users should be your guide in defining the best tool to use. If the tool is cumbersome and prevents a complete, high quality inspection from occurring in the field, then consider another option. Efficiency gains should always be considered. But saving time entering data does not add value if the quality is compromised or the process fails to identify non-conformant items. These misses compromise the PSSR process and may introduce associated risks into the operation. Always consider practical field constraints that impact usage. Wherever possible, single sheet checklists, front and back, that fit on a clipboard are preferred when using paper tools.

4 Follow-Up Action Items

Reviews and checklists are only useful to project success if items found as unacceptable receive follow-up attention and are corrected before the designated deadline. Small jobs may incorporate follow-up item closure on the original PSSR form. But frequently, follow-up action items are captured on a separate master list or in a separate database. These lists may be referred to as punchlists, follow-up items, action items, corrective action items, incomplete work list items, recommendations, or other descriptions. This list may include follow-up items from other stages of the project including design reviews, technical reviews, EHS reviews, PHA reviews, as well as the PSSR. The PSSR process will contribute follow-up action items through both onsite inspection and testing as well as through reconciliation of administrative requirements.

Follow-up action item descriptions should be clear, complete, and stand-alone so other reference material is not required to understand the unresolved issue. Every follow-up action item should be assigned a deadline, an owner, and closure requirements. It may also be advisable to indicate all drivers for the action item, such as: safety, operability, compliance, policy/procedure, and/or better practice. In many cases, all drivers may apply. But as noted in Section 2.2.5, certain pre-startup requirements may be driven by regulation rather than safety. The reverse may also be true where follow-up items are driven by safety, operability, or other needs rather than by regulation.

The deadline should be defined relative to project approval stages. For design reviews and PHA reviews, completion in advance of design approval may be specified. During PSSR, the deadlines will include pre-startup required, during startup, or post-startup allowed. All items that are assigned a deadline of before startup, regardless of the project stage in which they were generated, must be reconciled, closed, and signed off before operational startup occurs. Items indicated as during startup must be acknowledged and addressed during startup. Post-startup items allow for closure during ongoing operation. Closure of all post-startup items allows for closure of the overall project and MoC.

Pitfall: Post-startup action items are notoriously hard to give attention to and closure deadlines are notoriously hard to meet.

Solutions: If your company experiences these problems, conduct a problem-solving exercise, such as by cause mapping, to investigate why and implement solutions. Consider dedicated resources, use scorecards and metrics, and/or ensure that triggers are in place to prevent missing important deadlines. Management support of these efforts is essential.

Each follow-up item should be assigned an owner with primary responsibility for addressing and completing the item. Each follow-up item must also define sign-off, or approval, requirements for closure including if the item must be inspected in the field by operating company personnel. Indicating whether functional testing will be required for follow-up items is useful to note.

Critical items, including safety related items, will require PSSR Lead and Operations validation and sign-off. Whereas less critical items may allow for validation and sign-off by a construction team member including contract representatives. All tracking and closure of follow-up items should be dated and indicate the name or initials of the approver and include closure comments. Any items considered for closure by dismissal must be closely scrutinized. In order to dismiss, the drivers for the follow-up item must be defined and PSSR team members that identified and prioritized the follow-up item should be consulted and in consensus on the cancellation.

Red flag: Each pre-startup item must be addressed and closed before startup. Closing an item on one list and placing the item on a separate list with a new deadline does not constitute an acceptable closure mechanism.

Solution: In rare instances where the item deadline is reconsidered as being acceptable post-startup allowed after previously being assessed as pre-startup required, clear rationale must be given. This reassessment requires characterization of the drivers for the action item, i.e., safety, operability, compliance, policy/procedure, and/or better practice. Deadline reassessments may require risk assessment and should be signed off by the same team members that established the original deadline. All deadline reassessments should receive post-mortem investigation into the causes and include further corrective actions.

Any follow-up items that result in changes to approved design require an MoC equivalent review and approval process, including process hazard review. Follow-up items of this scope should be rare during the PSSR stage but must be handled appropriately when or if identified.

5 Continuous improvement

As with all tools and processes, every aspect of the PSSR process must be reviewed, refreshed, and continuously improved including the work flow process, the role definitions, the tools (specifically the checklists), and the follow-up item categorization and management.

Are the Operators happy with the process? Are the installers pleased that they understand the requirements and are they able to deliver those requirements? Are the project engineers engaged? Do the managers understand the process and are they fully bought in?

- Are items being missed? Add them.
- Are items not understood? Modify and clarify them.
- Assure that training is adequate. But recognize, PSSR is a relatively simple process. Assure that self-explanatory items are just that. Avoid unnecessary complexity in the

process. Include critical instructions within the tools used day-to-day. Do not bury important instructions in procedures and training materials.

- Is part of the process not working? Approach the gap from a problem-solving standpoint. Ask why. Cause map the problem and develop solutions.

No tool and process is ever perfect and every tool and process can always be improved. These improvements are accomplished through end-user feedback, client feedback, and auditing. Internal audits should be conducted to assess not only compliance but effectiveness of the PSSR process. The focus of the internal audit should be on quality and assurance that the PSSR is serving the intended purpose.

<u>Anecdote B</u>: Company "B" used two separate processes to complete PSSR. One process was used for large projects and a separate process for small to mid-sized MoC projects. Neither PSSR process was functioning well. A cross-functional team was created to improve the overall PSSR process and create common standards and expectations.

The small and mid-sized MoC PSSR process was driven by internal audit findings that identified gaps in conducting PSSR sufficiently or at all. To close the gaps, the requirement for PSSR was stated in the MoC procedure and several very generalized equipment checklists were made available; but no discussion was included regarding the who, what, when, where, or how it would be conducted. The generic checklists contained one-word item descriptions and included no acceptance criteria. No guidance was given regarding specifically who needed conduct inspections or how the process would work.

The large project process stemmed from legacy processes carrying other names. These work processes were extremely complex. The processes did support functional testing of equipment, but the onsite inspection process was not defined or clear. A binder full of very specific equipment checklists existed but none of those checklists were being used for the onsite inspections. Scheduling was generally insufficient to conduct individual inspections ahead of the turnover date. Expectations for these inspections were unclear.

The cross-functional improvement team defined a standard work flow process that could be scaled to projects of various sizes. The work flow process more clearly defined each element of the process, who needed to participate and approve, and how the process would work. The team also generated replacement checklists that were more useable, complete, and contained acceptance criteria. The checklists were not specific to the project team owning the work but were specific to the equipment involved in the project

Although the equipment checklists that were improved were part of MoC process toolkit, the QC department fully adopted the new checklists. QC touched a wide range of jobs and this new tool was applied broadly throughout the field. The QC department also adopted the practice of continuously improving and updating these checklist tools.

A primary learning from this effort was to apply consistent standards regardless of the project group owning the new installation, modification, or repair work. The equipment, not the project team, governs the requirements. A secondary learning was to strive for balance between too much detail in processes or tools and not enough.

6 Summary

PSSR or operational readiness reviews are key elements of PSM. Each operating company must develop and refine processes and tools that are appropriate for their organization, process, and personnel. However, learning through the experience of others may be useful. Field experience has highlighted the value in certain practices while illuminating certain pitfalls. As is often the case, certain pitfalls stem from practices that were implemented with good intent but may result in unintended negative outcomes. Table 1 below summarizes potential PSSR pitfalls and red flags along with possible solutions that have been captured in this paper. For additional details, refer to the Section that is referenced.

Table 1. PSSR Pitfalls/Red Flags and Solutions/Better Practices

#	Section	PSSR Topic	Pitfalls and Red Flags	Solutions and Better Practices	Key
1	2.2.2	Work Flow Process, What? Onsite Inspection	Deviations are identified between the approved for construction design and the installed equipment. 1) If approved during construction, beware of assuming that post-PHA changes were approved through a sufficiently rigorous MoC equivalent process. 2) If not previously approved, beware of assuming that changes found during PSSR should be approved. (Red flag)	Address deviations (i.e., design changes not appropriately approved) via a thorough MoC equivalent process before startup is allowed. When it occurs, reward rejection of unapproved changes and follow-on impacts in order to enable a fair, thorough review process. Address root causes of deviations through failure analysis study and follow-up after the project.	Yes
2	2.2.5	Work Flow Process, What? Procedures	An assumption that the initial version of startup procedures are final or "etched-in-stone."	After first use, startup procedures should be reviewed to determine if a better approach exists and then revised, as needed.	
3	2.3	Work Flow Process, Who?	Beware of perceptions that one entity, position, group, or department owns the MoC and/or PSSR process.	MoC and PSSR cross boundaries between functional groups. Ensure that all employees and contractors recognize their role in the integrated process.	Yes
4	2.2.8 and 2.3.1 and 5	Work Flow Process, Who? Operations	Lack of recognition of the field or plant operator as primary customer of the PSSR turnover process. Lack of inclusion of the field or plant operator due to oversight and lack of invitation. (from Anecdote A, Company "A")	Ensure that the field or plant operator is recognized as an essential participant and as the client in the PSSR process. Engage all participants, especially the Operator, in actively participating in and continuously improving the PSSR process.	Yes
5	2.3.5.1	Work Flow Process, Who? MoC Coordinator	Confusion regarding administrative authorities in the MoC and PSSR progression process with technical and operational approval authorities, such regarding MoC Coordinator role.	Ensure that MoC and PSSR protocol clearly define approval authorities. Ensure that tools used require all essential team members to sign off. Ensure that MoC Coordinator training is thorough.	
6	2.4	Work Flow Process, When?	An assumption that group onsite walkdowns are not needed	Group onsite walkdowns are a critical step in the equipment inspection portion of the PSSR.	Yes
7	2.4	Work Flow Process, When?	An assumption that a single group walkdown on location may be sufficient without upfront individual pre-inspections.	For equipment types not routinely inspected for startup, preliminary onsite inspection by individual team members is preferred to ensure adequate time is allocated for a detailed initial review.	Yes
8	2.6	Work Flow Process, How?	Beware of project fatigue, a desire and inertia to press towards startup and skim through the PSSR process. This inertia will negatively impact PSSR quality and completeness.	Conduct internal audits that are quality focused. Bring in dedicated inspection and testing resources for the PSSR stage. Hire detail-oriented people. Assure management support and incentives exist.	Yes

#	Section	PSSR Topic	Pitfalls and Red Flags	Solutions and Better Practices	Key
9	3.1.1	Effective Tools, Equipment Installation Checklist, Category specific	Overly specific equipment checklists. Checklists that replicate manufacturer information. Too many checklists to manage.	Use manufacturer guidance documents and installation procedures for detailed instructions. Use equipment walkdown checklists for verification of key equipment requirements, company-specific requirements, and items that are frequently missed.	
10	3.1.2	Effective Tools, Equip Installation Checklist, Standard Sections	An assumption that labeling is not an important nor required deliverable from the project team.	Establish requirements for labeling as part of the delivered project. Include labeling requirements on equipment installation checklists. Provide a label/tag maker and supplies to implementers.	
11	3.1.3	Effective Tools, Equip Installation Checklist, Acceptance Criteria	Checklists that refer to line items by a single word or phrase and have no acceptance criteria.	Ensure that the checklist includes acceptance criteria for each line item.	Yes
12	3.1.3	Effective Tools, Equip Installation Checklist, Acceptance Criteria	Having too many criteria on a single line item creates some risk that the inspector or auditor checks the line item as acceptable when all criteria have not been met.	Ensure that instructions on the checklist header clearly state that all criteria must be met. Emphasize during training. Audit the process occasionally. Identify the out-of- conformance item in comments and follow-up work list.	
13	4	Follow-Up Action Items	Beware of challenges associated with completion of post-startup follow-up items	Conduct problem-solving efforts to address causes of failure to close post-startup items in a timely manner and implement solutions identified, if this problem exists.	
14	4	Follow-Up Action Items	Cancelling or dismissing follow-up action items without PSSR team concurrence.	In order to cancel or dismiss a follow-up action item, the drivers for the action item must be defined and PSSR team members who identified the follow-up item should be consulted.	Yes
15	4	Follow-Up Action Items	Closing a follow-up action item designated as pre-startup required on one list and placing the item on a separate list with a new post-startup deadline does not constitute an acceptable closure mechanism. (Red flag)	Clear rationale must be given for reassessment of action item deadlines. Drivers must be defined, i.e., safety, operability, compliance, policy/procedure, and/or better practice. Any deadline reassessments should be risk assessed and approved by PSSR team members. All deadline reassessments should receive post-mortem investigation into the causes and include further follow-up action.	Yes
16	5	Continuous Improvement	Large project teams apply differing PSSR standards when installing equipment as compared with small projects or plant-led projects. (from Anecdote company "B")	The equipment-type or component, not the implementation team, governs the requirements for inspection and acceptance. Apply consistent standards specific to the equipment being installed, modified, or repaired.	Yes
17	5	Continuous Improvement	Overly cumbersome tools and processes will hamper practical implementation of PSSR. Overly simplified tools and processes will hamper quality. (from Anecdote B, Company "B")	Endeavor to find the correct balance between too much detail and not enough detail in order to ensure that the PSSR process is effective.	Yes

7 Conclusions

Your PSSR program should be framed by five key elements: defining a practicable work flow process; ensuring the correct resources are allocated; providing effective tools, in particular, functional checklists; ensuring follow-up on corrective action items; and making efforts to continuously improve the process. Management commitment and support enables the entire process. Consider the better practices and pitfalls discussed throughout this paper and summarized in Section 6 with respect to your PSSR program.

Success of the PSSR process will be achieved by using site-specific practices that find the right balance between too much information and not enough. The tools should be thorough but not cumbersome. Ensure that checklists are fit-for-purpose and are modified by end-users to meet site-specific needs. However, ensure that consistent standards are applied by equipment type regardless of the implementation team.

PSSR teams should involve a sufficient number of skilled resources without flooding the process with too many people. Team members must recognize that PSSR efforts occur at the boundary of project and operational work scope. The process is cross-functional and requires participation and ownership by multiple teams and individual positions. The process should also recognize the Operator as the ultimate customer.

Committing to quality and follow-through is critical for the success of PSSR. Part of ensuring quality comes through resisting pressure to skim through the process in order to achieve an earlier startup. Implementing successful PSSR is a combination of having strong processes, tools, resources, commitment, and follow-through. Failure of any of these components may cripple the overall PSSR process and potentially compromise the success and safety of an otherwise well-implemented project.

8 Acknowledgements

The author wishes to thank Donald Connolley, Fred Henselwood, Trish Kerin, Bruce Weiler, Angela Kuest, Bart Lewis, Mark Humphrey, and Mike Marshall for their review, comment, and contributions to this paper and presentation. The author sincerely appreciates having reviewers that include management, project, engineering, maintenance, operations, and regulatory perspectives.

9 References

9.1 Cited References

[1] S.W.Ostrowski and K.K.Keim, Tame Your Transient Operations: Use a special method to identify and address potential hazards, Chemical Processing, June 23, 2010, Available at https://www.chemicalprocessing.com/articles/2010/123/, Accessed on February 25, 2021.

[2] U.S. Department of Labor Occupational Safety and Health Administration. Process safety management of highly hazardous chemicals, 29 CFR. §1910.119(i) and (l), Jun 1, 1992, as amended Feb 8, 2013.

[3] CCPS. *Guidelines for Risk Based Process Safety*. Center for Chemical Process Safety. American Institute of Chemical Engineers, New York, NY, 2007.

[4] Cause Mapping®, ThinkReliability, Available at https://www.thinkreliability.com/cause-mapping/, Accessed on March 3, 2021.

9.2 Additional references not cited

[5] CCPS. *Guidelines for Performing Pre-Startup Safety Reviews*. Center for Chemical Process Safety. American Institute of Chemical Engineers, New York, NY, 2011.

[6] Marsh & McLennan Companies, Pre-Startup Safety Review: Risk Engineering Position Paper – 06, Available at https://www.marsh.com/uk/insights/research/pre-startup-safety-review.html, Accessed Feb 10, 2021.

SPRING21
+17TH GCPS
A Joint AIChE and CCPS Meeting

The Shift to Dynamic and Simulated Process Safety Risk Management

Simon Jones
Sphera Solutions, Inc
130 E. Randolph St., Suite 2900
Chicago, IL 60601 USA
simon.jones@sphera.com

Prepared for Presentation at
American Institute of Chemical Engineers
2021 Spring Meeting and 17th Global Congress on Process Safety
Virtual
April 18 - 23, 2021

The Shift to Dynamic and Simulated Process Safety Risk Management

Simon Jones
Sphera Solutions, Inc
130 E. Randolph St., Suite 2900
Chicago, IL 60601 USA
simon.jones@sphera.com

Keywords: Operational risk management. Frontline operational risk. Process Safety Management. Process safety barrier health. Dynamic risk management. Digital twin.

Abstract

This paper explores how companies are adopting digital twin technology to help close the gaps between process safety design intent and operational reality. A case study of an Operational Risk Management system implementation in a major refinery shows the data sources and underpinning models used to provide a dynamic, real-time view of risk status by surfacing critical barrier health. These systems aide frontline decision-making by simulating the risk impact of work prioritization options and are now evolving to model specific major accident hazard scenarios, associated risk trends and to provide assurance that critical barrier health is sustained in the operational phase.

1 Introduction

Studies in various industries indicate that for every accident that takes place, 50 to 100 near-misses occur, and these are underpinned by around 10,000 errors or failure conditions on manufacturing facilities[1]. Complex facilities constantly migrate toward higher-risk states, and it is often not noticed until a significant accident or incident occurs.

Hazard study tools and techniques help identify hazards and assess risk; they help specify prevention and mitigation safeguards to minimize the likelihood of experiencing a major hazard event. Once in place, these safeguards are monitored and maintained by numerous business processes and associated systems, including those Environmental, Health and Safety (EHS) management, maintenance management, data historians, asset integrity and asset performance management, control of work and more.

Yet these tools do not jointly analyze asset-level data. This leaves organizations vulnerable to major hazard risk. In fact, a 2020 Sphera survey[2] shows that just 37% of respondents classify their organization as taking a proactive approach to managing process safety risk, which means an incident or potential incident could be developing and they probably wouldn't even realize it.

However, the same research reveals that the challenge is recognized; 52% of organizations are now making investments in solutions to integrate data sources to be able to model and simulate real-time threats based on the health of their critical safeguarding barriers and minimize the possibility of a hazardous event occurring.

The next natural step is to use these systems to predict future risk status based on trends and to model the risk impact of potential changes that may be introduced to the facility.

2 Survey of process safety and operational risk engineers

In 2020, Sphera undertook its 5th annual survey[2] of process safety and operational risk management practitioners (Figure 1) in order to benchmark progress; understand best practices; and identify the future direction of risk management.

There were four main themes to the poll:

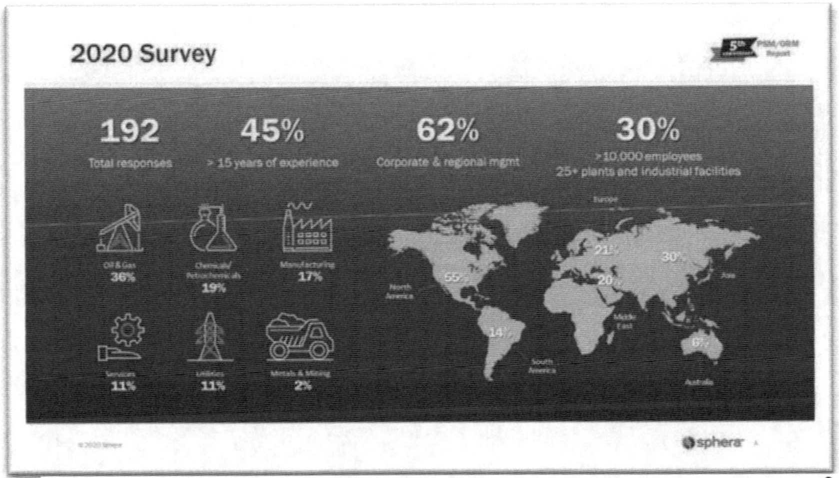

Figure 1 Sphera 2020 Survey of Process Safety Professionals[2]

- *safety culture*,
- the *reality of risk* in your industry,
- how *risk insight* is served; the methods organizations use for identifying, understanding and communicating risk, and
- to explore priorities and plans for *digital transformation*.

On the first theme of *safety culture*, the responses were positive in terms of safety intent (Figure 2). There is evidently a high degree of corporate commitment to:

- safety as a core part of the corporate value structure,
- commitment to continuous monitoring of general safety performance, and also
- a recognition of the need to strive to reduce operational risk and manage major hazards.

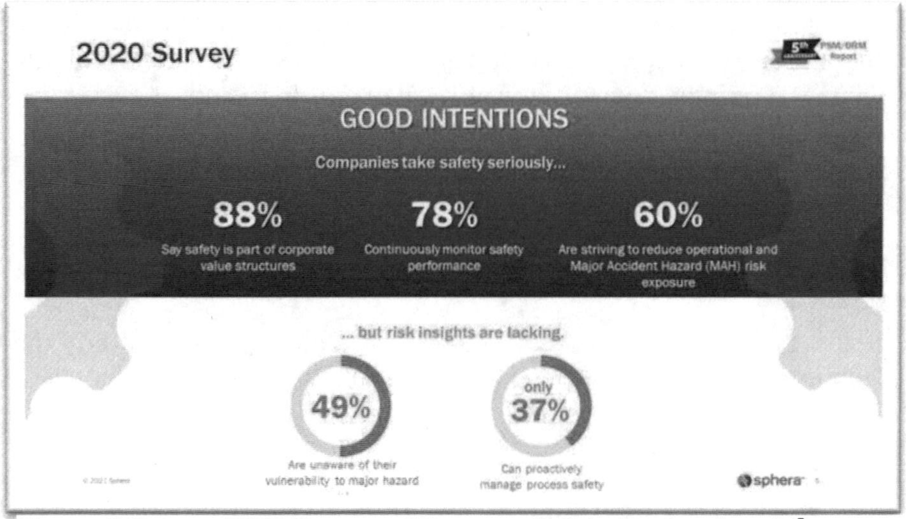

Figure 2 Sphera 2020 Survey of Process Safety Professionals[2]

However, despite good intentions, there is also evidence of some gaps when it comes to *risk insight*; 49% said most organizations are unaware of their vulnerability when it comes to major hazard risk, and only 37% characterized their organizations approach to managing process safety as “proactive”.

On this last point, it is noticeable that if we look back at the 5-year trend, the survey consistently shows between 35-40% of organizations categorizing themselves as “pro-active”.

To drill into this theme of *risk insight* a little further, we asked companies, how they are working to identify risk (Figure 3). Audits and inspections, HAZOP, What If studies, and other Process Hazard Analysis (PHA) techniques are the most common methods to identify hazards or non-conformances and analyze risk. In addition, 46% of respondents indicated they have deployed software-based business system to support their risk assessments. But they also highlighted concerns about the effectiveness of those studies and there is evidently a great deal of regional variation in terms of both the frequency and consistency of audit, inspection and hazard study revalidation.

By necessity, these kinds of studies are periodic. For example – around half of companies said they carry out risk reviews through PHA revalidation every 1 to 6 years. 53% said they are only assessing a portion of their facilities and half are concerned with the consistency and quality of their assessment and audit practices.

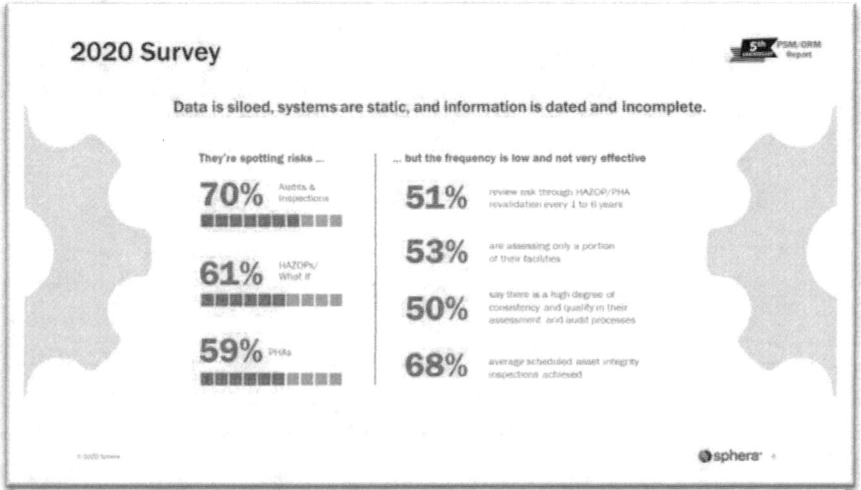

Figure 3 Sphera 2020 Survey of Process Safety Professionals[2]

Another interesting statistic is that, on average, companies said they are only able to complete 68% of their planned safety-critical maintenance and asset integrity inspections every year. Only half of respondents said they thought it was actually practical to achieve 100% of plan. The top 3 reasons given for this included limited resources (72%), conflicting priorities (71%) and limited budget (49%). These same top-three challenges to plan attainment have been seen every year for the last 5 years running.

So this paints a picture of reliance on periodic audit, inspection and hazard study revalidation which at best provides a retrospective snap-shot of the risk status on our facilities. In fact, we asked if any organizations were taking a more frequent look at risk status - just 8% of companies said they look at risk on a daily or weekly basis. If this is a reasonable goal, this suggests that perhaps the right tools are not available to make proactive safety decisions based on true risk status.

3 The journey to Dynamic and Simulated Process Safety Risk Management

3.1 Process safety gaps revealed by the 2020 survey

When asked about the *reality of risk* specifically at the frontline, 81% said there was a gap between process safety intent and what actually happens on the ground – for example, 58% said that when undertaking periodic process safety reviews – such as the PHA revalidations, they find that the risk on the facility has changed in the intervening period.

The main factors that drive the reported gaps between process safety intent and reality include,

- an exodus of experienced workers, according to about half of those surveyed, and
- 43% said there are conflicts between procedures and policies and frontline working practices,
- 39% indicated that poor process execution and management of change, and
- 43% said a lack of operational risk visibility is a key factor

There is increasing recognition of this gap between process safety intent and the reality of frontline operations. In 2020 both International Association of Oil and Gas Producers (IOGP) and the European Process Safety Centre (EPSC) published guidance on process safety fundamentals which highlighted the critical importance of frontline practices to sustaining good process safety performance.

Analysis by IOGP[3] based on data reported by members showed that for the period 2007-2017, nearly all (91%) of fatal process safety events were linked to one or more of a core set of 10 process safety fundamentals shown in Figure 4.

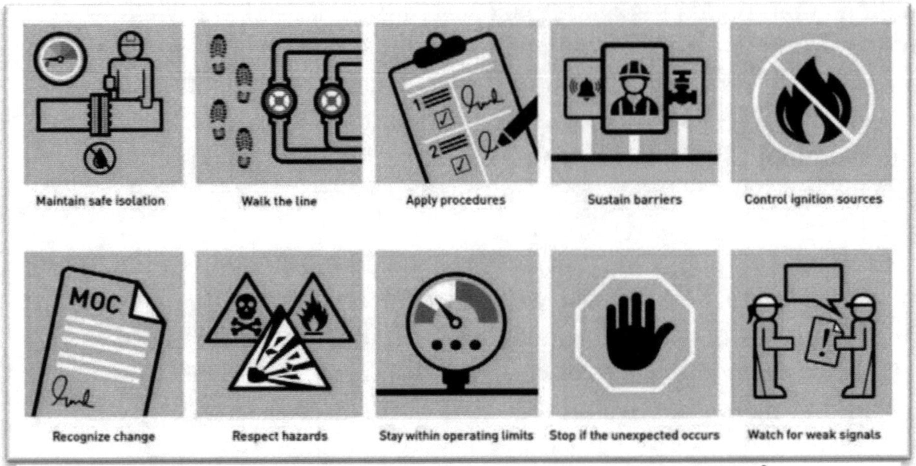

Figure 4 Ten IOGP Process Safety fundamentals[3]

A similar analysis[4] on approximately1000 process safety near-misses, high potential events and accidents in an EPSC member benchmark study showed that 51% of root causes identified were classed as Operational (with 14% Design and 35% Integrity). This drove the Centre to publish its own guidance on Process Safety Fundamentals (Figure 5).

These guidelines are designed to help those working in front-line operations and maintenance to start conversations and draw attention to the situations that are most likely to lead to process safety events and potential escalation.

Figure 5 EPSC Process Safety fundamentals[4]

Common to both sets of Fundamentals is the idea that it is critical to sustain barriers and safeguards on the facility whenever undertaking any activity. Indeed, the publications highlight activities that have the potential to cause a process safety accident, with the goal of increasing awareness and focusing attention and scrutiny of these activities.

The IOGP focus on the importance of watching for weak signals is a theme that is a core driver to the development of digital twin technology to highlight a dynamic real-time view of risk status.

3.2 Moving towards a dynamic real-time view of risk status

Perhaps un-surprisingly in this age of Industry 4.0 and Digitalization, 6 out of 10 respondents in the Sphera survey[2] said they are investing in technologies to create data-driven business processes, with particular focus on integrating EHS and sustainability data. When it comes to operational risk, half said they are investing in solutions to begin to track dynamic real-time risk and are also exploring digital twin technology to drive performance improvement, in the expectation that it will:

- drive safer operations,
- improve plant productivity, and
- help plan and deliver safer, more efficient shutdowns.

According to the Industry Analyst Verdantix[5], a digital twin is a concept spawned from the idea that computing power has evolved to the point today where we can create a fully digital replica of a complex facility. It models real-world systems and processes, using big volumes of data flowing from the physical asset to create its virtual twin.

The primary use-case for digital twins are aimed at managing the reliability and uptime of capital-intensive assets from an Asset Performance Management perspective. However, focusing on the critical barrier or safeguard health and the critical frontline practices, as suggested by the IOGP and EPSC guidelines, is the basis for the creation of an Operational Risk Management (ORM) digital twin (Figure 6).

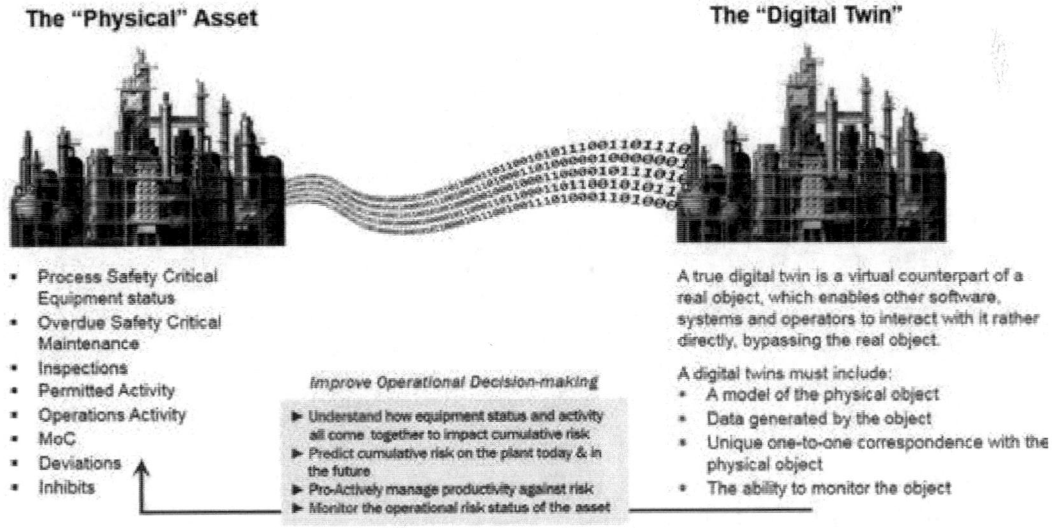

Figure 6 Modelling for an Operational Risk Management digital twin

From Verdantix' research, there are four key elements needed for a true digital twin:

1. Information gathering capability – a way to collect relevant information from sensors, human activities, and equipment status on the real-world asset;
2. Modelling capability - to pull the data together from multiple source systems to model the asset;
3. Analytical horsepower to model and analyze the impact of current status data, and also to predict future status based on trends or the planned or simulated changes that may be introduced to the system
4. Finally, we need innovative user interfaces to make the information accessible and enable people to understand and interpret the information to make informed decisions

For the information gathering capability required, it is clear from the IOGP and EPSC guidance that key business process and data sources fall into two broad categories:

- We are interested in all the operational activities that are planned or ongoing – whether that is permitted maintenance activity or routine operations activities; and
- We need to know the status or health of critical equipment or safeguards that exist at the site.

Information for both these categories is available in the application environment of a modern facility – examples of the kinds of systems are illustrated in Figure 7.

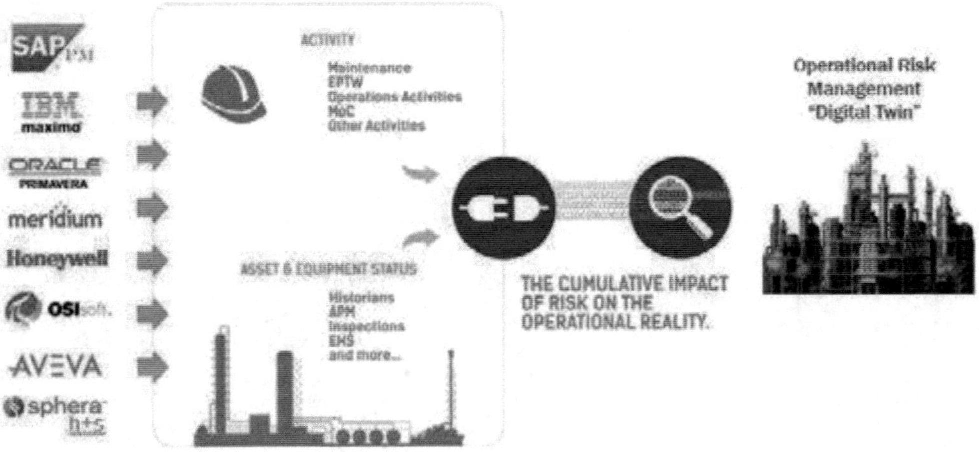

Figure 7 Data sources for an Operational Risk Management digital twin

It is important to recognize the value of pulling these two categories of information together – many accident investigations indicate that that these are interdependent – it is the combination of work activity and degraded - or absent - barriers that drive losses. In addition, silos of data and siloed business process exist – for example Enterprise Asset Management systems don't typically consider how human work practices can impact critical safeguards, and our safe systems of work – permits and job procedures - rarely have insight into the health of equipment barriers or safeguards that may impact the person executing a planned job.

The key point is that this is critical data that already exists in the business application environment of a modern industrial facility – the silos of data need to be connected to unlock process safety insights, as illustrated in the following case study.

3.2.1 Case study: developing an Operational Risk Management digital twin

A major national oil company has built and operates a world-scale refinery in the Middle East. During the greenfield phase, the operator commissioned a significant Industry 4.0 initiative to develop and deliver a technology-driven approach to integrate a suite of business systems to better support asset and operations processes and management. The operator has a sophisticated safety management system and clear corporate standards and practices. The operator's goal is to make a real-time view of operational and process safety risk a central element of decision-making from plant start-up onwards.

The operator has implemented an Operational Risk Management (ORM) system to manage all permitted activity and all deviations on the facility.

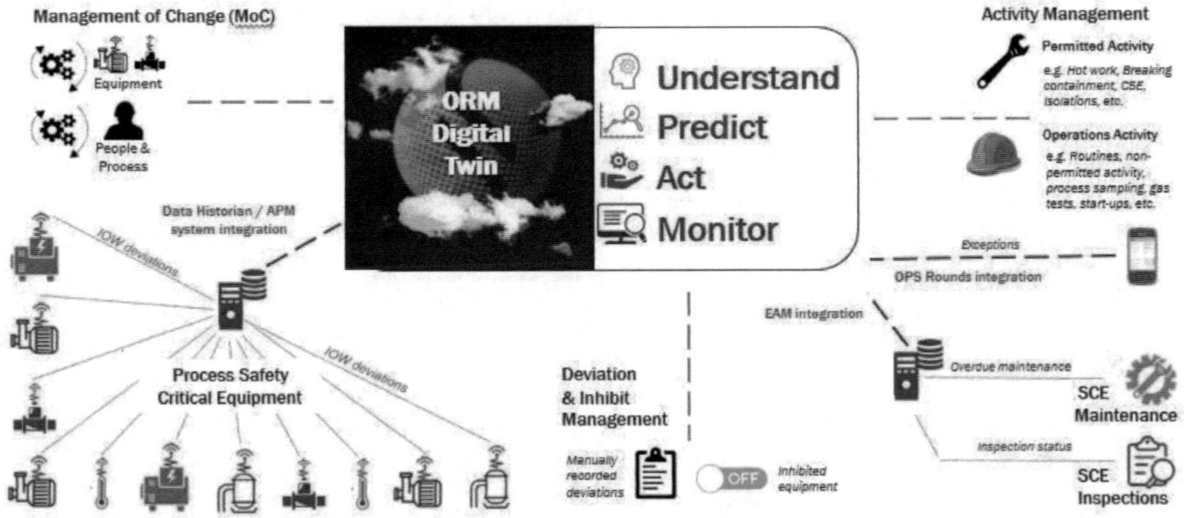

Figure 8 Case study: data sources and key business processes

The data in the ORM digital twin comes from four principal sources (Figure 8):

- The ORM system has integral business applications that manage the following key processes end-to-end: Management of Change; Permit to Work; Isolation Control; shift handover; and emergency preparedness. In addition risk assessments, approvals and monitoring associated with deviations or non-conformances on the facility are managed in the system.
- Other business applications that manage inhibits and operator rounds frontline business processes
- Technical system integrations to Enterprise Asset Management systems connect to the maintenance and inspection data – for this client the maintenance management system was SAP-PM – focusing on the maintenance and inspection record of specific critical barriers
- Similar technical system integrations to the data historians for the facility – in this case OSI-Pi – allows near real-time health status of critical equipment to be tracked.

These four sources ensure sufficient data capture capability to provide a near real-time feed of the changing risks and hazard exposure on the physical asset.

During design phase, the project team was able to access design package materials from the EPC contractors responsible for each refinery unit to identify the critical equipment. This includes bringing together a variety of information in useful formats:

- From design phase Hazop and asset integrity studies
- Health parameters associated with specific items of identified safety-critical equipment
- Records representing all critical equipment was set up in the ORM software
- Through integration, the ORM system "listens" for the health of critical equipment, based on the above parameters, from
  - Near real-time status from the data historian
  - Inspection records for critical equipment from operator rounds and inspection management system
  - Deferred planned maintenance for critical equipment from MMS
- Non-conformances are mapped to a fundamental process safety barrier model

In order to model and help visualize the risk impact, a fundamental barrier model, based on IOGP guidance[6] is used: where-ever an indication of weakness is found in the data, its impact on the barrier model for the location or system impacted is shown (Figure 9). Note these are not singular barriers, but a collection of equipment, instrumentation or people-driven processes which collectively fulfill the function of the barrier for that location.

Figure 9 Groupings to illustrate relative barrier health at location

Role-specific dashboards were created to support discipline-based use-cases, but there was also an effort to ensure all users could access three common screens (Figure 10) to show: Location - with a risk heat-map overlaid; Barrier impairments on the Barrier Group Model; and a Time - or risk-forecast – view, based on the upcoming work schedule.

Figure 10 Summarizing the Operational Risk Management status of the asset

From these common dashboards, users can drill down to understand what is driving the risk on the facility, to help support decision making in their function.

3.3 Focusing on major hazard scenarios, risk and assurance

The case study illustrates how a combined view of risk can be modelled, based on barrier or safeguard health data and the schedule of operational or maintenance activities for a location. From a technical safety perspective, the same system integrations for barrier health data can also be used to model risk status for the defined major hazard scenarios for the facility.

From a process safety point of view, the best starting point for modelling major hazard risk are the facility hazard studies – the HAZOPS, PHAs, LOPA studies and bow-tie diagrams. In some jurisdictions, these studies underpin a Safety Case document that identifies the major hazards (Major Accident Hazards – MAH) under management at a facility, and also specifies the system design and operational risk management intent to manage those hazards. This often takes the form of *specific* barriers – which may be hardware barriers or human barriers that are in place to minimize the risk of a defined top event[6] (Figure 11).

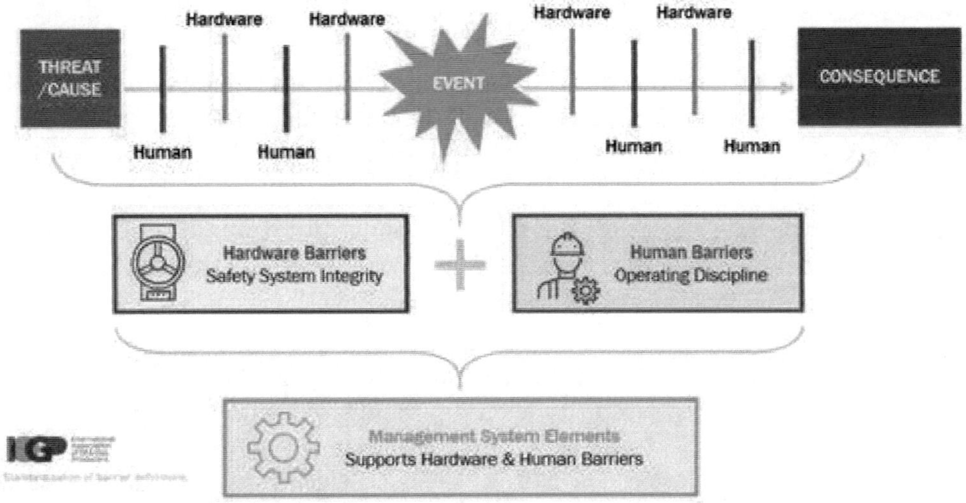

Figure 11 Major accident hazard event and defined hard and soft barriers[6]

These hazard studies are reviewed to establish the critical equipment tags and processes specified – this truly represents the inventory of tags or processes for which the asset has taken a risk reduction credit. As discussed in the case study, constructing an Operational Risk Management twin delivers the integrations to listen for the health of these tags or processes, wherever they appear in the business system environment.

The same source data and integrations are now used to monitor and visualize the health of "credible risk pathways" specific to the major hazard scenarios for the facility, illustrating the strength of each specific layer of protection along each pathway.

This drives the kind of visualization shown in Figure 12 – instead of focusing on location risk where we look at the cumulative impact of degradation to the Barrier Groups (Figure 9) for that location, the same data is presented in terms of risk driven by each *specific* individual barrier and its health for the major hazard scenario for which it was specified. Where problems are identified, user can drill in to see the details of the emerging risk pathways, understand the status of barriers and the relationship to the major hazards under management and intervene.

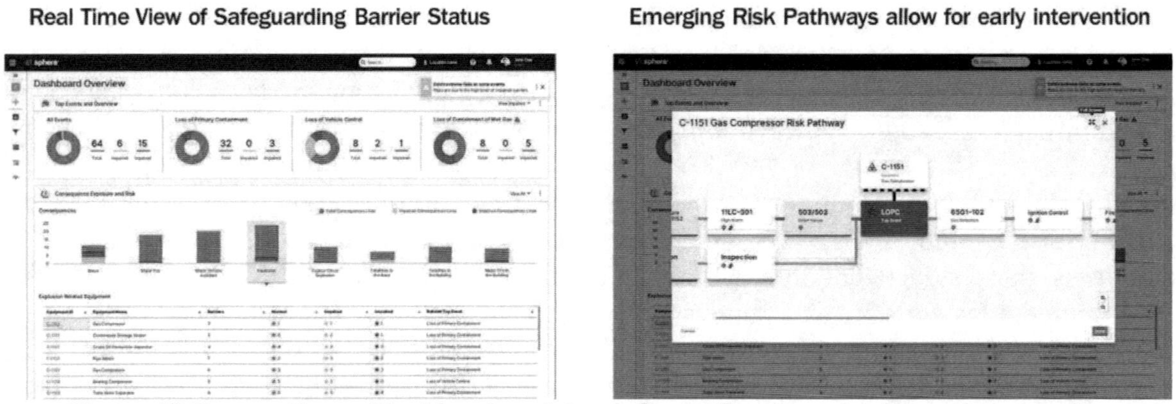

Figure 12 Connecting critical barrier status information to major hazard scenarios

By clicking on any barrier in the bow-tie like model, the system provides information on current assurance status, and why the barrier is impaired – e.g. maintenance or verification or testing not current, or a defective or absent device, audit finding, or even specific process conditions.

This allows the Technical Safety user to quickly understand the strength of specific safeguarding barriers against specific major hazard scenarios, and gives assurance that the hazards are under effective management.

Using this process safety source as the baseline set of data ensures laser-like focus on the major hazards and critical safeguards, which provides a practical way to connect process safety to the frontline.

4 Conclusion

There are gaps between process safety intent and what happens at the frontline. The dynamic nature of frontline operations represents a challenge from a risk management perspective. Companies are increasingly investing in so-called Industry 4.0 technologies to help close some of these gaps.

This represents an opportunity to rethink safety and risk management in operations: digital tools can be transformative and drive effective, efficient processes. These is an abundance of business process data sources in the application environment of a modern operating facility which represents a rich source of relevant safety and risk information. From a process safety perspective this data contains key information about the health of the critical barriers or safeguards that were specified in hazard studies and analysis

There are emerging Operational Risk Management systems which connect these data sources to the frontline, with the aim of providing a dynamic, real time view of risk to support practical decision-making. These systems deliver timely, relevant information to protect process safety barrier health and reduce exposure to major hazard risk.

5 References

[1] W.G. Bridges. "Gains from Getting Near Misses Reported", 8th Global Congress on Process Safety, Houston, Texas, USA, Apr 2012.

[2] Sphera, *2020 Survey on Process Safety and Operational Risk Management*, Sept 2020, https://sphera.com/research/2020-psm-orm-survey-report/.

[3] International Association of Oil and Gas Producers, *Process Safety Fundamentals*, Report 638, Oct 2020, London, UK.

[4] European Process Safety Centre, Process Safety Fundamentals, Nov 2020, https://epsc.be/Products/PS+Fundamentals.html.

[5] Verdantix, *Buyer's Guide: Digital Twins for Industrial Facilities*, May 2020, London, UK.

[6] International Association of Oil and Gas Producers, *Standardization of barrier definitions*, Report 544, Apr 2016, London, UK.

621786 Sponsored Technology Workshop - Benefits of Procedure Clarity and Effective Knowledge Transfer on Field Operator Performance (incl. Live Demo) - Voovio

Abstract:

Voovio Data collected from across North America shows "Procedure Clarity" is a major source of error that is overlooked by the industry. Clear and concise procedures, standardized on plant operation that have objective assessments, are foundational building blocks to ensure field operator competency and deliver predictable process performance. Yet operators are consistently challenged by a lack of clarity and sometimes correctness of their procedures, combined with inconsistent training. We will take a deep dive into the problem and show a live demo that solves these challenges across refining and petrochemicals.

SPRING21
+17TH GCPS
A Joint AIChE and CCPS Meeting

The e-SafetyCase – Electronic or Effortless?

Gareth Ellor
Director - Clean Energy & Innovation
Risktec Solutions Ltd
Wilderspool Park
Greenall's Avenue
Warrington
WA4 6HL
UK
gareth.ellor@risktec.tuv.com

Prepared for Presentation at
American Institute of Chemical Engineers
2021 Spring Meeting and 17th Global Congress on Process Safety
Virtual
April 18 - 22, 2021

The e-SafetyCase – Electronic or Effortless?

Gareth Ellor
Director - Clean Energy & Innovation
Risktec Solutions Ltd
Wilderspool Park
Greenall's Avenue
Warrington
WA4 6HL
UK
gareth.ellor@risktec.tuv.com

Keywords: Electronic, Safety, Case, HSE, HSSE, ALARP, Risk, Management, Digitalization.

Abstract

An Electronic Safety Case (e-SafetyCase) is an alternative structured approach to safety. The purpose of an e-SafetyCase and the information contained within it remains the same as a traditional Safety Case; to demonstrate risks are reduced As Low As Reasonably Practicable (ALARP). However, instead of being a lengthy and complex document, the e-SafetyCase is highly interactive, using clickable links as a means of connecting and navigating the information quickly and easily. Not only does this make it simple, intuitive and user-friendly, but by integrating it with day-to-day operations, it is kept current and evolves with the facility. This ensures there is no disconnect between the safety case and operations; they become one and the same! As a result, the e-SafetyCase not only demonstrates risks are reduced ALARP but it ensures risks are managed ALARP throughout the facility lifecycle; the ultimate goal of any safety management system. If done properly, an effective e-SafetyCase will be so useful and helpful to the workforce that it will naturally evolve hand-in-hand with the facility as routine business. Effortless safety – is that not too good to be true?

This paper introduces the concept of an e-SafetyCase outlining why, despite being a concept which has been around for some time, it has struggled to gain widespread traction across industry. It debunks the myths about what an e-SafetyCase is, in an attempt to knock-down the barriers we consider are blocking its widespread use. The majority of the paper is devoted to demonstrating what an e-SafetyCase looks like using a practical case study, highlighting the significant benefits it delivers to the safe and reliable operation of the facility throughout its lifecycle. In addition, it demonstrates the flexibility and adaptability of the e-SafetyCase approach to allow the considerable benefits to be realized by any major hazard facility. Finally, it will touch on ground-breaking work being done to further improve the functionality and hence benefits of an e-SafetyCase by blending with wider technological innovations. For example, the integration of virtual and augmented reality to really bring the Safety Case to life through modelling of accident scenarios and emergency response arrangements. Although ambitious, this kind of forward thinking could ultimately change the landscape of safety management in major hazard industries.

1 Overview

This paper introduces the concept of an Electronic Safety Case (e-SafetyCase), outlining why, despite being an aspiration which has been around for some time, it has struggled to gain widespread traction across industry. It debunks the myths about what an e-SafetyCase is, to help knock-down the barriers blocking its widespread use. The majority of the paper is devoted to demonstrating what an e-SafetyCase looks like using a practical case study from an Offshore Wind Farm. This case study illustrates the significant benefits it can deliver to the safe and reliable operation of the facility throughout its lifecycle. In addition, it demonstrates the flexibility and adaptability of the e-SafetyCase approach to allow the considerable benefits to be realized by any major hazard facility. Finally, it touches on the ground-breaking work being done to further improve the functionality and hence benefits of an e-SafetyCase by blending with wider technological innovations. Although ambitious, this kind of forward thinking could ultimately change the landscape of safety management in major hazard industries.

2 What is a Safety Case?

Before we start to explore the e-Safety Case concept, we must firstly remind ourselves of what a Safety Case actually is.

A Safety Case is generally accepted as being the written demonstration that a plant, facility, operation, etc. has been designed, conceived and executed in such a way that the major Health, Safety, Security and Environmental (HSSE) risks incurred during its lifecycle are reduced to a level that is As Low As Reasonably Practicable (ALARP).

They are virtually always written documents and despite best intentions from some to break-down established precedents and norms, and perhaps a lazy approach from others, they often end up being lengthy and complex. Whilst serving their core purpose, they can easily become impenetrable to anyone without direct Safety Case experience, heavily limiting wider benefits.

They cover the major HSSE risks thereby ensuring that most effort is focused on managing the risks that could cause greatest impact. They rely on the wider HSSE Management Systems to effectively manage less significant hazards on a proportionate basis.

They are generally associated with a legislative regime and, if one were to be cynical, have evolved in such industries to be a document issued to a regulator to show risk levels are ALARP to obtain a licence to operate. However, in reality, a Safety Case is independent of regulations. It represents good practice. It shows all stakeholders that risk is being managed effectively.

Myth	Reality
✗ Complex and theoretical.	✓ Proportionate to risk.
✗ A thick, weighty document that no one reads!	✓ Fit for purpose, simple, clear and effective. No baggage or precedents! Useful!
✗ Something to tick a box and get a stakeholder off your back and then sits on a shelf gathering dust!	✓ Provides a central focal point to maintain and improve safety throughout the life-cycle of an asset. A live document!
✗ Symptomatic of a legislative regime.	✓ Independent from legislation. Good practice. Shows stakeholders you are managing risk effectively!
✗ Very expensive!	✓ Effective management of risk proven to reduce overall project costs.

Figure 1. Safety Cases - Dispelling the myths [Ref 1]

Figure 1 summarizes the myths we hear time and time again about what Safety Cases are perceived to be (on the left) with the reality of a good Safety Case on the right. This has prompted us to use the Marmite analogy to inject some humor into what is a very dry subject ("you either love it or hate it!"), to try and break down these myths and allow industry to embrace the Safety Case approach and realize the huge benefits it offers. People are rarely ambivalent to a Safety Case. They either love it or hate it. In our experience, the haters usually come with a skewed view of what a Safety Case is. Figure 1 shows the huge untapped potential of a Safety Case and the e-Safety Case is ready to deliver on this.

3 What is an e-SafetyCase?

The purpose of an e-SafetyCase, and the information contained within it, remains exactly the same as a traditional Safety Case; to demonstrate risks are reduced ALARP. It is how this information is presented that is the key difference. Instead of being a lengthy, complex written document, the e-SafetyCase is highly interactive and intuitive, using clickable links as a means of connecting and navigating the information quickly and easily. As a result, the e-SafetyCase is accessible and helpful to anyone involved in the safe operation of the facility.

The best way to understand what an e-SafetyCase is, and get an appreciation of the benefits it offers over a traditional paper-based Safety Case, is to take a look at a practical example.

We have developed an e-SafetyCase for an offshore wind farm being built in UK waters. Although not a regulated industry, the client showed a strong appetite for a Safety Case approach as they understood the benefits it delivers.

Before we explore this, we must firstly understand how it was developed as this has a significant bearing on how it is presented.

3.1 Risk Management Process

A Safety Case should represent the output of a proportionate and practical risk management process which is fully integrated with the project development process for a particular design. Figure 2 illustrates the process adopted as the foundation of this Offshore Wind e-SafetyCase and shows how this meets the three fundamental challenges of effective risk management (see Figure 3).

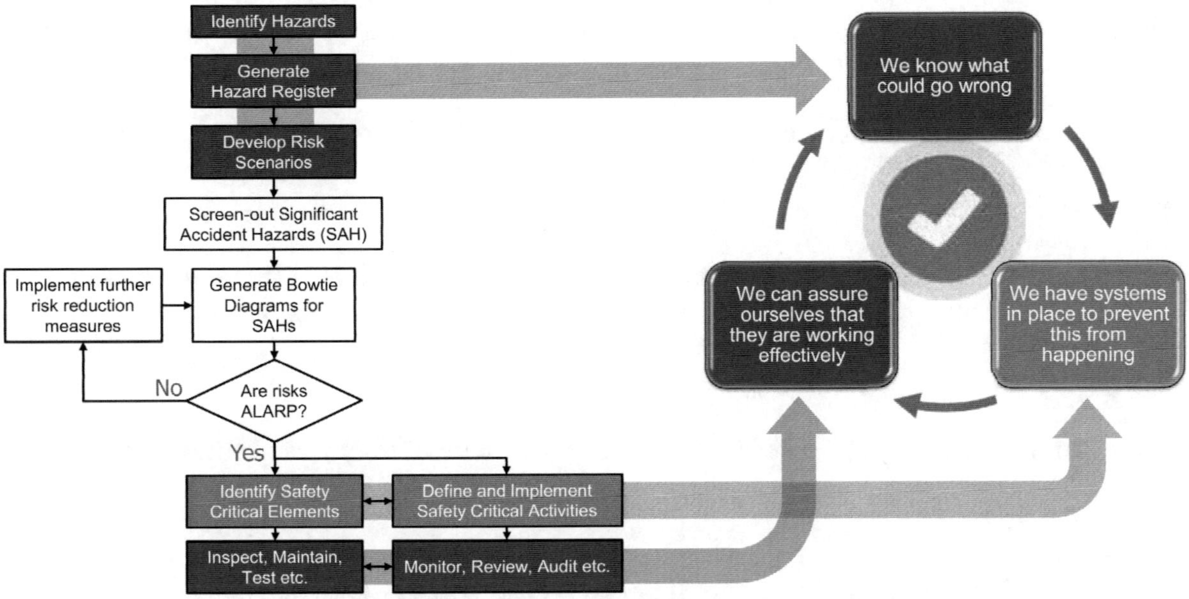

Figure 2. Risk Management Process – the foundation of the e-SafetyCase

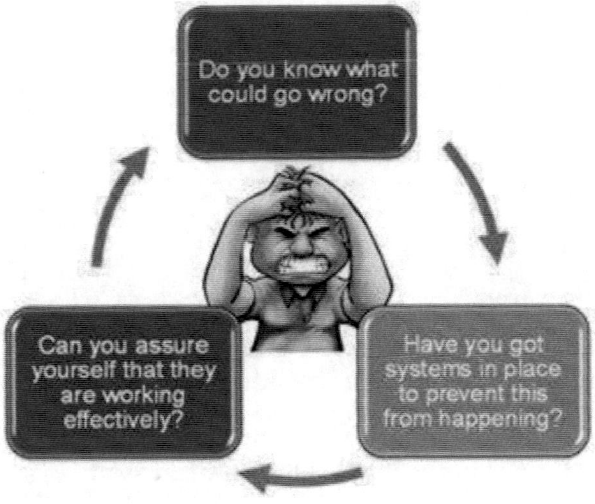

Figure 3. The challenges of effective risk management [Ref 2]

3.2 Definitions

Significant Accident Hazards (SAH) – Although not comparable to, say, the oil and gas industry in terms of ultimate consequences, if the risks associated with offshore wind are inadequately managed, they could still result in fatalities and loss of reputation. Whilst the term "significant" was used instead of "major" to define the hazards screened-out for detailed assessment, the approach is equally relevant to major hazard industries. The only difference being that in applying this approach to major hazard facilities, more hazards are likely to be identified for detailed analysis and/or the screening threshold may rise to ensure a proportionate approach.

Safety Critical Elements (SCE) are defined as those items of equipment or structures whose failure could cause or contribute to a SAH, or the purpose of which is to prevent or limit the consequences of a SAH.

Safety Critical Activities (SCA) are defined as those activities which are critical to ensuring safety and continued integrity of design, or ensure SCEs continue to function as intended.

3.3 Bowtie Approach

As illustrated in Figure 2 on te previous page, the Bowtie methodology was used to assess all SAHs.

The Bowtie technique is an extremely powerful way of clearly illustrating how risk is being managed within an operation, business, etc. It helps to ensure that risks are managed rather than just analyzed, partly by going beyond the usual risk assessment "snapshot" and highlighting links between the risk controls and management systems.

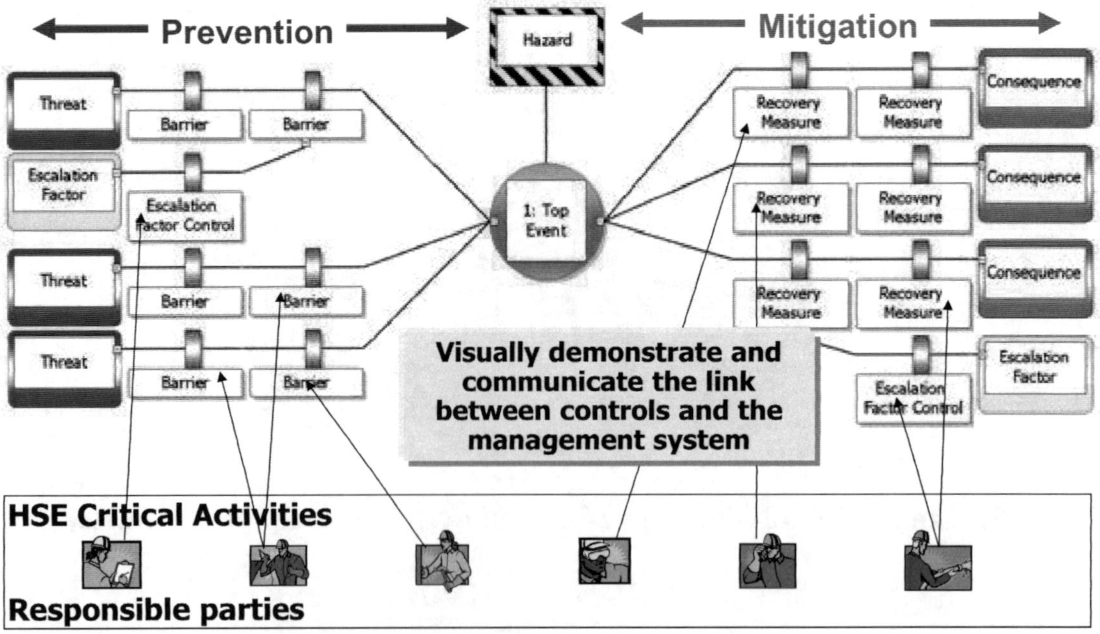

Figure 4. Bowtie Diagram

Figure 4 illustrates the basic structure of a Bowtie diagram, the key elements of which are:

- The **Hazard** which has the potential to give rise to the unwanted event you are concerned about. This is illustrated in the yellow/black striped box.

- The unwanted event you are concerned about if the hazard is "released". This is called the **Top Event** and is the red circle which forms the "knot" of the Bowtie.

- The credible causes of the unwanted even, termed **Threats,** and are illustrated in blue on the left hand side of the diagram.

- The unmitigated **Consequences** of the unwanted event. These are illustrated in red on the right hand side of the diagram.

- An **Escalation Factor** (or "degradation factor") which could compromise the integrity of a barrier.

- The **Barriers** in place to prevent the unwanted event occurring or minimize its consequences. Those on the left hand side of the diagram prevent the hazard being realized and are categorized as **Prevention**. Those on the right hand side limit the ultimate consequences of the hazard if realized and hence represent **Mitigation**. Barriers can also be included to eliminate or reduce the impact of an Escalation Factor on the integrity of a specific barrier, either prevention or mitigation.

- The specific systems, processes, procedures, etc. which must be in place to ensure the integrity of these barriers. This is achieved by identifying the Safety Critical Elements (SCE – plant, equipment, etc.) and their associated Safety Critical Activities (SCA – operations, inspections, checks, etc.), along with the roles, responsibilities and competencies of the people who ensure these work effectively.

This latter point represents the real benefit of the Bowtie methodology. It provides a clear, auditable trail from hazards right through to the specific systems, processes, procedures, etc. implemented, with performance standards and responsibilities clearly assigned, to reduce the associated risks ALARP. It therefore illustrates the complete safety assurance process in a clear picture which can be easily understood by everyone within the organisation.

3.4 Significant Accident Hazards & Bowties

The following Significant Accident Hazards (SAH) were identified for the offshore wind farm and Bowtie diagrams were generated for each of these:

1. Electrical hazards
2. Fire hazards
3. Loss of structural integrity
4. Personnel transfer
5. Personnel at height
6. Hydraulic and mechanical systems

7. Hydraulic systems – pressure hazards
8. Vessel transport
9. Helicopter transport
10. Suspended loads
11. External hazards
12. Explosion hazards

3.5 e-SafetyCase Overview

The e-SafetyCase we have developed is presented in a highly visual web-browser style. To illustrate just how simple this concept can be, the version we're going to show was developed in PowerPoint and turned into a .pdf file. It is therefore accessible on any computer and can be developed and updated using standard software on any PC. Furthermore, it is sized such that it can be readily saved on a laptop or tablet allowing it to be accessed by anyone, anywhere, anytime.

It can be navigated via two separate home pages which the user can toggle between.

Firstly the user can explore the e-SafetyCase from a process perspective (see Figure 5). This is based on a more simplified version of the risk management process illustrated in Figure 2. This allows the user to quickly and easily access the information based on the process through which it has been generated. This allows those familiar with the underlying risk management process to access the information they require in a very quick and highly intuitive way. It also brings the overall safety case together to show how all the elements combine to present a robust justification that risk is being managed ALARP. Finally, it raises awareness of the underlying risk management process to people less familiar with this, promoting an understanding of how risk is being managed across the business.

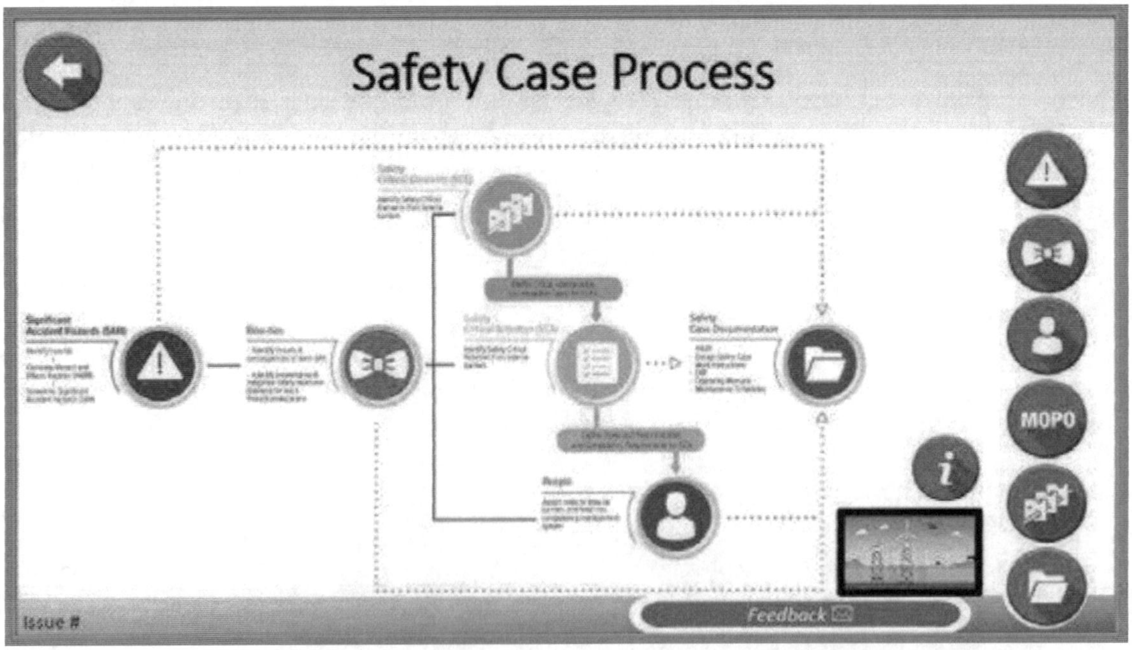

Figure 5. Process View Homepage

Alternatively, the user can switch to an infographic showing all the key elements of the overall wind farm to explore the e-SafetyCase from an asset perspective (see Figure 6). This is particularly useful to someone less familiar with the risk management process or perhaps wants to immediately dive into the detail behind how risk is being managed within a specific element (e.g. turbine, foundation, offshore substation, helicopter, etc.).

Figure 6. Asset View Homepage

On each homepage the user can click on the same set of buttons to access the following information:

Significant Accident Hazards (SAH)

Presents definition of a SAH, lists all SAHs and provides information on each individual SAH (description, video, training material, etc.).

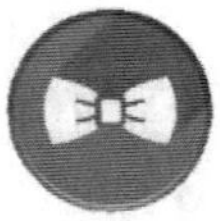

Bowtie Diagrams

Lists all Bowtie diagrams relating to each SAH and allows user to view individual Bowties to varying levels of detail (overview to complete).

People

Identifies all safety critical roles claimed within the Bowties and assigns these to a responsible person, whilst listing all of the SCAs each person is responsible for.

Matrix Of Permitted Operations (MOPO)

An interactive tool to determine whether it is safe to undertake key operations, recognising the conditions which may elevate the associated risk and/or other simultaneous operations.

Safety Critical Elements

Lists all SCEs, rolled-up into groupings linked to the principle of hierarchy of risk reduction.

Documentation

Provides links to all associated Safety Case documentation (studies, analysis, etc.) which underpins the position of safety set-out within the e-SafetyCase. This represents the supporting evidence.

The "Feedback" button, when clicked, opens up an e-mail to allow the user to provide feedback directly to the e-SafetyCase owner (e.g. something unclear, something wrong, something which can be improved, etc.). The e-SafetyCase owner collates this feedback and folds this into periodic updates where appropriate.

Whilst the presentation of the Safety Case in such a simple, intuitive and user friendly format is a quantum leap from traditional paper based Safety Cases, the real beauty of the e-SafetyCase is that all of this information is connected together to allow the user to access it via whatever route (or "journey") they so choose, via quick and easy links. In very simple terms, the e-SafetyCase brings the Safety Case to life through the navigation triangle illustrated in Figure 7.

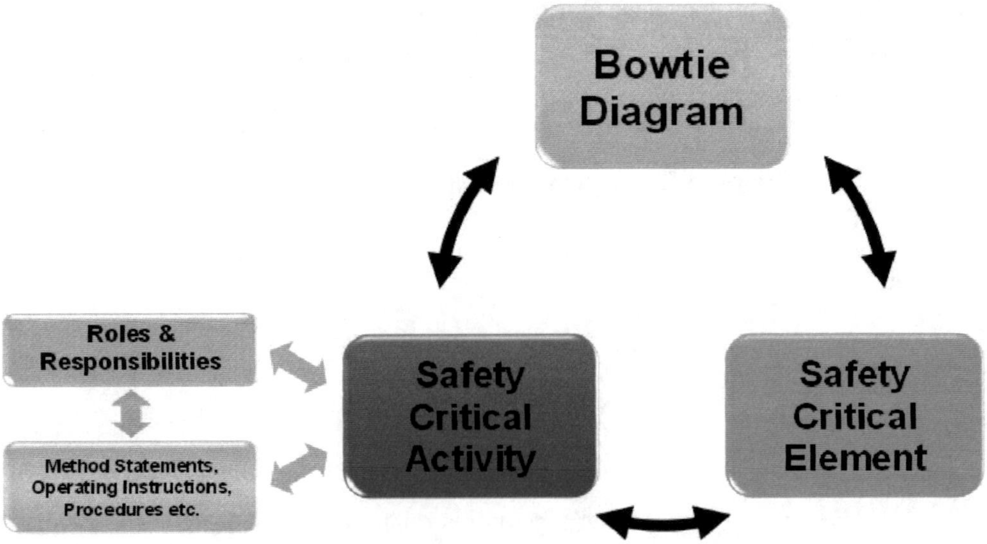

Figure 7. e-SafetyCase navigation triangle

For each SAH the user can, via single clicks, jump between the Bowtie, SCEs and SCAs from any entry point. This allows the user to:

- Gain an understanding of the role each SCE or SCA plays in ensuring risks are reduced ALARP by linking directly to the Bowtie and showing how each fits into the big picture.

- Link directly from a SCE to view the SCA(s) which must be performed to ensure that feature continues to perform its safety function correctly.
- Link directly from a SCA to the associated SCE to understand the feature the activity relates to.
- From a SCA, click directly to the method statement, operating instruction, procedure, etc. which must be followed to complete that activity correctly and identify who is responsible for completing that SCA.
- Quickly and easily identify all SCAs a specific individual is responsible for.

But this simple and easy navigation is not confined to individual SAHs. The links cut across all SAHs, Bowties, SCAs and SCEs and can be accessed both from a process and asset perspective.

The example user journeys described in Table 1 illustrate the flexibility and benefits an e-SafetyCase offers.

Table 1. Example e-SafetyCase user journeys

No	User	Requirement/Benefit
1	Service Technician	Familiarize himself/herself on the important factors in executing a safe transfer from marine vessel to offshore wind turbine whilst transiting out to the wind farm using the e-SafetyCase downloaded onto a mobile device.
2	Marine Controller	Refresh himself/herself of all the SCAs they are responsible for to ensure he/she performs these roles effectively.
3	Maintenance Technician	Find the method statement to allow him/her to complete one of their designated SCAs and watch an instructional video on how to complete that task.
4	HSE Manager	Assess potential vulnerability to a recent incident on a similar wind farm by interrogating the resilience of the barriers included within the relevant Bowtie diagram based on the findings of the incident investigation.
5	Regulator	Understand how the risk of fire is being managed within a wind turbine to form a view as to whether this risk is being managed ALARP.

<u>What it is not….</u>

It is important to recognize that the e-SafetyCase described here does not present a live risk monitor or risk “dashboard”; something which, confusingly, is often termed an e-SafetyCase.

This is something industry clearly has an appetite for but is a very different challenge to that being explored here. Work is ongoing to develop tools to do this and hopefully these could ultimately be blended with this e-SafetyCase concept to provide a complete, holistic solution to managing risk within a major hazard facility.

4 What benefits does an e-SafetyCase offer?

It goes without saying that an e-SafetyCase has to do what a traditional paper based Safety Case does; demonstrate risk is reduced ALARP. The real benefits of an e-SafetyCase come from the extra things it can do. It's a Safety Case, but not as we know it!

As previously described, this e-SafetyCase provides multi-dimensional links to allow any individual involved or interested in the safe operation of the offshore wind farm to quickly and easily access the information they need to perform their role effectively. It houses all of the information to ensure, demonstrate and support safe, reliable operation in one place and puts this at the fingertips of the right people in a format that suits them. This delivers three key benefits:

- It brings the information within the Safety Case to life in a simple, interesting and engaging way promoting a better understanding of safety and better safety compliance across the facility. As opposed to a traditional paper-based Safety Case, which tends to sit on a shelf gathering dust, an e-SafetyCase can actually promote and drive safe operations by encouraging feedback from the field that allows it to be improved and optimised to the benefit of the overall facility.

- By providing access to all of this information in a very simple and intuitive way, it becomes helpful and useful to anyone involved in the operation of the facility, not just those concerned about safety. When information is useful, it is well used, and so there is a natural incentive to keep it up to date. As a result, the e-SafetyCase evolves with the facility, ensuring there is no disconnect between the Safety Case and operations; they become one and the same (see Figure 8, with apologies to London tube stations). This overcomes the huge limitation of traditional paper-based Safety Cases; they are a burden to maintain and if they do not evolve with the facility, safety is inevitably compromised.

Figure 8: An e-SafetyCase bridges the gap between operations and safety

- By linking directly to traditional management system documentation such as procedures, method statements, job descriptions, etc., the e-SafetyCase provides a seamless link from the arguments as to why something should be safe right through to the documentation provided and responsibilities discharged to make sure it is actually safe. This is a significant improvement over traditional paper-based Safety Cases which tend to stop with the "why?" – the snapshot at that moment in time when the case was written.

So, in summary, the key benefit of an e-SafetyCase is that it:

- provides an accessible, engaging and intuitive basis of safety which proactively drives and encourages safe operations;
- naturally evolves hand-in-hand with the facility as routine business; and
- provides a seamless link between hazards and the actions people need to take to ensure risk is reduced ALARP.

Effortless safety – is that not too good to be true?

5 So why aren't all Safety Cases electronic?

The e-SafetyCase approach has been around for some time, but has struggled to gain widespread traction across industry, potentially due to misconceptions about the complexity and hence cost of development or the need for bespoke software. In reality, it's just about presenting information in a different and more accessible way, created using simple software found on most computers. What's more, this format can be applied to any industry for any facility. The cost of generating an e-SafetyCase is no greater than a paper-based Safety Case. All the underlying analysis, studies, etc. remain identical, it is just the final presentation that is different. If the decision to present the Safety Case in an "electronic" format is taken early, then it can be developed at no cost or programme penalty compared with a traditional paper-based version.

Another possible reason for not developing an e-SafetyCase is the concern that maintaining and updating it is too difficult and time consuming. Although updating the e-SafetyCase may seem more onerous than a traditional Safety Case, with a good management of change log and guidance, updating the e-SafetyCase should be no more taxing than updating a traditional Safety Case. It may just need someone with a different skill-set. Indeed, as previously demonstrated, an effective e-SafetyCase will be so useful and effective it will naturally evolve hand-in-hand with the facility as a natural part of routine business.

There's a danger the e-SafetyCase grows into a catch-all document due to the extra functionality it offers, however the intended focus of the e-SafetyCase should not be lost; it should not replace existing management systems or document systems but rather act as a link and interface into these complementary existing systems.

6 Further development and future potential

6.1 e-SafetyCase App

The offshore wind farm e-SafetyCase we have described here was built in PowerPoint with interactive links and then converted to a PDF. Whilst very simple to develop and use and low-cost to deliver, primarily because it uses standard software everyone has access to, all of the logic built into the Bowties had to be recreated within PowerPoint. This is obviously time consuming, introduces the potential for errors and makes updates more difficult.

To provide a more effective, long-term solution, we have developed an App which we can use to build an e-SafetyCase for an asset, facility, etc. The App works on PC, tablet and mobile phone (PC/Android/iOS). The App "sucks-in" the Bowtie native files which cover the major accident hazards for that facility (e.g. Offshore Wind Farm, Oil & Gas Offshore Platform, Onshore Refinery Plant, etc.) and provides a simple, engaging and intuitive way of navigating around this information in the same way as we have demonstrated here for the offshore wind farm.

So, in the analogy below, the Bowtie files are the image carousel and the e-SafetyCase App is the viewer. This therefore provides all of the benefits of the PowerPoint e-SafetyCase that we have demonstrated here, but with huge time and cost savings as it is automatically generated by the Bowties, thereby avoiding errors and making future updates and changes effortless as they simply evolve out of the Bowties which are maintained through the life of the asset/facility.

Rather than being a "product" which we sell, we use this App as an internal tool to help create an e-Safety Case for our clients. All Safety Cases are ultimately bespoke and are tailored to present a "Case for Safety" for an individual asset, facility, etc. Our VIVA App provides this functionality, enabling a bespoke e-SafetyCases to be built for each client and which is delivered to them in App form to use and maintain as they wish.

VIVA

BRINGING SAFETY TO LIFE

6.2 Blending with Digitalization

The e-SafetyCase concept has significant scope for further development and improvement by embracing wider technological innovations. A good example of this potential is the work we are doing to blend the e-SafetyCase with Virtual Reality (VR) to truly bring the Safety Case to life by immersing the user within the actual facility and overlaying Safety Case information on the plant itself, thus removing the disconnect between the facility and its Safety Case.

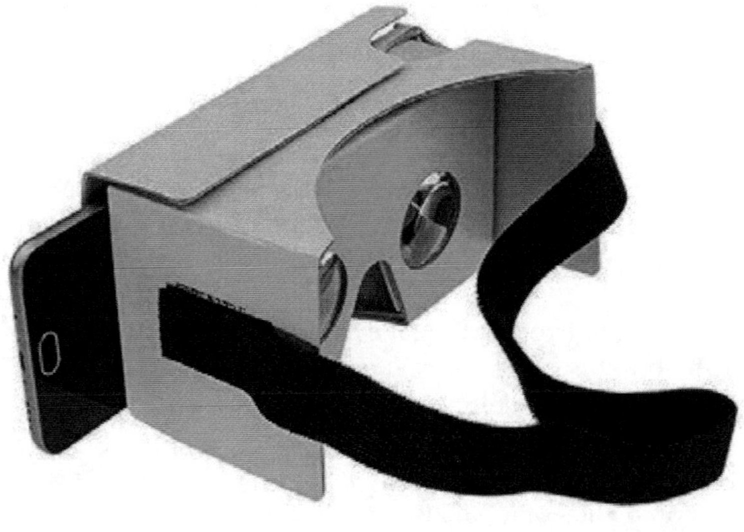

Figure 9. Google's Cardboard VR Goggles

VR technology has evolved significantly over recent years and is now available in very affordable and accessible formats such as Google's cardboard VR goggles (see Figure 9). These goggles incorporate special lenses and accommodate a standard mobile phone. The lenses translate moving images played on the phone's screen via compatible Apps into a 3D VR experience. The positional sensors within the phone are synchronized with the image being played on the device such that when the user tilts their head or turns around, they see the corresponding view creating a highly realistic, 360° immersive experience. In addition, a button on the side of the goggles taps the mobile phone screen allowing the user to select options built into the video footage played on the mobile device providing an interactive experience.

We have developed a prototype e-SafetyCase augmented with VR for a Wind Turbine Fire Safety Case. The Safety Case is viewed via an app downloaded onto a standard tablet device. Where VR enhances the Safety Case, a VR icon appears, which, when clicked, launches a VR snippet on a separate mobile device located within the Google goggles. The user can then put on the goggles and actually experience that element of the Safety Case in real, absolute terms. Some examples of the enhanced functionality this delivers include:

- A VR icon appears next to a SCE within a Bowtie diagram. Clicking the icon within the Bowtie launches a VR snippet which allows the user to see where that feature exists within the wind turbine in 3D terms and to understand its size, scale, configuration, etc. just as if they were physically there within the turbine itself.

- A VR icon appears next to the SCAs assigned to a responsible person. Clicking the icon launches a VR snippet which allows the user to see that feature so that they can experience in real terms how they must go about completing that activity.

- A VR icon appears next to the key stages of the various Escape, Evacuation and Rescue (EER) routes from the wind turbine in the event of all credible fire scenarios. Clicking these icons launches VR snippets which allow the user to experience the complete escape route, including;

    - the physical route they must follow;
    - the location of EER equipment (e.g. immersion suit, harness and descent device to allow an external escape from the nacelle of the turbine through the rear escape hatch as a last resort);
    - the time they are expected to take to progress through the route to ensure they can escape before it is compromised (based on findings of the EER study, supported by fire and smoke modelling);
    - the things they need to do at each stage (e.g. put on harness, connect to decent device, clip device onto anchor point, open escape hatch, etc.), and;
    - the extent to which visibility is likely to drop as they progress through the escape route by simulating the results from the supporting fire and smoke modelling.

  This delivers two key benefits:

    - It allows operators to familiarize themselves with EER arrangements – providing an excellent training tool.
    - It provides confidence (or otherwise) in the adequacy and acceptability of overall EER provision by allowing users to test, in a safe, virtual environment, whether escape can be completed safely before the overall route is compromised, thereby reinforcing the integrity of the case. Don't just read about why something is safe, experience it!

However, the direction of travel is not just one way. When wearing the goggles to view the VR snippets, the user can click the button on the side of the goggles to access related information which can be viewed back on the tablet. For example, SCEs are highlighted within the VR model and when clicked (via the button on the goggles) a simple menu opens-up allowing the user to select and then view (back on the tablet) either:

- The Bowties where that SCE is claimed;
- the SCAs associated with that SCE; or
- instructional videos associated with that SCE.

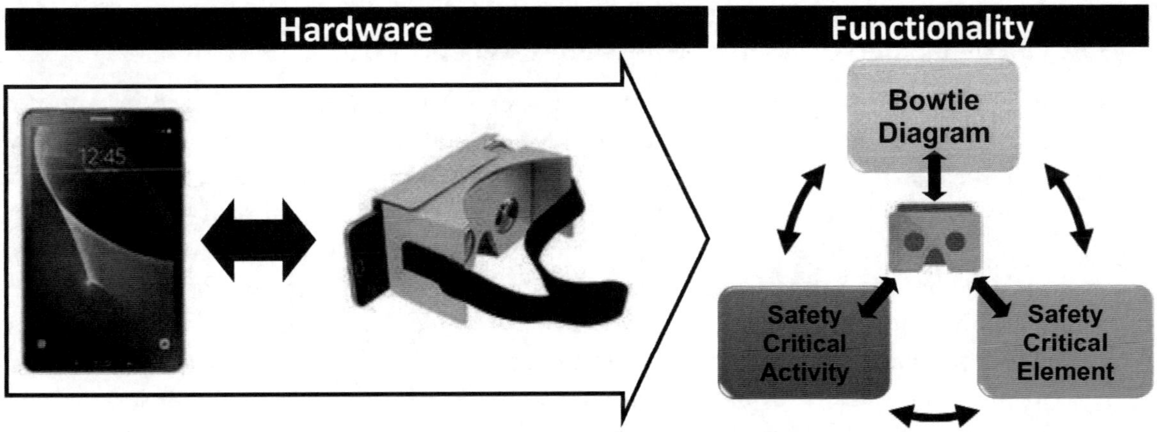

Figure 10. e-SafetyCase augmented with Virtual Reality

As illustrated in Figure 10, this two-way connectivity greatly improves the e-SafetyCase user experience from the version previously demonstrated (see Figure 7). The VR complements the e-SafetyCase, further bringing it to life, whilst also being a real life navigational tool for the Safety Case itself making it even more accessible, intuitive and easy to navigate.

The Holy Grail for any Safety Case is for the plant itself to show why it is safe. An e-SafetyCase augmented with VR goes a long way towards achieving this.

7 Perceived limitations of e-SafetyCase versus paper-based Safety Case

As the saying goes; "the proof of the pudding is in the eating!". As with any other innovation, e-SafetyCases will only become the established norm if they are better than what already exists. They must never be championed as an academic or theoretical crusade - they will rise or fall based on this very practical test. So, before significant time, effort and money is spent on the development of e-SafetyCases, as any good Safety Case author should always do, we must critically challenge whether they are indeed better than traditional paper-based Safety Cases by considering the following potential limitations:

1. **Ease of accessibility – can they really be "accessed by anyone, anywhere, anytime"?**

   By making the e-SafetyCase available as an App which can be downloaded onto a standard mobile device, yes it can. Most people have an electronic device which they could download the App onto. Also, the price of this technology is dropping all the time and so the cost to the company of providing all Safety Case users at a facility with an intrinsically safe tablet would be small in comparison to the cost of developing the Safety Case itself.

2. **An e-SafetyCase provides lots of peripheral benefits, but does it actually deliver on its core function – demonstrating a safe position?**

   An e-SafetyCase is founded on the same risk management process, studies, analysis, assessments, etc. as a traditional paper-based Safety Case. The only difference is how this information is presented, with these differences delivering the wider, peripheral benefits described elsewhere within this paper.

3. **A regulatory Safety Case is a strictly controlled document and for very good reason. How can an e-SafetyCase deliver the same level of control?**

   One thing a traditional paper-based Safety Case and an e-SafetyCase have in common is they are both deliverables that have a defined status and identity. As such, change management can be controlled in an identical way using existing processes. These will ensure revisions are developed, implemented, reviewed and approved by competent people, with updates strictly controlled to ensure that people only use the latest, approved version. Whether the Safety Case is presented on paper or in electronic format is immaterial in this regard. Indeed, the days of a SafetyCase being a printed book issued to a small number of approved users are (largely!) long gone. The traditional paper-based Safety Case has morphed into an electronic document which can be downloaded off servers and e-mailed around organizations. An e-SafetyCase presents no greater challenge to that which is already being managed.

   Making the e-SafetyCase available as an App does introduce additional challenges, particularly around security, but other businesses, with similar sensitive information which is version critical, manage these challenges effectively, e.g. finance/banking and e-commerce. In these examples, these challenges are managed in the public domain (the App is publicly available). There is no need for an e-SafetyCase to be publicly available with access only granted to approved users, thereby introducing another level of control.

4. **Regulated industries expect a traditional paper-based Safety Case. This is what they are used to and, indeed, the regulator has invariably shaped the format, structure, content, etc. of these. Will regulators accept an e-SafetyCase?**

   An e-SafetyCase does exactly the same job as a traditional paper-based Safety Case; it provides a robust demonstration of safety. The difference is in how this information is presented. With an e-SafetyCase being more accessible, intuitive and easy to navigate than a traditional paper-based Safety Case, it should make a regulator's job easier and so should be welcomed as a positive development. Indeed, it is a fundamental expectation of any regulator that operators embrace ways of reducing risk. With an e-SafetyCase providing an accessible, engaging and intuitive basis of safety which proactively drives and encourages safe operations, it should be viewed favorably, provided the challenges are adequately managed. Further support, engagement and buy-in would be gained by actively involving the regulator in the development of the e-SafetyCase concept, allowing it to be developed to meet their specific needs.

   Regardless of the above position, it is worth remembering that an e-SafetyCase is founded on the same risk management process, studies, analysis, assessments etc. as a traditional paper-based Safety Case. It would therefore be simple to also present it in the established, traditional format accepted within that industry. An e-SafetyCase which can be viewed in "traditional" and "electronic" modes would therefore provide the best of both worlds.

5. **Software projects are notorious for mushrooming in scope and complexity and blowing budgets and programmes. Would the development of e-SafetyCases suffer these problems?**

   An e-SafetyCase template would be developed as a tightly controlled project based around a detailed specification developed by the Safety Case users. This template would then be used to generate all future Safety Cases ensuring a consistent, proportionate approach. Clearly the scope and complexity of the e-SafetyCase could mushroom, but this can be readily managed through this one-off project. From that point onwards, the established blueprint will prevent individual e-SafetyCases being overly complex. This is actually an improvement over traditional paper-based Safety Cases where experience shows that they can tend to grow and expand unnecessarily from project to project, and also over the years as they often pass from one owner to the next and more information is bolted on in an ad-hoc way.

It will ultimately be the users (the designers, operators, managers, regulators etc. of major hazard facilities) who will decide whether the e-SafetyCase has a future.

There is clearly evidence that an e-SafetyCase can be at least as good as a paper-based Safety Case in terms of its core purpose of demonstrating a safe position. Adding in the wider, peripheral benefits it offers, there would appear to be a strong case to offer these users the chance to trial the concept so that they can make the ultimate decision. Let the people decide!

8 Conclusions

In today's world where answers to most questions can be Googled instantaneously, isn't it time the safety world caught up? The development of an e-SafetyCase is no more onerous than a conventional Safety Case, it's just different. Maintenance is no more onerous either. Indeed, an effective e-SafetyCase will be so useful and effective it will naturally evolve hand-in-hand with the facility as routine business.

Climbing the learning curve is a small price to pay for the considerable benefits this concept offers; providing an accessible, engaging and intuitive basis of safety which proactively drives and encourages safe operations.

Taking things further, the e-SafetyCase can be blended with technological innovations such as Virtual Reality to further bring it to life. This kind of ambitious approach has the potential to change the landscape of safety management in high hazard industries.

9 References

[1] The Nimrod Review
An independent review into the broader issues surrounding the loss of the RAF Nimrod MR2 Aircraft XV230 in Afghanistan in 2006
Charles Haddon-Cave QC
https://assets.publishing.service.gov.uk/government/uploads/system/uploads/attachment_data/file/229037/1025.pdf

[2] Buncefield: Why did it happen?
The underlying causes of the explosion and fire at the Buncefield oil storage depot, Hemel Hempstead, Hertfordshire on 11 December 2005
http://www.hse.gov.uk/comah/buncefield/buncefield-report.pdf

SPRING21
+17TH GCPS
A Joint AIChE and CCPS Meeting

Procedure Simulators That Ensure Your Procedures Are Accurate and Operators Are Fully Competent

Presenter/Principal Author: Francis Montemurro
Affiliation: Voovio Technologies
Address: 1888 Stebbins Dr, Ste 100, Houston TX 77043
Email address: fmontemurro@voovio.com

Other Presenter/Author: Christian McDermott
Affiliation(s): Voovio Technologies
Address: 1888 Stebbins Dr, Ste 100, Houston TX 77043
Email address: cmcdermott@voovio.com

Prepared for Presentation at
American Institute of Chemical Engineers
2020 Spring Meeting and 17th Global Congress on Process Safety
Virtual
April 18 - 22, 2021

Procedure Simulators That Ensure Your Procedures Are Accurate and Operators Are Fully Competent

Presenter/Principal Author: Francis Montemurro
Affiliation: Voovio Technologies

Other Presenter/Author: Christian McDermott
Affiliation(s): Voovio Technologies

Prepared for Presentation at
American Institute of Chemical Engineers
2020 Spring Meeting and 17th Global Congress on Process Safety
Virtual
April 18 - 22, 2021

Keywords: Field Operations, Human Error, Procedure Clarity, Procedure Execution, Procedure Simulator, Execution Errors, Unplanned Downtime, Training, Simulation, Operational Discipline, Competency.

Abstract

It is well-documented that many manufacturing unplanned events could have been prevented if procedures were executed properly. Ensuing investigations cite 'failure to follow procedure' as the event root cause resulting in next step recommendations, including additional operator training and discipline. However, based on data we collected from across North America, we consider "Procedure Clarity" a major source of error that is overlooked by the industry.

We believe there are three things that need to happen in order to drive real procedure clarity for operators of all levels:

1. You need clear and concise procedures;
2. You need the ability to visualize the procedure action steps;
3. You need the ability to practice and learn the procedure in a standardized way (learn by doing).

1 Procedure Clarity versus Operator Execution

ARC Advisory estimates that downtime costs the process industries somewhere in the area of a trillion dollars a year.[1] At Voovio, we believe the cost of inadequate procedure clarity and field operation execution errors is substantial, based on our own experience of working with customers across the petrochemical, specialty chemical and refining industries. An ASM Consortium study put the losses associated with unplanned downtime in the US Petrochemical

Industry at $10-20 billion annually.[2] Indeed, another ASM Consortium study[3] on human error and root cause of incidents identified causes of abnormal events as:

- procedure not followed : 51%
- procedure wrong: 40%
- procedure followed incorrectly: 6%
- Other: 3%

It is well-documented that many manufacturing unplanned events could have been prevented if procedures were executed properly.[4] Ensuing investigations cite 'failure to follow procedure' as the event root cause resulting in next step recommendations, including additional operator training and discipline. However, based on data we collected from across North America, a more systemic failure could be at the root of these unplanned events.[5] When an operational execution error occurs during a process or equipment startup, shutdown, changeover, process loss out of primary containment or other event, how often is "Procedure Clarity" considered the root cause? We consider it as a major source of error that is overlooked by the industry.

Operating procedures originate during capital project installation and are developed using the technical process design basis. Procedures include safety warnings, specific instructions, and details required to complete the desired outcome of the titled procedure. Following a successful project startup, the initial operating procedures are redlined, updated based on startup and practical experience, and institutionalized as part of the site's management of change (MOC) process.

We have identified two major reasons for suboptimal procedure clarity:

1.2 Procedure does not reflect operator practice

We analyzed 300 procedures from Customers across the US and Canada (between 2018-2020), broken down into the categories below in Figure 1. Interviews with customers and our procedure analysis shows a significant number of procedures do not match operator practice, a foundational element for Operational Discipline. 90% of the procedures analyzed contained generalizations.

Figure 1 – distribution of ~300 procedures analyzed by Voovio[6]

Type of Procedure	%
Abnormal	1.40%
Emergency Operations	2.46%
Emergency Shutdown	0.35%
Equipment Swaps	0.35%
Initial Startup	1.05%
Loading/Unloading	4.91%
Maintenance Overhaul	1.05%
Normal Operations	3.16%
Normal Shutdown	1.75%

Shutdown	27.37%
Startup	38.60%
Startup and Emergency Shutdown	3.16%
Startup and Normal Operations	2.81%
Startup and Shutdown	5.61%
Startup following a Shutdown	4.91%
Temporary Operations	1.05%
TOTAL	**100%**

When surveying manufacturing process subject matter experts (SMEs), the written procedures vary from the unit's established practice with some practices varying from shift to shift either due to training inconsistencies or a lack of standardized execution. This finding was previously documented as the cause for 23% of manufacturing's downtime losses by the Ron Moore Group, in Figure 2 below.

Figure 2 – Categories of downtime losses in process industry.[7]

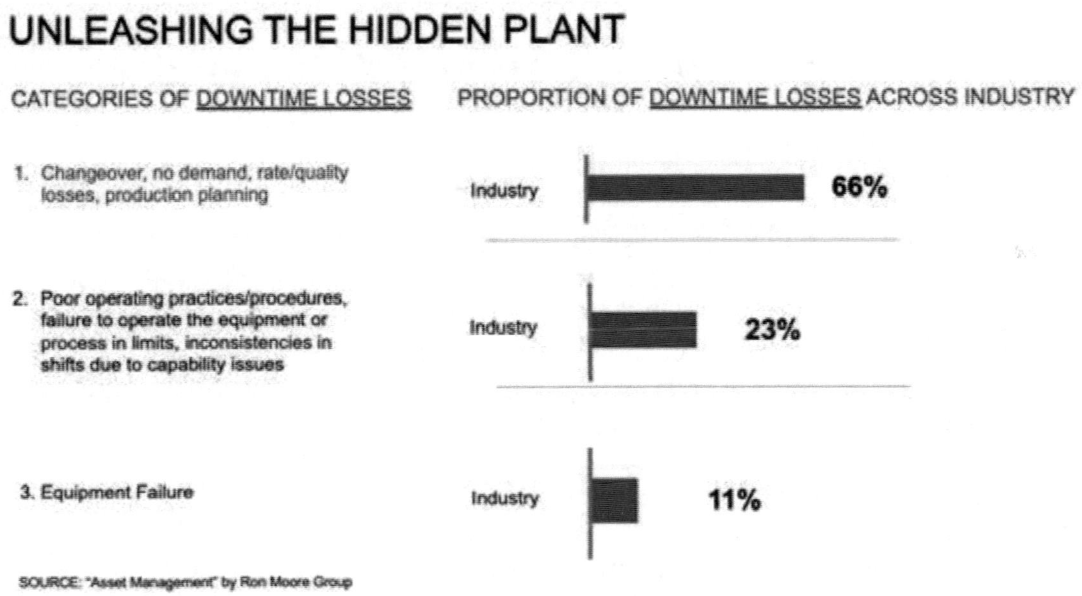

1.3 Ambiguities and generalizations

Using procedures, manufacturing operators train on how to operate the process, unit or plant safely and effectively. Operators are challenged to learn and qualify on hundreds and hundreds of pages across a range of procedures. Often, these procedures are written with a lot of technical content, generalizations and ambiguities that assume clear understanding by the operator. Procedures are typically written by engineers or experienced operations personnel (SMEs), individuals very familiar with the process, who tend to overlook the basic details of execution believing the details are largely understood. Below are some examples from actual manufacturing procedures (we have changed the equipment tag numbers to ensure anonymity).

Example: Procedure line with additional peripheral information and options on how to execute the procedure;

"Level indicator transmitters (ABC-111 and DEF-222 at A-C-9 and GHI-871 and GHI-872 at A-C-8) provide an initial inventory in these vessels. To provide initial inventory in A-C-7 and A-C-6, 30-psig condensate control valve (FV-4447) can be manually opened or put in Automatic at 30%. The steam regulator for A-C-6 is located beside the CO2 plant." [8]

The first and third sentences are not needed for procedure execution; thus, may add uncertainty to the operator. The second sentence gives the operator options; leaving to question what the preferred best method is to inventory the system.

Examples: General, ambiguous information in procedures;[9]

- *"Close all bleeders and open-ended lines"*
- *"Blow product return lines and control valve loop with blend gas from A-123 take-off manifold to ensure lines are clear then isolate from all equipment"*
- *"Outside operator will verify manual valves on feed system are lined up"*
- *"Ensure cooling water is lined up to all coolers and exchangers that require cooling water"*
- *"Manually close all condensate controllers outside"*

An operator is expected to follow the above procedural statements and remember where 'all' the components are located while executing the one procedure line item. In the 1990s and early 2000s, this problem was masked by workforce experience, shift teams had many individuals with 20 to 30+ years of operating experience, thus procedure clarity was not important.[10] With the gap in experience, the lack of procedure clarity, and few operator aids to ensure execution accuracy, incidents and unplanned events will continue to occur.

2 Impact of procedures during operator training

Typically, once the operator reads the required procedures, shadows an experienced operator for the designated time and completes some form of assessment, he or she is 'signed-off as qualified' in the job. However, industry data indicates the operator has only actually executed or practiced around 40% of the procedures due to the fact that equipment is always running and plant operation priorities.[11] Post qualification, the operator is released to work the job. Thus, procedure quality and, more importantly, specificity is the building block for developing operator competency.

Procedure formats and templates vary from company to company and even from site to site. Some companies have implemented company or site-wide digital solutions, involving many technical writers, significant company involvement and subsequent procedure conversions. Other companies maintain procedures manually with internal document control systems. Depending on the implemented solution, sites require operators to print procedures prior to execution and sign or check off steps as they are completed, or with digital formats, operators carry tablets in the field to use as a reference tool to read the procedure prior to executing each line item. Both formats require operators to read procedures for understanding. It is well documented in learning and development forums that individuals learn and retain ~90% using simulation and doing versus ~10% when reading procedures; see below. The industry challenge is to move operator training, proactive development, down the pyramid to increase individual retention and reduce an operator's time to full competency.

Figure 3 – the Learning Pyramid.[12]

The Learning Pyramid or Cone of Learning

Ultimately, to see a step-change improvement in operational discipline, procedures should be developed, implemented, and maintained based on successful execution by the end-user… the operator; and they should be clear, concise and action-based.

We believe there are two best practices during project development to ensure accurate procedures and standard operator execution. First, a Forward Engineer (creator of the procedure simulator)[13] performs a Line Walk for selected customer procedures with the resident SME to document procedure execution accurately using photographs and creating an action-based checklist. This visual walkthrough identifies gaps in the written procedure and the actual operating practice. All deviations from the documented procedure are reviewed and approved using the site MOC process. One extreme example, but not uncommon, is the following **one** procedure line step:

Pre-Startup Checklist SOP: "1. *Using 70-psig nitrogen, purge all equipment and process lines to remove oxygen and to ensure there are no leaks or bleeders left open before introducing feed to the unit.*"

Once the procedure had been simulated (following the Line Walk in the field with the SME and Forward Engineers) this single step consisted of 112 actions that the outside operator needs to perform across a sizeable section of the unit including:

- Interacting with bleeder valves 46 times
- Interacting with inlet/outlet valves 26 times
- Interacting with supply valves 8 times
- Performing line purges and other tasks

Operators confirmed that there is no 'agreed' way to perform this step, and there are variations in the way this step is performed from shift to shift. Typically, procedures as ambiguous and generic as in this example would be updated to better reflect the reality of what happens in the field, therefore providing a 'standard' for the unit. An action checklist detailing these specific actions is printed and submitted to the Operations Manager and Procedure Owner for review, with a red-lined version of the original procedure. The site's MOC process is followed.

Figure 4.

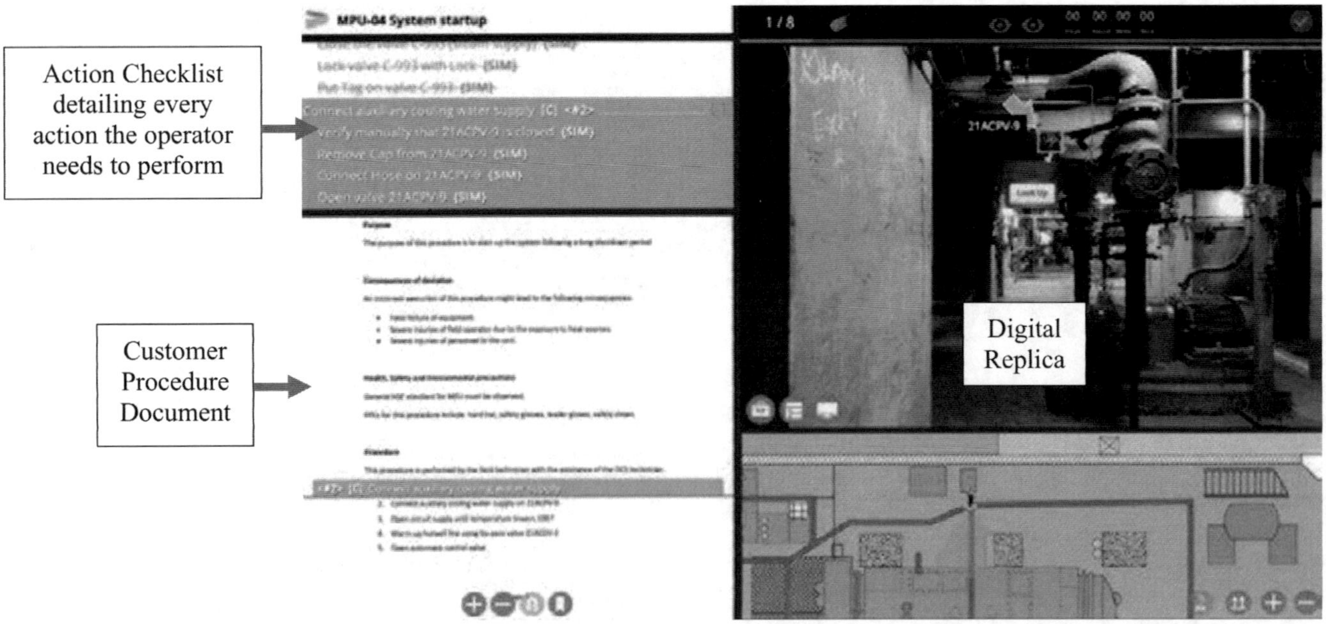

Additionally, using Voovio's patented technology, users can perform the newly documented procedure in an immersive photo-based environment (called the 'Digital Replica') and interact with and manipulate (open, close, push, etc.) on-screen components. This method of simulating operation allows operators, engineers, production specialists and others, the chance to **visually**

verify procedure accuracy. A simple and extremely effective way of validating and certifying the written procedure.

The second best practice is the addition of action steps, better known as a checklist. This well-documented methodology has significantly improved safety performance in air travel, medical procedures and other work environments.[14] During the Line Walk with the SME, the Forward Engineers confirm each action the operator is expected to complete. We regularly discover there are multiple actions within a single procedure line. Below are two examples to highlight this.

- The procedure line item, *"Close all bleeders and open-ended lines"*[15] resulted in 12 action steps because there were 12 bleeders and open-ended valves to be closed to ensure no contents left primary containment and/or caused an environmental incident.

- The procedure line item, "*Manually close all condensate controllers outside*"[16] resulted in 20 action steps to close 16 valves.

The process of simulating a procedure, with the Line Walk performed by an SME and Forward Engineer, results in (on average) 2.4 times more action steps than the number of procedure line items. Procedures are complemented by an action-oriented checklist, thereby making them clearer and improving the human reliability component when executing them in the field. A summary of past projects is depicted in the graph below (Figure 5).

Figure 5 – Procedure steps before and after a line walk for simulation.[17]

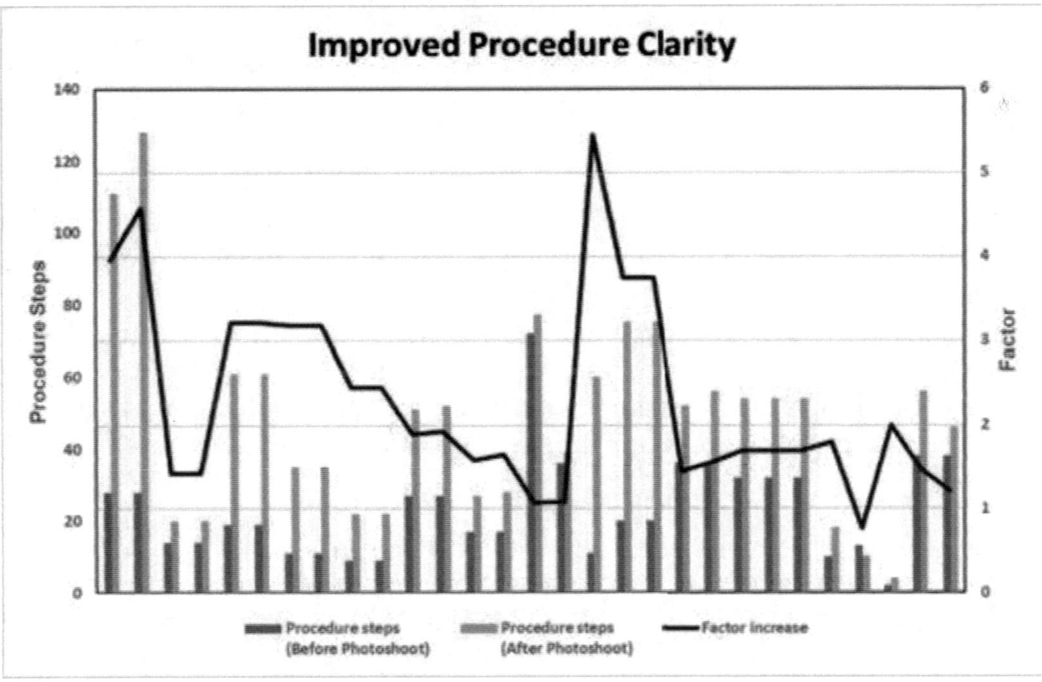

Voovio's interactive and comprehensive Forward Engineer (FE) Line Walk process eliminates procedure generalities and ambiguities resulting in clear and concise action steps. On average, procedure steps increase by a factor of 2.4

3 Digital Replica: environment required to 'simulate' and 'verify' procedures

Simulation requires a visual representation of the assets, which is displayed on a screen. Simulation technologies available include CAD-based technologies and Enhanced Reality technology. This paper considers simulation and procedure line walks that use Enhanced Reality technology, which was specifically designed for the manufacturing environment.

The Digital Replica is the basis of the simulated environment. It is a collection of standard photographs that are arranged in such a way that the user feels as though they are walking seamlessly throughout the process environment and able to interact with the equipment. In the Procedure Simulators, written procedures and the actions in the checklist are linked to the Digital Replica. Users are able to review and perform each action in the Digital Replica as if they were in the real plant. The Procedure Simulator Test Module monitors performance as users walk around and interact with the equipment, providing immediate feedback on incorrect and correct actions.

Operations, engineering, maintenance and other critical personnel can utilize the Digital Replica and Simulators to verify if the procedure is correct. It is not uncommon that a procedure is updated and improved after SMEs have utilized the simulation. Another simple example of procedure improvements thanks to using the Digital Replica and Simulators is an inspection procedure for a chlorine railcar. In order to perform the checks, operators would walk around the railcar almost 20 times with multiple inspection points. Following the simulation, which highlighted this unnecessary inefficiency to the senior trainers and SMEs, the procedure was updated to produce a more efficient and effective route for inspections.

4 How is this approach economically viable?

The criteria for selecting procedures to simulate include the frequency of incidents, including safety, environmental or product quality losses which can be traced back to operator error. In addition, procedures that are not frequently executed are also good candidates, because operators may not have the opportunity to experience or train on them unless they happen.

The Line Walk process with a Forward Engineer ensures that a procedure is correct by walking step-by-step with the Subject Matter Expert. In surveys conducted with end users (operators) between 2018-2020, 100% of respondents believed the Line Walk process and Simulated Procedures makes it easier to learn and understand procedures than their current (traditional) training methods, provided clarity on aspects of their job that they did not have before, and will increase compliance with operating procedures in their area.

In Figures 6-8, data taken from Voovio customers show the financial impact of downtime losses associated with specific equipment and procedures. This is a very small sample from different operating units in both Refining and Chemical plants. A single refinery or plant may have thousands of similar procedures. These incidents happen frequently across the industry. Having accurate procedures is foundational to having good operational culture and discipline. While clearer procedures will help reduce incidents, organizations need to provide operators with the tools and training to be successful.

In each example below, the cost of simulating the procedures (including line walks) was far less than the cost savings realized, resulting in payback in less than one year.

Figure 6[18] - Data from a support process for refinery, providing technical gases. The site suffered from high attrition and unplanned downtime related to boiler feedwater (BFW) pump failures related to mis-operation. Thanks to improved procedure clarity and the ability for operators to practice and learn prior to execution, it increased organizational productivity and reduction of unplanned failures with an annual benefit of $ 88,000 thanks to simulating one procedure.

ASSET AVAILABILITY	TECHNICAL GASES
Annual Downtime Costs, $	220,000
Plant Trips, #	1
Cost per Plant Trip, $	100,000
Critical Equipment Failures, #	1
Critical Equipment Failure Costs, $	150,000
labor costs per event, $	15,000
contractor costs per event, $	-
Other Costs, $	5,000
Estimated Downtime Reduction, %	40
Net Program Yearly Benefit, $	88,000

Figure 7[19] – Data from a Petrochemical plant (Nonwovens). The unit suffered from incorrect process start-ups, low product yields due to operator execution errors. Several startup procedures were simulated, leading to improved procedure clarity and knowledge for the operators resulting in reduction of product losses, saving over $ 500,000 annually.

ASSET AVAILABILITY	PETROCHEMICAL
Annual Downtime Costs, $	1,142,000
Unplanned Events/year, #	1
Cost per Plant Event, $	1,116,000
Critical Equipment Failures, #	-
Critical Equipment Failure Costs, $	-
labor costs per event, $	15,000
contractor costs per event, $	10,000
Other Costs, $	1,000
Estimated Downtime Reduction, %	50
Net Program Yearly Benefit, $	571,000

Figure 8[20] – Data from a refinery, truck loading operations. The refinery aimed to reduce overtime to train contract drivers and reduce unplanned downtime on loading racks due to damage and spills/resulting in costly repairs and clean-up (~2.5 per month). 3rd party truck drivers are remotely trained, with clear actions, and assessed prior to loading leading to a reduction of incidents.

ASSET AVAILABILITY	TANK TRUCK HAZMAT
Annual Downtime Costs, $	2,675,700
Plant Events, #	27
Cost per Plant Event, $	99,100
Critical Equipment Failures, #	-
Critical Equipment Failure Costs, $	-
labor costs per event, $	-
contractor costs per event, $	-
Other Costs, $	-
Estimated Downtime Reduction, %	10
Net Program Yearly Benefit, $	267,570

In addition to the reduction of unplanned downtime due to poor procedure clarity and mis-operation, time savings are a key contributor to savings in operations training. Figure 9 reflects average speed to competency gains by using Digital Replicas and Procedure Simulators. Units see an average of 50% reduction in Subject Matter Expert (SME) time dedicated to training new operators and when onboarding operators, on average it takes 30 to 50% less time to be fully qualified.[21]

Figure 9 – Traditional Methods of Training versus Digital Replicas and Procedure Simulators

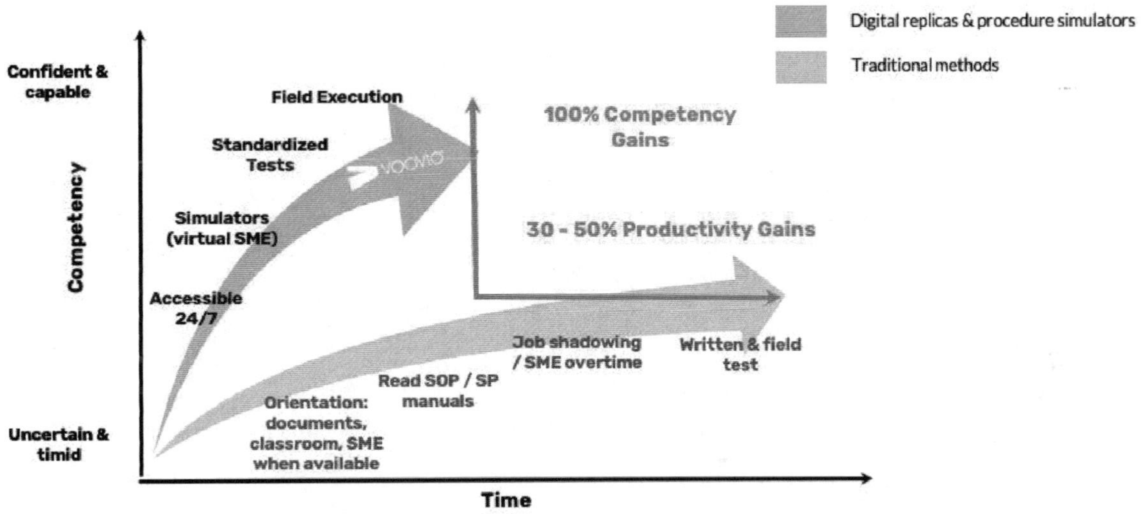

5 Conclusion

ARC Advisory concludes that "*the human role is paramount in reducing unplanned downtime… To avoid an incident, operators must receive better information and training to make the right decisions.*"[22] Whether this is for control room operators or field operators, the approach must be the same and the investment required to achieve this makes economic sense, based on the immediate payback achieved by preventing failures.

The data outlined in this paper, collected from Customers in North America, cite procedure clarity as a major source of error. The data shows there is a strong case for improving procedure clarity, so that the industry can improve how the 'knowledge' within those procedures can be transferred more effectively to operators themselves, thereby creating a standardized system for training and procedure execution, leading to a reduction in unplanned downtime.

Clear and concise procedures, standardized on plant operation that have objective assessments, are foundational building blocks to ensure field operator competency and deliver predictable process performance. Furthermore, operational discipline and speed to operator competency are and will remain competitive advantages in the Petrochemical and Refining industry due to the current and projected attrition slated for plant sites.

The road to execution excellence is complex, but by having the three foundational pieces we set out in this paper, manufacturing plant operators will have the tools to build their competency and deliver business results safely and successfully:

- Clear and concise procedures
- Ability to visualize procedure steps in the field
- Ability to practice standardized execution (*learn by doing*)

6 References

[1] *Process Industry Downtime and Key Performance Metrics, ARC Advisory, Larry O'Brien, April 2017.*
[2] *Process Industry Downtime and Key Performance Metrics, ARC Advisory, Larry O'Brien, April 2017.*
[3] *Improving Procedural Operations, ASM Consortium, Chris Morse / Andrew Ogden-Smith, Honeywell, September 2014.*
[4] *Based on reports from the Chemical Safety Board (CSB) and Francis Montemurro's own experience with investigations.*
[5] *Data collected from Voovio Technologies Customers between 2018 and 2021.*
[6] *Data taken from Voovio Customers, 2018-2020.*
[7] *Ron Moore is the Managing Partner of The RM Group, Inc. in Knoxville, TN. He is the author of Making Common Sense Common Practice: Models for Manufacturing Excellence (now in its 4th edition); and of What Tool? When? A Management Guide for Selecting the Right Improvement Tools (now in its 2nd edition), both from MRO-Zone.com; as well as Business Fables and Foibles and Our Transplant Journey: A Caregiver's Story; and over 50 journal articles.*
[8] *Extracted from a Standard Operating Procedure from a Voovio Technologies Customer, 2019.*
[9] *Extracted from a selection of Standard Operating Procedures from Voovio Technologies Customers, 2018-2020.*
[10] *Based on evidence collected from Francis Montemurro's 30+ year career with DuPont, in roles of plant management and operations leadership, including industry peers.*
[11] *Based on feedback from Voovio Customers, 2018-2021.*
[12] *Edgar Dale, The Learning Pyramid,* https://en.wikipedia.org/wiki/Learning_pyramid, *accessed on January 27th 2021.*
[13] *A Forward Engineer is an engineer by background, with experience in manufacturing who walks through procedures with an SME and captures the photographs required to produce the Procedure Simulators and documents the actions required to perform each step of the procedure.*
[14] *Data includes "The Checklist Manifesto – How to Get Things Right", Atul Gawande, 2010.*
[15] *Extracted from a Standard Operating Procedure from a Voovio Technologies Customer, 2018.*
[16] *Extracted from a Standard Operating Procedure from a Voovio Technologies Customer, 2019.*
[17] *Data collected from Voovio Technologies Customers between 2018 and 2021.*
[18] *Data taken from Voovio Customers, 2018-2020.*
[19] *Data taken from Voovio Customers, 2018-2020.*
[20] *Data taken from Voovio Customers, 2018-2020.*
[21] *Data extracted from Voovio Customers, 2018-2020.*
[22] *Process Industry Downtime and Key Performance Metrics, ARC Advisory, Larry O'Brien, April 2017.*

2021 AIChE Annual Meeting – Extended Abstract

Adsorption Kinetics of N-nitrosodimethylamine (NDMA) onto Activated Carbons

Kailey Garland, David Amidei, Paul A. Andersen, Catherine E. Brewer

According to the U.S. Environmental Protection Agency, N-Nitrosodimethylamine (NDMA) is classified as a probable human carcinogen. A risk assessment indicated that a concentration of 0.7 ng/L (parts per trillion) of NDMA in drinking water represents a 1 x 10^{-6} increased cancer risk level. NDMA is highly mobile in soil, and was derived from the treatment of water from rocket fuel testing, which resulted in the contamination of ground water near rocket testing sites. Additionally, NDMA can be an unintended byproduct of the chlorination of wastewater, making it a significant concern for wastewater treatment plants that use chloramines for disinfection. The treatment standards for NDMA typically range from 1 to a few ng/L, which creates analytical challenges for monitoring and development of remediation methods.

The purpose of this research was to test the ability of biochars made from local biomass wastes to adsorb NDMA from water at treatment-relevant concentrations. The adsorption kinetics of biochars produced from pecan shells and local forestry residues were tested and compared with a commercially available coconut shell-derived activated carbon. Further testing of the commercially available carbon was conducted in a lab-scale column for adsorption of NDMA to be used as a basis for future work. Concentrations of ^{14}C-labeled NDMA in well water from the NASA White Sands Test Facility (Las Cruces, NM) were measured at regular time intervals over the course of the flow through a biochar-filled column. Liquid scintillation counting was used for NDMA quantification.

Freundlich isotherm models were used to estimate column dimensions, flowrate, and breakthrough times. The resulting column was 18 inches (45.72 cm) long with a diameter of 0.5 inch (1.27 cm), and used an initial flowrate of 240 mL/min for a breakthrough time of 6 hours. Experiments showed breakthrough occurring immediately, however, despite changing experiment parameters such as flowrate and carbon preparation.

Calgon coconut shell carbon (CSC) provided the fastest and highest capacity for NDMA adsorption. Synthesized pine bark char adsorbed more NDMA by the end of the kinetics study than the commercial pecan shell char. Synthesized pecan shell char adsorbed the least NDMA. The initial mass transfer rate was controlled by film diffusion for all the tested carbons. Intraparticle diffusion was negligible for CSC but controlled a noticeable segment of adsorption for the other carbons. Column breakthrough trends did not follow expected patterns: breakthrough occurred much more rapidly than models indicated, and faster flowrates resulted in slower breakthrough and more apparent adsorption.

619077 Characterization of Cooling Water Blowdown in District Cooling and Determination of Its Effect on Wastewater Treatment Plants

Abstract:

The District Cooling (DC) process requires high volumes of water to operate cooling towers (CTs). In the State of Qatar, the use of Treated Sewage Effluent (TSE) in DC processes is mandated by the government for the purpose of sustainable management of limited water resources, efficient use of resources and to reduce Greenhouse Gas (GHG) emissions. The operation of CTs generates cooling water blowdown (CWBD), which is a contaminated liquid with very high concentrations of solids and chemicals. To adopt appropriate treatment and disposal strategies of CWBD, a full characterization of this wastewater is required. Th quality of CWBD depends on many parameters, including the quality of feed and make-up streams . Hence, the work presents a full characterization and analysis of the CWBD and TSE streams, including solids concentration, hardness, alkalinity, COD, pH, EC and toxicity. In addition, a bench scale study has been used to assess impacts of CWBD constituents on TSE and sludge. The results of the analysis and characterization work can be used in future research to simulate and validate the effect of CWBD constituents on the full-scale wastewater treatment process and, ultimately estimate the permitted volumes of CWBD that can be discharged to the WWTP without affecting the treatment process performance.

SPRING21
+17TH GCPS
A Joint AIChE and CCPS Meeting

Slaying the Dragons that Sabotage Protection Measures to Feed on the Ash of Your Plant

Denise Chastain-Knight
Exida Engineering, LLC
Sellersville, PA
dchastainknight@exida.com

Prepared for Presentation at
American Institute of Chemical Engineers
2021 Spring Meeting and 17th Global Congress on Process Safety
Virtual
April 18 - 22, 2021

Slaying the Dragons that Sabotage Protection Measures to Feed on the Ash of Your Plant

Denise Chastain-Knight
Exida Engineering, LLC
Sellersville, PA
dchastainknight@exida.com

Keywords: Hazard Identification and Risk Analysis, Safety Instrumentation Systems, Measurement and Metrics

Abstract

Dragons are legendary winged monsters that breathe fire and feed on ash. Or so we think. This is a tale of three dragons that have evolved to hide in plain sight and quietly feed on the underpinnings of protection layers, eroding the effectiveness of process safety mitigation measures. These dragons are mischievous and feed off process upsets, near misses and accidents. The Invisible dragon roams around the plant breaking devices and causing people to make mistakes that set off alarms and causes Safety Instrumented Systems (SIS) trips to initiate. Its special joy is creating initiating events that were not considered in design. Tiny and Giant dragons specialize in defeating the SIS protection layers, so that all three dragons can feed on the ash of an accident. The Tiny dragon baits SIS designers with devices having extremely low failure rate data causing the humans to believe Safety Instrumented Functions (SIF) are more reliable than they actually are. The Giant dragon works alongside the Tiny dragon to further undermine the SIS with its song of near perfect proof testing. This paper will illustrate the impacts of underestimating SIF demand rate, using overly optimistic failure rate data and overestimating proof test coverage. We will discuss techniques to identify and correct these issues and suggest metrics to monitor SIS performance thus slaying the dragons.

1 Introduction

Safety Instrumented Functions (SIF) are intended to mitigate a specific amount of risk based on the potential consequences of a specific hazard scenario as determined by the hazards risk assessment. The design of the SIF involves careful selection of components and operating and maintenance parameters, in order to deliver the required risk reduction. Verification and validation of the SIF performance is measured against the stated design specification. Failure to maintain any of the specification parameters can result in SIF performance that does not meet the design intent.

The risk reduction requirements for a SIF, in low demand, are stated in terms of Safety Integrity Level (SIL) target and a specific Risk Reduction Factor (RRF). For example, a low demand SIF with a target SIL 2 RRF of 225 must achieve an average probability of failure on demand (PFD_{avg}) of 4.4E-3 and meet SIL 2 hardware fault tolerance and systematic capability. Hardware fault tolerance is the physical redundancy as required in IEC 61511[1] or IEC 61508 [2] standards to address random hardware failures. Systematic capability is essentially the systematic fault tolerance and is assessed through certification or Prior Use justification. The discussion in this paper addresses factors that primarily impact the PFD_{avg} (PFD_{avg} = 1/RRF). The three dragons represent input parameters of SIL verification calculations where PFD_{avg} is assessed. This paper will illustrate how PFD_{avg} is influenced by demand rate, component failure rate and proof test coverage. It will also discuss how failure to maintain operation consistent with these parameters can result in SIFs that do not achieve the intended risk reduction target.

2 Invisible Dragon – System Demands

Design and verification rules for a SIF are determined based on the mode of operation. SIFs may operate in Low Demand, High Demand or Continuous mode. Low Demand mode is quite common in the process industry and random failure rate is assessed by PFD_{avg}. High demand and Continuous mode failure rate is defined as probability of failure per hour, or PFH. For a SIF to be operating in Low Demand mode, the initiating event frequency must be less than 1/(2*PTI)[3] where PTI is the proof test interval. A SIF with an initiating event frequency that exceeds the Low Demand threshold is considered High Demand and must be designed per the High Demand mode criteria, thus there is no benefit derived from proof testing. The effect of proof test interval on SIF demand mode is illustrated in Figure 1.

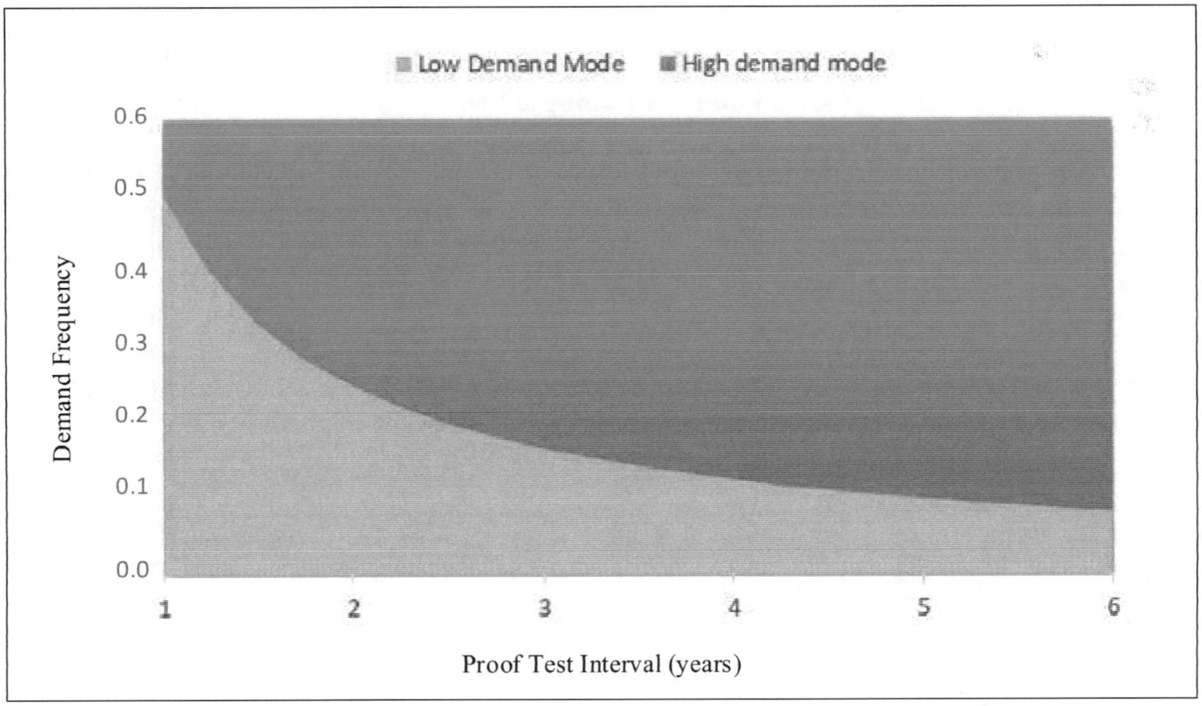

Figure 1 - SIF Demand Mode Threshold

Consider the process illustrated in Figure 2. During the LOPA, a scenario was developed resulting in a SIF to protect against low flow from V-100 to V-200.

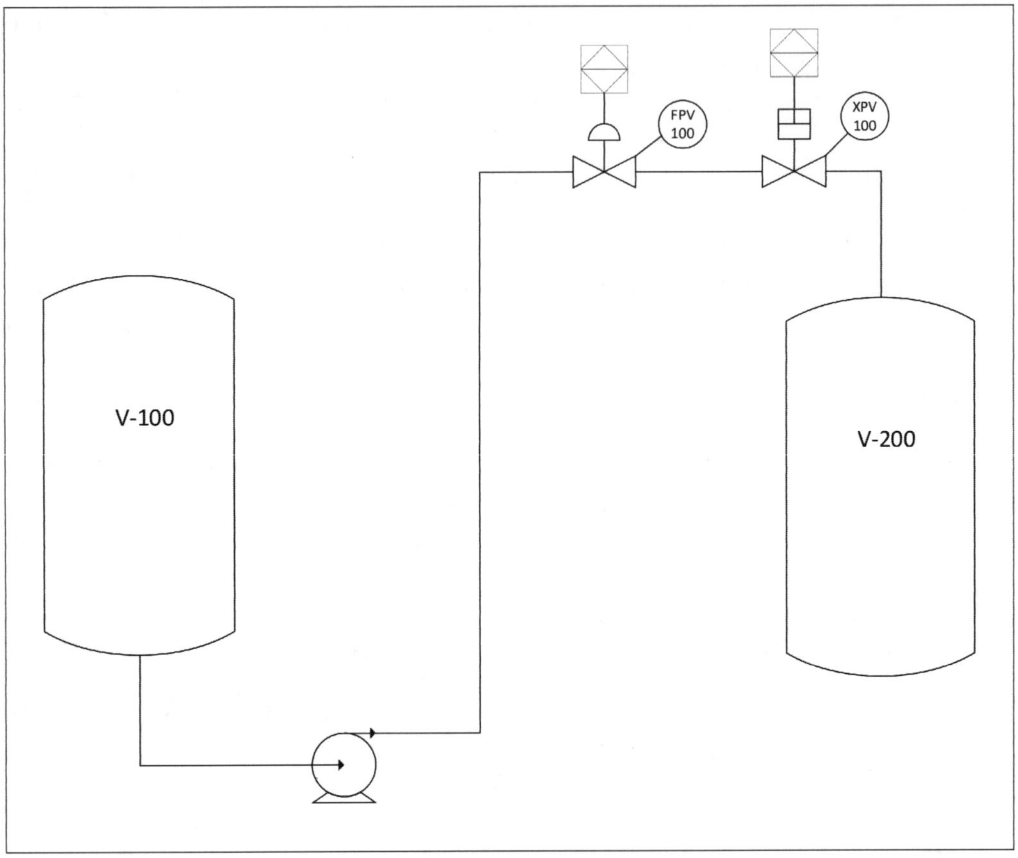

Figure 2 – Example Process

The LOPA given in Figure 3 considers pump failure, FPV failure and XPV failure, as initiating events for the SIF. Demand on the SIF is reduced for the pump failure case by a fault alarm with operator action. The SIF demand was determined to be 0.21/year. Proof test interval for the SIF is 2 years. The SIF demand threshold is calculated to be 1/(2*2) or 0.25/year so the SIF is operating in low demand mode.

The original design for the SIF is very near the threshold and could easily be pushed into high demand by an increase in proof test interval or the initiating event frequency. For example, if the facility decided to delay a turnaround by 6 months, the proof test interval is then increased to 2.5 years changing the high demand threshold to 1/(2*2.5) or 0.2/year. With this change, the SIF is now effectively operating in high demand mode.

1. Low of flow from V-100 to V-200		IPLs			
Initiating Event	Frequency [per year]	Pump Fault alarm	Loss of flow SIF	Intermediate Frequency [per year]	Comments
Pump failure	0.1	B 0.1	B 0.01	B 1.00E-4	
		E 0.1	E 0.01	E 1.00E-4	
		S 0.1	S 0.01	S 1.00E-4	
FPV-100 failure	0.1	B NA	B 0.01	B 1.00E-3	
		E NA	E 0.01	E 1.00E-3	
		S NA	S 0.01	S 1.00E-3	
XPV-100 failure	0.1	B NA	B 0.01	B 1.00E-3	
		E NA	E 0.01	E 1.00E-3	
		S NA	S 0.01	S 1.00E-3	

Figure 3 – Example Process LOPA

Because a decision to delay a turnaround is a conscious choice, an engineer can pause to perform SIL verification using high demand criteria quantifying the risk that is assumed during the delay. They can then implement additional protection and/or mitigation measures to mitigate that risk as needed.

But what if the initiating event frequency is higher than estimated in the LOPA? This is the Invisible Dragon causing additional risk to be accepted without the benefit of foreknowledge, so mitigation measures are not considered. For example, presume the FCV sticks periodically causing low flow and operators are accustomed to unsticking the valve when needed. If the actual failure frequency for the FCV is slightly higher, say 0.15 failures/year, than the assumed literature value (0.1 failures/year), the overall demand rate is actually 0.26/year and the SIF is always operating in High Demand mode.

How can the Invisible Dragon be made visible? Consider implementing metrics to monitor SIF demand rate or use forensic analysis to confirm that it is within the design basis. If the SIF demand is found to be borderline, similar to the example process, consider implementation of additional independent protection layers (IPLs) to reduce the demand so that the SIF remains in Low Demand mode at all times.

3 Tiny Dragon – Unrealistic Failure Rate Data

The SIL verification calculation equation for PFD_{avg} can become very complex as terms are included to address factors such as proof test effectiveness and diagnostics. The dangerous failure rate λ_D, whether detected (λ_{DD}) or undetected(λ_{DU}), appears in every term. If the verification calculation is to be realistic, then it is critical that the underlying failure rate data be realistic. The Tiny dragon baits SIS designers with devices reporting extremely low failure rate data causing the people to believe SIFs are more reliable than they actually are.

Figure 4 illustrates the structure of a SIF having a SIL 2 RRF 100 target. The SIF is designed to stop catalyst feed to a reactor when high pressure is sensed in the reactor. The sensor group includes three transmitters with two out of three (2oo3) voting. The final element group includes two identical valve assemblies consisting of a solenoid, actuator, and valve. The valve assemblies are placed in series in the catalyst feed line, so flow is stopped if either valve is successfully closed (1oo2 voting). A generic SIL 2 certified PLC is the logic solver. On the bottom left are the verification results (RRF 121.4) for the model using device data that has been vetted and is considered realistic. On the right, are the results (RRF 177) when the solenoid, actuator and valve reliability data is replaced with data from a certificate that appears to be optimistically low. The difference in the overall RRF achieved is significant. The significance of the difference is most obvious when the final element part of the SIF is compared in detail. The PFD_{avg} given by the optimistic data (8.7E-5) is less than 4% of that given by the realistic data (2.69E-3). Notice also that the Mean Time to Fail Spurious (MTTFS) for the final element group increases significantly from 40.03 to 159.77 suggesting the final element group is more resistant to spurious trips than it actually is. Finally note that the architectural capability (AC), or hardware fault tolerance (HFT), determined by the optimistic data suggests the final element arrangement could be used up to SIL 4, where the realistic data limits use to SIL 3.

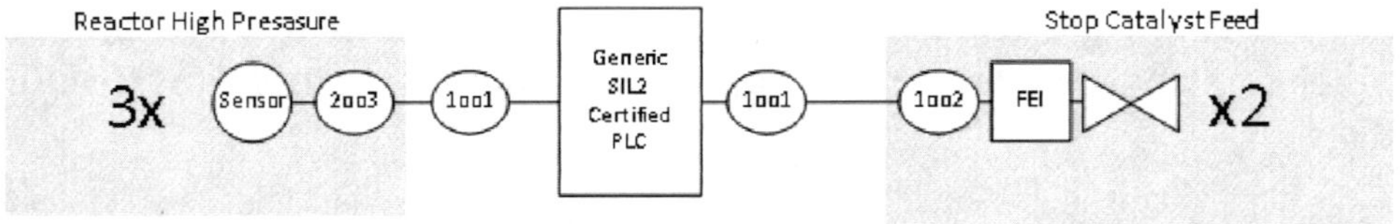

Vetted Valve Data

Safety Instrumented Function Performance	
Target SIL	**2**
Target RRF	100
Achieved SIL	**2**
PFDavg	8.23E-03
SIL (PFDavg)	2
SIL (Arch. Constraints IEC 61508)	2
SIL (Systematic Capability)	N/A
Achieved RRF	121.4
MTTFS (years)	7.42

	PFDavg	MTTFS	SIL Limit	
			AC	CAP
Sensor Part	4.03E-03	549.13	2	N/A
Logic Solver Part	1.54E-03	9.26	2	N/A
Final Element Part	2.69E-03	40.03	3	N/A

Unrealistic Valve Data

Safety Instrumented Function Performance	
Target SIL	**2**
Target RRF	100
Achieved SIL	**2**
PFDavg	5.65E-03
SIL (PFDavg)	2
SIL (Arch. Constraints IEC 61508)	2
SIL (Systematic Capability)	N/A
Achieved RRF	177
MTTFS (years)	8.62

	PFDavg	MTTFS	SIL Limit	
			AC	CAP
Sensor Part	4.03E-03	549.13	2	N/A
Logic Solver Part	1.54E-03	9.26	2	N/A
Final Element Part	8.70E-05	159.77	4	N/A

Note: Contribution of common cause failure is not included

Figure 4 – SIF Results Vetted vs Unrealistic Failure Rate Data

Clause 11.9.3 in IEC61511 states "the reliability data used when quantifying the effect of random failures shall be credible, traceable, documented and justified and shall be based on field feedback from similar devices used in similar operating environment.". Failure rate data is obtained from industry data sources, provided by the vendor with device certification documents, or collected from within the operating facility. Unfortunately, one does sometimes encounter a certificate or vendor literature that includes questionable failure rates. Figure 5 is a plot of failure rate data for solenoids, which includes data points from three industry databases, calibrated Failure Modes Effects and Diagnostics Analysis (FMEDA)[4] assessment results, and data reported on several certificates. While the industry data and the FMEDA are reasonably consistent, the four datapoints in red would appear to have overly optimistic data.

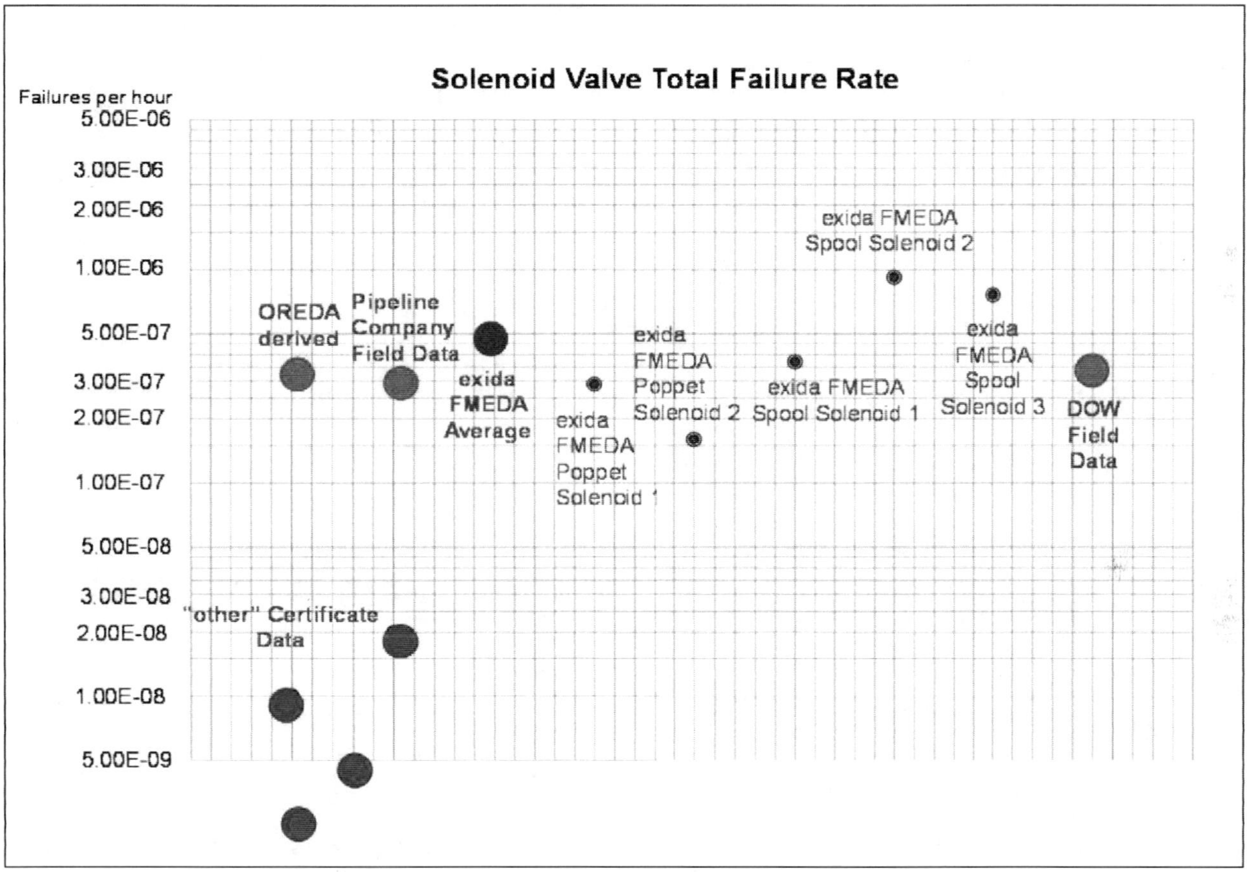

Figure 5 – Solenoid Valve Total Failure Rate

So how does one know if data presented by a vendor is realistic or not? The SILSafeData tool, http://silsafedata.com/, provides upper and lower boundaries for dangerous undetected failure rates for devices commonly used in process industries based on over 200 billion unit operating hours of field failure data. Designers should vet the failure rate information provided by vendors against industry data as part of the device selection criteria before selecting for use in a SIF.

4 Giant Dragon – Perfect Proof Testing

SIFs operating in low demand mode benefit from proof testing because there is opportunity to detect failures between demands and effect repairs. Proof test coverage (PTC) is the percentage of failures that can be identified with the proof test. While it is very tempting to assume that a proof test is perfect (100%), the reality is that proof tests are not always executed correctly or completely, or that the test may not effectively identify all failures[5]. The Giant Dragon undermines functional safety goals by using PTC values that are too large.

Figure 6 illustrates the impact of assuming perfect proof testing. The SIF is the same as described in Figure 4 above. The results shown on the left of the table include proof test coverage, which is based on careful analysis of the failure modes and the capability of the proof test to identify failures. The results (RRF 121.4) shown on the right are for the same SIF with the perfect proof test coverage in sensor, logic solver and final element groups.. The latter predicts a much higher overall RRF (181.3) and suggests the SIF is far more reliable that it actually is. The difference is less obvious in the final element group details, so one must review the PFD graph for the final element group shown in Figure 7 to appreciate that the higher PFD_{avg} is achieved through larger 'recovery' at proof testing.

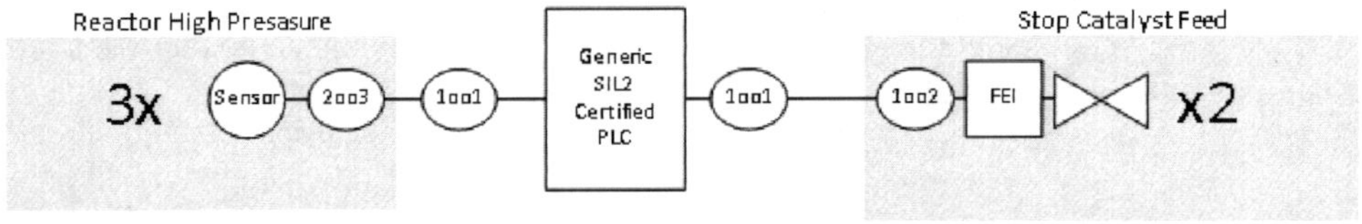

Realistic PTC Assumed

SAFETY INSTRUMENTED FUNCTION PERFORMANCE	
Target SIL	2
Target RRF	100
Achieved SIL	2
PFDavg	8.23E-03
SIL (PFDavg)	2
SIL (Arch. Constraints IEC 61508)	2
SIL (Systematic Capability)	N/A
Achieved RRF	121.4
MTTFS (years)	7.42

	PFDavg	MTTFS	SIL Limit	
			AC	CAP
Sensor Part	4.03E-03	549.13	2	N/A
Logic Solver Part	1.54E-03	9.26	2	N/A
Final Element Part	2.69E-03	40.03	3	N/A

Perfect PTC Assumed

SAFETY INSTRUMENTED FUNCTION PERFORMANCE	
Target SIL	2
Target RRF	100
Achieved SIL	2
PFDavg	5.51E-03
SIL (PFDavg)	2
SIL (Arch. Constraints IEC 61508)	2
SIL (Systematic Capability)	N/A
Achieved RRF	181.3
MTTFS (years)	7.41

	PFDavg	MTTFS	SIL Limit	
			AC	CAP
Sensor Part	2.64E-03	561.18	2	N/A
Logic Solver Part	1.54E-03	9.26	2	N/A
Final Element Part	1.35E-03	39.69	3	N/A

Note: Contribution of common cause failure is not included

Figure 6 – Imperfect Proof Testing Comparison

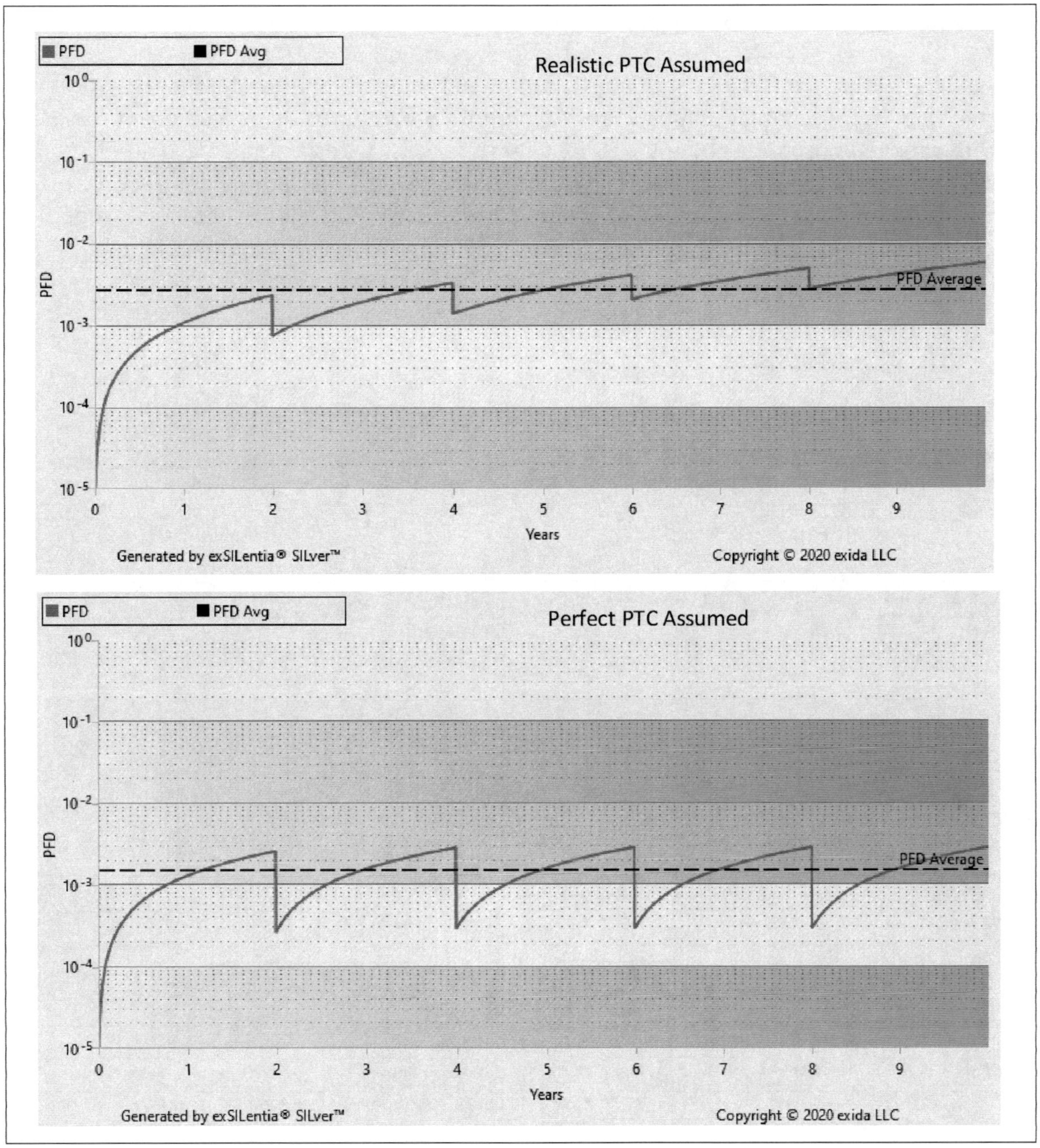

Figure 7 – Imperfect Proof Testing Comparison Final Element Group Graph

The analysis for the realistic model in Figure 6 and Figure 7 used PTC values from the software database that ranged from 51% to 99% for individual components. The proof test coverage of a device can be determined by FMEDA analysis[6] considering the proof test steps and the physical construction of the device. Many vendors will provide recommended proof test procedures with associated coverage in a product's safety manual.

5 The Tiny and Giant Dragons Team Up

When the Tiny Dragon and the Giant Dragon team up, the combination of unrealistic failure rate data and perfect proof testing compound results. Figure 8 compares the model using realistic failure rate data and proof test coverage (left) with the model using unrealistic final element data and perfect proof test coverage (right). The model with realistic failure rate and PTC basis delivers RRF 121.4 and the final element group is limited to SIL 3 architecture. The model using optimistic information delivers an overall RRF of 237.1 and suggests the final element group is acceptable for SIL 4 service.

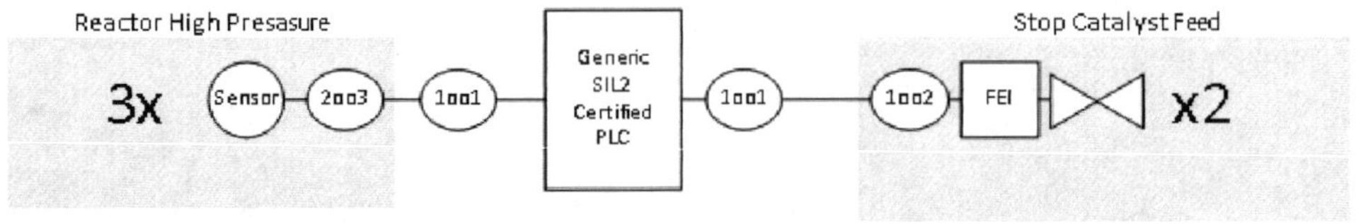

Realistic Failure Data and PTC

SAFETY INSTRUMENTED FUNCTION PERFORMANCE	
Target SIL	**2**
Target RRF	100
Achieved SIL	**2**
PFDavg	8.23E-03
SIL (PFDavg)	2
SIL (Arch. Constraints IEC 61508)	2
SIL (Systematic Capability)	N/A
Achieved RRF	121.4
MTTFS (years)	7.42

	PFDavg	MTTFS	SIL Limit	
			AC	CAP
Sensor Part	4.03E-03	549.13	2	N/A
Logic Solver Part	1.54E-03	9.26	2	N/A
Final Element Part	2.69E-03	40.03	3	N/A

Unrealistic Failure Data and Perfect PTC

SAFETY INSTRUMENTED FUNCTION PERFORMANCE	
Target SIL	**2**
Target RRF	100
Achieved SIL	**2**
PFDavg	4.22E-03
SIL (PFDavg)	2
SIL (Arch. Constraints IEC 61508)	2
SIL (Systematic Capability)	N/A
Achieved RRF	237.1
MTTFS (years)	8.62

	PFDavg	MTTFS	SIL Limit	
			AC	CAP
Sensor Part	2.64E-03	561.18	2	N/A
Logic Solver Part	1.54E-03	9.26	2	N/A
Final Element Part	4.83E-05	159.71	4	N/A

Note: Contribution of common cause failure is not included

Figure 8 – Unrealistic Failure Rate Data and Imperfect Proof Testing Comparison

Comparing the overall PFD charts in Figure 9 one can observe that the model using realistic PTC and failure rate information predicts periods where the actual RRF falls below 100 where the 'sawtooth' peaks cross the 10^{-2} line. With a mission time of 10 years, the chart shows that the SIF is achieving less than a RRF of 100, increasingly significant periods of time in years 6-10. To mitigate the risk, a designer might recommend that the mission time of the SIF be reduced and schedule refurbishment or replacement of components at year 6 instead of 10. The model utilizing unrealistic information does not indicate these increased periods of risk so the need for reduced mission time would go unrecognized.

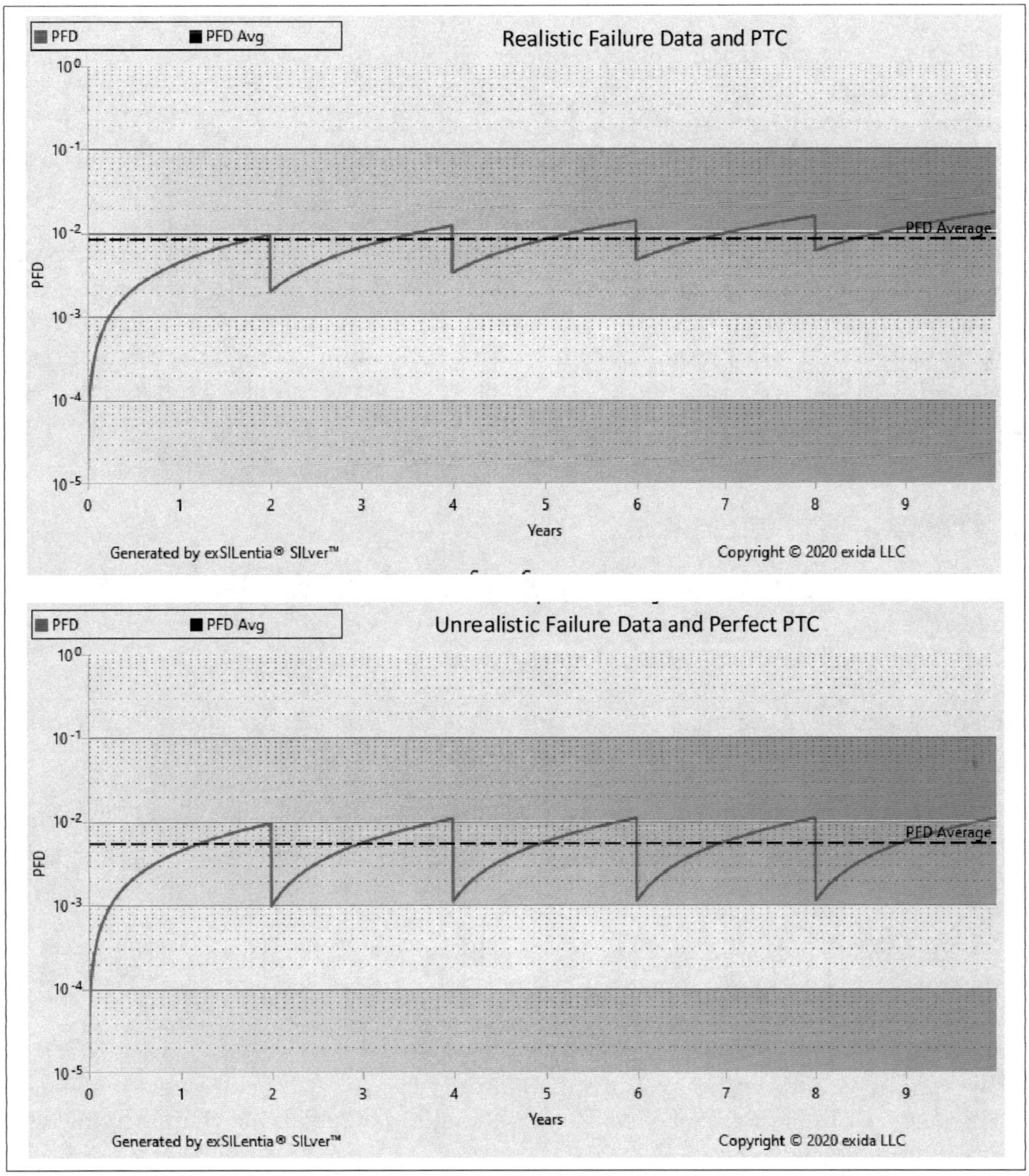

Figure 9 – Unrealistic Failure Rate Data and Imperfect Proof Testing Comparison Graph

6 Conclusion

What can designers do to slay the hungry dragons that sabotage protection layers, leading to the unknowing acceptance of additional risk?

Invisible: Remember that extending proof test interval of a SIF can move the operating mode from low to high demand. When adjusting PTI, designers should revisit SIL verification calculations in order to understand the risk that is assumed. Also, designers should periodically review the failure rate information for initiating events to confirm that the assumptions made during the PHA and LOPA remain valid.

Tiny and Giant: Don't fall for the lie of low failure rate data or the lure of near perfect proof testing. During device selection confirm that the data is credible, traceable, documented and justified per the requirements of IEC 61511 clause 11.9.3. Use of unrealistic failure rate data or assuming perfect proof test coverage can result in an overly optimistic PFD_{avg} calculation. When completing SIL verification calculations, avoid the use of overly optimistic failure rate information by vetting against industry data. Finally, use realistic proof test coverage for SIL verification based in the capability of the proof test procedure.

7 References

[1] IEC 61511-1 Ed 2.0, *Functional Safety: Safety instrumented systems for the process industry sector – Part 1: Framework, definitions, system, hardware and application programming requirements,* IEC, Geneva, Switzerland 2016

[2] IEC61508 Ed 2.0, *Functional safety of electrical/electronic/programmable electronic safety-related systems,* IEC, Geneva, Switzerland 2010

[3] Iwan van Buerden, William M. Goble, *Safety instrumented System Design, Techniques and Design Verification*, Instrument Society of America (ISA), Research Triangle Park, NC, 2018

[4] *Getting Realistic Failure Rate Data*, exida white paper, Sellersville PA, 2020. https://www.exida.com/Resources/Whitepapers/Getting-Realistic-Failure-Rate-Data (accessed March 1, 2021)

[5] Bukowski, J.V. and van Beurden Iwan, *Impact of Proof Test Effectiveness on Safety Instrumented System Performance*, Reliability and Maintainability Symposium (RAMS), Fort Worth, TX, January 2009.

[6] Goble, W. M., *Strategic Proof Testing*, exida white paper, Sellersville PA, 2020. https://www.exida.com/Resources/Whitepapers/strategic-proof-testing (accessed March 2, 2021)

SPRING21 +17TH GCPS
A Joint AIChE and CCPS Meeting

"I Didn't Inhale, and Never Tried It Again."

Proper Handling of Toxic Inhalation Hazard (TIH) Chemicals in Plants and During Transportation

Michael P. Broadribb
Baker Engineering and Risk Consultants, Inc.
3330 Oakwell Court #100, San Antonio, TX 78218
mbroadribb@bakerrisk.com

Rafael Callejas-Tovar
Baker Engineering and Risk Consultants, Inc.
11011 Richmond Ave. #700, Houston, TX 77042
rcallejas@bakerrisk.com

Prepared for Presentation at
American Institute of Chemical Engineers
2020 Spring Meeting and 17th Global Congress on Process Safety
Virtual
April 18 - 22, 2021

"I Didn't Inhale, and Never Tried It Again."

Proper Handling of Toxic Inhalation Hazard (TIH) Chemicals in Plants and During Transportation

Michael P. Broadribb
Baker Engineering and Risk Consultants, Inc.
3330 Oakwell Court #100, San Antonio, TX 78218
mbroadribb@bakerrisk.com

Rafael Callejas-Tovar
Baker Engineering and Risk Consultants, Inc.
11011 Richmond Ave. #700, Houston, TX 77042
rcallejas@bakerrisk.com

Keywords: toxic hazards, process safety, refinery, chemical plant, transportation

Abstract

Numerous fatalities and serious health effects have occurred in the process industries through personnel exposure to toxic chemicals and materials. A Toxic Inhalation Hazard (TIH) chemical can be particularly hazardous as a release of some chemicals cannot be detected through sight or smell. TIH chemicals are widely handled in process units and also transported by road, rail, marine, and pipeline. Fatal incidents involving chlorine (Graniteville, SC, 2005), hydrofluoric acid (South Korea, 2012), and methyl mercaptan (MeSH) (La Porte, TX, 2014) (Ref.1) emphasize the risks of TIH chemicals if they are not handled with a sense of vulnerability and proper care. Most elements of CCPS Risk Based Process Safety (RBPS)(Ref. 2), plus some engineering design considerations, are important for the proper handling of TIH chemicals.

Transportation of TIH chemicals poses additional risks as an accidental release can potentially expose a large public population. In addition to other RBPS elements, hazard identification and risk analysis (HIRA) can be used to quantify the consequences and risks associated with transportation operations. A comprehensive quantitative risk analysis (QRA) involves the development of an event tree to identify scenarios, dispersion simulations to calculate vulnerabilities along the route, and a thorough frequency analysis to pair the scenarios with their likelihood to calculate risk. The results are typically presented in terms of FN curves[1], societal risk, and location specific individual risk, which are compared against risk tolerance thresholds to facilitate the decision-making process. These results have helped companies understand if the

[1] FN Curve is a plot of cumulative frequency (y-axis) versus consequences (x-axis), often expressed as number of fatalities. The value plotted on the y-axis is the cumulative frequency of experiencing N or more fatalities.

benefits of their transportation operations justify the associated risks, and to develop comprehensive risk mitigation strategies.

This paper describes the authors' experiences of applying: (i) good practices for handling TIH chemicals, (ii) a RBPS approach for assessing TIH risks at chemical plants, including loading/unloading (railcars, road tankers, ISO tanks), (iii) use of quantitative methods for TIH transportation risk decision-making, and (iv) real case examples of potential TIH risk reduction strategies.

1 Introduction

Toxic chemicals pose a potential risk if people are exposed to the chemicals, which includes not only the workforce within chemical plants and oil refineries, but also the local communities. The transportation of toxic chemicals by road, rail, marine, and pipeline also pose potential risks to the public and often travel through high density population areas. As a result, it is paramount that hazards are properly managed.

Toxic Inhalation Hazard (TIH) chemicals are a class of toxic chemicals that act on the respiratory tract and can lead to pulmonary injury. There are many industrial TIH chemicals that are widely handled in process units and transported, and some, such as chlorine bleach and ammonia, are commonly found in households. A TIH chemical can be particularly hazardous as a release of some chemicals cannot be detected through sight or smell. A fatal incident in South Korea that released a colorless volatile TIH chemical (hydrofluoric acid) is described in the case study below.

Hydrofluoric Acid Release, Gumi, South Korea, 2012 (Ref. 3)
Two workers at the Hube Global chemical plant climbed onto a road tanker to hook up an air hose intended to unload the hydrofluoric (HF) acid cargo by air pressure into a storage tank. While one employee connected the air hose, the other, presumably to save time, began unscrewing the bolts of the end plate on the acid outlet valve. He inadvertently bumped the manual outlet valve arm, which partially opened. When the valve opened, 8 tons of HF was released under pressure. During transportation, the cargo had become pressurized due to thermal expansion. Neither worker was wearing appropriate personal protective equipment (PPE) and were killed, as were three others on a neighboring facility. Another 18 workers were hospitalized, and 4,195 people, including 416 firefighters were treated for nausea, skin rash, and breathing difficulties. A significant area of arable land downwind of the release was contaminated, resulting in the loss of crops, over 3,900 cattle, and 1,200 corroded vehicles.
Lessons learned:

- Industry procedures were not followed. The tanker should have been de-pressured and appropriate flexible hoses deployed before efforts to commence unloading.
- Workers should wear proper acid suits with supplied air hoods.
- Municipal fire-fighters arrived without proper PPE and sprayed water on the release that intensified the rate of HF release.
- Local authorities were slow to evacuate the residents in the vicinity of the release, and deployed inadequate resources, such as neutralizing agents (lime).

2 Practices for Handling Toxic Chemicals

The first practice for handling toxic chemicals is to understand how to maintain primary containment of the chemical. One of the short definitions of process safety is:

Keep it in the Pipe!

This is especially true for TIH chemicals. It is essential to know if a chemical is corrosive, unstable, reactive with other chemicals, or has any other physical property that can impair primary containment. A Safety Data Sheet (SDS) should provide the necessary information. Obvious safeguards against a loss of primary containment (LOPC) include selection of appropriate materials of construction, avoiding inadvertent mixing with specific other chemicals, and appropriate storage and operating conditions. These and other safety measures are discussed below.

The next practice for handling toxic chemicals is to understand their physical properties in case of a LOPC, which should also be provided on the SDS. TIH chemicals can have acute or chronic health effects, so it is important to follow the required personal protection equipment (PPE) specified. Other key physical properties to understand are:

- If a release occurs, will it be a liquid or vapor/gas or aerosol?
  - Process conditions (temperature, pressure), ambient temperature, and if the TIH chemical is mixed with other chemicals, will determine which phase.
- If a liquid, know whether the TIH chemical is volatile, heavier or lighter than water, and if soluble.
- If a vapor or gas, know if the TIH chemical is heavier or lighter than air, and what happens when it cools.

When implementing capital projects and small management of change (MOC) projects that involve TIH chemicals, consideration should be given to inherently safer design (ISD) principles. These principles are:

- Intensification - minimize inventory of hazardous materials including TIH chemicals,
- Substitution - eliminate hazards by substituting less hazardous materials where possible,
- Attenuation - minimize operating severity by avoiding extremely high pressures, high temperatures, or low temperatures, and
- Simplification - avoid unnecessary complexity/redundancy to minimize potential release sources and facilitate operator understanding.

Other design considerations in addition to ISD involve risk management and reliability hierarchies, as detailed in Table 1 and Table 2 below.

Table 1. Risk Management Hierarchy

ELIMINATION is better than
PREVENTION is better than
CONTROL is better than
HAZARD MITIGATION is better than
EMERGENCY RESPONSE

Elimination is not always practical as it involves selection of alternative processes that do not handle TIH chemicals.

Some examples of PREVENTION include, but are not limited to:

- Corrosion – use of corrosion-resistant alloys,
- Joints – use of welded piping to prevent leaks/releases at flanges, screwed fittings and couplings,
- Rotating Machinery – use of canned pumps and double seals,
- Unmanned facility – automated remote operation to prevent injury/fatality,
- Secondary containment,
- Good operating, maintenance, and process safety practices (but may not be 100% successful – see reliability hierarchy in Table 2).

Examples of CONTROL to reduce size or frequency of the hazard include, but are not limited to:

- Emergency shutdown (ESD) systems – remotely operable, excess flow valves,
- Blowdown/dump systems – remotely operable,
- Process quench/neutralizing systems,
- Fire protection – deluge/sprinklers/monitors/foam - remotely operable,
- Restriction orifices,
- Minimize surface area of spilled liquid release,
- Manual intervention/shutdown – needs PPE,
- Good operating, maintenance and process safety practices.

Examples of MITIGATION of the impact on people include, but are not limited to:

- Secondary containment – dikes, water curtains,
- Toxic release general alarm – site and local community,
- Evacuation – escape packs / respirators, other PPE,
- Shelter-in-place – building tightness (doors/windows/louvres), pressurized (elevation of heating ventilation and air conditioning (HVAC) inlet), HVAC ESD (gas sensors, automation).

Table 2. Reliability Hierarchy

PASSIVE controls are more reliable than
ACTIVE controls are more reliable than
ADMINISTRATIVE or PROCEDURAL controls

Examples of PASSIVE controls include, but are not limited to:

- Containment dikes and kerbs,
- Restriction offices,
- Fire resistant cladding on structural steel,
- Smoke/fire dampers, fire walls, and fire doors,
- Blast resistant buildings.

Examples of ACTIVE controls include, but are not limited to:

- Emergency shutdown systems,
- Blowdown/dump systems,
- Process quench/neutralizing systems,
- Active fire protection – deluge/sprinklers/monitors/foam.

Administrative and procedural controls tend to be the least reliable as they involve human interaction, and human intervention in the case of an emergency or abnormal situation. Process units should have clear expectations for operator intervention in the event of a TIH emergency, who will require appropriate training and PPE. Similarly, if the facility has its own emergency response team or relies upon mutual aid and/or the local fire department/emergency medical technicians (EMT), they must have toxic release capability (knowledge and training/drills in headcount, personnel rescue, and stopping a release, plus all of the necessary equipment and PPE to do so).

3 Transportation of Toxic Inhalation Hazard Chemicals

Transportation of TIH chemicals is a key activity in the supply chain, which has been increasing due to market globalization. The transportation of TIH chemicals might involve long distances and different modes between the origin and destination, such as truck, rail, crane, ship, etc. Since the transportation operations of TIH normally use the existing infrastructure, the routes might pass through locations with high population, which poses additional risks as an accidental release can potentially expose a large public population. An incident in South Carolina released a volatile TIH chemical (chlorine) during transportation and impacted the local community resulting in fatalities, hospitalization, and evacuation. This event is described in the case study below.

Chlorine Release, Graniteville, South Carolina, 2005 (Ref. 4)

Many trains traveled through Graniteville each day. A rail spur in the center of the town serviced the Avondale plant, which received daily deliveries of chlorine by rail for mill operations. On January 5, 2005, the regularly assigned conductor and engineer were both off duty, and standby employees operated a local train that was parked in a siding near the Avondale plant. The switch connecting the spur to the main line had to be turned manually for deliveries. Two track warrants protecting the local train's use of the tracks were cleared. Although the railroad switch for the Avondale siding was supposed to be set for mainline operation before the train crew departed for the night, the train's brakeman later told the NTSB that he was "not 100 percent sure" he had aligned the switch for mainline operation. No train operated through the area or used the affected tracks for the rest of the day.

At 2:38 am on January 6, 2005, a freight train approached Graniteville at 48 miles per hour. When the engineer saw the improperly aligned switch, he emergency braked, but there was insufficient distance to stop. The freight train was diverted by the improperly lined switch onto the siding and collided with the local train. The collision derailed 3 engines and 18 freight cars.

The freight train was transporting chlorine, sodium hydroxide, and cresol. A railcar loaded with 90 tons of chlorine ruptured and the liquefied chlorine rapidly vaporized. Nine people died of chlorine inhalation (eight at the time of the accident, one later), and at least 554 people were treated for chlorine exposure. About 30% of the chlorine was recovered by industrial responders. While the fire chief ordered shelter in place and initiated a major evacuation, these actions did not take place immediately. In total, 5,400 residents within one mile of the crash site were forced to evacuate for nearly two weeks while hazardous materials (HAZMAT) crews decontaminated the area.

Lessons learned:

- The incident could have been avoided if railroad employees had been alerted to the switch error by an electronic signal.
- The Federal Railroad Administration (FRA) sets guidelines and standards for railroad companies and lines. However, the railroad company, main line, and rail siding located in Graniteville complied with these standards that were inadequate in 2005.
- Regular shipments of hazardous materials through populated areas potentially expose the public to significant hazard.
- The incident illustrated the difficulties of achieving adequate and efficient evacuation for hazardous materials events. The local community did not understand how to react (shelter in place, evacuate) to a chlorine release.

In addition to other RBPS elements, hazard identification and risk analysis (HIRA) can be used to quantify the consequences and risks associated with transportation operations. A comprehensive quantitative risk analysis (QRA) can provide the risk profile along the length of the route and help to identify potential high-risk locations and operations. This powerful insight can be used to develop cost-effective risk mitigation, if needed, and facilitate the decision-making process.

The main steps involved in the QRA are:

1. Identification of the scenarios to be considered.
2. Consequence analysis (dispersion and vulnerability calculations).
3. Frequency analysis.
4. Risk calculations.
5. Comparison of results vs. risk tolerance thresholds.
6. Risk mitigation strategy and ALARP[2] demonstration.

The QRA is a screening level analysis that requires a comprehensive review of the potential failure modes to identify the ones that can cause loss of containment scenarios during the transportation operations (Ref. 5). Typically, the scenarios can be categorized as:

- External impacts (e.g., crashes, derailments, dropped objects, etc.)
- Spontaneous mechanical failure of the container device or its components
- Escalation scenarios (e.g., pressure relief device releasing due to external fire)

The identification of scenarios to be considered in a transportation QRA is done through the development of an event tree. Figure 1, an example event tree for a TIH transportation QRA, shows that scenarios from the categories above are included. Also, the event tree includes the different parameters of the scenarios to create a comprehensive mapping of the potential outcomes. Representative release sizes, weather conditions, and wind directions must be included in the QRA.

Once the scenarios are identified, their consequences can be calculated using dispersion modeling. Figure 2 shows example toxic dispersion contours for a scenario at a given point of a route using BakerRisk's proprietary software SafeSite$_{3G}$©. The concentrations calculated from the dispersion calculations are then used to calculate the lethality of the toxic cloud, typically using probit models. Since the scenarios can occur at any point of the transportation route, the consequence analysis must consider the different population numbers along the route as well as the relative location of the populations with respect to the potential scenarios. Population data is typically derived from census data or other population estimates to calculate the potential number of fatalities of the scenario. This thorough approach is computationally intensive and requires software tools able to handle thousands to millions of scenarios, depending on the characteristics of the route.

[2] As Low As Reasonably Practicable (ALARP), UK Health and Safety at Work etc. Act 1974.

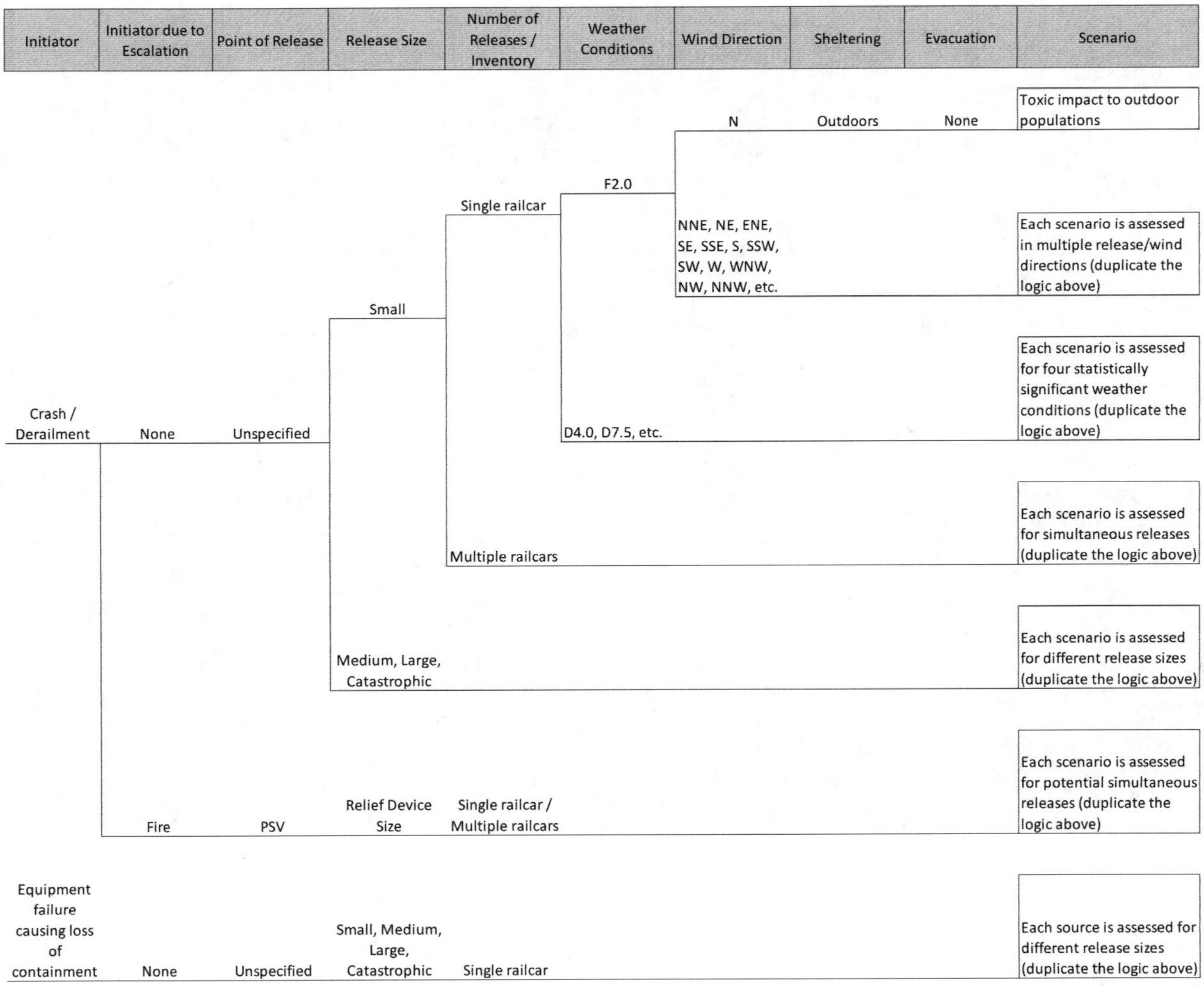

Figure 1. Example Event Tree for TIH Transportation QRA

Figure 2. Example Toxic Dispersion Contours

A frequency analysis is required to calculate the risk of each scenario once the consequences are calculated. In the frequency analysis, statistics on transportation incidents are used to estimate the likelihood of each scenario identified in the event tree and conditional probabilities are applied to the different parameters of the scenario (e.g., release size, weather condition, wind direction, etc.). However, it is important to understand the basis of the statistics reported as well as accounting for historical data specific to the operation analyzed.

The risk results can be presented in many ways, such as FN curves, societal risk, and location specific individual risk. However, the most typical way of reporting transportation risk is using FN curves based on the risk tolerance criteria lines for transportation of dangerous substances provided by the UK Health and Safety Executive Advisory Committee on Dangerous Substances (HSE ACDS) (Ref. 6). Risk tolerance criteria allow the owner/operator to classify the risk results based on published risk tolerance from which it can be determined if risk mitigation is required.

Figure 3 shows example FN curve results. It is observed that there is intolerable risk (above upper criteria line) for the baseline transportation operations; this requires the implementation of risk mitigation to decrease the risk to tolerable levels. Once the baseline risk results are obtained, the QRA model can be used to quantify the effectiveness of risk mitigation options. The risk mitigation strategy should be developed following the risk management hierarchy in Table 1. For example, alternative routes through locations with less population, enhanced design of the containers that reduce the likelihood of a release upon impact, controls above the industry norm, etc. can be evaluated using the QRA model. Risk reduction of each mitigation option is calculated and compared. In the example in Figure 3, the Risk Mitigation Option 2 reduces the risk more significantly than Option 1. Using the risk mitigation results with the estimated cost of implementing each of the risk mitigation options allows for a cost-benefit analysis to assist in the decision-making process.

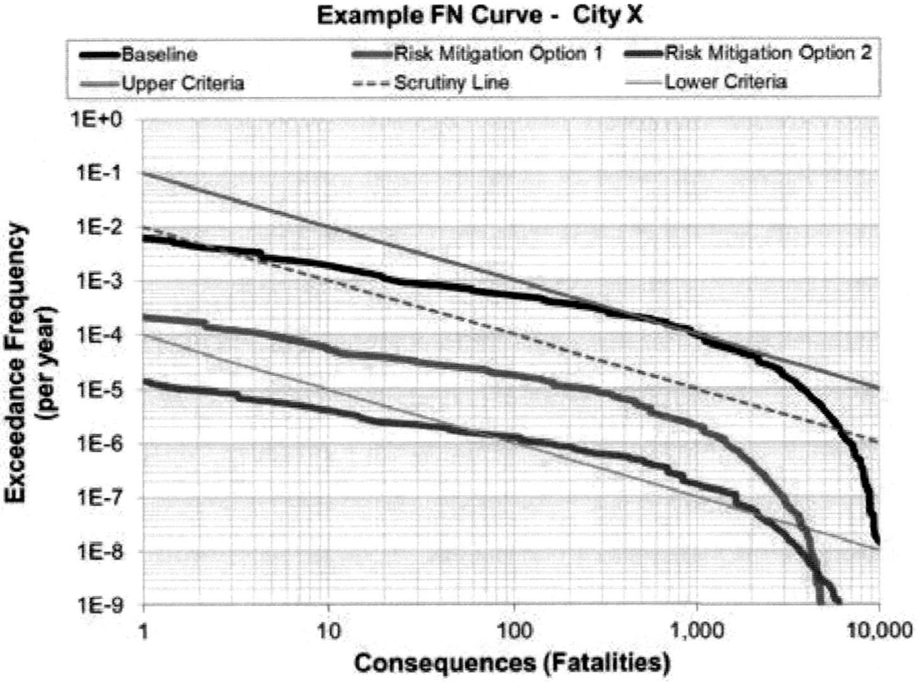

Figure 3. Example FN Curve Results[3]

4 Audit Protocol for Toxic Inhalation Hazard Chemicals

Auditing is a critical evaluation of a facility's management system to identify and correct weaknesses in the management system design and implementation. Auditing can be applied to process units handling and transporting TIH chemicals to identify opportunities for reducing risk and improving the effectiveness of the management system. To truly manage TIH chemicals it is necessary to look beyond local regulations, such as OSHA PSM[4] in the USA. Compliance with regulations is no guarantee that process safety incidents will be prevented. Facilities need to (i) identify the hazards, (ii) understand the risks associated with the hazards, and (iii) manage the risks by "*doing what is right*". This invariably means going beyond regulatory compliance and a risk-

[3] Risk tolerance lines based 0n HSE ACDS guidance (Ref. 6).

[4] Occupational Safety and Health Administration, Process Safety Management of Highly Hazardous Chemicals standard (29 CFR 1910.119).

based approach (i.e., 20 elements of CCPS RBPS) to assess all hazards and risks that could lead to an onsite and/or offsite TIH incident may be necessary.

Not all hazards and risks are created equal. Therefore, the emphasis of the auditing should be on what is truly safety critical, and dig deeper where the risk is high. While all RBPS elements should be addressed, in the case of TIH chemicals the main audit focus may be on:

- Operations personnel's understanding of hazards and risks,
- Operations and maintenance personnel competencies (knowledge, skills, abilities, and experience – not just training),
- Maintaining primary containment in pressure vessels, piping (including cross-country pipelines), tankage (including ISO tanks), road tankers, and rail cars,
- Safety culture and conduct of operations.

The author has developed a detailed and comprehensive audit protocol specifically for process facilities (process units and transportation) that handle TIH chemicals. This protocol addresses all 20 elements of RBPS, plus two additional elements for design and sharing learnings that are not explicitly covered by the CCPS Guidelines for Risk Based Process Safety (Ref. 2), see Table 3. For facilities that are subject to periodic compliance audits under OSHA PSM, the protocol is designed to remove unnecessary duplication.

Table 3. Example of Audit Protocol for Design of Facilities Handling TIH Chemicals

Question	Thought Prompts
When designing new projects and MOCs, does the facility adopt an Inherently Safer Design (ISD) approach?	• Intensification – minimize inventory of hazardous materials? • Substitution – eliminate hazards by substituting less hazardous materials? • Attenuation – minimize operating severity by avoiding extremely high pressures, high temperatures, or low temperatures? • Simplification – avoid unnecessary complexity / redundancy to minimize potential release sources and facilitate operator understanding? • Risk reduction measures – preference for Prevention before Control before Mitigation? • Reliability / availability measures – preference for Passive before Active before Administrative? • Other?
Do existing and new facilities have prevention systems to avoid hazards from being realized?	• Corrosion – use of corrosion-resistant alloys? • Joints – use of welded piping to minimize flanges, screwed fittings and couplings? • Rotating machinery – use of canned pumps? double seals? • Other?

The intent is that prior to the audit, the client will determine jointly with the auditors on which elements the main emphasis and focus will be placed. For example, the audit may address all elements, omit some elements, or alternatively devote less time to some elements deemed less relevant to the client's operations. These decisions on the scope of the audit should be based on risk and past experience (incidents, near-misses, prior audits, metrics, citations, etc.). As indicated above, it is likely priority/weighting could be placed on certain elements associated with understanding hazards and risks, managing risks, and softer issues related to culture and conduct of operations.

This protocol has been successfully used at several client facilities, and the client's employees have also been trained in the topic and proper use of the protocol. In fact, the author's colleagues were also trained to ensure a consistent application and interpretation of the protocol before auditing a client's facility. Like any review or audit, a written report is produced to cover the observations, findings, and recommendations where potential weaknesses have been identified. As this is not a regulatory compliance audit, some clients prefer to conduct the audit confidentially under legal privilege.

5 Conclusions

TIH chemicals are widely handled in process units and transported by road, rail, marine, and pipeline. Fatalities and serious respiratory impacts can occur through personnel exposure if these chemicals are not managed properly with a sound safety culture and sense of vulnerability.

While the workforce in process facilities is at risk, transportation poses additional risks as an accidental release can potentially expose a large public population. A comprehensive quantitative risk analysis allows identification of potential high-risk locations and operations, as well as to calculate the risk reduction of mitigation options, if needed. This powerful insight can be used to develop cost-effective risk mitigation and facilitate the decision-making process using a cost-benefit analysis.

Auditing both process facilities and transport operations with a specialized RBPS protocol can also be a powerful tool to identify management system weaknesses and improvements *before* a serious incident occurs. When the RBPS audit is combined with a comprehensive QRA, the information is available to proactively manage, prevent, and mitigate potential hazards associated TIH chemicals.

So, are you going to inhale or are you going to assess and reduce your risks? Then you'll never try it again!

6 References

[1] CSB (2019), "Toxic Chemical Release at the DuPont La Porte Chemical Facility La Porte, Texas", Report No. 2015-01-I-TX, U.S. Chemical Safety and Hazard Investigation Board, Washington D.C.

[2] CCPS (2007), "Guidelines for Risk Based Process Safety", Center for Chemical Process Safety, American Institute of Chemical Engineers, New York, NY.

[3] ARIA (2015), "Hydrofluoric acid, the worst of all acids?", ARIA News Flash, The Analysis, Research and Information on Accidents (ARIA) database, Bureau for Analysis of Industrial Risks and Pollutions, France.

[4] Dunning, A. E. & Oswalt, J.L. (2007), "Train Wreck and Chlorine Spill in Graniteville, South Carolina, Transportation Effects and Lessons in Small-Town Capacity for No-Notice Evacuation", Transportation Research Record, Journal of the Transportation Research Board.

[5] CCPS (1999), "Guidelines for Chemical Process Quantitative Risk Analysis", 2nd Edition, Center for Chemical Process Safety, American Institute of Chemical Engineers, New York, NY.

[6] HSE (1991), "Major Hazard Aspects of the Transport of Dangerous Substances", Health and Safety Executive, Advisory Committee on Dangerous Substances (ACDS), UK.

617617 Means to Achieve Backflow Prevention of Hazardous Chemicals from Process Vessels to Utility Pipelines and Its Importance during HAZOP Studies

Abstract:

Design of a chemical manufacturing facility is a challenging task because one has to ensure that every design feature has been carefully analyzed for Hazard and Operability (HAZOP). Along with the HAZOP studies of the main process components like reactors, crystallizers, distillation columns, etc., it is imperative to study the possible failures of the utility connections, which may get overlooked. A system is as safe as its weakest point of failure. A common failure is the back flow of hazardous chemicals into the utility connections. Utilities like nitrogen, steam, plant air, and deionized water are often directly connected to process vessels. The utility systems are generally common to the entire plant and so depending on the reaction conditions in an equipment, backflow may occur and contaminate other equipment. Different backflow prevention instrumentation is required based on the utility service. For instance, gaseous and liquid utilities require different backflow preventors. While choosing the material of construction of pipelines, specification brakes are placed at junctions where utilities are connected to the process vessels. This talk will outline various backflow preventor setups that have been successfully implemented in different chemical manufacturing facilities. Based on the severity of failure, the back flow prevention appropriate for different utility services will be discussed. Different kinds of back flow preventors, including differential pressure measuring instruments, flow transmitters, liquid trap with level switch, on/off valve with interlock activation etc., will be explained in detail and compared for their function and safety features.

619079 Technical Review: Suitable Conventional and Emerging Water Treatment Systems for Cwbd Management

Abstract:

Cooling water blowdown (CWBD) is wastewater effluent from cooling towers that are used in industry for various purposes including cooling systems[1]–[3]. In the State of Qatar, district cooling facilities are used to supply large-scale cooling needs in high density developments such as business districts, commercial communities, university campuses, etc.Water quality of CWBD changes and depends on the quality of make-up water, the number of cycles of concentration (COC) and chemical additives used for reducing scaling and corrosion [4]. High concentrations of these contaminants in CWBD streams deem them unsuitable for discharge into surface water and wastewater treatment plants; and in some cases, may require pre-treatment. Therefore, the discharge regulations of the CWBD become more stringent to control the amount discharged to the eco-system, reduce makeup water requirements and to recycle and reuse as much water as possible. Hence, in the presented paper we investigate the existing and emerging water treatment technologies for the treatment of CWBD. We present the contaminants in CWBD and their corresponding impacts on the ecosystem; and we compare membrane and nonmembrane based conventional and emerging treatment technologies based on performance, cost, and limitations along with other factors. The membrane-based technologies reviewed in this article include membrane distillation (MD), reverse osmosis (RO), nanofiltration (NF), and vibratory shear enhanced membrane process (VSEP). For the nonmembrane-based technologies, electrocoagulation (EC), ballasted sand flocculation (BSF), and electrodialysis (ED) processes will be also discussed. EC and VSEP processes are emerging technologies with high performance and can be considered as the most suitable technologies for the treatment of CWBD compared to other processes. However, they are energy intensive and that can be controlled by utilizing renewable sources of energy. The study also compares international and regional (Gulf states) policies and regulations related to the standards for water contaminant discharge levels into waterbodies including surface water. Such regulations will help in delineating the limits of contaminants concentrations, consequently determining the most suitable technology for the treatment of these contaminants. The paper findings will guide the development of suitable policies for the management and discharge of CWBD into surface water and wastewater treatment plants.

References

[1] H. I. Abdel-shafy, M. A. Shoeib, M. A. El-khateeb, A. O. Youssef, and O. M. Hafez, "Electrochemical treatment of industrial cooling tower blowdown water using magnesium-rod electrode," *Water Resour. Ind.*, vol. 23, p. 100121, 2020.

[2] O. M. Hafez, M. A. Shoeib, M. A. El-Khateeb, H. I. Abdel-Shafy, and A. O. Youssef, "Removal of scale forming species from cooling tower blowdown water by electrocoagulation using different electrodes," *Chem. Eng. Res. Des.*, vol. 136, pp. 347–357, 2018.

[3] E. A. Anwer and B. A. Abdul Majeed, "Different Electrodes Connections in Electrocoagulation of Synthetic Blow down Water of Cooling Tower," *Iraqi J. Chem. Pet. Eng.*, vol. 21, no. 1, pp. 1–7, 2020.

[4] J. Ahmed, Y. Jamal, and M. Shujaatullah, "Recovery of cooling tower blowdown water through reverse osmosis (RO): review of water parameters affecting membrane fouling and pretreatment

schemes," *Desalin. Water Treat.*, vol. 189, no. June, pp. 9–17, 2020.

SPRING21
+17TH GCPS
A Joint AIChE and CCPS Meeting

Keeping Process Safety Management Intact during Global Pandemic

Muhammad Salman Shafique
Fatimafert Limited
28-KM, Lahore- Sheikhupura Road, Lahore, Pakistan
Salman.shafique@fatima-group.com

Prepared for Presentation at
American Institute of Chemical Engineers
2020Spring Meeting and 17thGlobal Congress on Process Safety
Virtual
April 18 - 22, 2021

Keeping Process Safety Management Intact during Global Pandemic

Muhammad Salman Shafique
Fatimafert Limited
28-KM, Lahore- Sheikhupura Road, Lahore, Pakistan
Salman.shafique@fatima-group.com

Keywords: COVID-19, Process Safety Management, Crisis Management

Abstract

Process Safety Management (PSM) is the backbone of any process industry and is fundamental to its businesses, plant reliability, safe operations, employee well-being and prevention of catastrophic events.

In order to grow or sustain business and be responsible for its employees' health, good organizations never compromise on the foundation of PSM in their business and operational decisions. But what if a deadly disease alters the way how organizations and systems typically work? Can we quickly adapt to such unprecedented challenges? Are we able to fulfill all requirements of a process safety management system in this extraordinary scenario?

The paper will cover this topic in detail and it will shed light on the encountered challenges and tested best practices in order to maintain PSM during a global pandemic. Effects of a pandemic on key important factors related to PSM will be discussed like management of change, employee motivation, process hazard analysis, job supervision, process monitoring, competency levels, training & development etc.

Building on similar personal experience, the paper will also cover many interesting methods and suggestions for safety professionals and line managers to successfully adapt to the changes in this scenario while not compromising their process safety management systems.

1 Background

1.1 Introduction of Fatimafert Limited (FFT)

Fatimafert Limited is one of the three manufacturing plants of Fatima Group, which is the largest conglomerate of Pakistan producing Nitrogenous fertilizers along with businesses in Energy, Cement, Mining, Packaging and Commodities.

Fatimafert Limited (FFT) is located in the Lahore region of Punjab Province of Pakistan and produces Urea under the brand name of "Babarsher". Commissioned in 1971, it is one of the oldest plants of Pakistan producing Ammonia and Urea (prilled). It has 1000+ employees on permanent, company contract and third party contract basis.

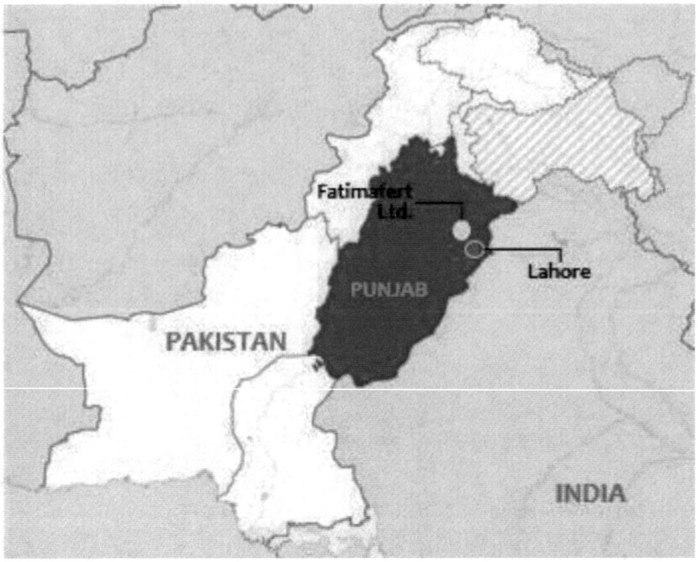

Figure 1. Location of Plant Site and Lahore in the map of Pakistan

1.2 Background of Process Safety Management at Fatimafert Limited

Fatimafert Ltd. has implemented DuPont's Process Safety Management (PSM) system which is a 22-element program consisting of 10 risk-based elements along with 12 cultural elements. All elements are driven by Visible Management Commitment and are connected with each other. Failure of any one element can result in a failure of the whole program. Fatimafert Ltd. has been declared on "compliance" level through a comprehensive assessment by DuPont.

1.3 COVID-19 Pandemic in the Region

COVID-19 pandemic rocked the world in early 2020 and was spread to Pakistan very quickly due to close proximity and continuous trade with China. First case of the disease was reported in Pakistan on February 26, 2020 and the country was swift in reacting to the pandemic. Lockdown was imposed and stringent SOPs were implemented by the government of Pakistan.

Fatimafert Ltd. plant site is situated near Lahore which is the 2nd largest city of Pakistan and is a sprawling metropolis with a population of around 12 million. As with all other densely populated cities of the world, Lahore remained the epicenter of the COVID outbreak in the region with highest COVID positivity ratio in all of the country.

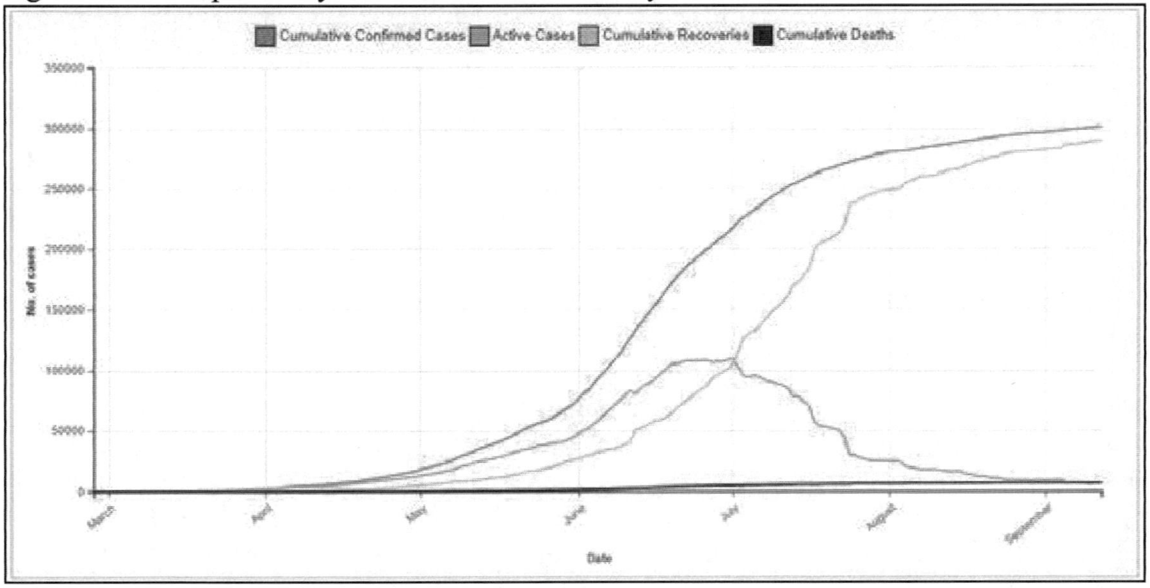

Figure 2. COVID-19 cases (cumulative) in first wave in Pakistan

2 Challenges due to Pandemic

Due to very close proximity with Lahore and almost 90% of workforce traveling daily between the plant site and different areas of Lahore, Fatimafert Ltd. faced a huge challenge to ensure the health and well-being of the employees and also to keep Process Safety uncompromised during operations of a highly hazardous fertilizer manufacturing plant. It was believed that any lapse in Process Safety Management system due to impacts of a very surprising and unusual situation can result in a catastrophic process safety incident and organization geared quickly to adapt according to the situation. Below are major challenges related to Process Safety which were encountered at Fatimafert Ltd. due to the pandemic;

1. Health & Safety of Employees

Keeping each and every member of the team safe from the outbreak remained one of the most pressing challenges as team consists of management, staff and contractual employees (skilled, semi-skilled and unskilled laborers).

2. Mental Well-being

A global pandemic and its rapid outbreak in the region seriously threatened to affect mental health of workers, which directly related with the cultural part of Process Safety Management and could have increased the process safety risks due to human errors.

3. Chaos due to Rumors and Inauthentic data

Due to the unique nature of the pandemic and many conspiracy theories associated with it, managing chaos and stresses due to inauthentic and over-inflated sensationalized statistics was a major challenge.

4. Operational Sustainability linked with National Food Security

Fertilizer production is directly linked with food security and keeping the national food supply chain intact in harsh circumstances remained a big challenge, which could only be achieved through uninterrupted and safe manufacturing operations.

5. Compliance to Process Safety Management

Usual 100% compliance of Process Safety Management program by all cadre of the workforce posed the biggest challenge as compliance levels in such unusual situations vary substantially from the routine safe circumstances. Along with the cultural elements, keeping the risk-based elements (e.g. Process Hazard Analysis, Process Safety Information, Management of Change etc.) intact was the biggest challenge due to abrupt changes to site dynamics including manning limitations, social distancing protocols, ban on physical meetings etc.

6. Minimum Competency Levels across functions

Management of the Personnel Change element of PSM requires every department to maintain a certain level of competency, which was challenged due to Work-from-Home or under-care employees and serious effects on conventional capability development program.

7. Field Supervision and Governance

Field presence and job supervision by operating personnel and maintenance supervisors were challenged due to work-from-home requirements and social distancing protocols.

8. Manpower Management during Lockdowns

Major and minor lockdown imposed by the government in severely affected areas of the region prohibited critical work force to travel and join plant duties putting safe plant operations in jeopardy.

9. Low Literacy related to COVID-19 amongst workforce

Overall literacy levels of the region are low and a new, unknown, widely-spreading disease further challenged comprehension capabilities of grass-root workers, making it tough for them to understand the risk completely and respond in a safe manner.

10. Gadgets and hardware for protection

Low comprehension of the disease, availability of personal protective equipment (PPE) and required protective gadgets remained a challenge especially during the first wave of the pandemic.

11. Risk Assessment & Hazard Analysis Studies

As risk assessment and process hazard analyses are a routine part of the hazardous process plants, organization of these multi-disciplinary team activities in restrictive conditions was very challenging.

12. Social distancing during operational and maintenance activities

Every facility, offices and buildings are designed to accommodate a specific number of people. But when the pandemic hugely limited occupancy levels, it put a question mark on safe execution of almost all operational and maintenance activities.

13. Budgeting and Funds for Safety

Extraordinary situations required extraordinary finances to overcome the challenges related to competency drainage, protective gadgets, testing, hospitalization without compromising routine safety related budgeting and expenses.

14. Securing the Employee Bubble instead of employee only

Due to high transmission rates of COVID-19, only putting efforts on keeping our employees safe was not sufficient unless the same level of efforts were put on awareness and security of families and social circles of the employees to curtail the spread.

3 Best Practices at Fatimafert Ltd. against Pandemic

Early realization of the extent of the threat the region faced and general risk awareness at Fatimafert Ltd. resulted in proactive planning and implementation of preventive actions including development of SOPs, protocols, employee education and handholding all segments of employees. COVID crisis management was formally triggered at Fatimafert Ltd. in last week of February 2020 after reporting of the first COVID positive case in Pakistan.

All efforts at Fatimafert were divided in *two major categories* as described below;

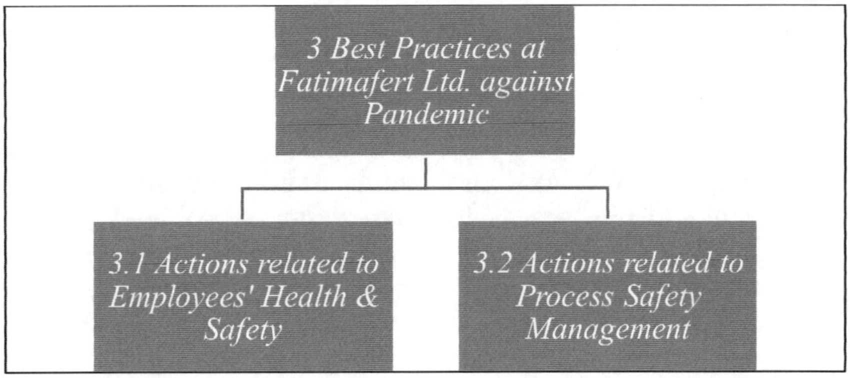

Figure 3. Major Categories of Best Practices at Fatimafert Ltd. against Pandemic

3.1 Actions related to Employees' Health & Safety:

Cultural elements of PSM are directly dependent on protection and well-being of workforce so dedicated efforts were initiated from the start to keep essential workers protected, safe and healthy from the disease. The COVID Crisis Management at Fatimafert Ltd. consisted of following four major elements;

3.1.1 Proactive Policies:

- **No Handshake Policy** – Handshakes and all types of physical contacts were immediately banned on company transport and within site premises.
- **Risk Profiling of Employees** – Medical risk profiling of all employees was carried out in order to clearly identify employees vulnerable to COVID infection including those with respiratory diseases and ischemic heart conditions.
- **Work-from-home for vulnerable employees** – Employees identified as "Vulnerable" through risk profiling were immediately barred from traveling to the site and were facilitated to work from their homes instead.
- **Shorter Working Hours for Immunity Boosting** – To facilitate employees (both traveling to site and on work from home) in having sufficient rest and to build immunity, duty hours were shortened significantly with same pay/wages.
- **Special Focus on Mental Well-being** – Emergence of COVID brought along with it huge anxiety and stress due to fear of the unknown. Therefore, the organization was quick in initiating special efforts to keep employees happy and relaxed. Special "Fun Breaks" were designed and incorporated in routine duty timings which included jokes, games, story sharing, physical exercises etc. In addition, a dedicated fitness program "Fit for Life" was launched during the first wave of the pandemic. Through this program, employees were coached and facilitated to develop a fitness mindset through steps tracking, weight loss, healthy diet, nutrient support etc. These efforts yielded surprising results and employees remained calm, relaxed and cheerful throughout the time of panic and despair.
- **Isolation of essential workforce in colony** – Fatimafert Ltd. maintains a small colony adjacent to the plant for accommodating critical workforce of all departments (around 50 families). Colony premises were locked down to ensure 100% security of the last line of defense; they were a critical workforce.
- **International/Domestic Travel and visitors handling protocol** – International and domestic traveling of employees were banned. A stringent health declaration and medical checkup protocol implemented for extremely unavoidable circumstances. All unnecessary incoming visitors and Vendor Services Manager (VSMs) were prohibited. Strict controls were implemented for extremely crucial VSMs only. These protocols

contained Declaration Forms (covering travel history, interactions history and family health status) and screening by medical professionals at the site's entrance.

- **Online meetings and trainings** – All team meetings and classroom training sessions were immediately prohibited and virtual meetings and training sessions were introduced. Required tools (software, licenses, user training) were immediately arranged for smooth execution.
- **Termination of gatherings & shutdown of Café** – All gatherings and team events were cancelled and café operation was terminated to eliminate the chance of cross contamination during meals preparation and consumption. All incoming employees were encouraged to bring along their meals from their homes.
- **Benchmarking of Policies with National & International Guidelines** – All COVID Crisis Management policies were benchmarked with best national and international guidelines including CCPS (Center for Chemical Process Safety), OSHA (Occupational Safety and Health Administration), government of Pakistan Guidelines & other national level manufacturing industries to learn from others.
- **Progressive Motivation** – In order to appreciate and encourage front line warriors like health care workers from medical center, plant operators, HSE (Health, Safety and Environment) Inspectors etc., a dedicated campaign by the name of "OUR HEROES" was launched. In addition, awards and recognition programs were also launched to acknowledge the efforts, reward the best contributors and keep the morale high.

3.1.2 Gadgets and Hardware:

- **Face masks and gloves for all employees** – N95 and surgical masks were arranged immediately for the whole workforce including contractual laborers and the use of masks in company transport and plant premises was made strictly mandatory.
- **Hand sanitizers & hand washing** facilities – Hand sanitizers were installed at entrances of all buildings and offices. Similarly, portable hand wash facilities along with soaps were placed at all entrances and places of usual commute of the workforce.
- **Extra vehicles for transportation** – Fatimafert Ltd. provides transportation to all outstation employees in the form of shuttle vans. In order to ensure social distancing (at least 3 feet) during commuting, vans occupancy limit was reduced to 50% of the total capacity and extra vans were arranged to ensure social distancing.
- **Full body disinfection at entrance** – Full-body disinfection enclosure were installed at all entrance gates of the plant site and colony.
- **Thermal screening of entrants** – Thermal screening devices were arranged and all incoming personnel were screened in two steps while entering plant premises. Anyone

having higher than normal body temperatures was denied access and escorted to the medical center for further checkup.

- **Installation of Door Openers** – Door openers were immediately provided to doors of all offices, washrooms etc. to keep them open and reduce chance of transmission through door handles.
- **Isolation Facility** – A secure isolation facility was established at a safe location to immediately isolate any employee showing symptoms during duty.
- **Anti-body test kits** – More than 100 Antibody test kits were procured for on-site verification of COVID infection in employees showing symptoms related to the disease.
- **Full body protection kits** – Full body protection kits were arranged for Health Care Workers of Plant Medical Center for safe medical screening of visitors and handling of employees with symptoms.

3.1.3 Employee Education:

- **Innovative Use of Social Media for Education of Employees** – In order to spread the awareness on authentic statistics and preventive measures to all level of employees down to grass root level, WhatsApp Broadcasting Groups were created and all company and contractual employees were added in those groups. Through it, only well-researched and reviewed information was shared with the workforce, which remained hugely effective and beneficial to avoid spread of rumors and panic situation.
- **Families & public education through Social Media Page** – A social media (Facebook) page was created specifically targeted to impart COVID education to employee families and the public. Education material shared through this page reached to residents of 70+ countries demonstrating a contribution of Fatimafert towards educating the community and the public in general.
- **Trends & Statistics** – Trends and statistics extracted of COVID information related to the region was compiled in easy-to-understand format and circulated across the site on a daily basis to inform employees on the ever-changing situation of the pandemic.
- **Communication circulars** – Communication circulars containing awareness data, preventive measures and company COVID related SOPs were developed in national language (Urdu) and circulated via emails as well as in form of pamphlets for distribution to employees and their families. These awareness brochures were also distributed in the nearby community.
- **COVID management certifications through online platforms** – COVID related awareness and management online training and certifications were arranged for employees from all departments who acted as COVID warriors in their respective domains.
- **Visualization of Preventions and Protocols** – All COVID preventive measures (e.g. hygiene etiquettes, hand washing method, relevant symptoms etc.) and company

protocols (where to report, how to report, SOPs, rules etc.) were designed in visually appealing patterns and clearly displayed at strategic locations in form of awareness boards, banners and posters.

- **Webinars on mental health** – Special webinars and coaching sessions were arranged for employees and their families by company medical officers on mental health, stress and anxiety management due to the COVID pandemic. It helped a huge deal in ensuring the mental well-being of employees and their loved ones.

3.1.4 Auditing:

- **Behaviors & awareness checks through Management Audits** – During MSAs (Management Safety Audit Program), special focus was put on verification and immediate correction of behaviors, compliance and awareness related to the COVID crisis management (e.g. masks compliance, social distancing compliance, physical gathering etc.). It helped greatly in aligning behaviors of the workforce with COVID protocols.
- **Reporting of Non-Compliances thru Observations Reporting Program** – Fatimafert Ltd. has a well-established and efficient medium for reporting of HSE observations by the name of "Mukhlis". All employees were encouraged and empowered to use this tool for reporting of any non-compliances of COVID protocols in company vehicles and plant premises. Best reporters were rewarded on their care for the colleagues.
- **Daily audits of face mask & social distancing compliance** – As face mask compliance was made 100% mandatory along with social distancing , a formal, round-the-clock auditing mechanism by a dedicated team to check compliance levels were successfully implemented.
- **Catering hygiene audits of workers canteen** – A workers' canteen was kept open under strict compliance watch to provide meals for shift working teams. Strict auditing regime was developed to audit COVID protocol compliance including social distance during meals collection, meals consumption, meal handling staff, etc.

3.2 Actions related to Process Safety Management:

Maintaining same compliance level of all Process Safety Management protocols and requirements during the pandemic was one of the biggest challenges faced at Fatimafert Ltd. Below is explanation of alternate methods taken against major elements of PSM;

- **Risk Assessment & Process Hazard Analysis**

Process Hazard Analysis is a team exercise and a discussion based brainstorming activity. Fatimafert Ltd. conducts the PHAs as per an approved PHA plan and PHA studies of three different process sections were due in 2020 as per the plan. It posed a great challenge to conduct the studies in COVID circumstances due to restrictions on physical meetings and most of the team members working from home. As a solution, remote PHA / HAZOP studies were introduced successfully. Area visits for facility siting survey and hazards identification were made shorter to reduce exposure time and were conducted by smaller group of PHA team members, staff interviews were conducted remotely and HAZOP study brainstorming sessions were carried out through virtual mediums (Skype for Business). Special trainings were arranged for PHA leader and team members for efficient utilization of time, resources and virtual mediums for this extremely important process safety activity.

- **Process Safety Information (PSI)**

Keeping Process Safety Information (PSI) include Hazards of Material, Process Design Basis and Equipment Design Basis. Keeping PSI documents of all units up-to-date and providing easy access to these packages for all employees in the office or working from home posed a major challenge and most of the documents (e.g. manuals, diagrams, flow-charts etc.) were kept in the Project Library. Therefore, after COVID arrived, extensive efforts were made to scan thousands of PSI documents and creating an online portal in which soft files of complete PSI packages were uploaded, sorted, catalogued and marked for quick and easy access by all employees from office and homes which they could use during PHAs, MOCs and various studies.

- **Management of Change (MOC)**

All non-critical and non-urgent modification jobs at the plant site were prohibited during the COVID pandemic. Only emergency changes were allowed after stringent reviews. Instead of paper-based forms and manual review process (office-to-office circulation of MOC documents posing risk of cross-contamination), a 100% digital, ORACLE-based platform was launched for virtual data entry, review and approval steps completion from laptops and PCs thus completely eliminating the need for physical review meetings and virus spread through paper-based files and documentation.

- **Mechanical Integrity & Quality Assurance (MIQA)**

Due to vintage design and aging of the plant, quality assurance, preventive maintenance of safety critical machines/equipment and their inspection becomes extremely important in order to ensure asset integrity and plant reliability. Due to this reason, the organization was not able to afford any delay or lapses in procurement/testing of safety critical devices and implementation of preventive maintenance/inspection plans. To assure quality of safety critical equipment being procured, all the correspondences were made online, virtual verifications during FATs (Factory Acceptance Test) were conducted instead of physical visits to the suppliers. A detailed SOP was developed for maintenance and inspection activities which chalked out stringent protocols related to tools sanitization, maximum technicians for each job, separate scheduling of each job etc.

• Incident Investigation & Reporting

In order to facilitate employees in immediate reporting of any Near Miss Event or any other type of safety incident and quality investigation/Root Cause Analysis by individuals and investigation teams, access of available online incidents portal Safety Management Information System (SMIS) was expanded, new licenses were purchased and reporting through paper-based forms were immediately prohibited to curtail virus spread. Through SMIS anyone could report an incident. Investigation teams were able to work together from their separate offices or homes through virtual mediums and conduct quality investigations. All interviews were conducted virtually and visits of incident locations were made shorter and by a maximum of two members of the investigation team following COVID protocols including masks and social distancing.

• Permit to Work (PTW) System

At Fatimafert Ltd. Permit to Work (PTW) system is a paper based system containing nine different types of permits including cold job, hot job, electrical job, instrument job, confined space entry, excavation, lifting over process lines, vehicle entry, and radiography permits. All kind of activities are covered through these permits and, due to complexity of the system and certain requirements, making the permits online was not possible. Therefore, alternate administrative controls were implemented instead. Instead of physical meeting of permit issuer and acceptor in field rooms, permit desks were set-up in open areas to ensure 100% ventilation, social distancing, and a queuing mechanism was introduced to reduce physical interactions. Instead of direct exchange of paper-based PTW forms between issuer and acceptor, PTW baskets were introduced to reduce chance of contamination and virus spread.

• Integrated Organizational Structure

Fatimafert Ltd. manages it Process Safety Management program through an integrated organizational structure of 13 committees including Executive PSM committee, contractors' management committee, behaviors and audits committee, occupational health committee etc. Each committee consists of 8-12 multi-disciplinary employees who meet at an identified frequency. Keeping in view high risk of contamination during these meetings, physical gatherings were immediately banned and virtual mediums (Skype for Business) were provided to all committee members for smooth meetings along with an online tool for recording, communication and stewardship of minutes of meeting.

• Audits & Observations

Process Safety Auditing Program consists of behavioral audits by site management called Management Safety Audits (MSAs) (60 audits per month), internal PSM audits (annual), second party audits by Fatima Group Corporate HSE team (annual) and external PSM audits by DuPont (usually after three years). As per plan all audits, except external PSM audit, were due during the pandemic and were conducted safely and successfully. SOP of MSAs was revised to incorporate guidelines for making safe safety contacts, shorter area visits and online audit report submission portal. On the other hand, preparatory, kick-off, discussion and closing meetings of all PSM audits were conducted virtually instead of physical sessions. All required documents and evidences were exchanged through shared folders instead of files and paper-based documents to eliminate risk of cross contamination.

- **Emergency Response & Planning**

Fatimafert Ltd. has defined roles and maintains a crisis handling team through two Emergency Control Centers (ECCs) for effective handling and control of any process safety emergency. Due to limits on site occupancy and work-from-home protocol implementation, the organization faces challenges related to availability and presence of key ECC members which could have affected gravely in case of emergency. To counter that challenge, back-up membership of all control centers was identified and socialized. Emergency role statements were revised accordingly and targeted training on these updated role statements were provided to new members. Moreover, weekly and the quarterly emergency drills/mock exercises mechanism was modified to meet COVID protocols including sanitization of emergency handling gadgets, fire and safety articles, identification of safe distances, group activities with a smaller workforce and critique meetings in open air instead of critique rooms.

- **Training & Development**

Due to the ban on physical / classroom training sessions, development of employees and their capacity building was a huge challenge which could have direct consequences on process safety of the plant. The company invested in purchasing well-recognized online training programs including Harvard ManageMentor, Linkedin Linda etc. Along with that, a dedicated "Capability Development Program" was launched and successfully executed through which employees of all cadres and groups achieved trainings and certifications on topics as diverse as COVID management, process safety, HSE, digital tools, remote working and functional / technical topics etc. It resulted in completion of 850 online courses (certificates & diplomas) resulting in 2230 training man-hours in addition to routine training man-hours derived from annual training calendars.

4 Conclusion

Despite of the very susceptible situation of Fatimafert Ltd. plant site due to proximity with the epicenter of the COVID outbreak in the region and very high exposure of 90% of employees to the affected areas, so far there have been a total of nine COVID positive cases in over 350,000 man-days. There has been no process safety incident at the site during the COVID pandemic, all process safety studies, audits and protocols are being followed as per normal conditions.

All of this had been possible just due to a proactive approach, flexible policies related to employee well-being but very strict protocols related to COVID crisis management, innovative methods of employee education (especially the use of social media), alternate methods to ensure compliance of Process Safety elements, digital tools, stringent auditing and, most importantly, care for employees and the community which reigned supreme during this time of need and mutual support.

SPRING21
+17TH GCPS
A Joint AIChE and CCPS Meeting

Pandemic Risk Management; Protecting people while ensuring business continuity

James Sneddon
Principal Consultant, Risktec Solutions (Canada) Ltd.
Suite 1000, Bankers Hall West, 888 3rd Street SW, Calgary, AB
James.Sneddon@risktec.tuv.com

Prepared for Presentation at
American Institute of Chemical Engineers
2021 Spring Meeting and 17th Global Congress on Process Safety
Virtual
April 18 - 22, 2021

Pandemic Risk Management; Protecting people while ensuring business continuity

James Sneddon
Principal Consultant, Risktec Solutions (Canada) Ltd.
Suite 1000, Bankers Hall West, 888 3rd Street SW, Calgary, AB
James.Sneddon@risktec.tuv.com

Keywords: Pandemic, COVID-19, Risk Management, Bowtie, LOD

Abstract

The COVID-19 pandemic swept across the globe in the latter half of 2019, throughout 2020 and into 2021. In response, many companies implemented work from home policies, while others stopped operations entirely in an effort to limit the spread throughout their workforce and supporting communities.

This containment strategy was not universally viable; long-term shutdowns impacted the economic viability of companies, and some industries were designated as an 'essential service' and thus continued operations. These employers faced the proposition of balancing the needs of the business and the community with a continued responsibility to provide a safe workplace for employees.

This paper, and accompanying presentation, will demonstrate how the application of bowtie analysis, a commonly adopted methodology in high hazard industries, can help the risks associated with continued operation in a pandemic to be better understood and managed, thus ensuring the safety of both your personnel and business.

1 Pandemic Risk Management; An Introduction

Since its appearance in December 2019, the SARS-CoV-2 virus (COVID-19) has continued to spread, touching almost all corners of the world. The progression from outbreak to pandemic was soon accompanied by the swift imposition of lockdowns in many countries.

In support of the various restrictions imposed upon their communities, many organizations implemented work from home policies, while others stopped operations entirely in an effort to limit the spread. There were, however, industries where this approach was not viable, either because they provided an essential service or because of economic necessity. Now, more than a year on from coronavirus initially upending daily life and work, this situation remains.

Employers requiring their employees to remain in, or return to, an office environment, work site or operating facility face the ongoing proposition of balancing the needs of the business and the community with a continued responsibility to provide a safe working environment for their employees.

Common risk management methodologies, such as bowtie analysis combined with the Lines of Defense (LOD) model, can be adapted to help employers better understand the risks involved with a resumption or continuation of in-person operations, ensuring the measures they have in place to manage this risk are appropriate and effective.

2 Bowtie Analysis

2.1 Introduction to Bowtie Analysis

Risk assessment lies at the heart of any form of risk management, and one of the most powerful of these techniques is the bowtie method. Its strength is that it goes beyond the usual risk assessment 'snapshot' and highlights the links between controls, assurance & verification activities, and the underlying management system; a valuable trait when assessing the constantly evolving nature of a pandemic.

Bowties originated as a method for assessing operational risk, with the earliest mention of such an approach appearing within an adaptation from the ICI plc HAZAN Course Notes 1979, presented by The University of Queensland, Australia [1]. The Royal Dutch Shell Group was the first major company to integrate the bowtie method into its business practices [2, 3, 4] and is credited with developing the technique which is widely used today.

The bowtie method provides a readily understood visualization of the relationships between the causes of business upsets, the escalation of such events, the controls preventing the event from occurring, and the mitigation measures in place to limit the business impact (see Figure 1).

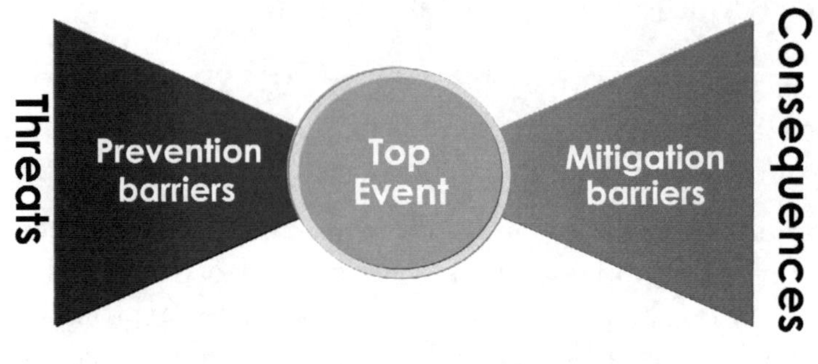

Figure 1: Bowtie Framework

A description of the different components of the bowtie, and the method for building such a diagram are well-documented [2, 3, 5], hence this paper focusses predominantly on its specific application within a pandemic risk management setting.

2.2 Application of Bowtie Methodology to Pandemic Risk Management

The application of bowtie methodology can be shifted from its more traditional use in high-hazard industries to assessing the risk to an organizations workforce during the COVID-19 pandemic in a relatively straightforward manner. Through graphical representation, bowtie analysis can map threats that may impact worker safety or the ability to maintain safe operation, identify and assess the safeguarding in place to prevent or mitigate different scenarios, and readily highlight any deficiencies or non-conformances.

A representative example of how the bowtie methodology can be applied in the assessment of 'worker safety' is provided in Figure 2, below.

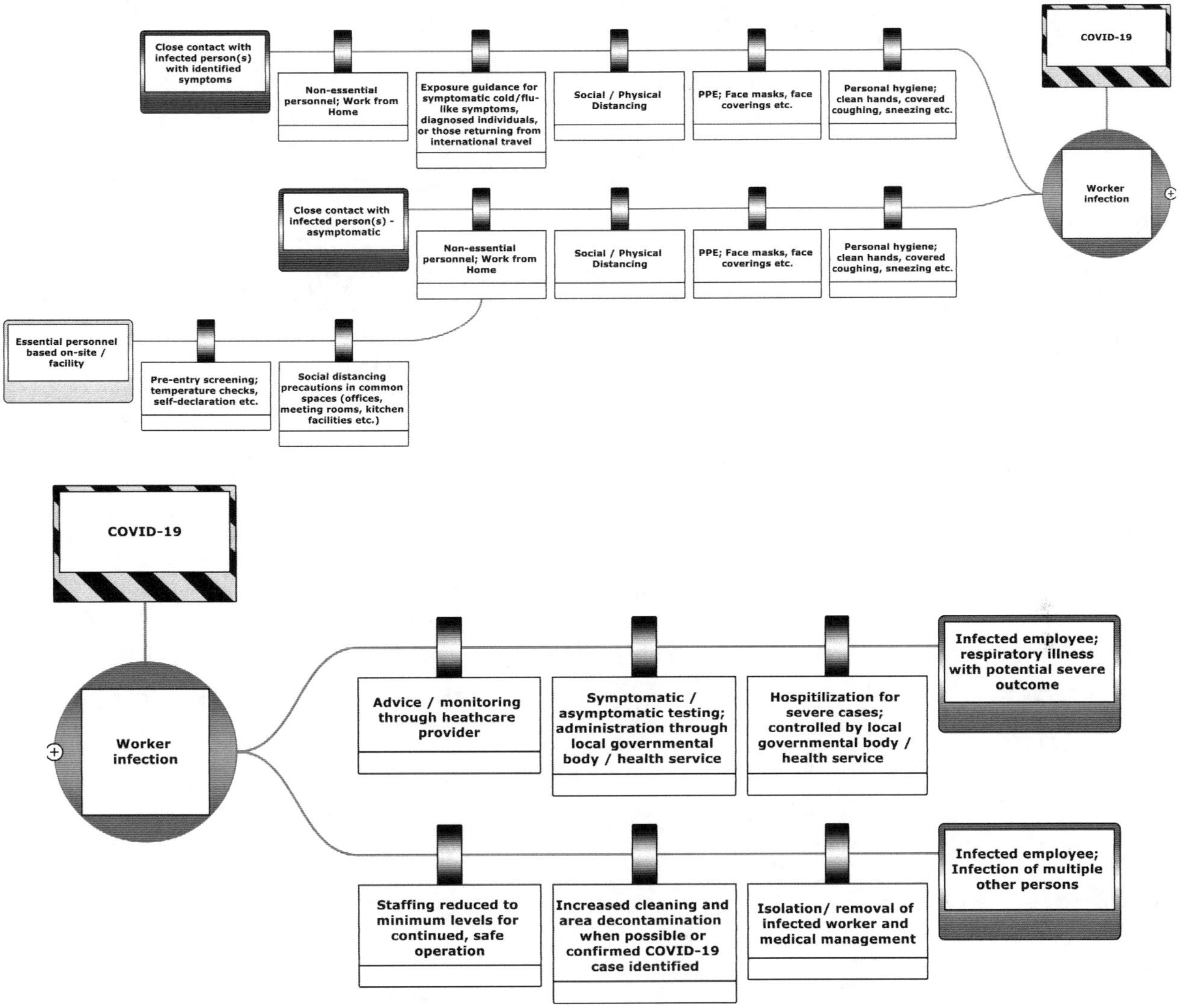

Figure 2: COVID-19 Worker Infection Example

3 Risk Management Framework

Bowtie analysis is not a panacea for managing risk in the current environment. It must be integrated within a wider risk management framework, enabling barrier implementation to be controlled and the risk profile for the organization to be consistently monitored.

The adoption of such a framework is of special significance to large organizations who face the following challenges:

1. How do we coordinate our response in a consistent manner across multiple facilities and offices?
2. How do we provide a large and geographically diverse workforce with readily available information on our response and the resources available?
3. How do we map and track compliance of safeguarding to ensure conformance and identify any deficiencies?

The following features of bowtie analysis, and its application within a robust risk management framework help address these challenges.

3.1 Pandemic Response Communication

The bowtie is an excellent communication tool for the coordination of response. Dissemination of key information to the workforce can be actively managed via targeted outputs from the bowtie diagrams.

Information for each barrier on the bowtie diagram can be easily and swiftly communicated to the workforce in a one-page summary (Figure 3), ensuring a coherent and well-informed response, covering questions such as:

- What is the barrier?
- What does it do?
- How does it perform?
- How is it tested?
- Where do I find documents with further information?
- Who should I contact for further details?

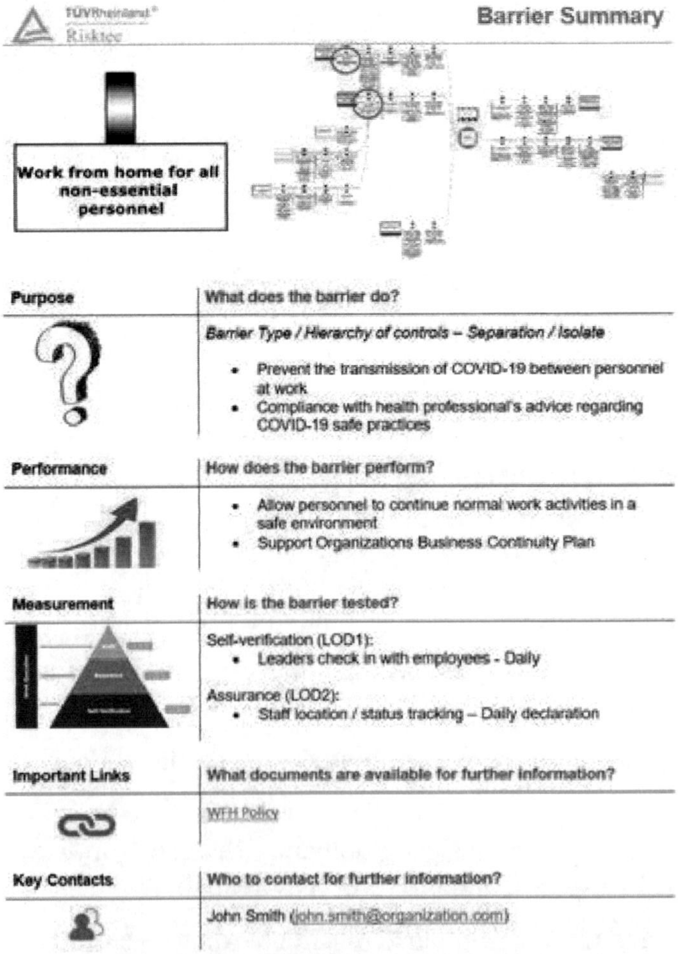

Figure 3: COVID-19 Safeguarding Summary

3.2 Barrier / Safeguarding Assurance and Verification

In order to track how well prevention and mitigation barriers are performing, assurance and verification criteria must be defined as benchmarks to measure success. Such criteria are commonplace within high hazard industries in the form of Performance Standards. However, a more appropriate approach for managing the COVID-19 response is the three lines of defense (LOD) model.

The LOD model is typically deployed in an internal auditing function [6]. LOD modelling is a method used to gauge performance, enhance clarity regarding risks and controls, and help improve the effectiveness of risk management systems. It is based around three broad concepts:

- LOD1: Self-verification that activities have been completed as prescribed. Barriers are effectively controlling risks and are delivering planned performance;
- LOD2: Independent functional assurance of conformance to requirements and quality of operating activities; and,
- LOD3: Internal, external and regulatory audits to confirm compliance.

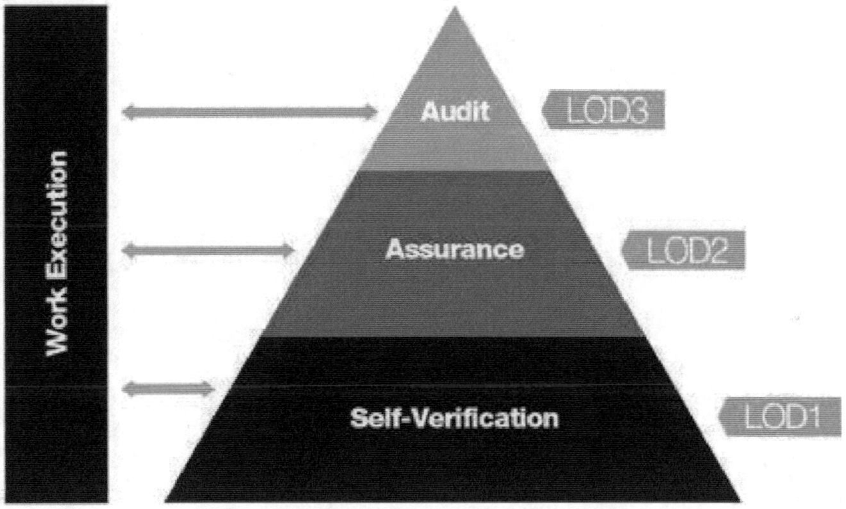

Figure 4: Lines of Defense Model

The LOD model can be applied to the prevention and mitigation barriers identified within the bowtie in the following manner:

1. Identify assurance and verification activities for each barrier;
2. Disseminate these criteria to operating locations, offices, entities in questionnaire format for completion by site representatives (see Figure 5);
3. Collate responses within the auditing function of the bowtie software used (e.g. BowTieXP); and
4. Display and monitor location performance using a Key Performance Indicator (KPI) dashboard (see Figure 6).

TÜVRheinland®
Risktec

Daily COVID-19 Checklist; LOD1 Self-Verification **Date:**

Barrier Ref.	Barrier	Self-Verification Action	Guidance	Responsible Person / Role	Check Status
W-P1	W-P1 Work from home for all non-essential personnel	Check-in with 90% of employees & contractors	Check-in can be by video, phone, or e-mail. The 90% value is a guide to indicate that most staff should be regularly engaged, but 100% compliance is not realistic every day.	Department / Team Leads	
W-P2	[illegible]	Check-in with 100% of employees & contractors who have self-declared and are isolating	Check-in can be by video, phone, or e-mail. These are people who are in self-isolation due to travel, exhibiting symptoms, or potential exposure to an infected individual.	Operations Manager	
W-M1	W-C3 Pre-access screening at facility; Temperature check	Verify temperature screening device is accurate and being used on 100% of all entrants	This check ensures the device being used to record temperatures of entering workers is properly calibrated as per the manufacturer guidelines. It also verifies that the temperatures of all entering workers were taken and those exceeding the upper limit were not permitted entry.	OH&S Lead	
W-M2	W-C7 Quarantine wing / area for isolation established	Cleaning schedule in place with 100% compliance	This check applies specifically to areas set aside for site quarantine of people experiencing symptoms of COVID-19. Such workers are restricted to this quarantine location while awaiting transportation from the site.	Facility Management Lead	

Figure 5: Self-Verification (LOD1) Checklist Example

The application of the LOD model to COVID-19 response enables a real-time view of the health of barriers, ensuring all non-conformances and deficiencies are highlighted, together with a risk profile for each facility, office, or entity under consideration.

KPI dashboards tracking compliance with the assurance and verification steps for the more common applications of equipment or activities deemed 'Safety Critical' can adopt a higher level of complexity than would be required for managing risk in a pandemic setting. A relatively simplistic traffic-light approach can serve to easily and effectively highlight how the identified barriers are being adopted across an organization.

A representative example of a KPI dashboard is depicted in Figure 6.

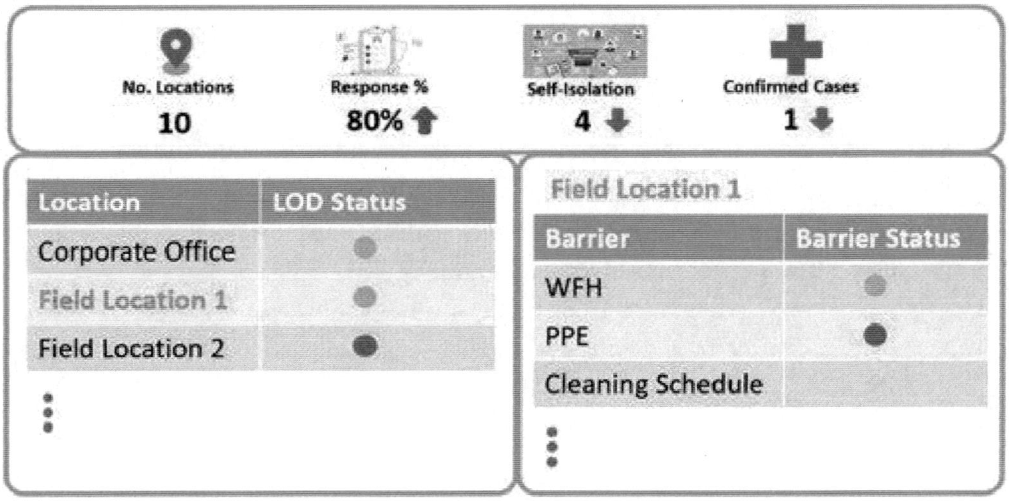

Figure 6: KPI Dashboard Example

4 Conclusion

COVID-19 continues to present a unique and challenging environment for society and industry alike. As organizations continue operations, begin return-to-work preparations, or simply prepare for the everchanging restrictions placed upon them, the need to manage the risks inherent in such activities is paramount.

Bowtie analysis, and its application within an integrated risk management process, can help us better understand and manage these risks by:

- Ensuring that the key hazards have been identified and appropriately assessed; including correct identification of threats, consequences, and both preventive and mitigative barriers;
- Developing, implementing, and tracking a set of assurance and verification tasks, in line with the LOD model, to ensure performance of the identified barriers;
- Highlighting any deficiencies in safeguarding, or non-conformances with the developed assurance and verification tasks, and
- Ultimately providing assurance that the identified risks are being managed / mitigated in a manner which is considered to reduce this risk to an acceptable (or As Low As Reasonably Practicable – ALARP) level.

5 References

1. University of Queensland, Australia, Minerals Industry Safety and Health Centre, National Minerals Industry Safety and Health Risk Assessment Guide.
2. Primrose, M.J., Bentley, P.D., van der Graaf, G.C., Sykes, R.M. Shell International Exploration and Production B.V., "The HSE Management System in Practice-Implementation," SPE 35826, 1996.
3. Primrose, M.J., Bentley, P.D., van der Graaf, G.C. Shell International Exploration and Production B.V., "Thesis – Keeping the Management System "Live" and Reaching the Workforce," SPE 336034, 1996.

4. Gower-Jones, A.D., van der Graaf, G.C., Milne, D.J. Shell International Exploration and Production B.V., "Application of Hazard and Effects Management Tools and Links to the HSE Case," SPE 36031, 1996.
5. Centre for Chemical Process Safety (CCPS), Bow Ties in Risk Management, A Concept Book for Process Safety, 2018.
6. The Institute of Internal Auditors (IIA), IIA Position Paper: The Three Lines of Defense in Effective Risk Management and Control, January 2013.

619345 Relief System Design Revalidation Warranted during Current Pandemic

Abstract:

Numerous operating companies globally are faced with major unparalleled challenges during these disruptive times. Companies are taking radical actions to reduce capital and operational expenditure. One common action taken to limit the spread of the virus is reduction in number of on-site personnel or "only essential to operations" personnel to be on site. As the current wave is looking more ominous compared to what we saw few months back, one can expect these measures getting further harder and stricter. But at the same time, impact of such actions on process safety must be understood. API Standard 521 allows credit for operator intervention and administrative controls to reduce likelihood of overpressure scenarios from occurring. The actions taken during current pandemic can certainly overburden operators jeopardizing administrative controls and compromising overpressure protection. This presentation dives deep into relief system design where administrative control and operator intervention credits have possibly been taken during original design and sheds light on the scenarios needing re-evaluation.

HIRA in Mining Companies Is Necessary and a Big Challenge

Presenter/Principal Author
Américo Diniz Carvalho Neto
RSE CONSULTING
Brazil
americodcn@rsem.com.br

Other Author(s)
Afonso Costa
RSE CONSULTING
Brazil
Afonso.costa@rsem.com.br

Prepared for Presentation at
American Institute of Chemical Engineers
2021 Spring Meeting and 17th Global Congress on Process Safety
Virtual
April 18 - 22, 2021

HIRA in Mining Companies Is Necessary and a Big Challenge

Presenter/Principal Author
Américo Diniz Carvalho Neto
RSE CONSULTING
Brazil
americodcn@rsem.com.br

Other Author(s)
Afonso Costa
RSE CONSULTING
Brazil
Afonso.costa@rsem.com.br

Keywords: HIRA Project,

Abstract

The implementation of a risk-based process safety management system is a turning point for any company that intends to have a better understanding of the risks associated with their process. However, full implementation of all elements takes time, and in the meantime, catastrophic scenarios should be managed accordingly, without wasting time.

To fulfill both criteria, having long and short run implementation strategies, HIRA (Hazard Identification and Risk Analysis) process was developed for mining companies worldwide. Based on the Critical Control Management (CCM), HIRA process is focused on "quick wins" of process safety, management gaps that must be eliminated shortly, to keep high consequences scenarios from happening. Those scenarios are known as MUE (Material Unwanted Event).

HIRA for mining companies is based on the methodology developed by ICMM (International Council of Mining and Metals), which is comprised of 9 steps, 6 of them are only to identify scenarios and their causes, consequences and mainly the critical controls associated. The methodology intends not only to identify critical controls, but also to check their status (implemented or not implemented), identify items that can make them more reliable and to assign responsibility for the management of the control, throughout its lifetime, to guarantee they will perform their function when demanded.

To accomplish such tasks, HIRA process widely uses Bow Tie Analysis, a very intuitive and visual risk management tool, which allows even people who do not have a lot of experience in the subject to easily understand the graphic representation of the MUE, such as cause, consequence, top event and mainly the controls and their status. Bow tie sessions are followed by field verification of each critical control identified, to check coherence between records (data), integrity and reliability, of the controls.

As a result, both existing and possible controls (known as Future State) are identified, according to the MUE analyzed and, comparing information from field verification, a qualitative risk assessment is carried out for three different categories – Inherent, residual and forecast risk.

By the end of the HIRA process, many types of controls are identified, such as procedures, safety valves, inspection plans, guard rails, alarms, interlocks…and so on. They now will be managed differently, as they are classified as critical controls – items that are crucial do prevent the MUE form happening or mitigate its consequence.

1 Main section 1 heading (Introduction)

Many companies have improved their performance on safety-related issues using a wide range of rules, procedures, and standards; however, Material Unwanted Events, such as fatal accidents, rare catastrophic events and circumstances of over-exposure that are harmful to health, are still subject to occur. For this reason, the ICMM (International Council for Mining and Metals) developed an approach called Critical Controls Management (CCM), aiming at the continuous improvement of management over rare events, but potentially catastrophic, emphasizing critical controls. The methodology to improve the management control of these types of events, called Material Unwanted Events (MUEs) follow below:

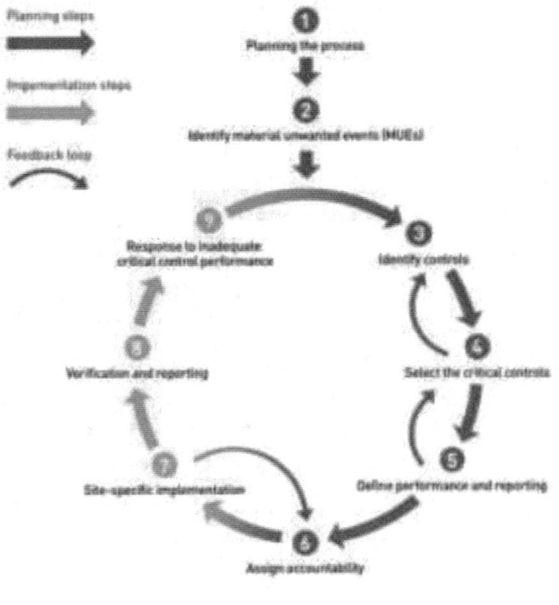

ICMM Critical Controls Management Process (CCM)

CCM's approach is based on some STEPS as follow:

Step 1: Develop a project plan that contemplates the organizational context, objectives, team responsibility and the individuals and business areas affected in the implementation.

Key actions for STEP 1

• Develop a plan that describes the scope of the project.

This includes:
– organizational context
– project objectives

– responsibilities
– business sections involved.

• Develop methods to:

– identify potential hazards and unwanted events
– assess risk
– review MUEs
– select critical controls
– assess objectives and performance of critical controls
– investigate critical control underperformance
– measure impact of the project
– identify ownership and accountability.

Step 2: Identify the MUEs to be managed and summarize the main information for each MUE. In order to identify, verify if they present material risk, evaluate the opportunities to eliminate it by improving the designs and describing the MUE completely (hazard, materialization mechanism and possible consequences).

Key actions for STEP 2

• Understand major hazards and identify potential MUEs.
• Apply selection criteria to MUEs with a focus on the consequences.
• Identify design opportunities to address the hazard, reducing the potential consequences and eliminating the MUE from the CCM process.
• Describe the identified MUE, including the relevant hazard, mechanism of release and nature of the consequences.

Step 3: Identify the controls, developing Bow Ties diagrams and evaluating them.

Key actions:

• Identify the controls.
• Prepare a bowtie diagram.
• Assess the adequacy of the bowtie and the controls.

Step 4: Select the critical controls for each identified MUE, summarizing its main information.

Key actions For STEP 4

• When identifying critical controls, apply the critical control definition and guidance in this section.
• Consider the performance requirements of the potential critical controls and how they could be verified.

• The final set of critical controls for an MUE should represent the critical few that, when managed using CCM, can effectively manage the MUE risk.

Step 5: Define the objectives and Critical Controls performance requirements, how they will be checked and the activities that will be affected by them.

Key actions for STEP 5

• Define objectives and performance requirements for each critical control.
• Identify current activities that affect the critical control's performance.
• Describe activities to verify performance and reporting requirements.
• Identify what would trigger immediate action to stop or change the operation and/or impose the performance of the critical control.

Step 6: Develop a list of those responsible for the MUE, for the Critical Control and for its verification, including the preparation of the final information report.

Key actions for STEP 6

• Assign owners for MUEs, critical controls and verification activities.
• Describe reporting plan for the health of critical controls.
• Assign owner for review of reports.

Step 7: Define a verification plan and reporting plan for each MUE, implementing it according to the specific needs.

Key actions for STEP 7

• Critical control information must be specific to a site or asset.
• Adjust the critical control definition, performance information and verification requirements as necessary to suit the local context.
• Site-specific planning for implementation may involve an iterative process.
• Site-specific planning should include establishing a foundation for CCM that includes leadership, communication and appropriate development of knowledge and understanding related to the critical controls.

Step 8: Implement verification activities and report the process, defining the status of each critical control.

Key actions for STEP 8

• Undertake verification activities for critical controls as described in MUE/critical control verification and reporting plan (developed in Step 5).
• Report a summary of verification activity results to the critical control owner.

• Report critical control verification status to the MUE owner.
• Reports should highlight priority information succinctly using traffic light system.
• Action initiated if critical control performance drops below the defined triggers (established in Step 5).

Step 9: Performance investigation if critical controls are performing below expectations.

Key actions for STEP 9

• Take action when critical control performance is inadequate (below the defined trigger threshold).
• Investigate the causes of unacceptable critical control performance.
• Information and data from the investigation should be used to continuously improve the CCM.

The companies risk map, in most cases, does not fully reflect the main risks to which the operation and the external population are exposed. This is due to the lack of a unified methodology to identify, quantify and measure risks. In view of this reality, one company can initiate the HIRA Project (Hazard Identification and Risk Assessment), to conduct a quick technical scan of risks in its operations to identify relevant and unmitigated situations.
HIRA was designed to quickly identify and verify the main risks of the company units operations. This is a global project to be implemented and will be at the center of the operational risk agenda in the coming months. The HIRA Project will focus on the very critical risks of severity in the company operations and its subsidiaries and its results, will be reviewed frequently by the Executive Committee for Operational Risks, the Executive Board and the Board of Directors. His results will guide the main actions to reduce the organization's risk profile.

GOALS

The current study aims to identify and analyze operational safety risks with a very high-risk magnitude or very critical severity of a company site. In addition, the identification, definition of performance criteria and the establishment of criteria to guarantee the performance of associated critical controls are part of the scope of this work. From this, it will be possible to help the company to produce a complete and accurate inventory of the relevant unwanted events of the organization (MUEs) and, later, based on the identified MUEs, identify critical controls, associated performance requirements and means of verification.

For this purpose, this study will include:

1. A consolidated list of operational safety risks assessed with a very high risk magnitude or very critical severity (MUEs) for the unit;

2. Bow Ties based on the identified MUEs and a corresponding list of critical controls identified for each (following the suggested ICMM decision tree and based on the group discussion);

3. One-Pager for each critical risk control.

4. List of action plans to be developed based on the prioritization of risks.

5. Updated MUEs risk matrix of the unit.

METHODOLOGY

The Bow Tie Analysis (BTA) can be used to identify and assess hazards with the potential to cause major accidents involving company site. This study can be conducted through presential or remote meetings through the TEAMS videoconferencing platform, by multidisciplinary teams composed of corporate company members and company site members and can be supported by external consultants.
Initially, happens a Workshop session, jointly between the consultants of RSE Consulting and company employees present at the meeting. The site MUEs were defined taking into account that the chosen MUEs would be among the most at risk in the facilities analyzed. In these sessions, MUEs were selected and evaluated from a previous list of more than 170 different MUEs, from different units of the company, including units in different countries which the HIRA project will be developed. Other new scenarios were also discussed and included as discussed with the local technical team. After determining the likely MUEs applicable to the site, these were then categorized as follow:

• Category 1: Scenarios in which, all the information presented during the Workshop, was sufficient for validation as a MUE. These are considered priority scenarios;

• Category 2: Scenarios in which, despite the information presented during the Workshop, these were not sufficient to validate the MUE in remote mode, thus requiring field visits and more information for validation. Also, in this category are included scenarios that can be considered as developments of scenarios classified as Category 1;

• Category 3: Scenarios in which, despite the information presented during the Workshop, these were not enough for validation as a scenario that meets the HIRA criteria, to be considered MUE. An on-site visit would be necessary to complement information and analysis. Thus, the team present at the MUE Workshop session with different specialists concluded in a list some (%)

MUEs classified as Category 1, that is, priority to the units of the complex. These meetings can be remotely, and the list was subsequently validated by company leaders. It should be noted that the identification of MUEs needs to consider the historical events (Accidents and Near Miss) of the site, company and sector, in addition to predictable events, given the operations and activities at individual locations. Therefore, the identification of MUEs needs to include personnel with adequate experience and a review of relevant data.

Then, meetings were held remotely to analyze the MUE and generate project deliverables for the Complex. The analyzes were composed by members of the company corporate members, site regional members and external RSE consultants, where the consequences, threats and the respective existing controls for each MUE of the site were identified. During these meetings Bow Ties from other company units were analyzed with similar processes as a reference. Details about the Bow Tie technique are presented of this paper.

To identify the critical controls and controls of the MUEs, the proposal decision tree was followed:

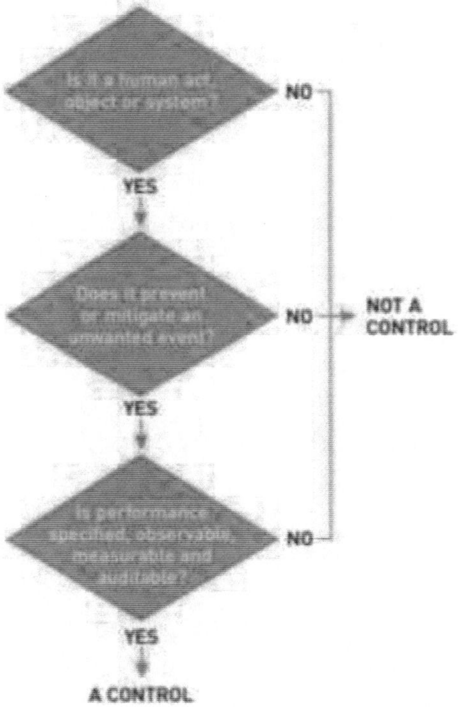

Parameters for defining a control

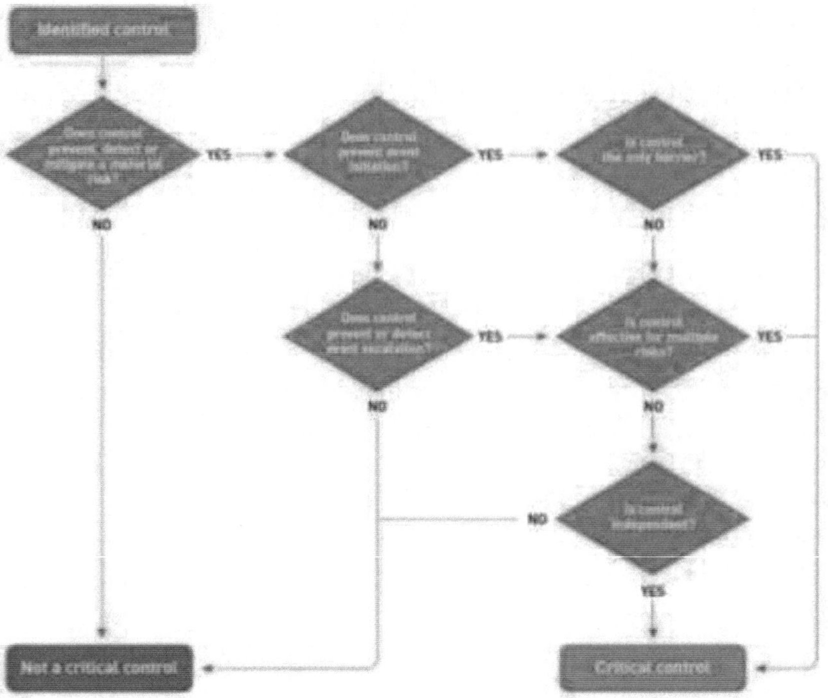

Parameters for defining a Critical Control

Once the critical controls were identified, a "One Pager" was generated for each one, that is, a quick briefing of just one page for each critical control (except "Future State" critical controls) containing the following information:

• Objectives of related critical control
• Performance requirements to meet objectives
• Activities that can enable or strengthen control
• Activities that can be verified to attest to the performance of a critical control
• Performance for critical control
• Trigger to stop the unit, revision or investigate, depending on the performance of the control
• Critical Controls attributes

The current levels of effectiveness (status) of performance of these critical controls were classified initially without field verification, based on the opinion of company specialists present at the meetings, to generate previous actions for the company, according to the methodology established for carrying out remote HIRA, during the COVID-19 pandemic. However, these classifications must be rectified / ratified in another phase of HIRA, when the field visit is done.
Remote risk classification meetings were also held, categorized as Inherent, Residual and Forecast, classifying them as follows:

• Inherent risk: premise = the existing critical controls to prevent each cause or mitigate each consequence are null (there are no controls);

• Residual Risk: premise = the existing critical controls to prevent each cause or mitigate each consequence are fully implemented or not implemented, depending on the classification of the status of each critical control determined through remote meetings by the Vale specialists team

• Forecast Risk: premise = critical preventive controls, mitigating controls, future controls (Future State) and all actions and recommendations captured during the meetings are fully implemented and functional, that is, they are compl ete and reliable.

It is important to note that risk assessments were carried out, remotely, without the field verification by the HIRA teams to validate, based on the One Pagers, the critical controls. Therefore, critical controls were evaluated in future scheduling, considered the next phase to be carried out at HIRA on site.

For the risk analysis of each MUE, the risk acceptability criteria of the company must be adopted. Below follows a brief explanation of Bowtie methodology selected in this Project.

BOW TIE TECHNIQUE

Bow Tie is a qualitative risk assessment method used to assess how high-risk scenarios develop. The essence of Bow Tie consists of plausible risk scenarios surrounding a certain hazard, and ways in which the organization prevents these scenarios from happening. The method takes its name from the shape of the diagram you create, which looks like a male bow tie.

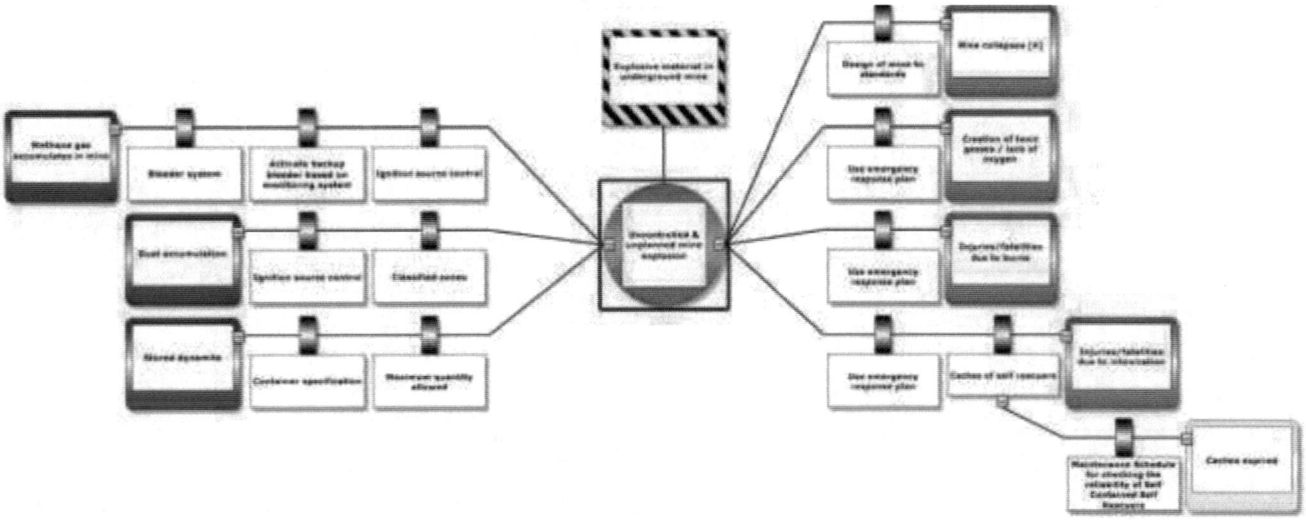

The Bow Tie methodology has several objectives:

• Provide a framework for systematically analyzing a hazard.

• Helps to decide whether the current level of control is sufficient (or, for those who are familiar with the concept, whether the risks are as low as reasonably practical or ALARP).

• Helps to identify where and how to invest in resources that could have the greatest impact.

• Increases communication and risk awareness.

General step by step construction of the Bow Tie diagram:

Hazards Identification The purpose of hazard identification is to find, recognize and describe the risks that can help or prevent an organization from achieving its goals. Hazards can be operations / activities (operating rotating machines, driving a car),

I. substances (chemicals, hot fluids, etc.) or situations (a suspended load) that we deal with in the normal processes of our business. As long as these hazards are under control, they will not cause harm, but introduce the potential for harm.

II. Definition of the Top Event / MUE When control over a hazard is lost, it is usually possible to identify the exact moment when a normal situation changes to an abnormal situation, that moment is called Top Event. The Top Event is not yet a catastrophe, but the company is now exposed to the potential for harm from hazard.

III. Identification of Consequences The consequences are classified as the possible negative results directly linked to the top event, which can affect people, the environment, the facilities and the company's reputation.

IV. Threats Identification Factors that can lead directly to the top event are called threats. These are capable of causing the Top Event independently and are not associated with barrier failures.

V. Identification of Preventive Controls Risk management is about controlling risks. This is done by implementing barriers to prevent certain events from forming. A barrier (also called control) can be any measure taken that acts against some unwanted force or intention, in order to maintain a desired state. Preventive barriers, identified on the left side of the Bow Tie, can prevent the top event from happening or can eliminate the threat.

VI. Identification of Recovery Controls Controls or recovery barriers identified on the right side of the Bow Tie can prevent the top event from resulting in unwanted consequences or mitigate the consequences.

VII. Define Escalation Factors Once the controls have been identified, it is recommended, according to the Bow Tie method, to identify specific conditions or actions that reduce the effectiveness of the control. These are called escalation factors.

VIII. Identification of Escalation Factors Barriers Once the escalation factors have been identified, the last step is to look at what barriers we have to manage these factors. These barriers have the same concept as all the barriers previously discussed, but, now they do not prevent / mitigate a top event or consequence, they will prevent a barrier from failing. The BowTieXP software, version 9.2, developed by CGE Risk, was used to construct the Bow Tie diagrams. BowTieXP is the most widely used risk assessment software based on the Bow Tie methodology that allows you to easily create risk assessment diagrams.

ANALYSIS RESULTS

Table attached shows the Material Unwanted Events (MUEs) that were hipotetical identified and jointly defined by RSE Consulting and the company team. As defined by the company, for an event to be considered as MUE it must have the potential to cause one or more fatalities or the potential to cause a major environmental impact that goes beyond the limits of the company.

In addition to these available scenarios (Category 1), the MUEs Workshop, conducted with a multidisciplinary team, generated potential scenarios for MUEs whose information presented was not sufficient to validate them in the remote mode, requiring additional information through field visits for validation (Category 2 and 3).

In total, 221 existing controls were defined to prevent and / or mitigate MUEs and 138 of these were classified as critical controls for these scenarios. These numbers consider the proposed future controls and the repetition of the same control between the scenarios.

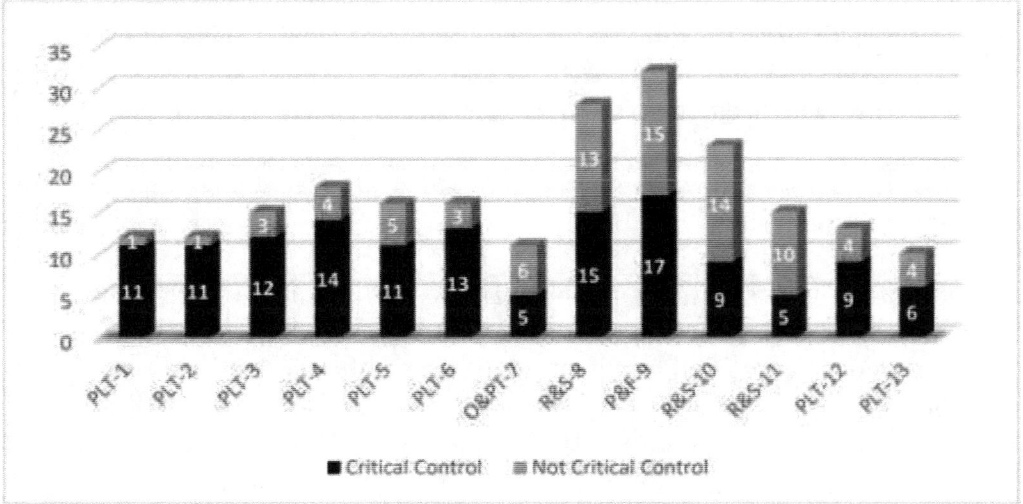

Controls and Critical Controls

According to the criteria adopted for the status of critical controls, the following were identified:

✓ 98 Critical controls classified as "Implemented"

✓ 09 Critical Controls classified as “Not Implemented”

✓ 31 Critical Controls classified as “Future State”

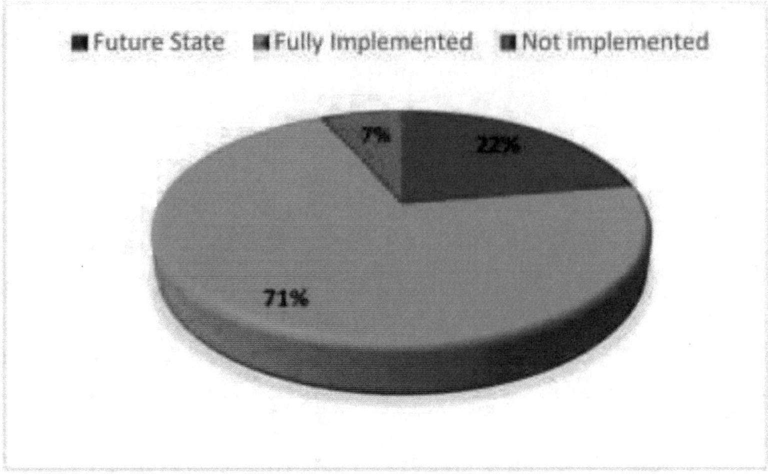

Percentage distribution of the status of Critical Controls

Regarding the type of critical controls, the following were identified:

✓ 104 Critical Controls classified as "Preventive";

✓ 34 Critical Controls classified as "Mitigatory".

This distribution demonstrates the greater effort of the professionals who participated in the assessments and identification of risks / controls in the Search for preventive solutions instead of mitigating unwanted scenarios, which satisfies the initial premise of HIRA.

Additionally, considering a larger sample of 19 sites, where 252 MUEs were analyzed, one can have a better understanding of the results.

• Critical controls - for the 252 MUEs evaluated in this sample, 4283 controls were identified, of which 2487 were considered critical controls. In this number are included the controls considered Future State, that is, the proposals identified by the group to reduce the risk of the assessed scenario. This means that 58% of the controls were considered crucial in preventing the occurrence or mitigating the consequences of the MUE, according to the evaluation team. This is always a point of attention when evaluating controls because it is not uncommon for members to tend to classify a large percentage of controls as critical. This is because companies prioritize resources to ensure the reliability of their critical controls, and the Operation team can use this criterion as a factor for prioritizing resources, going in the opposite direction to the methodology, which is a big mistake. It is up to the consultant to ensure that the flow of classification of critical controls is properly followed.

A rule used, which has served for the present project, is to consider as critical controls something between 45% to 55% of the controls. If the numbers indicate something greater than this, a ratification with the evaluation team is recommended.

For the same project, 1115 controls considered critical were classified as administrative controls, that is, they are basically controls that are highly dependent on human action to fulfill their required function. Examples of these controls are procedures and inspections. The latter is one of the main points of attention because it presents many opportunities for improvement as inspections are common not only in mining but in any company. However, many operational inspections carried out do not have the required reliability, that is, they are not able to identify deviations in integrity that can lead to the MUE.

These inspections are characterized by having a high frequency (sometimes daily) of execution and low level of details. On the other hand, there is the integrity inspection, which has a low frequency and a high level of details. The latter is usually performed by certified inspectors, has an associated procedure and inspection report issued at the end of each inspection. That is, depending on the type of inspection carried out, it is possible that this process is very effective in identifying structural damage or just a simple routine daily verification process, without the required level of detail.

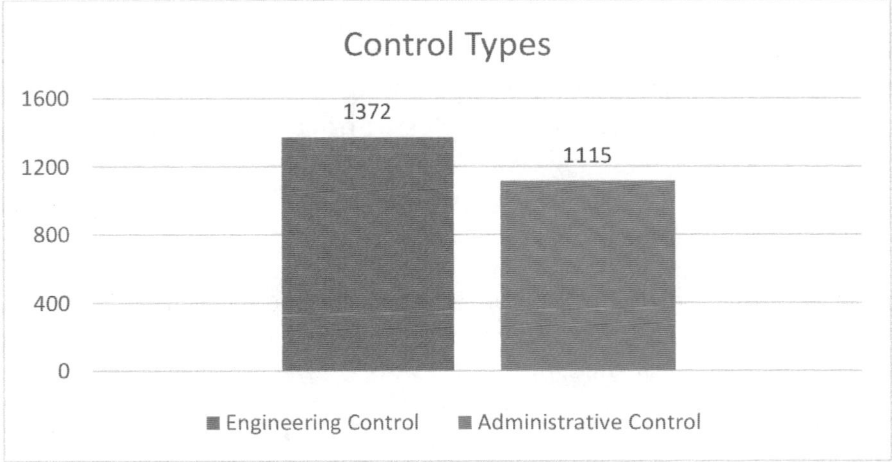

Graph 2 – Types of Critical Controls

It is highly recommended that this information about the items that can enable or strengthen critical control, in the case of operational inspection, are clearly described in the One Pager.

Preventive or mitigating controls - Considering the type of control as preventive or mitigating, again for the project considered, it was found that 1874 (75%) were preventive and 613 (25%) were mitigating. The mitigating controls are basically the emergency action plan and fire fighting system. Less frequently, dykes also appear as important controls for the MUE of loss of fuel containment.

It is important to say that, although they represent only 25% of the identified critical controls, mitigating controls are responsible for a higher percentage of recommendations. One of the most recurrent is the failure to consider the MUE in emergency pre-plans and drills.

Conclusion:

Throughout HIRA´s remote work of the company site, risk scenarios considered more critical were evaluated, with the potential for multiple fatalities or significant environmental impact or level 1 process safety event. Although the result of critical controls considered as implemented was in higher percentage, deviations were identified in the management of some pillars of Process Safety, such as Mechanical Integrity, and Risk Management. For a better understanding of the deviations listed, the list of actions is done.

A larger number of "automatic" or engineering critical controls type were identified in relation to those with human interface, such as operational procedures, inspections, maintenance plans, among others. For some of these controls, actions were proposed to ensure greater reliability on demand, activities such as systematic inspections, maintenance, calibration plan, looping control tests, fail-safe mode, etc.

About the critical controls identified with human actions, it is necessary the following requirements to the reduction of the failures: to reach deadlines of procedure's review, elaborate training plan/program with knowledge retention (evidence with records), recycling and periodic evaluation of the team performance, with actions for "under the established standard", as well as investigation records of near miss incidents.

About company site, it is important to highlight:

1. Treat the critical controls as critical items into the SAP or the specific management software.

2. Ensure proper inspection and maintenance for the automatic controls (safe interlock systems);

If the field verification of critical controls is performed in the company site made it possible to diagnose opportunities for improvement, strengthening and implementation of Process Safety Risk Management. Considering the concepts of hazard identification, risk analysis, change management, implementation of Process Safety Management System will be essential for the company site to manage its risks associated with the MUEs, to keep them at an acceptable level.

2 References:

[1] RSE Consulting HIRA methodology process, Americo Diniz Carvalho Neto, 2017.

[2] RSE Consulting Barriers Audit methodology, Americo Diniz Carvalho Neto, 2007.

[3] Health and safety critical control management Good practice guide, ICMM.

[4] CCPS. *Guidelines for Bowtie in Risk Management.* Center for Chemical Process Safety, American Institute of Chemical Engineers, New York, NY, 2018.

[5] CCPS. *Guidelines for Hazard Evaluation Procedures.* Center for Chemical Process Safety, American Institute of Chemical Engineers, New York, NY, 2008.

616590 Title: Silencing a Noisy Agitator Seal By Managing Stress

Abstract:

Title: Silencing a Noisy Agitator Seal by Managing Stress

Abstract: Did you know the sound that comes when you run your finger on the rim of a wine glass is referred to as "partial boundary layer lubrication"? Did you know this same phenomenon also occurs commonly on buffer and media tank agitators in bio-processing facilities? However, on agitator seals the sound can amplify to over 90 decibels, which exceeds OSHA's personal exposure limits.

Using an FEA model, John Crane found, that the noise from partial boundary layer lubrication is a result of vibration caused by the application of an even or consistent stress to the seal face, which is consistent with field trials since the seals rarely squealed at startup. Instead, the seals squeal after some run time when the right amount of stress is applied. This stress could come from the pressure of the vapor space in the tank or vessel, which also explains why the noise level and pitch can come and go depending on the pressure in the tank, agitator speed, as well as other scenarios.

This led to the discovery that the squealing can be reduced, or even eliminated, by managing the stress imposed on the agitator seal faces. Subsequently, John Crane developed a solution that results in an active seal face, meaning the degrees of flatness change with the stresses applied during operation. In other words, the rim of the wine glass changes as you run your finger around it preventing the noise condition from ever taking place.

Take Home Message from Poster: The noise from a media or buffer tank agitator seal can be reduced, making a quieter and safer workspace.

619172 Use of Data Analytics to Obtain Cost Reductions and Performance Improvements in a Regulatory Application

Abstract:

All operating nuclear power plants (NPPs) in the United States are required to monitor and assess the performance of plant structures, systems and components (SSCs) as specified in the Maintenance Rule (10CFR50.65). This rule requires that all licensees identify the occurrence of any SSC failure that results in a loss of safety function, often referred to as a Maintenance Rule Functional Failure (MRFF). All MRFFs are subsequently evaluated through the plant corrective action program (CAP) as required by 10CFR72.172. To ensure these requirements are met, all NPPs document events that occur which are evaluated to identify impact on the plant. A typical dual unit generating facility generates as many as 15,000 of these incident reports (IRs) per year which each must be evaluated by experienced plant staff. Of these IRs, only 0.1-0.2% typically are MRFFs. NPPs have found the monitoring of plant SSCs for compliance with the maintenance rule to be expensive and resource intensive.

An alternative approach which has the potential to reduce these costs is to automate the review of IRs through the use of data analytics and machine learning (DA/ML). Advances over the past several decades in computational capabilities and computer programming techniques / algorithms now permit application of DA/ML techniques to analyze the large historical data sets that contain information on the performance of SSCs over years of plant operation. Unfortunately, these techniques have not been widely adopted within the nuclear industry due to its highly regulated nature. A consequence of this structure is that any DA/ML models used in applications subject to regulatory oversight must provide a high degree of confidence in the predictions due to potential regulatory impacts associated with an incorrect determination.

The MRFF Analyzer is a DA/ML application which significantly improves the efficiency of the maintenance rule screening process. The model uses a combination of natural language processing (NLP), Bayesian statistics, and machine learning to identify the likelihood that the incident described by an IR represents a MRFF. The software was trained, tested, and validated on four years of historic incident reports (2015-2019) generated across Exelon Generation's fleet of operating NPPs. This data set consisted of approximately 650,000 IRs. Approximately 420,000 IRs from 2015 through the summer of 2018 were used to train the model, and 80,000 IRs from summer-winter 2018 were used to test the model. Model performance in the field was validated during 2019 with each IR generated by the sites analyzed by the software with review performed by on-site subject matter experts as specified in the plant program implementing procedures. Once the performance of the MRFF Analyzer software was verified as adequate to meet the requirements of the Maintenance Rule, the parallel use of review by human experts was discontinued.

This paper presents the development, training, and deployment of the MRFF Analyzer software across Exelon Generation's fleet of operating NPPs. The algorithms in use in the software are discussed in detail. The challenges dealing with highly skewed data are presented and the approaches used to address these challenges discussed. The results of a detailed sensitivity assessment are presented and used to estimate the expected statistical performance of the software. The overall performance of MRFF Analyzer since its deployment at Exelon Generation is discussed and estimates of the cost savings are provided.

617819 Search: A Symptom-Based Expert for Advanced Response to Chemical Hazards

Abstract:

Chemical exposure accidents pose a risk of enormous personal injury and property damage if initial response is not appropriate. In Korea, from 2014 to October 2019, 1,000 chemical accidents occurred, and 20 people died and 444 were injured. In addition, the number of chemical accidents shows a trend of increasing from 233 in 2014 to 277 in 2018. Existing chemical substance detection technology is focused on the development of response technology for leaks at known sites, and research on the detection at dynamically changing environment and knowledge service for hazardous chemicals considering symptoms is insufficient. Therefore, there is a need for research on a real-time hazardous chemical detection and diagnosis system and knowledge service that protects workers or exposed people from unexpected chemical accidents in the field, by detecting exposed chemicals quickly and helping to cope with dangerous situations early.

In this study, we develop an AI-based system for chemical identification from actively collected exposure symptoms, for real-time response to hazardous material incidents. In order to analyze the symptoms of chemical exposure accidents or contact with substances, information related to chemical substances (symptoms, toxicity, structural information, etc.) is collected and extracted from WISER, PubChem, MSDS, etc. to systemize the related information. The knowledge base of 1,000 core substances is built by designing ontology for symptom knowledge of accident-involving chemical substances. In case of WISER developed by NIH, chemical substances are estimated by simply matching symptom words manually inputted, but the proposed system actively collects the symptoms and predicts the candidate chemical substances (with probabilities) through the knowledge base inference considering the exposure situation. The knowledge base inference uses a knowledge graph as knowledge base that is easy to search, and uses a reasoning method such as the dynamic RDFS++ reasoner, which is easy to infer dynamic data provided by AllegroGraph, and OWL2RL materializer, which is easy to infer static data. In the version 1 of SEARCH, the dynamic RDFS++ reasoner is mainly used to proceed with inference.

Using Machine Learning (k-NN, Decision Tree, SVM, etc.) and Deep Learning, it also predicts toxicity from chemical substance information (structure, etc.) or symptoms from toxicity [1,2] for building expanded symptom KB. The accuracy of the chemical symptoms estimated from the structure and toxicity of the chemical substance is derived from the model with good performance among the previously used models, developing a system with an accuracy of 80% or higher that can estimate 43 symptoms for 1,000 chemicals. Through the combination of two network models that predict toxicity from the structure of chemical substances or predict symptoms from toxicity, it is possible to provide symptom knowledge to be used for symptom detection even for new chemical substances.

SEARCH provides a wide range of information on hazardous substances, including substance identification support, physicochemical characteristics, human health information, and containment and suppression advice. The prototype system is serviced with AI speakers as the main interface and demonstrated at the Gyeonggido Disaster and Safety Headquarters, and plans to combine the established knowledge base and chemical substance-symptom estimation model. The voice recognition interface using AI speakers enables effectively acquire information by saving time and effort rather than directly inputting and matching information manually.

617983 Application of Machine Learning Algorithm for the Analysis of OSHA Accident Database to Identify Trends in Performance Indicators of Process Safety Incidents

Abstract:

Many workplace injuries and fatalities occur in facilities dealing with hazardous chemicals due to unexpected leaks, exposure to dangerous process conditions and other loss of containment events. The U.S. Occupational Safety and Health Agency (OSHA)'s Process Safety Management (PSM) program has been in effect for almost three decades in the United States, requiring companies to establish comprehensive management programs to deal with such hazardous chemicals with the target to assure safety and health in the workplace. Analysis of the trend of process safety incidents in these industries can be a good performance indicator for the effectiveness of the PSM program.

For a long time, OSHA has been maintaining and publishing accident data collected from various industries since the 70's and currently has recorded information of more than 130,000 incidents involving fatalities and injuries arising from various causes. Analysis of such a large database to identify process safety incidents manually is extremely difficult and time consuming. Current study focuses on the development of a machine learning algorithm to screen the database and identify process safety incidents through use of the keywords and incident descriptions associated with each accident provided by OSHA.

In order to validate the machine learning process, the large OSHA database is filtered to analyze incidents from 1984 to current time and only chemical manufacturing and petroleum industries are considered through use of industry classification codes. More than 2,200 incidents are screened out in this manner and are analyzed to determine those that were process safety related incidents. The analysis was done both manually and through use of the developed machine learning algorithm to validate the capability of the latter. The keywords are classified into various categories to aid the algorithm in the learning process. This is the beginning step to the application of artificial intelligence to analyze big databases such as the OSHA database to determine common patterns behind injury and fatal incidents and provide industry with performance indicators that will provide an understanding of where the current laggings are and enable proactive measure to be taken to prevent future incidents.

617448 Game Based Learning – Making the Message Memorable

Abstract:

A perennial frustration for organizations working within high hazard industries is that despite going to great time, cost, and trouble to communicate important process safety messages across the entire workforce, incidents continue to happen. Making these messages "stick" is essential. This presentation will outline a novel approach to developing process safety knowledge and competence through Game Based Learning (GBL).

Based on research that suggests people remember 90% of what they 'do', GBL takes your message out of PowerPoint and broadcasts it interactively to the trainee. A GBL session typically lasts 30-60 minutes, is hands-on and, by using every-day analogies for technical subjects and grounding them in the real world, it is accessible to all personnel across an organization. Indeed, GBL for major accident awareness has been run for school children as part of Science, Technology, Engineering and Mathematics (STEM) teaching for years. GBL comes into its own in high hazard industries where, generally speaking, improvement is about small margins. This presentation will include some examples of games developed by Risktec, and cover a wide range of common process safety themes, to help prompt the audience's ability to replicate similar ideas in their own workplace.

SPRING21
+17TH GCPS
A Joint AIChE and CCPS Meeting

Process Safety Cards – A Good Deal Safer

David Hatch
Process Safety Integrity
Derby, UK
david.hatch@psintegrity.com

Prepared for Presentation at
American Institute of Chemical Engineers
2021 Spring Meeting and 17th Global Congress on Process Safety
Virtual
April 18 - 22, 2021

Process Safety Cards – A Good Deal Safer

David Hatch
Process Safety Integrity
Derby, UK
david.hatch@psintegrity.com

Keywords: Process Safety Cards, Bowties, Gamification, Microlearning, Workforce Engagement.

Abstract

To improve the awareness of less technical personnel to potential Loss of Containment (LoC) scenarios and increase their vigilance to such events & their potential causes so they can be promptly addressed; memorable images are shown on familiar playing cards representing typical LoC events.

1 Introduction

Process Safety Cards have a simple goal – to improve the awareness of a broad range of personnel to potential major accident scenarios and increase their vigilance to such events & their potential causes so they can be promptly addressed. These memorable images represent typical LoC events to deliver knowledge literally into the hands of those who might contribute to and/or be affected by Process Safety Incidents.

With an ageing workforce and increased outsourcing of maintenance and operations activities, it is vital that time and knowledge poor stakeholders (who typically fall outside corporate process safety training strategies) are engaged and equipped to ensure they play their part in sustaining the integrity of assets.

2 Challenge

Process Safety Incidents, Major Accidents or Loss of Primary Containment Events continue to occur in the 21st Century despite better technology and bigger data. For example, on 28th January 2021, a liquid nitrogen release at the Foundation Food facility in Gainesville, US resulted in 6 fatalities and multiple injuries.

A review (by the author) of the 100+ CSB Investigations [1] indicates 200+ fatalities occurred of which approximately 40% were Contractors. This does **not** suggest or imply that the Contractors were at fault, however it does highlight that 3rd parties are often involved in major accidents.

Although Process Safety initiatives, typically referencing Process Safety Management frameworks including OSHA [2], CCPS [3] or the Energy Institute [4], are now commonplace, these may exclude external resources including outsourced Operations & Maintenance personnel who are therefore not as equipped as their internal counterparts to recognize and address potential Process Safety incidents.

We must also acknowledge that wisdom is being lost with the ageing workforce and that assets are getting older whilst those responsible for them are getting younger.

Are we possible focusing on the wrong people and/or protection measures?

How do we put knowledge into the right hands?

3 Workforce Engagement

In order to gain and sustain the attention and participation of the workforce, which may range from novices who need to be informed to veterans who need to be reminded, it is becoming more common to use Gamification methods. These employ game design elements e.g., point scoring, competition with others, rules of play to improve user engagement, learning and knowledge retention.

However, many games require considerable time to setup and play or often cannot be played in the workplace due to the cost and complexity of deploying electronic devices e.g., laptops or tablets to host the games safely (flammable atmosphere ignition potential).

Process Safety Cards offer an alternative approach which offers the benefits of gamification without the time and cost burdens and is based on Microlearning e.g., small learning units absorbed in small time units.

Players are encouraged and helped to learn and apply simple principles:

- REMEMBER Events,
- RECOGNISE Causes (Threats) of Events,
- RESPECT Controls (Barriers) which prevent or mitigate Events.
- REPORT Concerns about the presence or performance of the Controls/Barriers,

Thereafter there is an obligation on the duty-holder or owner/operator to:

- RESPOND to Feedback (concerns, questions, or suggestions).
- RECOVER Protection to its original state.

Using this simple & memorable 'R' model aims to embed these principles into the minds & memories of frontline personnel. It is critical that the RESPOND & RECOVER steps are performed otherwise there can be no faith that contributions are respected and that improvements are being made.

4 Deployment

Process Safety Cards represent 52 different LoC events that occur in the Process industries which are presented as familiar playing cards split into 4 separate suits.

- DEGRADATION Failure within design conditions or "under-strength"
- DISCHARGE Open route to atmosphere
- DAMAGE External impact
- DEVIATION Failure beyond design conditions or "over-stress"

These events are drawn from industry publications including the HSE [5] and CCPS [6] and whilst they are not claimed to be an exhaustive list, they can be attributed to familiar major accidents across a variety of industries.

Cards make it easier to communicate in all directions, top-bottom, side-to-side and inside-outside the organization by providing a common visual language. This gets and keeps the workforce onboard with the Process Safety or Asset Integrity mission.

Sample cards are shown in Figure 1 below:

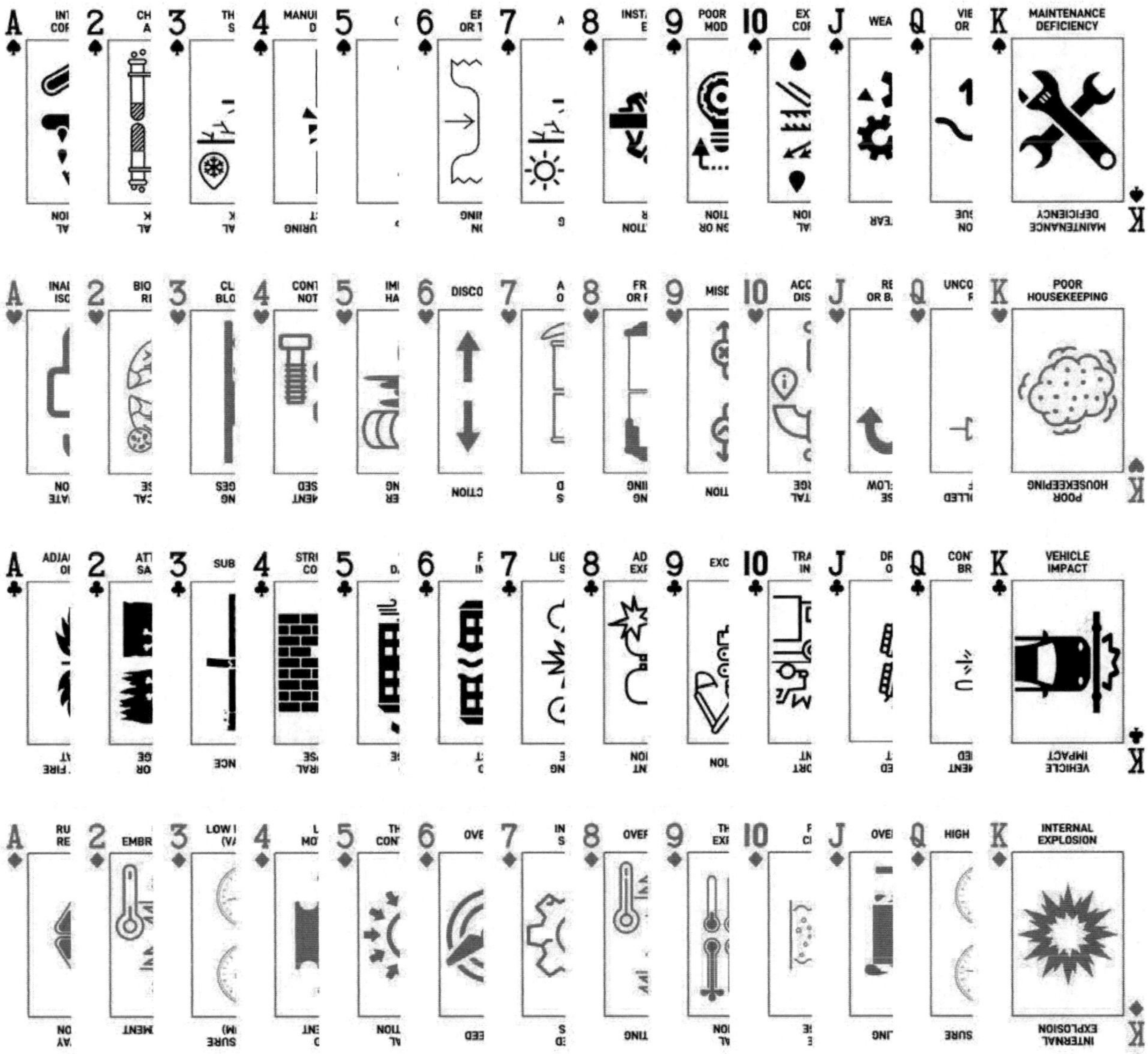

Figure 1 Sample Playing Cards

One of the core concepts of this approach is to “Know your Enemy” [7] where hazards are not the enemy (these are part of doing business) – it is Loss of Containment that is everyone’s opponent.

There are several different opportunities to deploy the cards in operational arenas, because they are easy to carry & exchange:

- Simple passive learning (by osmosis) via familiar, informal card games.
- Toolbox talks, e.g., explain your experience of the card you’ve been dealt.
- ‘Pop’ or quick quiz e.g., name the event on the card (without title)
- Hazard spotting or surveys e.g., punch hole in cards and tag/tie them onto the plant at vulnerable locations

- Hazard reminder e.g., attach relevant cards to Permit to Work.
- Inductions e.g., visitors, suppliers or short-term contractors during outage, turnaround etc.

The cards are not limited to frontline personnel as they can also be used to engage engineering or management personnel.

- Safety moments e.g., discuss your role related to the card you've been dealt.
- PHA/HAZOP validation e.g., ensure all deviations (cards) have been discussed.
- Performance dashboards e.g., incidents categorized according to LoC type (or card)
- Hazard potential e.g., sum (add) the values of each card i.e., Ace = 11, King/Queen/Jack = 10 & other cards numbered
- Campaigns e.g., each week showcase specific events or cards.
- Relate cards to client or industry incidents e.g., CSB investigations [1].

Just as a pack of cards can create a variety of different games e.g., poker, blackjack, bridge etc. Process Safety Cards utilize the same playing cards elements (or 'pips') which offers an organic approach to gamification that can grow with peer or community feedback i.e., facilities, organizations or industries can develop their own games based on their technology and culture.

5 Scenario Summaries

Playing cards increase awareness of the potential LoC events, however in order to address these, an appropriate level of insight is necessary. Visualizing the LoC scenarios in bowtie format provides a clear, concise summary which is more easily understood and retained than textual or numeric data.

A one-page representation of each type of LoC event i.e., playing card, provides typical:

- Cause (Threat)
- Controls (Barrier)
- Effectiveness or Vulnerabilities (by barrier type)
- Degradation factors (by barrier type)

These are not intended to replace existing analyses and should not be considered sufficient to properly monitor & manage barriers, however they do provide memorable context with a particular focus on the Degradation Factors which create or enlarge the holes in the Swiss Cheese model and are controlled via asset and/or safety management systems e.g., competence, information, maintenance change etc.

A sample is provided in Figure 2 below:

INTERNAL CORROSION

Typical **CAUSES** (Threats)

- ◇ Contamination
  - Process e.g. carbon dioxide, water, acids & other chemicals.
  - Microbiologically Influenced Corrosion (MIC)
- ◇ Stress corrosion cracking
  - Chlorides, sulphides, alkali & nitrates
- ◇ Chemical embrittlement
  - Hydrogen assisted/induced or Liquid metal (mercury)

Typical **CONTROLS** (Barriers) & **VULNERABILITIES** (Degradation Factors)

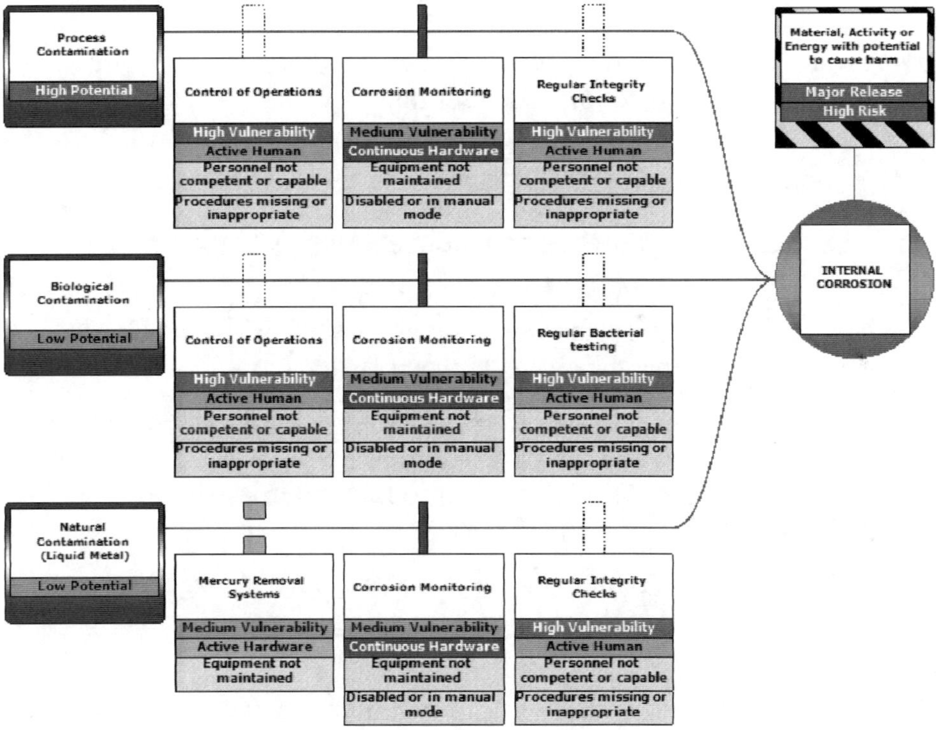

Version 2.1 Process Safety Cards © Process Safety Integrity 2020 www.ProcessSafetyCards.com 1

Figure 2 Sample Scenario Summary

Scenario summaries can be posted inside or outside populated or high traffic areas such as control rooms, canteens, workshops, or meeting rooms and rotated on a campaign basis e.g., a different LoC per week. They can also be published online e.g., intranet, SharePoint, or newsletter.

6 Online Collaboration

Although Process Safety Cards are designed to be physically handled, the current COVID-19 pandemic prevents or discourages such personal interaction, therefore digital solutions can be used to use or play the cards via collaborative platforms e.g., virtual whiteboards to connect remote or isolated team members.

7 Conclusions

Although based on gamification principles, the Process Safety Cards are not a game in themselves i.e., there are no rigid rules to follow. They simply provide a knowledge exchange toolkit to facilitate communications (in both directions) between management and frontline personnel to share learning using universal images and local languages.

These are not a new strategy (although they are a novel idea), they simply help deliver or reinforce established Process Safety (Asset Integrity) initiatives which are designed to 'Protect the Protection' i.e., look after the systems (hardware & humans) that look after you.

The cards increase awareness, the bowties enhance understanding, knowing your enemy improves vigilance which drives protection and asset integrity.

8 References

[1] U.S. Chemical Safety Board (CSB) Investigations (https://www.csb.gov/investigations/).

[2] U.S. Department of Labor, Occupational Safety & Health Administration (OSHA) Process Safety Management of Highly Hazardous Chemicals (29 CFR 1910.119) (https://www.osha.gov/process-safety-management).

[3] American Institute of Chemical Engineers (AIChE), Center for Chemical Process Safety (CCPS) Risk Based Process Safety (RBPS) Framework (https://www.aiche.org/ccps/resources/publications/summaries/summary-guidelines-risk-based-process-safety).

[4] Energy Institute High Level Framework for Process Safety Management (https://publishing.energyinst.org/topics/process-safety/leadership/high-level-framework-for-process-safety-management).

[5] UK Health & Safety Laboratory "Loss of Containment Incident Analysis" HSL/2003/07.

[6] American Institute of Chemical Engineers (AIChE), Center for Chemical Process Safety (CCPS) "Guidelines for Chemical Process Quantitative Risk Analysis".

[7] Sun Tze "The Art of War".

618386 Are Your Safety System Barriers Still Work for You?

Abstract:

Process Hazard Analysis is the tool to assess the systems in terms of the process safety scenarios and to determine the ALARP level of the risks that may arise. All the barriers in the Process Hazard Analysis is assumed to be working in the normal operations. However, how you manage to sustain the ALARP level? Do you know any deficiencies and by-pass during the operation? How do they effect the ALARP level? Are you still in the safe side?

All the barriers like Emergency Shut Down Systems, Pressure Safety Valves, Fire and Gas Detection Systems work for your system and you need to follow up in terms of any deficiencies or by-passes made on these. This process starts with reliability of these systems and continues to follow up during the operation. Emergency Safety Systems by-pass are undesired activities but during the operation there would be a need to do this. Here the important thing is how you manage the system with these by-pass and how you follow up these by-pass procedures. Industry and automation systems are improving and there are lots of ways to simultaneously cover all these changes and do the risk analysis to operate the systems safely during these by-pass times. So many companies would be using different ways of controls, in our refinery we have improved a system to follow up all the emergency safety systems with a hierarchy of control and approval automatically including the "Plan B" risk assessments for safe operation of this time periods.The important point here is the type of by-pass (conditional or non-conditional), duration of by-pass (approval hierarchy), what else do you have to operate safely, what other parameters should be followed up by the operator during the by-pass and does he know it?

These automated systems should be covered and controlled during the manual operation or by-pass time intervals in order to eliminate the human errors and not to have any kind of process related incidents. We do need to remember important Process Safety Incidents related with the by-pass of emergency shut down systems like Chernobyl. This solution and automated system would be a barrier not to have such incidents.

Effective response to a power loss event is critical to the safety and mechanical integrity of a manufacturing plant. Power failures can lead to accidents, injuries, loss of primary containment (LOPC), government recordable events, mechanical failure and loss of production. Due to the unpredictability of a power event, there is no single solution or response plan that can mitigate all of the potential issues that a power event creates.

A power event can cause alarm showers/floods which prevent the operator from seeing and reacting to high consequence alarms that could prevent a serious incident if responded to in a timely manner. Based on this, we performed investigation into the last several power failure events at our sites with a focus on the distributed control system (DCS) and its impact on response to power events. This presentation covers the findings from this investigation which include when power events occur, alarm counts/metrics during the events, the severity of each different power event, and findings during root cause investigations (RCI).

We then discuss the different tools that have been implemented into the DCS specifically to aid in operator response to a power event. Some of the tools that are covered in this presentation are DCS graphics "overview" screens for power failure events which provide one-click capability to layer 3/4 process graphics, alarm hiding/suppression concepts, bulk alarm acknowledgement/unlatching techniques, alarm silencing, and automated actions in logic/programming to build more resiliency against an event.

Development of Mobile App for Behavioral/Near Miss/PSE Reporting: Executive Summary

Ecolab, a global petro/chemical manufacturer wanted to improve their safety data reporting procedures, and create a consistent process for collecting/analyzing data.
The company also had the goal of incorporating more digital technology into their work practices and process production areas.

After a needs analysis, leadership recommended:
1. Development of a consistent Safety reporting procedure for all global plant sites to follow
2. Incorporate digital technology by use of a mobile app with shared database entry
3. Standardized analysis of safety data to quickly address trending safety events

Development of a consistent Safety reporting procedure

Global plant sites weren't required to use a consistent tool or process for safety data reporting. The project team analyzed numerous safety data reporting practices at sites, to look for consistencies and best practices. These were used to develop a consistent reporting procedure.
The procedure included types of events for reporting, as well as roles and responsibilities of employees at the plant sites. Required data would include Behavioral observations, Process Safety Events (PSE) and Near Miss Events. The overall goal of the procedure was to eliminate unsafe conditions and behaviors (leading indicators) before they escalated to more severe events (lagging indicators).

Incorporate digital technology by use of a mobile app with shared database entry

With the company goal of digital transformation, the project team looked for a solution that would eliminate paper-based practices at the sites. A mobile app solution was recommended, and pre-made options were explored. Nothing was found to meet the needs of the plant sites, so an internally developed app was recommended.
The project team used Microsoft Power Apps as a low-cost solution to develop an app in-house. The app included data entry fields for plant site, type of event, location event occurred, and many other important data points. The app was translated into several languages.
Safety events could be reported either using a mobile device, or on computer terminals located throughout the plant site. All reported safety events were downloaded from the app into a shared database for easy data analysis, and easy access by many different plant sites.

Standardized analysis of safety data to quickly address trending safety events

Part of the reporting procedure included roles and responsibilities for analyzing safety data reports. Safety data was analyzed to identify trends. Specific recommendations/actions were made to eliminate the unsafe conditions/behaviors found in the safety data trends.
PowerBI was used to quickly identify safety data trends during the analysis process. Trends were communicated through the PowerBI dashboard by using graphs and other easily interpreted visual elements.

Project Pilot

The pilot consisted of a 90 day "Safety Challenge" where a specific group of plant sites reported as many near miss and PSE observations as they could identify. Trends in the data were tracked through system generated reports, and displayed using a PowerBI dashboard. Plant site safety leaders immediately saw the benefits of a standardized safety reporting procedure, and ease of use with the mobile app.

The pilot has expanded to include additional plant sites in various countries. This has created a new set of challenges for deploying the Safety reporting procedure, expanding use of the app, translations, and identifying best practices for mobile devices. The pilot is set to complete in early 2021

617938 Using Technology to Improve Safety

Abstract:

As the world grows and changes, our business processes evolve, and we must change to meet our new needs. Recent technologies like smart phones, tablets, computers, smart televisions, smart homes, and applications that can manage every aspect of daily life are now widely available in any employee's personal life. These devices and programs reduce the time needed to do routine tasks, providing new opportunities to focus on things the individual may feel are more important or fun. In some cases, these processes learn from previous transactions and tasks and require less of the individual's input as it continues to operate.

However, in a person's professional life, it is common to have a significant portion of one's day composed of filling out paperwork and performing hands-on tasks. Although work instructions, procedures, and policies for everyday tasks are now available electronically, a field employee needs access to a computer to view, access and print documents. This limits the productivity and effectiveness of employees since it forces them to access all information at a single access point or having them carry around paperwork that can easily be damaged by work conditions and the environment. One limitation is unexpected equipment failure. In such cases, it is rare that employee would have all technical information readily available to help troubleshoot in the field immediately, except for minimal visual HazCom labels and warnings. This decreases overall process efficiency by inadvertently creating restraints – specifically, having to go back and collect all required paperwork- on the employee which can be overcome with everyday technology.

Understanding how so much information is instantly available outside of work is important to replicate that level of access and ease in the workplace. At Rain Carbon Inc., our mission is to be "a progressive industrial company committed to collaborative science-driven innovations that result in safe, reliable and efficient operations to ensure outstanding results for our customers, partners, employees and investors." With this as our guide, a new system was implemented to target several areas: safety, reliability, information availability and efficiency.

At our Candiac advanced materials site in Quebec, Canada, integrative software was implemented in 2017 as a pathway to achieving the company's mission. This software provides instant access to process information through phones, tablets and computers, including work instructions, policies, procedures, ability to complete round check sheets and work permits, action assignment and tracking, picture and video capability, report unsafe conditions and incidents, equipment malfunctions, and the management of workforce skills. The capabilities provided by the software have proven to be most effective when used on mobile devices as it allows for information to be accessed directly from any location within the plant helping us improve in all the below points:

- Enhancing site management by providing a live view of notices, work requests and unsafe conditions to immediately communicate and respond to the situation at hand;
- Increasing employee involvement in safety due to higher visibility and lower response times to address unsafe acts/conditions;
- Reducing redundancies of having to go back and forth between paper and electronic work; and
- Decreasing unplanned downtime by providing all relevant technical information instantaneously enabling employees to address deficiencies without delay.

Rain Carbon has pushed forward with innovation and technology by implementing QR codes for all equipment linked to the integrative software solution. This adds a significant amount of efficiency to work processes on the shop floor. When unplanned work is to be performed, the employee can take any mobile device with the software installed on it and scan the QR code on the respective equipment. This will take the employee to a screen with all necessary information regarding the respective equipment, including the work permit and list of materials needed to perform the work in question. A process that would, in most cases, take a matter of hours is reduced to a few minutes and a couple of taps.

Although reliability and efficiency have increased since the system was implemented at the Candiac facility, the most significant improvement has been observed in safety. Prior to implementing this software, safety walks were completed in an electronic or printed version of a Word document, safety incident notifications would be accomplished through phone calls and emails, and near misses would be recorded in a separate system. The centralization of reporting and the enhanced information availability highly encouraged employees to report all safety items with the backing of a more engaged management group, thus enhancing employee safety. Reporting increased with the introduction of the program by 772% on the first year and by 2,150% on the second year. Similarly, during that timeframe, near misses decreased by 40% and first aid cases by 33% compared with the current year.

Results:

From 2017 to 2019, there was a clear increase in reporting after the implementation of the software. Through management support and commitment, we can see a clear decline from 2019 to 2020 YTD of overall identified items, increasing the quality of work and safety across the site.

Implemented in 2017, the program has significantly improved the safety performance of Rain Carbon's Candiac site by engaging management and ensuring all safety items are properly communicated and addressed with as little downtime as possible.

617740 Facility Siting Revalidation - a New Tool

Abstract:

Facility siting studies require periodic review at specified intervals during the life of a facility. As stated in 29 CFR Part 1910.119 (e) (6), revalidation is required every five years based on the previous full study completion date. While revalidation is a regulatory requirement, little is often known about the process of undertaking the initial study or review and, in some cases, this process can become challenging.

A new and simple revalidation tool is available that helps operators determine if their facility siting study needs a complete redo, a retrofit, an update, or a simple revalidation. This paper introduces and explores this simple revalidation tool, providing examples of how it can be used in a plant environment.

617756 Quantifying the Conditional Dependencies between Layers of Protections in Event Tree Analysis

Abstract:

An approach to include conditional probabilities when calculating probabilities of disaster in quantitative risk analysis is presented.[1] Since not all layers of protection are independent of each other, using conditional probabilities leads to more accurate calculations of probabilities of disaster in a chemical process. Based on event tree analysis, probabilities of disaster are calculated assuming that events are statistically independent (independence assumption), and separately considering that events affect each other (conditional probabilities), in three cases with increasing complexity and number of layers of protection. First, a simple alarm – operator case is studied, then two cases involving a chemical reactor and operator with increasing number of layers of protection are analyzed. For all cases, we considered multiple scenarios to show that results do not depend on the input probabilities of particular events. For the latter two cases, the input probabilities of each event were randomly sampled with Monte Carlo simulations. Probabilities of disaster calculated with the assumption of statistical independence were compared to probabilities of disaster calculated with conditional probabilities. In most scenarios of the three cases, findings show that the probabilities of disaster calculated using the independence assumption are lower than those with conditional probabilities, which can lead to undercounting the layers of protection that are needed. Process engineers need to be aware that assuming that events are statistically independent can lead to the possibility of undercounting of layers of protection needed for a chemical process.

Reference: Mott TC, Kivistik PM, Panorska AK, Cantu DC. Comparing conditional probabilities and statistical independence in layers of protection analysis. Proc Safety Prog. 2020;e12215. https://doi.org/10.1002/prs.12215

617799 Comparison of Methodologies for Dropped Objects Analysis in Shallow Water Drilling Rigs in the Gulf of Mexico.

Abstract:

The reform of the energy sector undertaken in 2013 by the Government of Mexico to empower foreign investment in the hydrocarbon sector led to the arrival of private companies with new projects and technology to the Gulf of Mexico. The projects managed in this region move continuously from the exploration stage to the production stage. During this stage transition, Offshore Projects require development of safety evaluations for the design of the platforms that were not common in the Mexican market; among these evaluations there is "Dropped Objects Analysis".

The accidental dropping of objects is considered among the main risks on offshore production platforms since its occurrence can affect personnel (injuries and / or deaths) and damage the facilities (causing spills and / or fires). Thus, as part of engineering design development, there is a need to assess the potential risk associated with impact from oscillating objects and loads that could fall during routine and non-routine lifting activities.

The analysis focuses on the possible damage to the area of operation of a given platform, structures, process equipment and subsea pipelines, evaluating the frequency of dropping combined with the type of damage that would result from the impact. However, for Dropped Object Analysis there are different methods that involve complex calculations, including solving integral equations, which for practical purposes can result in a large time investment.

This work addresses the use of a numerical method that allows a complex calculation routine to be carried out towards a solution with simple arithmetic, thus facilitating the evaluation of the frequency and impact of dropped objects on the platform and reducing the solving time.

As an application case, the Dropped Objects Analysis is presented in two offshore platforms for the exploitation and transportation of hydrocarbons installed in southeast Mexico by a private operator. During the study, the frequency and consequence of dropped objects were evaluated to issue protection recommendations in the design of both platforms and were also compared with the use of traditional calculation methods.

617993 CFD-Based Ventilation Analysis - Basis for Conducting Consequence and/or Risk-Based Assessments

Abstract:

The combination of powerful capabilities in the development of Computer-Aided Design (CAD) drawings; i.e., 3D models; and the performance of formidable computer processors confirm the execution of robust Computational Fluid Dynamics - CFD-based simulations in manageable short-duration periods of time; i.e., minutes. Using CFD tools allow considering the fluid-geometry interaction, key factor for identifying specific features of the site section being analyzed. This information is not possible to be foreseen when using screening tools based on experimental data correlated and/or phenomenological methods.

The present paper illustrates a case study that provides examples of how a CFD-based ventilation analysis can provide insight information of areas that require special attention when a consequence or risk-based assessment is to be performed. The identification of regions with air recirculation, stagnant flow areas with low wind speeds, congestion volumes, etc. verify leak locations and orientations to be evaluated when performing advanced dispersion, fire, and explosion analysis.

SPRING21
+17TH GCPS
A Joint AIChE and CCPS Meeting

SYSTEMATIC APPROACH TO PROCESS SAFETY MANAGEMENT ON ROAD LPG TRANSPORTATION

Presenter/Principal Author
Américo Diniz Carvalho Neto
RSE CONSULTING
Brazil
americodcn@rsem.com.br

Other Author
Adriano Nonato
RSE CONSULTING
Brazil
ext.adriano.nonato@rsem.com.br

Prepared for Presentation at
American Institute of Chemical Engineers
2021 Spring Meeting and 17th Global Congress on Process Safety
Virtual
April 18 - 22, 2021

SYSTEMATIC APPROACH TO PROCESS SAFETY MANAGEMENT ON ROAD LPG TRANSPORTATION

Presenter/Principal Author
Américo Diniz Carvalho Neto
RSE CONSULTING
Brazil
americodcn@rsem.com.br

Other Author(s)
Adriano Nonato
RSE CONSULTING

Keywords: LPG transportation Riks; PSM in LPG distribution, LPG Transport

Abstract

Process Safety Management is systematically adopted in sectors of Brazilian industry, such as oil and gas and petrochemicals, with clear operating safety gains, based on performance indicators. In common, they all use fixed structures and are siting centered to carry out their operations. Also important in the energy chain in Brazil, bulk LPG transportation supplies industrial chain, and business and residences as well. With all this importance requirements for robust process safety management are still missing in LPG distribution sector.

This work aims to implement a safety management system within a company currently responsible for the largest share of the bulk LPG market, considering the particularities of this type of operation, totally dependent on road transport for LPG distribution. To achieve this the first step was identifying the main activities that represent the bulk system's operation, involving bobtail trucks, as follows: supply through road routes; unloading, which is the supply to the customer and the operation of the bobtail truck itself. The strategy of implementing the safety management system considered sequential application of the following steps: 1) Reassessment of the bobtail truck project, aligning it with the latest technological solutions; 2) Conducting a hazard identification study for supply and unloading activities, proposing a customized hazard identification model for supply; 3) barriers auditing to evaluate unloading components reliability; 4) Auditing operating activities based on CCPS RBPS in order to identify existing gaps. The fifth and final step is the definition of the Process Safety Management approach aligned with road transportation best practices worldwide and in compliance with legal and corporate requirements.

The studies already developed demonstrate that a careful analysis of the main normative reference identified gaps in the bobtail project, with emphasis on passive safety systems, absent in the original project. In the supply risk assessment stage (routes assessment), a specific model for route risk analysis was developed using a customized Excel spreadsheet for risk analysis through predefined menus. This model reduced the assessment time required and the level of subjectivity, allowing the company to develop hundreds of analyzes with the use of video conferencing tools, essential in this case, due to its nationwide scope.

This is a work in progress and the next steps, when completed, will make the company achieve a degree of excellence in Process Safety Management through monitoring for the implementation of all RBPS elements that have been identified as absent or deficient. Many other industrial sectors that carry out transport of dangerous products can benefit from this methodology for implementing RBPS, through the use of the described models, adapting them to the specific modal.

Introduction

It is recognized that chemical transportation risk management is more than just a service; but rather is a wide-ranging system of activities that involves numerous stakeholders with varying backgrounds and interests. To effectively manage transportation risk the entire supply chain, spanning the delivery and handling of raw materials through the offloading and unpacking of finished products, needs to be included and evaluated. In addition to safety and business continuity, the evolving threat of security concerns are now included as part of a comprehensive transportation risk management program. Given the complexity and global reach of many transportation networks, as well as the tradeoffs between safety and security, the management of transportation risk can challenge even the best business plans and programs.

Considering all these aspects and risks involved in LPG transportation, RSE Consulting developed a PSM project with one of the biggest gas distribution companies in Brazil, o order to develop a diagnostic, and create a PSM system to be implemented in the company.

1 Bulk LPG transportation risk assessment

This stage aimed to evaluate a pilot project of adequacy of LPG transport vehicle. The use of common risk analysis methodologies and verification of compliance with international benchmarks were applied.

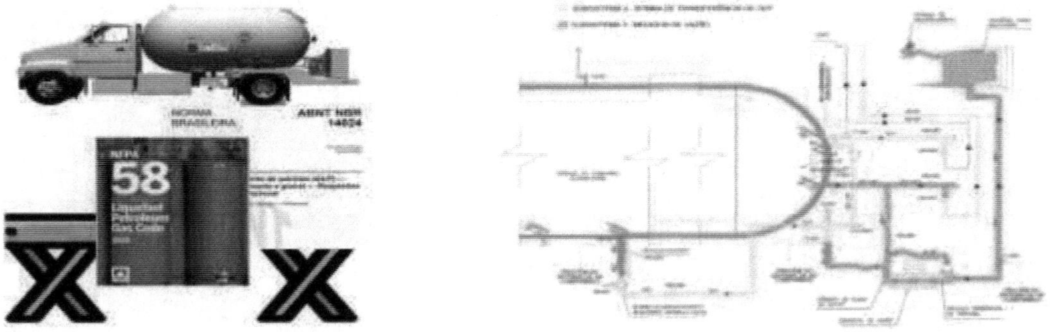

A HAZOP of the system was carried out to identify operational weaknesses

A revised project was obtained considering improvements in management and safety elements, some highlighted below:

Include the identified critical components into the inspection and maintenance plan (Bobtail). Ensure that inspection and maintenance plans have effective tests and tasks for identified and potential damage mechanisms.

Conduct training with operators on the new operating demands (Bobtail), reinforcing the concept of critical components, defining the points that should be evaluated during inspection routines, in order to ensure that the trailers transporting LPG meet the appropriate conservation conditions in relation to the Preventive Barrier of Periodic Inspection of Trailers.

Ensure that truck fleet will have a supply pump interlock system when the minimum level of 5 % of the volume of the LPG tank is reached, to avoid cavitation of the pump.

Every operational unit across the country should comply with managerial based actions concerning standardized execution of maintenance and inspection of critical components.

Include as managerial based actions the adoption of indicators for assets of integrity and reliability as stated on API 754

Define a maximum and minimum speed range to be adopted during the hose rewinding activity on the reel in order to ensure operation compatible with operator safety.

Ensure that a relief valve is installed on the swivel joint. This valve has the function of relieving pressure due to LPG expansion that can occur between two shutoff valves. Ensure that its construction and pressure class meet NFPA 58 requirements

Evaluate the feasibility of deploying a portable remote device for activate emergency activation system.

Modify emergency activation system so that the entire system is shut down or otherwise enforced to fail safe state (pump, pneumatic valves and engine and truck electrical system).

Reevaluate the removal of LPG liquid supply line check valve, which, as designed, will be replaced by a pneumatic-driven shutoff valve.

2. EVALUATION OF TRANSPORTATION ROUTE

The company under assessment supplies LPG for the vast majority of brazilian states

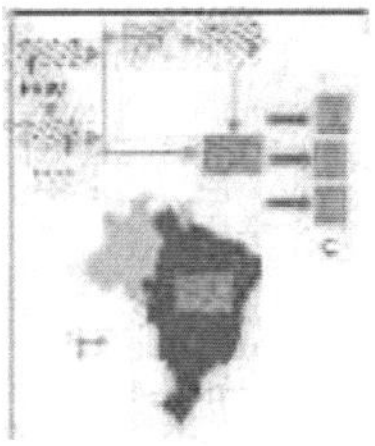

Process safety management on this company is still on its infancy ant still being circumscribed to the staff on Brazilian southwest headquarter. The staff engaged on conducting the evaluation had to overcome the lack of any previous knowledgeabout Risk Analysis of the operating team further hampered by mobility limitations imposed by the pandemic.

A method was developed for route risk analysis as recommended by the CCPS.

All relevant attributes that impacted the risk of the routes were identified and assigned weights for them.

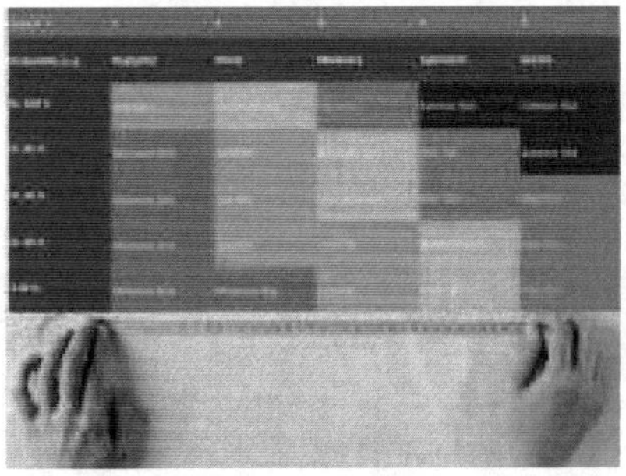

The weights have been calibrated to ensure consistency with the normative risk matrix.

An Excel spreadsheet was developed to assess risk, deploying the afore mentioned risk attributes, allowing teams to engage with maximum objectivity and time savings. Another advantage of the spreadsheet was its smooth learning curve, enabling the use for different teams of the company and possibility of future improvements to evaluate management elements

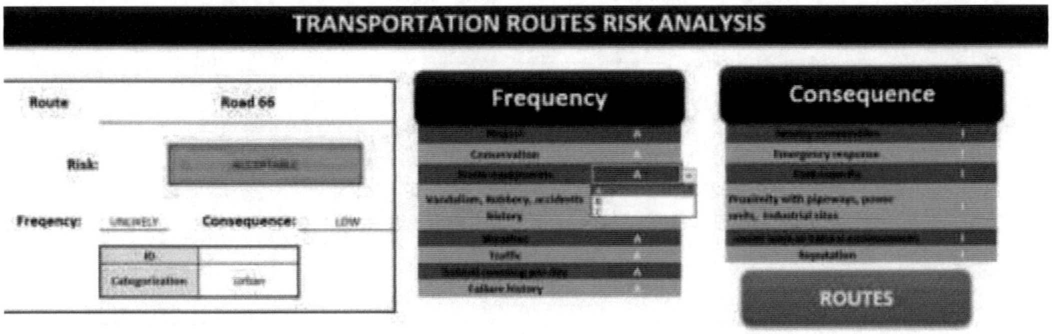

A total of 22 routes were analyzed. It is expected that teams can perform a much higher number of analyses with a minimum of tutoring, given the smooth learning curve for using the tool. Improvement points "appear" from the analysis of the result graphs, allowing recommendations to be properly addressed, as follow:

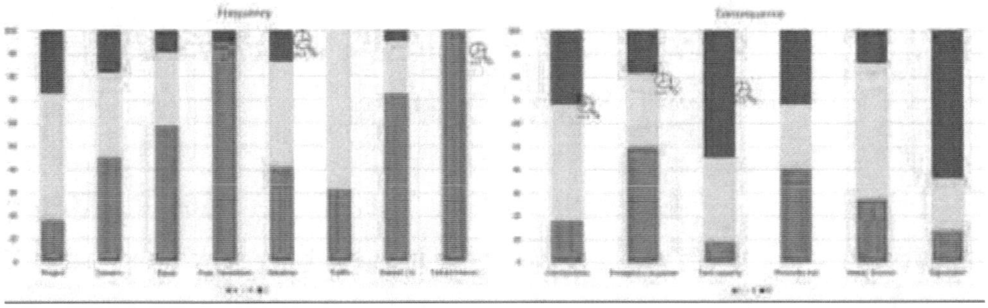

3. CUSTOMER RISK ANALYSIS

The 3rd stage of the project (in progress) will be the application of a bow tie diagram to the customer facilities.

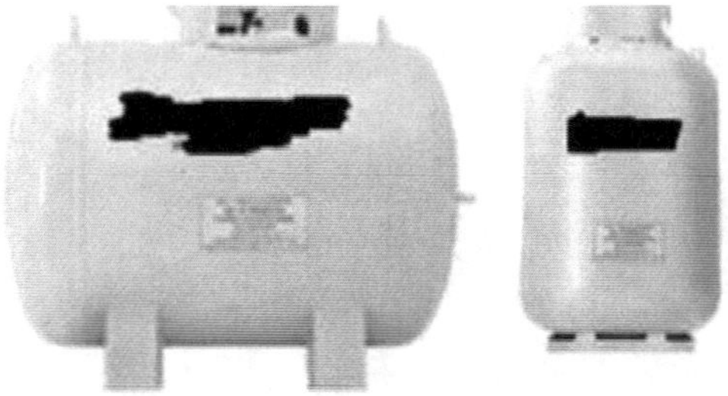

The bow tie will facilitate the process of identifying barriers for subsequent analysis.

4. BARRIER AUDIT

Step to be performed as follows below:

- ➢ Definition of critical risk scenarios for the preparation of protocol d andaudit;
- ➢ Preparation of audit protocols of critical scenarios and preventive and mitigating barriers;
- ➢ Audit of barriers / critical controls of the scenarios defined in the 1st step
- ➢ The systems to be analyzed are those of customer facilities. CCPS Bow Tie Approach was the chosen method.

5. IMPLEMENTATION OF PSM

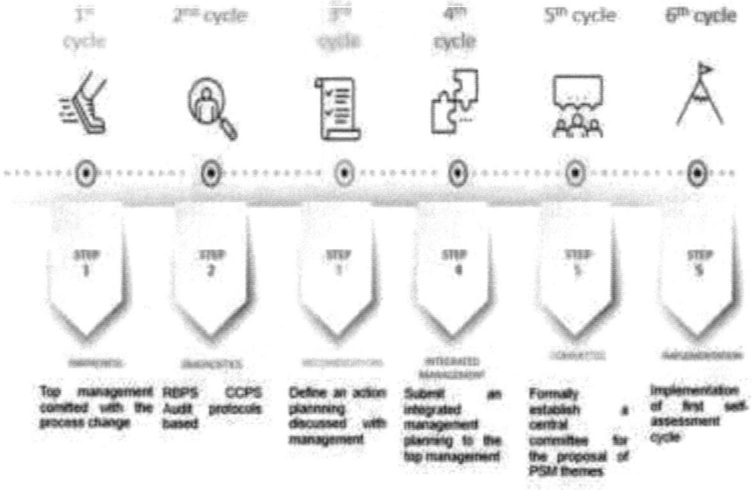

The PSM system implementation strategy is structured in the following steps:

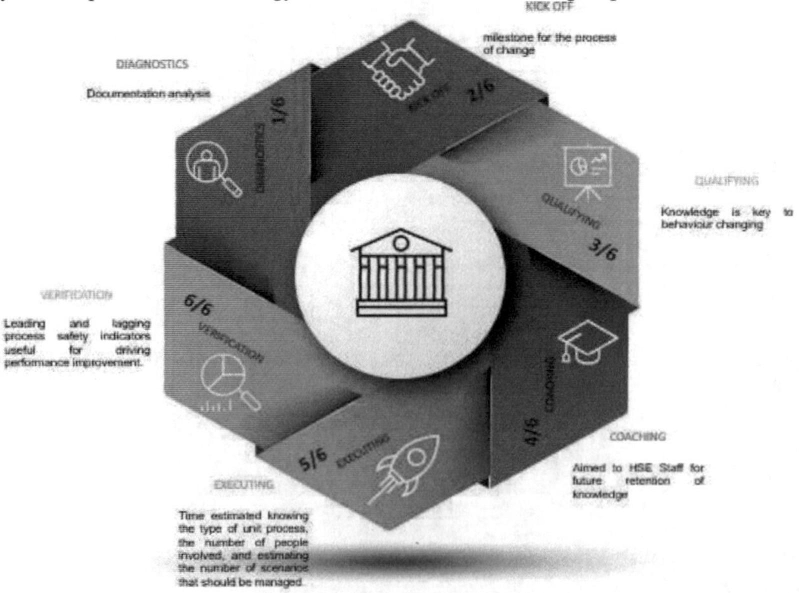

Conclusion

The area of transport of dangerous products, specifically the transport of LPG in large cities, presents many positive contributions to society, however this logistical process has many risk scenarios with high severity in all stages, from loading to the process bases, distribution of the LPG via tank trucks, and refueling at customers' different facilities. This Project allows the organization to not only make a technical diagnosis of the risks of the entire process, but also to set up a Risk Management System that, through the implementation of the requirements of the Management System and procedures, will be allowed to achieve a high performance with the elimination of leakages. gas with the potential for catastrophic accidents.

2 References (examples given to show format).

[1] J.A. Klein and B.K. Vaughen. "Implement an Operational Discipline Program to Improve Process Plant Safety," June 2011.

[2] CCPS. *Guidelines for Chemical Transportation, Safety, Security and Risk Management.* Center for Chemical Process Safety, American Institute of Chemical Engineers, New York, NY, 1995.

[3] RSE. RSE Consulting procedure for Risk transportation qualitative and quantitative risk analysis, 2019. Developed by Professor Americo Diniz Carvalho Neto.

618466 The Evolution of Emergency & Crisis Management in Petronas

Abstract:

Managing emergency and crisis effectively is a paramount aspect in Process Safety Management, especially in the complexity of oil and gas business. Optimum readiness is aimed to efficiently address potential impacts arising from incidents; immaterial of size; to contain the extent of exposure as quickly as possible, with the primary intent of saving lives and protecting the environment. Often we pay too much attention on the left hand-side of the bowtie (control barriers) that we often overlook the right hand-side of the bowtie (recovery barriers). The objective of this paper is to share the evolution of emergency and crisis management underwent by PETRONAS, with the ultimate aim to ensure our interests (people, environment and asset), and reputation are protected.

PETRONAS operates globally in more than 50 countries, hence effective establishment and implementation of emergency and crisis plans are vital. Historically, emergency and crisis management and preparedness has been a reactive science. We had our own share of incidents, and multiple lessons learnt had been consolidated to further improve our systems and barriers. The integration of emergency and crisis management as part of the HSE Management System (HSE MS) and Mandatory Control Framework (MCF) built a strong foundation and governance, and ensures continual improvement, taking into account all the internal and external lessons learnt. Conducting and monitoring the number of emergency drills and exercises, specifically based on the Major Accident Hazard (MAH), is imperative to ensure effective response in case of real emergency and crisis. Learning from real cases raised the vital need of coordinated response and putting the rights interventions and controls in place to minimize the impact of incident. Effective emergency and crisis management is crucial to support process safety implementation throughout the asset life-cycle covering design, operation and maintenance stage.

Throughout the years, we are also continuously enhancing our capability in terms of resources and people. In a nutshell, this paper will talk about the evolution of emergency and crisis management in PETRONAS, where it started with system establishment and capability building, whilst moving towards digitalization to better manage emergency and crisis. This paper will share some real incident cases and how we learn from these incidents to further intensify our emergency and crisis management.

617018 Pressure Relief System Force Balance Assessment – a Review of Its Effectiveness

Abstract:

Design documentation requirements for spring-loaded pressure relief systems are very detailed, based on the definition in API Standard 521, 6th edition. The Installation Documentation section of API Standard 521 recommends provision of inlet and outlet piping pressure drop calculations as part of a complete analysis.

Inlet pressure drop is an important factor, since it is considered that excessive inlet pressure drop will affect pressure relief valve performance, and potentially result in chatter which may damage the valve and piping. In situations where inlet pressure drop exceeds 3% of set pressure for spring loaded pressure relief valves, previous editions of API Standard 520 Part 2 have allowed for an "engineering analysis" to be conducted for these systems - however the content of the engineering analysis has never been well defined.

The latest version of API Standard 520 Part II provides a more detailed description of an engineering analysis and allows a Force Balance Assessment to be conducted for spring loaded pressure relief valves exceeding 3% inlet pressure drop. This method can predict valve stability during relief flow, and identify whether chatter, flutter or stable flow is likely to occur. Valve characteristics such as valve opening and closing times are critical input parameters for this analysis. If stable operation can be demonstrated, then expensive mitigation efforts, such as modifications to piping or the valve itself, can be avoided.

This paper presents an overview of the force balance assessment in a step-by-step manner. Additionally, a sample of spring-loaded pressure relief systems which exceed the 3% inlet pressure drop criteria is reviewed. The aim of this review is to determine the effectiveness of conducting a force balance assessment, and to identify which systems show stable operation, and which systems were identified as likely to chatter.

618390 Chemicals Management inside a Refinery - Does Everyone Knows the Hazards of the Chemicals on Site?

Abstract:

Chemicals are widely used in Oil&Gas industries. Each of them has different phsyical and chemical properties and these differences make it hard to store and use them in a proper way.

In the refineries, those widely used chemicals require a good storage and transportation standards. In order to create a safe environment for those areas, Tüpraş has created its own "Chemical Interaction Matrix". In this matrix, various aspects of chemicals are handled and each Unit has its own matrix.

As a starting point, each of the chemicals are checked as per their stability and reactivity properties from their safety data sheets. The chemicals that do not have detailed information are reacted to each other in the laboratory in order to see if there is a dangerous reaction.

Chemicals that may cause a dangerous reaction to people and environment are stored seperately and the properties of each are shown in the matrixes. These matrixes are hanged on the unit boards so that everyone can see and check if the storage is provided correctly.

It is important to keep these matrixes up-to-date. The matrix is updated when a new chemical arrives in the unit and everybody is informed about update. The update and information is achieved wit the actions that we tracked via our MoC (Management of Change) system.

617653 Operational Risks Involved on Integrating Flare Gas to Fuel System

Abstract:

Flaring is a safe industrial process to dispose waste or excess gas from a plant. But, it has limitations in energy conservation and environmental loss. Flare gas integration to cogeneration plant benefits from utilizing high calorific value waste gases as supplemental fuel to cogeneration plant or gas turbines. Earlier studies on flare gas integration have shown environmental incentive and economic viability. However, the impact of flare gas quality on the operation of cogeneration plants are yet to be identified.

This study delves into the operational hazards and risks arising from the variability of flare gas stream. A novel framework is proposed which can be applied to new or integrated processes which has inadequate information on its operational performance and associated hazards. The study identifies the technical hazards through design data collection, process simulation and sensitivity analysis. Followed by a systematic approach to evaluate the critical operational event occurrences and their impacts through scenario development and quantitative risk assessment.

An important outcome from this study is the identification of critical flare-fuel gas parameters affecting the cogeneration operation through process simulation. High hydrogen content and rich fuel-air ratio in the boiler can affect the boiler operating conditions. Increase in the hydrogen content in fuel system can increase the boiler gas temperature, combustion mixture and flame stability inside the firebox. Risk analysis through Bayesian Network showed a significant risk escalation. With a presumed frequency of 12 hrs/year of high and medium hydrogen gas flaring, there is a substantial rise in the occurrence of boiler gas temperature exceeding design limit and rich fuel mixture formation in the firebox. These top events has major influence on the operational time as well as loss control events - flame impingement and tube rupture.

618388 Process Hazard Analysis Methodologies and Refinery Applications

Abstract:

Process Safety Management System starts with managing the process related risks and from the first step of establishing a new process unit that is needed to evaluate the risks and define the necessary measurements. Process Hazard Analysis methodologies are the systematic tools which help to evaluate the risks in a process. For different needs there are several methodologies like DOW FETI, HAZOP, What if, HAZID, Fault Tree and Event Tree Analysis, Scenario Modelling, Facility Siting etc. Some of these methodologies are semi quantitative some are quantitative methodologies. In order to define and prevent major industrial accidents in our refinery we use DOW FETI, HAZOP, FTA and ETA and Scenario modelling. Moreover for some kind of projects we may also apply What if and HAZID analysis instead of HAZOP depending on the complexity. Facility Siting is another tool for risk related studies which is used to evaluate the risks for buildings in order to protect people inside the buildings.

The workflow for a full unit PHA study starts with DOW FETI calculation which is used for risk ranking of al static equipment inside the unit. DOW FETI calculation is done in accordance with the equipment temperature, pressure, hold-up amounts, components and also reaction types if there is any. Then risk assessment process continues with HAZOP studies which is for scenario identification. In HAZOP, first step is chopping the unit into small parts to be evaluated which is called nodes. After that all nodes are evaluated using P&IDs and PFDs by keywords like temperature, pressure, flow, level etc. HAZOP helps to understand what kind of scenarios are possible inside the unit and these critical scenarios are taken through FTA and ETA studies. HAZOP is a semi quantitative risk assessment method where FTA and ETA are quantitative methods. FTA is applied for critical scenarios in order to find out the probability of top event then ETA is then applied in order to find out the probability of scenarios like fire, explosion, toxic dispersion etc. caused by this top event. FTA and ETA analysis all use the statistical PFD and failure rate values of all initiating events and barriers defined in HAZOP for the related scenario.

Scenario Modelling is the following step which is used for understanding the effect of the probable scenario. The most probable scenarios coming from ETA is modelled to see the effect zones. These effect zones are used to plan Emergency Response Studies for all these potential major accidents.

Facility Siting is another check point in a unit PHA study all potential scenario modelling results are used to evaluate the risks for the buildings and also if there are any in the effect zones of fire, explosion or toxic release then there should be actions defined.

619040 Identification of the Process Hazards By Way of Technical Integration:

Abstract:

Undisclosed threats can lead to **death** and **devastation** in the process industry. Right and constructive hazard recognition supports security procedures, employment health and risk control techniques. It thereby is essential to the health and security of all company employees and to effective business practices.

Without understanding of process hazards, risk cannot be controlled and risk assessment is compulsory for incident prevention. Managers want their workers to be able to recognize not only apparent but evolving threats effectively. Risk assessment needs various **strategies**, **skilled** and testing teams to identify risks and create obstacles to mitigation in their operating environment. Different approaches can enhance the detection of hazards to the risk assessment process.

This paper explains the approach used in the implementation of combined **traditional** and **non-conventional**, **qualitative** and **quantitative** approaches, to identify process hazards, risk assessment and risk management. An insight on the non-conventional techniques *(Temporary repairs reduction & effective permit to work system)* will enable conference audience from all sectors of process industry, in developing integrated process hazard identification and evaluation

617089 Keep the Lion inside the Cage

Abstract:

Keep the Lion inside the Cage

All Chemicals, that we utilize to produce our final Products, are considered as Lions. We have to maintain their healthiness in good condition & under control inside the Plant (Cage) elements (Pipe, Equipment, Vessel, Tank). Process Safety Engineer's Career is an adventure where all safeguards, Barriers & mitigations shall be designed, constructed & operated safely in order to achieve the Process Safety ultimate Goal: "Prevent Loss of Containment". When the chemical leaks, the Lion is free to have high Consequences. With integration for all Reliability, Process Safety & Asset Integrity, we can keep the Lion inside the Cage & eliminate all Probability to have any Consequence whatever its Risk Level.

In order to achieve our ultimate goal, several aspects shall be managed & steward-shipped by the own Organization. There are many Process Safety pillars starting from "hazards identification" till removing the Risk or reducing it to insignificant level; including The Process Safety Competency Development for all Organization levels, Process Safety Knowledge Enhancement, System Support, Technical Guidelines, Risk Management, Process Safety Performance Monitoring and end with the Emergency Preparedness & Response Capability.

The beauty of the Process Safety Concept is the alignment with the Emotional Intelligent that we feel during the dealing with Chemicals (Lions). They shall be always contained & maintained inside their containments. In the same manner of having a specific Process Safety Competency Development Plan for All members of Organization to maintain the Chemicals inside the containment, it shall be available for all members in order to have the Lions inside the Cage. Our ultimate goal is to complete the Journey in safe manner & to maintain the Healthy condition ... in order to keep "Chemicals" inside the pipe & to keep the Lions inside the cage.

617835 Beware of Blind Risk -What You Do Not Know Can Hurt You

Abstract:

This paper discusses the concept of blind risk, how it arises, its dire consequences, and how to best avoid it.

Blind risk is most often due to incomplete, inconsistent, or out-of-date asset and process safety information. Most often, these information quality issues are not obvious or visible, and as such they are a blind risk. They lie dormant, like a beartrap in the undergrowth of a forest until someone inadvertently steps on it. They become gaps in our "layers for protection" -the systems and processes we trust to stop incidents before they happen. But blind risk is also a choice, and it is a measure of a company's tolerance for risk.

Safety, engineering, capital, and operating decisions are only as good as the information they are based on. Most companies may not ***objectively*** know how good (read reliable and trustworthy) the quality of their underlying asset and process safety information is. It does not have to be this way.

There are two things you can do today that will have a significant impact now and into the future:

- Use standardized information quality metrics and tools to measure, monitor and report the true state of asset and process safety information across multiple repositories, applications, and formats. Let all stakeholders including executive management know the good, the bad and the ugly. Use this as a baseline to measure your improvement against.
- Monitor and quantify the cost of poor asset and process safety information. Use this information to justify funding for information quality improvement initiatives.

619136 Relief Systems Work Is a Money Pit: Change My Mind

Abstract:

As a relief system engineer, I revalidate and correct pressure relief system designs daily. Many times, the owner/operator views the pressure relief systems efforts as a 'bottomless money pit' that never returns value and is seemingly never complete. This presentation will dispel this notion and explain how to ensure your pressure relief systems efforts turn into cost-avoidance measures and high-value money-makers.

This three-part presentation will summarize the business purposes of pressure relief system efforts, describe how good relief systems design bases are converted into cost-avoidance measures, and finally share how targeted relief systems efforts can turn into net-positive returns for your facility.

Example content includes safety and regulatory obligations, keeping pace with industry, insurance impact, risk reduction, reputation, safe operating limit changes, and debottlenecking.

ISO 9001-2015 and Risk Mindedness in the Management of Heat Exchangers in Ammonia Synthesis Process

Festus Addo-Yobo, Adrian Lutchman, Sandra Maharaj, Devon Indar
The University of Trinidad and Tobago, Point Lisas Campus, Esperanza Street, Brechin Castle,Trinidad and Tobago.

Introduction

The ammonia synthesis process is one of the most important in the manufacturing Industry because ammonia is a feedstock for diverse products including fertilizer. This process is an interconnected system of production units which include: generation of hydrogen, separation of air to produce nitrogen, ammonia synthesis using catalysts, ammonia condensation and recycling. In the twenty first century efforts were concentrated on the optimizing of the processes either independently or as groups of processes providing inputs to various stages of the processing plant (CEP 2015). As integrated design has been largely successful in optimizing the production plant that is delivered to the producer, attention is now focussed on running the plant so as to optimize its performance. Plant safety, operational efficiency, effectiveness of operation, utilization of resources and productivity are desired outcomes. To achieve these ends, Process Control Engineers must necessarily have conversations with custodians of Quality Management Systems (QMS) in order to optimize the plant performance at a higher management level.

Bayer Technology Services offer Online Optimization & Operational Excellence solutions . Others utilize the TQM and Six Sigma quality assurance models. Nowadays, most plants are run by Distributed Control Systems (DCS) and have modelling software such HYSYS for carrying out such tasks. Thus the opportunity to optimize plant performance online has become feasible. BASF Services have teamed up with Haldor Topsoe to offer Ammonia producers new optimization opportunities online (Process Worldwide 2020).

Ammonia plants have optimized energy recovery through design. Waste heat in the process stream exiting the front end, exiting the secondary reformer, is recovered via heat exchange to produce high pressure steam. Similarly the heat produced by the highly exothermic conversion reaction in the backend is also recovered to produce high pressure steam. The heat exchangers used in these sections (shown in red boxes in the adjacent figure) operate under high temperatures (>400 ºC). They thus are susceptible to failure mechanisms not usually encountered in most petrochemical operations.

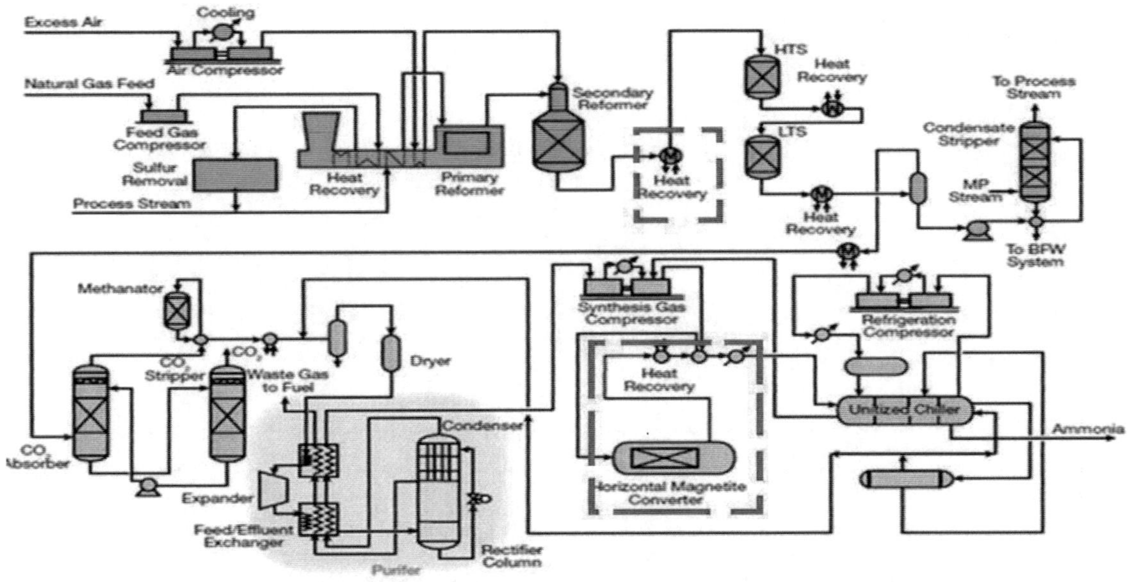

www.aiche.org/cep September 2016 **CEP** Copyright © 2016 American Institute of Chemical Engineers (AIChE)

The Quality Management System

An important development in the standard to which quality management systems for production processes must conform is the revision of the ISO 9001 Standard to the 2015 version which makes it a requirement to address risk in all aspects of the QMS and provide anticipatory remedial actions. The standard places the responsibility on top management. Addressing risk is identified in the Standard as having the potential to reduce cost and also providing opportunities. One important way to operationalize risk-mindedness as advocated in the standard is to model the plant performance or the performance of relevant components and investigate scenarios to identify the risks. Heat exchanger tube thinning In the nuclear power generation industry is well researched and reported (Gao et al – 2017, Jonas – 2004, Snachez-Carrera – 1984). Remedial action in the nuclear industry includes scheduled inspection and replacement of heat exchanger tubes prior to their rupture according to protocols using Computer Programmes e,g,CHECWORKS™ (Smith & Horowitz – 2014). In the Ammonia Industry, non-destructive testing can identify tubes that have thinned significantly, which are then blocked off as a preventive measure to avoid water leaking from them to the process stream.

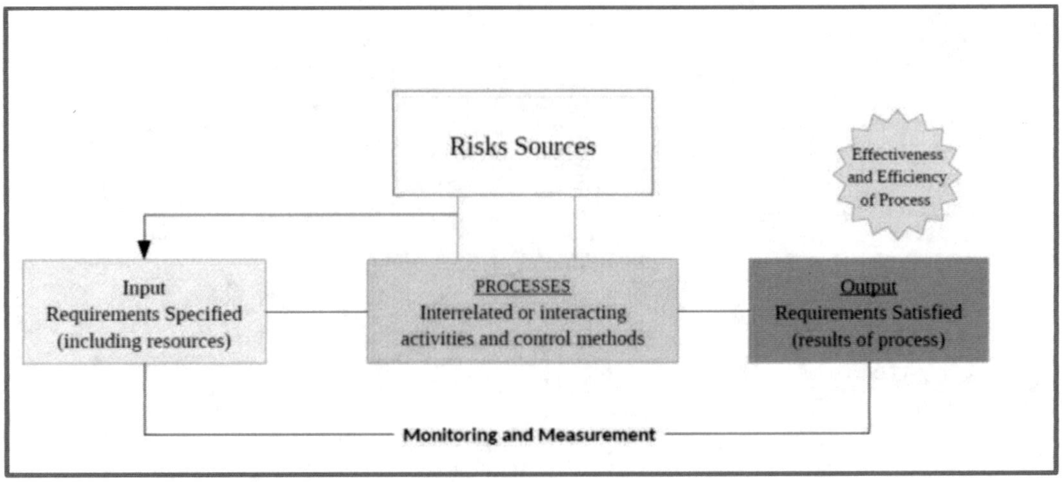

Juxtaposing the great effort and successes in ammonia process optimization to the experience of plant operators, demonstrates the existence of a gap. Even when plants are controlled at desired set points, heat exchanger tubes can and do fail leading to days and weeks of unplanned downtime. Given the high throughput of the plant (600 tonnes/day), this represents significant reduction in plant productivity. Such failures are risks that can be expected, though the events cannot be predicted with accuracy. Heat exchanger networks form an important part of the ammonia manufacturing process. Therefore, these risks should be managed by modelling the operation of the networks. In this work we use HYSYS to model the performance of the heat exchanger network used in typical Ammonia Plant (Chemical Engineering Progress 2015) before and after heat exchanger tubes are blocked off. We investigate numerically, the influence using Boiler Feed Water (BFW) to manage four of the interacting process variables which influence the risk of tube failure: process gas temperature, the velocity of the steam-water mixture in the tubes, the average temperature and the quality of the steam-water mixture in the boiler tubes.

Identification of Risk Factors

The mechanism of platelet erosion has been identified as leading to the thinning of ductile stainless such as the high carbon steel used in the nuclear industry. Abrasive particulates scrape the surface and take off the protective oxide which is swept away by the current of fluid flowing in the tubes, leading to further corrosion. This is known as flow assisted corrosion. Levy (1986) shows that above about 250ºC, temperature speeds up tube thinning process exponentially. Other factors which contribute to Flow Accelerated Corrosion (FAC) are the velocity of the particle, their hardness, the shape and the impinging angle. The geometry of the pipeworks is very important because bends and fittings serve as targets for high velocity dislodged protective oxide.

The HT exchanger tubes which carry BFW for cooling process gases are made of Stainless steel 304L whose mass fraction of chromium are as high as 18-19%. Chromium forms an oxide layer up to 300nm which protects the steel. Should this layer be removed, chromium would react with dissolved oxygen in the feed water to form Chromium Oxide or an even more protective layer comprising a solid solution of Fe, Ni, Cr oxide. Even though the BFW is purified by ion-exchanged to remove chlorides, the possibility of chloride contamination by downstream cooling seawater cannot be ruled out. Chromium can be leached from the protective layer by presence of alkali chloride, which could react with oxygen to release chlorine which further attack chromium oxide and leach out the chromium at temperatures above 500ºC. At lower temperatures, it is postulated that the chloride could react with the oxygen and chromium oxide according the eqn below.

Cr_2O_3 (s) +4KCl(s/l)+2H_2O +1/2 O_2 $\leftrightarrow$ 2K_2CrO_3 (s/l) + 4HCl.

The chromate formed would either remain in solid phase in the absence of liquid water or dissolve in the hot BFW water and be leached away. In either case, no contiguous layer of protective chromium oxide layer would remain and the iron underneath would rust. This mechanism would lead to thinning. Thus, the risk factors are: the contamination of the BFW with chloride, high temperatures and the oxygen levels in the BFW. The steam weight would also constitute a risk factor. The presence of nitrates is a risk factor as it known to corrode SS304L. These two mechanisms would apply to all boiler tubes in all the heat exchangers and thus all the tubes would be equally exposed to this risk.

In an effort to increase heat transfer coefficients in heat exchangers in the ammonia process, "static" mixers are often inserted in boiler tubes. The increase in the heat transfer coefficients is accompanied by the disruption of the laminar boundary sub-layer which also leads a higher mass transfer coefficient, thus increasing the potential for leaching chromium and iron ions from the metal. The additional risk factor associated with presence of static mixers in tubes is the higher velocities in the pipes they cause by occupying space.

The tube inserts pose additional risks introduced at the design stage. They are pendula bodies which will undergo complex vibrations and translational motions during plant operation. The amplitude of the vibrations will depend on the flexural rigidity of the metal. Their motion pose two additional risks of tube thinning: they can impose forces normal at the tube wall which could disrupt the formation of replacement protective oxide layer (F. Addo-Yobo Unpublished work); they could also make contact with the tube wall, peel off existing protective layer and set off vibrations in the walls which can continually weaken the replacement oxide layer and increase its porosity to migrating metal ions thus accelerating erosion and corrosion thinning of the steel underneath.

Thus the operational risk factors are: steam Weight, temperature in boiler tubes, velocity in the boiler tube. Design risk factors are: presence and motion of inserts for promoting turbulence, vibrations; and Utility risk factors are: contamination of process water with chlorides and nitrates after the ion-exchange stage.

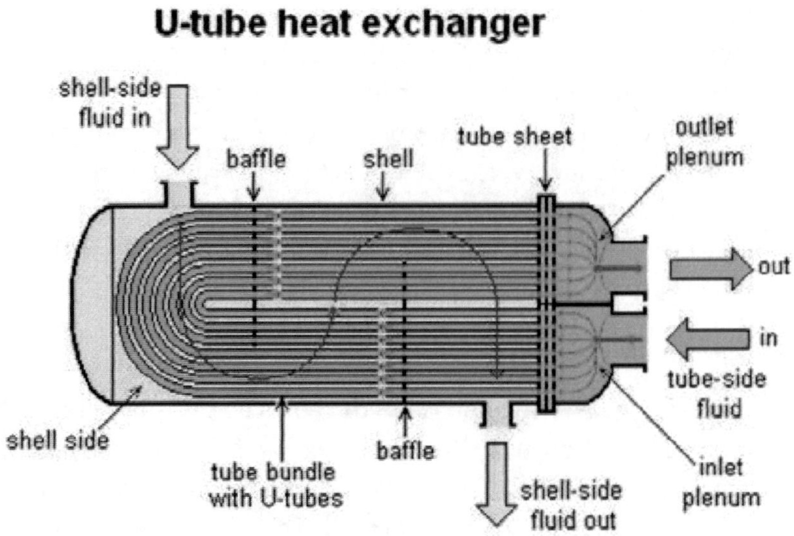

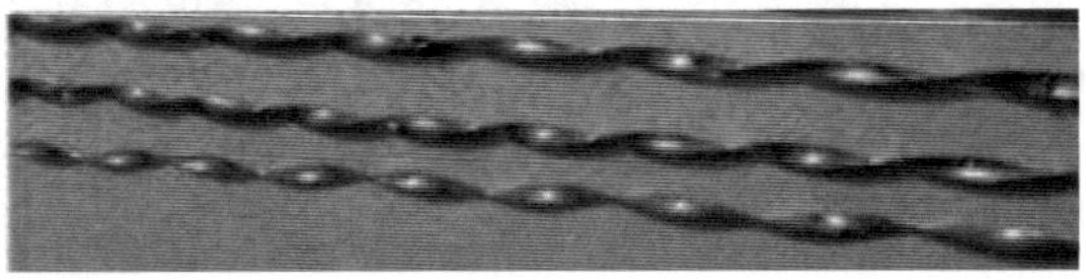

The Models in Hysys

he heat exchangers enclosed in the dotted red boundaries in the Process Flow Diagram above are modelled as a network. E2A is immediately downstream of the ammonia converter, E2B is the second in that stream. E111 is the exchanger in the front end stage of the process. The flow of Boiler Feed Water for cooling, in the boiler tubes in the heat exchanger E2A, E2B and E11, commence as single phase then becomes two phase with the generation of steam as heat is transferred from the process gases on the shell side. The dispersed two phase model in HYSYS was adopted to account for this.Whereas standard HYSYS models for heat exchanger could not accommodate finned and unfinned tubes in one exchanger, the rigorous EDR models were modified successfully to include this capability without changing the hydraulic and heat transfer characteristics. We were also able to incorporate variations in the process gas specific heat with composition without much computational expense.

From heat balance, the weight fraction of steam in the tubes (v_f) relates the process gas flowrates ($\dot{m}_G$) and temperature drop (T^G), BFW flow rates ($\dot{m}_G$) and changes in enthalpies (H^W). The steam weight was computed as shown in the adjacent equation.

$$v_f = \frac{\frac{\dot{m}_G}{\dot{m}_W}\left[C_{pc2}(T_1^G - T_2^G) + C_{pc1}\left(T_{1,in}^G - T_{2,out}^G\right)\right] + H_1^W - H_2^W}{H_2^v - H_2^W}$$

Thus confirming that the flow rate of the BFW is a suitable manipulated variable for controlling gas outlet temperatures. As stated above, the velocity of BFW in the tubes is important and its value can pose a risk. Its value varies with the weight of steam. The dispersed two phase model for steam-water mixes in HYSYS was used for modelling the flow and heat transfer to BFW in the tubes of the heat-exchanger. The shell side heat transfer coefficient in HYSYS are predicted by the well-known Dittus-Boelter equation.

The models built in HYSYS accurately predicted the output pressures and temperatures of the process gas entering and leaving the three heat exchangers, under standard conditions, within 3% percent of the expected values, as shown in the Table1. The simulations were carried out under standard operating conditions at 100% plant rate. The table also shows the risk values of these deviations from the TEMA values, which will be discussed later.

Results and Discussion

Often process Engineers find the need to increase the throughput of a Plant beyond the design values in order to improve productivity. However, this may carry additional risks of shortening the lifespan of the heat exchangers. In order to protect the heat exchangers and avoid overloading them, the product stream can be spilt and a percentage diverted around exchangers E2A and E2B. Table 2 compares output variables when model plant is operated with higher throughputs than the design levels. Columns 4,5,6 of Table 2 below show that the tubeside temperatures can be maintained within 0.3% of the design values with the bypass arrangement. However, steam weight increases beyond the prescribed TEMA values of 30%, thereby increasing potential operational risk. The weight fraction of steam increases beyond prescribed TEMA values with additional feed rate of 7% to 17%. Columns 5 & 6 of the above table show that at around 17% excess feed rate, increasing the cooling water rate by around 8% reduces the steam weight fraction to values close to the prescribed limit, thereby reducing operational risk and at the same time raising production rate.

The studies show that when 5% of tubes in E2A are blocked to mimic the response of industry to tube thinning beyond a certain level, to eliminate the risk spilling water from burst tubes wetting the product gases, the steam weight remains at reasonable levels close to the TEMA limit. However, the velocity in the tubes increase to values close to the prescribed TEMA limit because of the reduced cross sectional area available to the flow of BFW. The risk values of plugging tubes in exchanger E2A on exchangers E2B and E11 are negligible. Tables 3 summarizes the predictions of output variables caused by the plugging 10% of the tubes in E2A on E2A. The simulations show a reduction in the area available for heat transfer, in exchanger E2A by about 3% accompanied by a reduction in the heat duty of ca 4.3%. The reduced cross-sectional area available for flow on the tubeside results in 11% increase in the mean velocity through the tubeside and a 7% increase in the maximum velocity in the tube. The latter is close to the prescribed TEMA limit thereby increasing the operational risk posed to the plant by the blocking the tubes. The effect of this increase in velocity could be a shortening on the service life of E2A.

Table 1: Comparison of the predicted plant output variables with the design historic values.

	Mass Flow Rate	% Deviation of predicted Temperature from Standard TEMA values	Risk Value	% Deviation of predicted Pressure from Standard TEMA values	Risk Value
EE11 Gas Inlet	Standard	0.00	None	-1.85%	None
EE11 Gas Outlet	Standard	-0.21%	None		None
EE11 BFW Inlet	BFW rate	-0.39%	None		None
E2A BFW Inlet	BFW rate	0.18%	None		None
E2A Gas Inlet	Standard	0.00	None	1.31	None
E2A Gas Outlet	Standard	-0.05%	None	2.62	None
To ** (E2A BFW Outlet)**	BFW rate	-0.29%	None	-1.33	None
BFW	BFW rate	2.53			None
E2B Gas Outlet	Standard	1.57%	Negligible	2.80	None

Table 2: Effect of Increasing the Plant Throughput

KEY PROCESS PARAMETERS	Design Specification or Historic data	100% PLANT RATE	7% increase in feed process gas	17.4% increase in feed process gas	16.7% increase in feed process gas

Process Gas Rate (kg/s)	Standard	No deviation	+7%	+17.4%	+16.7%
Deviation in Water circulation rate (Wt/Wt)	Standard	No deviation	-17.2%	-21.9%	-15.7%
Deviation in steam weight $\frac{\boldsymbol{Predicted - Tema\ limit}}{\boldsymbol{TEMA\ limit}}$	-23.3%	-23.3%	+33.73%	+46.8%	+13.4%
Risk Status E2A due to steam weight	None	None	Highest	Highest	Highest
Deviation in tube-side outlet temperature of E2A	None	-0.29%	-0.14%	-0.143%	-0.15%
Risk value of tube temperature E2A	None	None	None	None	None
		Mean/Max			
E2A TUBESIDE % Velocity deviation	No Data	-61%/-20%	-64%/-3.7%	-63%/+3.6%	-63%/-3.7%
Risk Value E2A %Vel deviation	N/A	No Risk	/Localized Medium Risk	/High risk	/Medium risk
E2B % dev in TUBESIDE VELOCITY	-51%	-59%/-57%	-66%/-64%	-66%/-64%	-63%/-61%
Risk value	No risk	No risk	No risk	No risk	No risk
E11 TUBESIDE VELOCITY (m/s)	1.92	-60/-58	-66/-64	-66/-63	-64%/-61%
Risk value in E11 tube side velocity		No risk	No risk	No risk	No risk

Table 3: The Impact of blocking off 5% (10%) of the tubes Exchanger E2A of Performance of E2A

Heat Exchanger Characteristics	E2A with no plugged tubes at	E2A with plugged tubes	Percent Deviation

	theoretical conditions	at conditions from DCS	
Effective Surface Area available for Heat Transfer (m^2)	*	*	-3.07
Heat Duty (kW)	*	*	-4.35
Dev from Standard in Mean Temperature of Outlet of tubeside E2A	-0.15%	-0.16%	
Risk value	No Risk	No risk	
Corrected Mean Temperature Difference (°C)	*	*	7.34
Deviation in Heat Transfer Coefficient	*	*	-8.1%
Mean Shellside Velocity (m/s)	-56.2%	-56.1%	0.28
Maximum Shellside Velocity (m/s)	-53.2%	-53.2%	0.00
Risk value of deviation in maximum shell side velocity	No risk	No risk	
E2A TUBESIDE % mean Velocity deviation $\frac{mean\ velocity - Tema\ vel\ limit}{TEMA\ velocity\ limit}$	-63%	-59%	
Risk Value E2A %Vel deviation	None	None	
Dev maximum Tubeside Velocity (%) $\frac{Predicted\ max\ velocity - Tema\ vel\ limit}{TEMA\ velocity\ limit}$	-6.7%	-0.23%	
Risk value of Predicted maximum velocity	Medium risk	Medium risk	

Discussion, Conclusion & Recommendation for Future work

The studies have demonstrated the usefulness of building a rational model of the operation of the heat exchangers in the Ammonia process based on the diagrams presented in CEP (2015). They confirm the remedial actions taken by the Industry to improve productivity and maintain safe operation are sound. However, they point to the unfortunate fact that the quest to increase the heat transfer coefficients in the U-tube heat exchangers by inserting turbulent promoters are fraught with design risks that might not be possible to eliminate.

The simulations coupled with the present understanding of the mechanism of corro-erosion of high chromium stainless steel products (Folkeson et. al. 2011) suggest that once tube thinning is allowed to commence, caused by a risk factor introduced at the design stage, e.g. the inclusion of unfastened tube inserts, which increase heat transfer coefficients but which disturb the protective oxide layer, the exchangers may have to be eventually de-commissioned because of the difficulty of laying down a contiguous protective layer as shown in the schematic shown above right.

In order to firm up on the mechanisms that trigger tube thinning by corro-erosion of stainless steel boiler tubes in the Ammonia Process, further research into the risks posed by tube inserts and the careful weighing up of the cost of mitigating this factor as required by the Standard need to be carried out.

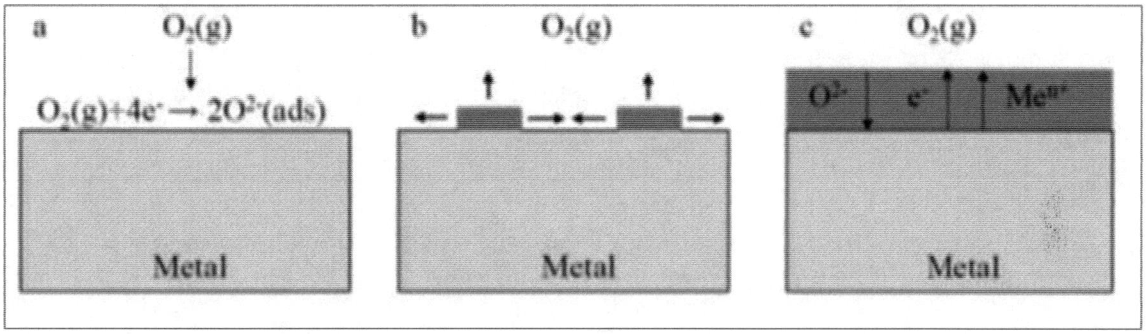

Schematic of the initial formation of an oxide scale by adsorption of oxygen, oxide nucleation and growth to form a continuous oxide scale (stages a–c).

References

1. Venkat Pattabathula & Jim Richardson 2016, Introduction to Ammonia Production, Chemical Engineering Progress pp 69-75
2. Process Worldwide: 12:08:2020 Editor Ahlam Rais. Haldor Topsoe, BASF to Offer Ammonia Producer new Optimisation possibilities. ID 46800904 https://www.process-worldwide.com/haldor-topsoe-basf-to-offer-ammonia-producers-new-optimisation-possibilities-a-955495/
3. The Role of CHECWORKS in an Effective FAC Programme, David Smith and Jeffrey Horrowitz , ICONE 22, 22nd International Conference on Nuclear Engineering, 2014 Prague
4. Gao, Weimin, Li, Yungang and Kong, Lingxue 2017, Numerical investigation of erosion of tube sheet and tubes of a shell and tube heat exchanger, *Computers and chemical engineering*, vol. 96, pp. 115-127, doi: 10.1016/j.compchemeng.2016.10.005.Boiler tube failure in Heat Exchangers for Power generation in the Nuclear Industry

5. L.E Sanchez-Carrera, "Mechanism of Corrosion-Erosion in Steam Extraction lines in Power Station", PhD Thesis Dept of Mechanical Engineering MIT 1984

6. Jonas, O. (2004). FAC and Cavitation: Identification, assessment, monitoring, prevention. *Combined Cycle Journal*.

7. Suez Water Technologies and Solutions. (2020). *Handbook of Industrial Water Treatment Chapter 11- Preboiler & Industrial Boiler Corrosion Control.* Retrieved from https://www.suezwatertechnologies.com/handbook/chapter-11-preboiler-and-boiler-corrosion-control

8. Md. Moniruzzaman Bhuyan1, Ujjwal K. Deb, M. Shahriar, Simul Acherjee 2017, Simulation of Heat Transfer in a Tubular Pipe Using Different Twisted Tape Inserts, Open Journal of Fluid Dynamics, 2017, 7, 397-409, http://www.scirp.org/journal/ojfd

9. Folkeson, N., Jonsson, T. Halvarsson, M. Johansson, L.-G., Svensson J.-E., The influence of small amounts of KCl(s) on the high temperature corrosion of a Fe-2.25Cr-1Mo steel at 400 and 500°C. *Materials and Corrosion*, (2011), 62(7), pp. 606–610

10. F, Addo-Yobo 2021, Motion of Pendula inserts in heat exchanger tubes in the Ammonia Process and their impact on exchanger risk, Unpublished work.

617202 Optimal Placement of Gas Detectors Using Dispersion Modeling

Abstract:

Historically, gas detector placement was performed by experienced designers using rules of thumb and judgment based on fundamental principles of gas cloud behavior. Unfortunately, designs using these approaches have been inconsistent and, in many cases, poorly performing. Historical data from the UK HSE shows that certain fixed gas detection systems routinely were only able to detect less than 70% of "major" gas releases. In the past decade, advances in computer modeling and 3D graphics has enabled the implementation of chemical process quantitative risk analysis (CPQRA) in the service of gas detection system design. The general process for using CPQRA to develop and verify a gas detection performance metric called coverage, which is essentially the fraction of the all release scenarios that are detected by a gas detector array, has been documented in the International Society for Automation (ISA) technical report ISA TR84.00.07 – Guidance on the Evaluation of Fire, Combustible Gas, and Toxic Gas System Effectiveness. This tutorial provides an overview level explanation of the workflow for the process of gas detection mapping, including the risk-based selection of performance targets and quantitative verification of the achievement of the coverage performance targets using dispersion modeling results and leak rate estimations that are similar to the proven approaches used in CPQRA. The tutorial goes through the entire process of gas detector placement for a sample facility where liquefied petroleum gas products are stored, beginning with risk assessment and going all the way through a completed gas detector layout.

619065 Addressing Aging Processes and Equipment

Abstract:

Number of the fertilizers plants across the globe have been installed and are in operation since many decades. The operation with the old equipment and conventional process techniques is being managed coupled with budget constraints. Hence the old aged plant equipment have high risk of failure under the shed of outdated processes, and may cause severe damage to the plant assets as well as to human. Consequently, one has to bear more and repeated maintenance cost and regular replacements of equipment.

Fatima Fert Plant (Previously Daud Hercules) also commissioned in 1969 and is in operation to date. Dealing with the challenge of old aged equipment and process techniques, very strict and robust reliability plans have been established and implemented to inspect the equipment with the advanced ultrasonic techniques and to make sure that the asset is in operation under controlled parameters. Based upon the finding of inspections, 10 year rolling plan have been developed to replace the equipment that are prone to failure.

The paper will be providing the details of the challenges and risks faced due to old age equipment and the way forward to mitigate and control that risk with optimized budget utilization.

619392 Why Is There a Regularity of Process Incidents Occurring, or Is It All Just a Re-Occurrence of the Past?

Abstract:

We attend conferences to hear the experiences of others, learn lessons from past events and take that experience and learning back to our place of work. Yet a simple analysis of suggest there is a plateau in the progress to reducing the frequency of process safety incidents.

So perhaps the question to address is, “Why are lessons not being learnt”?

This paper will highlight various data demonstrating the “plateauing” of performance; highlight particular recurring indicators that need to focused on and look at how learning may be distributed to those that need to increase their knowledge of PSM. There will be a particular emphasis of the use of case studies in this approach.

The paper will particularly interest those still on the learning journey in PSM and those questioning alternative approaches.

618397 How You Manage the Actions Coming through Process Safety Management Systems?

Abstract:

For improving process safety in the plant, we are putting great afford managing task of operators. Alarm management is important for control operator's tanks. Quantity of the alarms should not be more then control operator can manage.

However, there is gap in engineer's tasks management. Workload of an engineer is an important parameter in complex plants. This workload can be rapidly increase with HSE related tasks. As a result of Management of Change, Incident investigations, ... most of the tasks are given to operations engineers, projects engineers, maintenance engineers.

If the workload is too high it is easy for the engineer to miss the important actions. With HSE Action Portal, we aim to give management a view of task load for each department. The department responsible can focus on the overall tasks of the department

HSE Action Portal make possible to reach actions and tasks coming through PSM system elements such as Management of Change, Pre-Start up safety review, incident/accident investigations and process hazard analysis in the same platform. Users are able to analyse their action/tasks in terms of number and completion percent. In this way, users are triggered to follow closely their actions and complete them in a correct way.

It is an useful tool also for top management in order to collect data and create statistics periodically. It can be seen actions/task situations for each refinery and department.

617997 Guidelines for Risk Ranking Hazardous Scenarios When Conducting PHAs

Abstract:

Most of the companies in the process industry use corporate guidelines, or dedicated criteria and principles for conducting Process Hazard Analysis (PHAs). Examples of factors addressed are the following: nodes selection, process deviations to be brainstormed, level of detail when writing causes/consequences, safeguards categorization, risk matrix to be used, and how to risk rank the identified scenarios.

This paper collects different approaches from leading PHAs at several companies. An analysis comparing these approaches is provided and strengths and weaknesses are identified. Finally, based on safeguards and consequences categorization, guidelines are proposed for consistently risk rank hazardous scenarios.

619107 Practical Guide to Operationalize & Visualize Cumulative Process Safety Risks Perspective

Abstract:

The paper describes how Process Safety Management can be seamlessly integrated with Operational Risk Management. It demonstrates how operational processes like Permit to Work, MOC, Job Hazard Analysis, Incident Management and Process Safety Management can be brought together to support day-to-day decision making based on the actual cumulative risk status of the facility.

Key thinking is developed to answer questions such as *How do we make Process Safety Risk as visible as HSE* and *How can we prevent previously identified hazardous sequences of events from reoccurring?* , thus increasing the entire system safety.

This paper presents practical approaches in overcoming the challenges in collecting data from disparate sources and learnings on realising Dynamic Operational Risk Management (ORM) from various client implementations. In addition, it touches upon topics such as *How to analyze the data* and *Use of machine learning to predict future risk* .

The paper presents a risk calculation methodology that calculates barrier status and visualizes cumulative risk:

in a standardized barrier model, based on IOGP 544 standard, in Swiss-Cheese view

geographically by site and area/location using site graphics

per critical scenario in a BowTie view

It connects and visualizes process safety risk by automatically collecting data for Tier-1,2 and 3 safety indicators that are covered by various processes spread across the organization.

Furthermore, it shows how the cumulative risk status can be used in operational processes, potential conflict detection and planning of activities to manage and minimize the actual risk.

Keywords: Barrier Management, Process Safety Management, BowTie Methodology, Swiss-Cheese Model, Operational Risk Management, Project Implementation Methodologies, permit-to-work, MOC, JHA/RA, Machine Learning

ENGINEERING & OPERATIONS

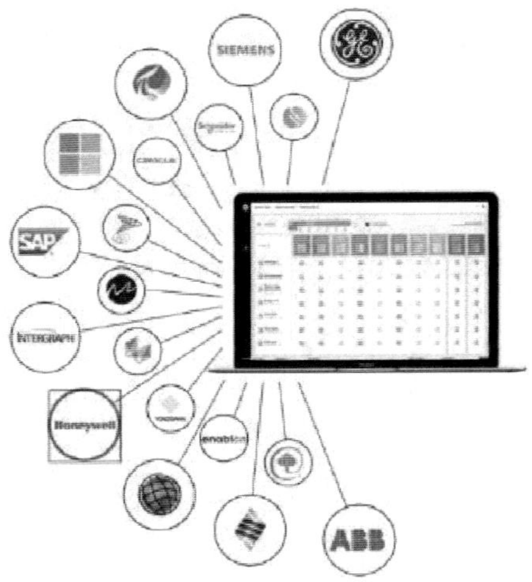

Wolters Kluwer Enablon Proprietary and Confidential

BARRIER MANAGEMENT INTEGRATION DATA POINTS

Integrity Data: CMMS, Audits & Assurance, Competence Management, Inspections, Engineering data.

Operational Intervention: Process Control System / Historian, Shift Logs, Operator Rounds, Permits, Temporary defeats.

Barrier Challenges & Failures: Incident Management, Management of Change, Process upsets, PSM events.

BARRIER MANAGEMENT – PROCESS SAFETY MANAGEMENT

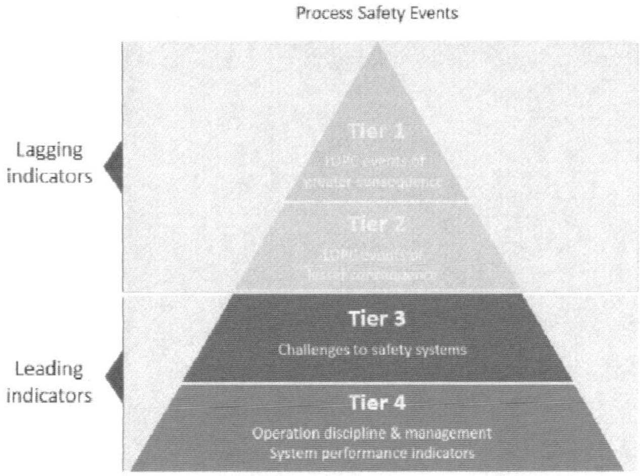

The focus of Barrier Management

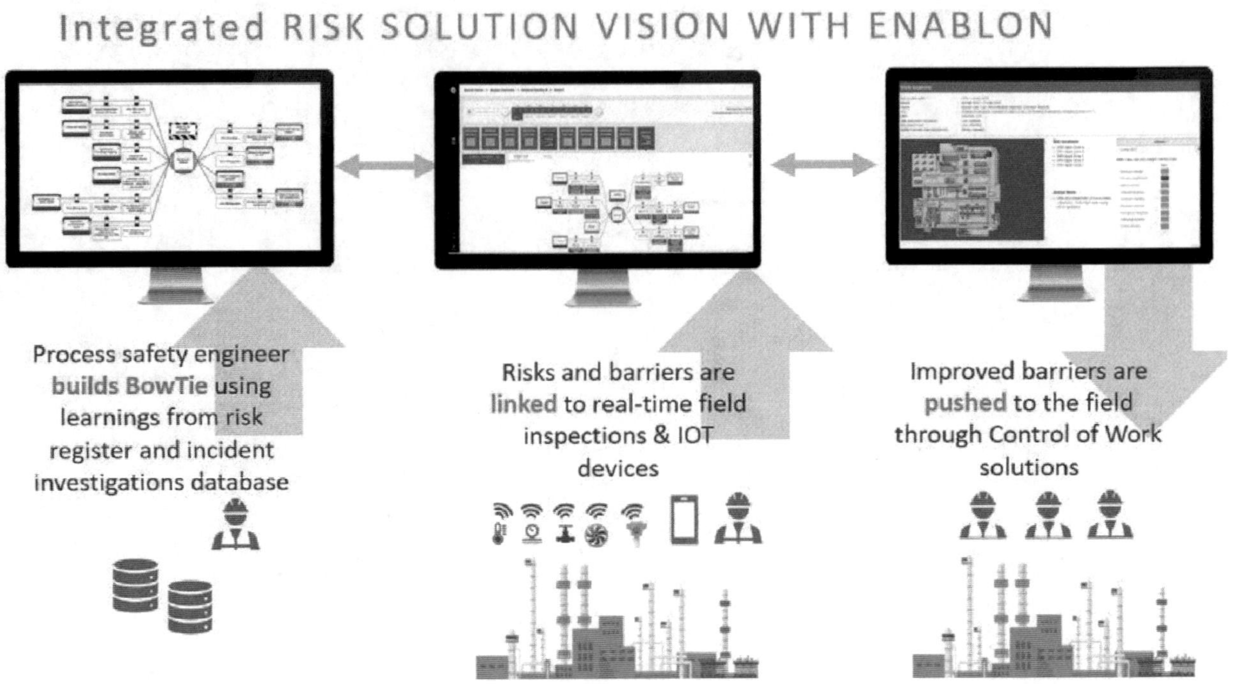
Integrated RISK SOLUTION VISION WITH ENABLON
Process safety engineer builds BowTie using learnings from risk register and incident investigations database
Risks and barriers are linked to real-time field inspections & IOT devices
Improved barriers are pushed to the field through Control of Work solutions

619126 Controlling Human Performance Between Both Unplanned and Planned Tasks within Abnormal Operation Mode

Abstract:

There is confusion in terminology used in the chemical-related industry for the class of procedures commonly referred to as Abnormal Mode of Operation and Abnormal situation management. This paper provides a clear definition of each mode of operation and gives examples of how the human performance is controlled for each.

1. Normal – either a continuous mode or a normal batch mode of operation.
2. Planned Non-routine (Non-Normal)– Startup, shutdown, are online maintenance are the main non-routine or non-normal modes of planned operation. But planned Temporary procedures, with time limits, are also part of these.
3. Abnormal– These include those activities covered by generalize procedures or guides (planned in a general sense) the activities that don't have written, step-by-step procedures (but noting that many companies have guides for most of these abnormal situations). A further breakdown is possible:
    1. Response to upsets using a Trouble-shooting Guides(TSG) for handling deviations from the operating window (these or normally triggered on an alarm); these may ultimately lead to a shutdown, safe park, or emergency shutdown (all of which are also proceduralized), if the deviation cannot be corrected in time.
    2. Response to failures using a Temporary procedure - such as how to run in bypass mode if the flow controller fails and you want to keep running in manual mode.
    3. Response to Unanticipated Events - The FIRST occurrence of this is handled by what is generally referred to as Emergency-MOC, which is really saying we will make a change immediately, and do the risk review of the change later, and then learn from this one case and proceduralize into a Temp Procedure or TSG for the next time it comes up.
4. Emergency Operations- a diminished or reduced operating plan; normally a Temporary procedure (similar to 3.b. above)
5. Emergency Shutdown - a TSG that fails to resolve the issue in time will go to this; or for some events such as a sudden loss of containment, we go straight to these
6. Emergency Response- normally in conjunction with 4 or 5, but focused on the protection of people, assets, environment, given the release or other imminent harm is in play.

This paper will provide a framework to ensure no mode of operation is overlooked and it will help sites understand what is needed to control risk during each mode of operation. Most of the time in the paper and presentation will be focused on classification 3, Abnormal Mode of operation.

619218 Proactive Lopc Indicator Critical Analysis through Power BI Dashboard

Abstract:

This work applies Power BI dashboard in order to proceed a critical analysis of proactive loss of primary containment indicator so that possible process problems can be detected before greater events occur by prognoses and decision-making to direct resource management in a more assertive way.

A process safety event is an unplanned or uncontrolled release of any material including non-toxic and non-flammable materials from a process, or an undesired event or condition that under slightly different circumstances could have resulted in a release of material (API 754). The pyramid model can represent the predictive relationship between lower and higher events.

It is important to guarantee the good register of pyramid base events (Tier 3) to achieve enough information, identify the fail installation's weakness, avoid higher process safety events, reduce losses and improve the facility performance. Nevertheless, the only register is not enough to achieve these improvements. Its's necessary to quantify and analyze the data to guide the decision-making. A critical analysis involves the following parameters: leaked product, fail causes, equipment that failed, engineer discipline, LOPC location, consequences.

The first step to make the critical analysis is having a reliable database. This step involves training the operational team in leading and lagging process safety indicators and process safety management. The communication must be direct and clear. Furthermore, it is necessary to provide a satisfactory registration platform with the necessary fields to carry out the subsequent critical analysis.

After that, it is possible to extract all the information to proceed the proactive LOPC indicator critical analysis. At this point Business Intelligence (BI) presents itself as an adequate tool. Power BI, from Microsoft, is one of the most prominent BI tools. It is a collection of software services, applications and connectors that work together to consolidate, make information from different data sources coherent, visually engaging and interactive (from an Excel spreadsheet to a cloud-based data collection).

Through the LOPC critical analysis using Power BI dashboard it is possible to evaluate the Tier 3 incidents, focused on developing a learning approach. Therefore, it is possible to prevent and monitor failures trends and effectiveness of implemented actions. The critical analysis permits to direct the resource management by identifying the equipment and systems with recurrent failure events. It is worth mentioning the effective contribution for Process Safety Culture, since the leadership, operation and related areas could see the cause-effect relationships between the minor events and the daily operations.

619342 A Case Study on Ensuring "ALARP" Demonstration to Justify Continued Operation with Respect to "Mah" By Integrating Process Safety Techniques for Existing Process Facility

Abstract:

A lot have been written on process safety. However, practical applications of all these theories & studies has its own challenges. At the end, after overcoming the challenges as well as developing, implementing & monitoring strategies, a risk to existing facilities can be reduced to ALARP/ acceptable level.

This paper will throw light & discuss about requirements to integrate different process safety strategies, challenges during implementation & their solutions to demonstrate that existing oil & gas facility is safe for operation, through a practical case study of one complete process train. The methodology includes;

- Identifications of process trains
- Data collection
- Data Review
- Review past changes & their implementation
- Update records & P&IDs
- Carry out/ Update Integrated PHA with process simulation
- Development of Failure Scenario
- Risk Calculation considering existing safeguard (preventive and mitigating measures)
- Risk submission and ALARP demonstration
- Compare risk
- Identification of safety critical elements (SCEs) & Safety Critical Tasks (SCTs)
- Development of Performance Standard of SCEs
- Developing management system of PLAN-DO-CHECK-ACT
- Integrating the same in Audit/Inspection
- Develop training through simulations

Challenges faced by team for existing facility to achieve optimum process safety & ensure ALARP level of risk, and their solutions will be discussed.

Conclusion

This exercise provides evidence that wherever practical, the design and operations of the existing oil & gas facility has attained a level of risk that Justifies Continued Operation with respect to major fire and explosion events. This demonstration of Justification for Continued Operation is based on the number of safety measures inbuilt at design stage, Quantitative Risk Assessment and various other process safety studies revalidated during operational phase of the oil & gas process facility as well as after implementing changes through MoC. It must be noted that a separate process of action tracking is in place to prioritize and monitor implementation of recommendations provided to reduce the risk further to As Low As Reasonably Practicable (ALARP). A systematic approach of Plan-Do-Check-Act is developed to ensure that, safety critical tasks are carried out in timely manner by competent personnel, to ensure safety critical elements are in place and functional in line with performance standard.

616668 Towards Management Systems Perfection – Successes with Implementing a Layer of Verification Analysis to Continually Treat Process Safety Risks

Abstract:

Trust, but verify, by continually checking everything! This paper will demonstrate the importance of verification as a critical step in continually improving the treatment of process safety risks, including implementing inherently safer designs and operating practices, and controlling residual risks.

A well-established design concept for management systems is the continual improvement lifecycle involving the steps: Plan-Do-Check-Act (PDCA). Similar to standards for managing quality, environmental, occupational health and safety, and security, models for effective process safety management (PSM) systems also utilize the PDCA cycle. At a higher level, the PSM elements of Mechanical Integrity, Incident Investigation and Compliance Audits address verification in the "Check" step; while every management process of the PSM System should incorporate a "Check" step with verification processes and procedures to monitor and assure compliance and performance. This paper will present a taxonomy of activities for meticulous verification against policies, objectives, plans and other requirements for a vibrant PSM system, showing the importance of examining, analyzing, assessing, confirming, inspecting, testing, and reviewing planned activities, desired conditions and achieved results.

Attendees will benefit from practical takeaways of how each management system element can be developed with "verification designed in" and a verification culture created to treat process safety risks at three opportune times: before the incident (prevention), during the incident (emergency management) and after the incident (incident investigation, root cause analysis and lessons learned).

A comprehensive case study with practical examples will be presented to demonstrate a proven technique of Layers of Verification Analysis (LOVA) as a best practice integrated into the conduct of a process hazard analysis (PHA). The case study will show the audience "how to" enhance the PHA facilitation wearing an "auditor's hat" to raise questions and verify the effectiveness of existing safeguards, all of which are managed by the PSM System. Borrowing from the basic principles of management systems auditing, LOVA asks these questions for each hazard scenario: are systems, processes and procedures associated with causes (of human errors and equipment malfunctions), with consequences (of physical phenomena and resulting impacts), and with safeguards (for both engineering and administrative controls) *(1) in place?, (2) adequately designed? and (3) effectively implemented?* LOVA has also been used for supplemental analyses within the PHA to address human factors, previous incidents, facility siting and changes (e.g. MOCs). Recommendations from the LOVA in PHAs can be used as lessons learned for improving the implementation of the PSM system on a daily basis, without having to wait on the next periodic audit or PHA revalidation. The LOVA technique has been shown to be equally suitable for any stage of the process lifecycle, including inherently safer design evaluations.

The paper will also present three incident case histories to show how missing, inadequately designed or ineffectively implemented verification processes and activities have contributed to process safety incidents.

This paper will draw upon the authors' extensive experiences in designing, developing, implementing and auditing PSM systems, as well as conducting PHAs, building siting evaluations, and incident investigations.

617112 Study and Experimental Validation of CO2 Blowdown at Cryogenic Conditions

Abstract:

CO_2 solidification is one of the critical issues being identified in cryogenic technology process, particularly during the blowdown process. This poses critical challenge in developing the technology for separation of CO_2 from natural gas as there are limited studies and data to support the design and simulation of the process. Depressurization of high CO_2 fluid at cryogenic temperature would lead to possibility of CO2 solid formation, hence potential blockage of process equipment and venting line.

The new cryogenic technologies are currently being tested for the proof of concept. Hence, a pilot plant, which is a scaled down version of the technology was developed. One of the major challenges faced during the pilot plant testing is the emergency depressurization philosophy as the process involves CO_2 solids handling which is uncommon to the industry standard. Lack of experimental data of CO_2 blowdown at these conditions presented a technical gap which needs to be addressed to properly predict the solidification and mitigations, hence avoiding over-design of the plant and chances of accidents and unscheduled downtimes can be minimized.

Therefore, this paper will focus on the experimental study which have been done on CO_2 blowdown scenarios at a pilot plant. The pressure and temperature profile provide real insight of the solidification of CO2 during several blowdown scenarios. Finally the paper would validate a simulation data conducted using the experimental data as one of the potential prediction tools for CO2 blowdown process.

617588 The Supervisors Next Big Challenge: How to Ensure Operational Discipline on the Work Fronts

Abstract:

The Supervisors Next Big Challenge:

How to Ensure Operational Discipline on the Work Fronts

There is a problem that the petrochemical and oil and gas industries have not solved effectively: how to transform work fronts in environments based on operational discipline.

All the commitment and effort that an organization mobilizes in its risk management, finally, is updated in what happens between supervisors and operators on the work fronts. That is why it is so necessary to work on the alignment of the workforce of operating companies and contractors.

In the last decade, Senior Management became more aware of its role: defining what were tolerable risks and assuring resources. Process Safety, Integrity and Environment ceased to be discursive themes, to effectively have a genuine place at the small table where strategic decisions are made. The biggest challenge has been getting it to operate according to those definitions.

Supervision is carried out on the work front, the true executive arm of the organizations in the field. And for this organizational segment to ensure a disciplined operation, they need to be able to manage the culture.

A common mistake in the risk management approach has been to think of the lack of operational discipline of supervision as a problem of isolated individuals, and not of the context in which these people work, that is, of the context that the activities provide. organizations, from the context produced by culture.

When analyzing the supervision segment, one should not think of undisciplined people, but cultures that tolerate them, or that reinforce risky behaviors and shortcuts to the established norm. If supervision does not ensure operational discipline on the work fronts, it is because the organization does not provide the conditions for this to occur. People are the reflection of the culture in which they operate. Individuals do not impose a personal style of managing risks in isolation, but adapt to group agreements, displaying safe or unsafe behaviors depending on what is collectively tolerated. Field supervisors only take risks that the culture allows them to take.

The change of mindset in the supervision segment should not be built on the basis of shortcuts or "crash plans", but rather in avoiding in a progressive and lasting way the repetition of incidents, through a transformative process: generating a culture of discipline, learning and mutual trust in the work fronts, based on group agreements between operators and contractor companies, managed by a trained supervision to make decisions based on risks, combat the normalization of deviance and keep in mind the sense of vulnerability.

The key is to generate the conditions for effective learning. In addressing group problems, and not individual ones. The challenge is to create the conditions so that operational discipline is not a declaration of principles but a reality in the field. For this they do not need new knowledge, but to deepen what they already know, in the right context.

The pandemic, paradoxically, provided an opportunity to improve the quality of supervision development, personalizing it. Although online self-training has been around for a long time, it is not the most appropriate tool to generate changes in operations. For that, culture needs to intervene.

The next decade seems to keep a central challenge: transforming supervisory teams into custodians of culture and guarantors of risk management, through work fronts based on operational discipline.

Presenter Information	Contact Information
Presenter Name: Bejoy K. Bharatiya	**Contact Name**: Bejoy Bharatiya Telephone No.: +61 413 098 008
Presenter Biography: Bejoy Bharatiya was Process Safety Advisor at DNV GL Offshore and Industrial Services (B) Sdn. Bhd. He has over 30 years of professional experience in the upstream oil & gas, refining, petrochemical, and mineral processing industry. Prior to joining DNV, he held various positions, Process Safety Group Lead at Saudi Aramco Shell Refinery Company (SASEF), Senior Process Safety Engineer at SANTOS (South Australian Oil Search) Pty Ltd. Australia, Senior Operations Engineer with Saudi Aramco Gas Operations, and the Principal Process Engineer of Kellogg, Brown & Root (KBR), Australia.	Fax No.: E-mail Address (Required): **bejoyb@hotmail.com** **bejoy2014@gmail.com** Your Company/Department:

Abstract (max of 1500 words)

Project Assurance Review and Audit Framework for Capital Project of a Petroleum Refinery - A Process Safety Approach

Bejoy K. Bharatiya

The petroleum industry has long since recognized that certain risks are involved in the major hazard industries. It becomes a major challenge to minimize risk by **comprehensive project assurance/audit framework** in the project execution of a major process facility in the hydrocarbon industry. Over the years, the oil and gas industry has learned lessons from a number of process safety events and various hazards such as loss of control and loss of containment top events that could result in: fires, vapour cloud explosions, toxic and asphyxiant exposure, which occurred in various parts of the world and are not limited to the country of origin. The design and operation of process facility for the refinery and petrochemical complex have been the subject of major hazard to the oil industry.

It has been the continuous effort of the industry to update the design and operating philosophies to improve the inherent safety of the processing units. The present paper is concerned with defining the major safety considerations of international oil companies, and Industry Standards in designing and operating the refinery and petrochemical process facilities. Inherently safe design and structured project assurance and audit program are mandatory to guide and drive the design intent of HSE critical systems. The main safety philosophies include: ESD and blowdown, fire and gas detection, toxic gas (H2S) detection, active and passive fire protection (including guidelines for firefighting systems and fire protection), layout HSE philosophy, process safety and containment etc.

This paper provides the **assurance/audit framework** for a major project but does not attempt to cover in detail every aspect of safety design. Reference should be made, therefore, to specific discipline philosophies and specifications as appropriate. (Refer to **Figure 1**).

Project audit or assurance program shall be applied throughout all the project development phases, engineering, construction, start-up, and operation via dedicated and well-structured efforts. The safety philosophy, and resulting protection of persons and assets, is combining all provisions undertaking during all project phases. (Refer to gated process, 1-6 **Steps** explained below).

HSE Design Philosophies

During the detailed engineering, and subsequent construction and commissioning, particular care will be taken by achieving:

- Good engineering practice
- Application of an all-encompassing QA system
- Development and implementation of a safety management system
- Design reviews at 30%, 60% and 90%, layout review, HAZID, Design Safety Review (DSR), accessibility reviews and other reviews.
- Qualitative and quantitative risk assessments and IPF review
- Design of installation for extreme weather conditions
- Execution of an extensive pre-start up review of the plant before introduction of feed (Hydrocarbon, gas) covering both hardware and, control systems and procedures.

- Conformity check with local regulations, project standards and other applicable industry standards and directives.

Gated Process During Project Development and Project Execution Phases

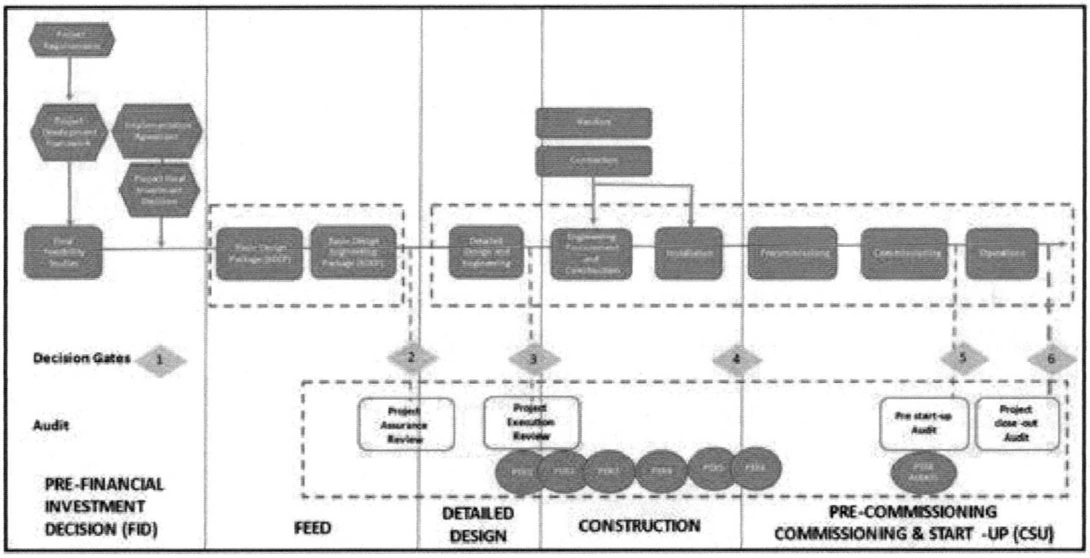

Figure-1 Gated Process for Project Assurance/Audit Framework for Project Execution

Project Assurance Review/Audit is required at all stages of project execution.

1. Pre-Financial Investment Decision (FID) stage:

1) Management: review critical contractor's capability, business license, & project experiences, PEP, schedule.
2) QA/QC: review critical contractor's ISO 9000 certificate, QA/QC plan, standards/policies/procedures (**SPP**), sample check of work records/reports etc. (see Table-1 International Codes and Standard)
3) HSE: review critical contractor's ISO 18000 certificate, project HSE plan, SPP, sample check of work records/reports etc.
4) Technical: review Feasibility Study, Environmental Impact Assessment (EIA) reports., compare critical design area engineering standards against international standards as well as best industry practices, etc.

Deliverable: review comments or audit findings

2. Front-end Engineering Design (FEED) Stage:

1) Management: review critical contractor's capability, business license, & project experiences, PEP, schedule
2) QA/QC: review critical contractor's ISO 9000 certificate, QA/QC plan, SPP, sample check of work records/reports etc.
3) HSE: review critical contractor's ISO 18000 certificate, HSE plan, SPP, sample check of work records/reports, including **PHAs/HAZOP/QRA** etc.
4) Technical: review FEED engineering design, & risk assessment e.g. HAZID, HAZOP, QRA, etc.

Deliverable: audit findings on Statement of Review for FEED, Statement of Review for FEED Safety Study.

3. Detail Design Stage:

1) Management: review critical contractor's capability, business license, & project experiences, PEP, schedule

2) QA/QC: review critical contractor's ISO 9000 certificate, QA/QC plan, procedures, standards/policies/procedures (**SPP**), sample check of work records/reports etc.
3) HSE: review critical contractor's ISO 45001 certificate, HSE plan, standards/policies/procedures (**SPP**), sample check of work records/reports etc.
4) Technical: review Detail Design engineering design, design variation, & risk assessment e.g. closeout of HAZOP recommendations, etc.

Deliverable: review comments or audit report on Statement of Review for Detail Design, Statement of Review for Detail Design Safety Studies.

4. Procurement Stage:
1) Management: review critical contractor's capability, business license, & project experiences.
2) Technical: review specifications of safety critical equipment and QA/QC process.
3) Engineering requirement e.g. specification, database, drawings etc.
   - Manufacturing documents e.g. drawings, inspection & test plan, witness Factory Acceptance Test (FAT), review FAT report & Manufacturing Databook/Spare parts, etc.
   - Inspection & Test Organization documents e.g. Inspection Report, 3rd party certificate etc.

Deliverable: review comments or audit findings.

5. Construction Stage:
1) Management: review critical contractor's capability, business license, & project experiences, PEP, schedule, etc.
2) QA/QC: review critical contractor's ISO 9000 certificate, QA/QC plan, standards/policies/procedures (**SPP**), sample check of work records/reports etc.
3) HSE: review critical contractor's ISO 45001 certificate, HSE management system, plan, procedures, standards/policies/procedures (**SPP**), sample check of work records/reports etc., construction environmental monitoring, waste treatment plan, emergency response plan
4) Technical: compare critical design area project engineering standards against international standards, review construction methodologies, check list, punch list, site construction/safety/environmental/QA/QC activities, design variation, mechanical completion, handover report etc. (**see Mechanical Completion Certificate Checklists** for conducting site audits

Deliverable: review comments or audit report, Statement of Review for Construction

6. Pre-Commissioning & Commissioning Stage:
1) Management: review pre-commissioning & commissioning plan schedule, integration plan etc.
2) Technical: review pre-commissioning & commissioning plan, isolation list, punch list, Site Acceptance Test (SAT), red-line markup drawings, design variation, witness pre-commissioning & commissioning activities, Prestart-up Safety Review (PSSR), ready for startup inspection etc.

Table – 1 International Codes and Standards

IEC 79	Electrical Apparatus for Explosive Gas Atmospheres
API RP 520	Recommended Practice for the Sizing, Selection and Installation of Pressure Relieving Devices (Part 1 and 2)
API RP 521	Recommended Practice for Pressure Relieving and Depressurizing Systems
NFPA 10	Portable Fire Extinguishers
NFPA 11	Low Expansion Foam
NFPA 13	Installation of Sprinkler Systems
NFPA 15	Water spray Fixed systems for Fire Protection
NFPA 16	Deluge Foam Water Sprinkler and Foam Water spray systems
NFPA 17	Dry Chemical Extinguishing systems
NFPA 24	Installation of private fire service mains and their appurtenances
NFPA 30	Flammable and Combustible Liquids Code
NFPA 72	National Fire Alarm Code
NFPA 101	Life Safety Code
NFPA 2001	Clean Agent Extinguishing System
NFPA	Fire Protection Guide to Hazardous Materials

IP Model Code of Safe Practice Part 3	Refinery Safety Code
IP Model Code of Safe Practice Part 19	Fire Protection at Petroleum Refineries and Bulk Storage Installations
BS 5839	BS 5839 Part 1988 for Fire Detection and Alarm Systems for buildings
IEC 61511	Functional Safety, Safety Instrumented Systems for the Process Industry
ISO 14692	Petroleum and Natural Gas Industries – Glass Reinforced Plastics (GRP) piping

This paper discusses industry experience and summarizes the salient design considerations and risk assessment to improve safety integrity in the execution of capital projects for refinery and petrochemical complex. An effort has been made to highlight the ***project assurance/audit framework*** *with HSE methodologies adopted by major oil companies using project standards, and industry best practices and standards like API, ASME, NFPA, IEC, ISO etc. Emphasis has been made to increase the awareness among the discipline design engineers and process safety engineers to develop safer designs of the refinery and petrochemical facilities*

619336 Case Studies – Beware of Common Errors during Risk Assessment of Powder Handling

Abstract:

Case studies – Beware of Common errors during Risk Assessment of Powder handling

"Single size shoe does not fit all," sounds so much in synchrony when we want to assess risk related to powder handling. Powders, unlike liquids have varied challenges right from it's transportation, receipt storage and handling.

Dust handling can have primarily 2 major hazards:

1. Dust explosion
2. Dust thermal decomposition

Few of the important parameters to understand for dust explosion are:

1. Fuel characteristics of the powder:
2. MIE: Minimum energy that can ignite dust cloud in the air.
3. MIT: lowest temp of a hot surface that will cause dust cloud (not dust layer) to ignite.

- Particle size: Smaller size poses higher risk

1. Kst: Max Rate of pressure increase. lower the kst number means the lesser severity 0 bar m/s means no risk while >300 bar m/s means high risk and bigger explosion severity.
2. Ignition source:
3. Hot Surfaces
4. Mechanical sparks/friction

- Electrical motors/contacts

1. Static electricity

Few important parameters to understand and control thermal decomposition in powders is:

1. Onset temperature of decomposition
2. Ageing of powders in vessels that the in process powders are handled such as Dryer, pipelines etc
3. Bulk storage stability of powders

Powders have challenge in terms of dissipation of heat, sampling of representative samples and uniform temperature indication and thus control.

The case studies will focus more on common errors during defining of controls for powder handling and will focus on how preventive controls can be built in to avoid major incidents.

Sr.No	Case Study Title	Controls in brief
1	Drying of Powders with detonation property	Selection of right type of Dryer to ensure metal tramp is not an issue Inert atmosphere
2	Aging of powders and thus lowering of thermal decomposition onset	Thermal studies post commissioning to ascertain periodic cleaning of process vessels such as dryers to avoid decomposition at lower temperatures
3	Handling of shock sensitive powders	Handling the powders in situ in inert atmosphere Ensuring that all process lines such as filter to dryer are flushed every batch Higher authority control for permit to work on such systems Avoid use of agitators for such processes.
4	Handling low MIE powder formulations including drumming out	Static charge generation is a concern. Some preventive controls to be considered are: Ensuring that drop height of powder is less than 3 mtr. Use of earthing continuity stem such as earthrite system. Ensure personnel use antistatic body protection and shoes. Use of conductive FIBC or drums

Good housekeeping

All the above studies are based on Author's 10 years of experience in process hazard analysis for bulk chemical manufacturing and is expected to benefit the audience at large.

SPRING21 +17TH GCPS
A Joint AIChE and CCPS Meeting

BS Team: Finding Competence Gaps

Salvador Ávila
UFBA
avilasalva@gmail.com

Luiz Fernando Lopes de Souza
COPENOR
lflsouza@globo.com

Gunther Josua Costa
UNIVASF
guntherjosua@gmail.com

Lucas Menezes Pereira
UFBA
lucasmpereira13@hotmail.com

Prepared for Presentation at
American Institute of Chemical Engineers
2021 Spring Meeting and 17th Global Congress on Process Safety Virtual
April 18 - 22, 2021

BS Team: Finding Competence Gaps

Salvador Ávila
UFBA
avilasalva@gmail.com

Luiz Fernando Lopes de Souza
COPENOR
lflsouza@globo.com

Gunther Josua Costa
UNIVASF
guntherjosua@gmail.com

Lucas Menezes Pereira
UFBA
lucasmpereira13@hotmail.com

Keywords: BS, Competences gaps, Decision models

Abstract

The academy form competences to work in the routine, without worries, without unknown hazards. The rules and objectives in routine activities are clear and require operational discipline to control the cost of production and product quality. The current educational model does not prepare society and groups for an emergency or even for decisions in a crisis, where resources are scarce mainly because the necessary knowledge about danger and its mechanism of action is lacking.
What really happens in the chain event scenario when the danger is unknown? What are the typical models of action of the emergency leader and his team? What are the required actions in crisis management required for a BS type event considering the scarce resources, the high level of stress and the direction of the danger energy? Which models are indicated for the situation of unknown danger and its intense energy flow with high impacts during these events? What is the appropriate mind map of those who lead and execute the actions considering the low level of visibility in the events and the dynamic of geopolitics in the BS scenario where, depending on the preparation for emergency, in case of intense demand for attention and action, possible modifications may occur in the map mental for decisions?
This work intends to review the discussions already performed about new concepts and tools to be used in the crisis prevention area. Between this works we will understand the effect of the stress level on emergency decisions valuating human error and observing team response using LODA tool. The Human elements designed in the organizational culture, can stop the high danger energy that can flow from some industries. The principal aspects to be constructed and monitored are technology (risk and complexity); management (leadership and stress); and behavior (cooperation, commitment, competence, and communication).

By the other hand, human factors are analyzed to avoid situations which hazard energy carriers during top events, as the effect of cultural aspects to chain reaction until the occurrence of disaster. Decision models for the emergency brigade (leader and team) indicate the motor and cognitive gaps that cause lack of control during contingencies. Finally, the analysis of simulation types, by rotating observation of field simulation allow find the deficiencies of the emergency team in a real situation in the future.
The methodology divided in following steps are dynamic: first, rescue the discussions about stress and the BS events and, in second step, recognize the intense hazard energy flow through failure events, antecedents, elements and human factors. In third step, we try to locate the gaps and environmental conflicts influencing losses from accidents. Finally, we assess the current event of covid-19 pandemics to suggest correct emergency preparedness actions in the preparation of leaders and their teams regarding cognitive and motor characteristics in situations of high demand. When comparing the characteristics already described by TALEB about BS with the event COVID-19, we intend to carry out a case study to indicate the possible causes this global impact crisis, or even, understand the technological disaster causes where uncontrolled events from chemical, nuclear, and oil industry begins chain reactions until the disaster. Unfortunately, a Crisis situation can be caused from several intricated dimensions, making difficult the contingency and mitigation actions.

1 Introduction

The academy form competences to work in the routine, without worries, without unknown hazards. The rules and objectives in routine activities are clear and require operational discipline to control the cost of production and product quality. The current educational model does not prepare society and groups for an emergency or even for decisions in a crisis, where resources are scarce mainly because the necessary knowledge about the hazard and its mechanism of action is lacking.
The knowledge learned to graduate the engineer is limited to logic and objective concepts. The students do not discuss different concepts, as social culture, legislation, workstation behavior at the same time, mixing objective aspects with subjective ones. We need to understand Cultural traits, geopolitical aspects, intrinsic variables from worker during routine to discuss about response models for emergencies, necessary to avoid disasters.

Unfortunately, we only look for new competences after the disaster happen! Why?

The metrics for Process Safety management are discussed in methods indicated in the CCPS books [1]. Some important issues such as: management and investigation of major incidents [2]; technological changes that cause accidents [3]; lack preventive actions for organizational changes [4] are exemplified and barriers are suggested, so as not to affect the safety of processes with big accidents.
These issues and solutions seek to avoid the accident through management actions, technological systems, and the attempt to control human factors [5] based on the lessons learned by various industries and associations when investigating the events that occurred. This investigation of the past, looking back to the facts, has a limit when external, dynamic environments threaten normality. The crisis scenarios can be very dynamic, not visible and of parallel occurrence, at the same time as those caused by climate change, by social conflicts, by conflicts arising from

degraded geopolitics, and by the effects of multiculturalism in the world of work. and in society - indicating low tolerance behavior.

In 2018, CCPS [6] with the leadership of Eastman, raises the issue of normalization of deviation, indicating repeated situations of acceptance of operational deviations, in the shift routine, are considered normal situations. The lack of research with the indication of bottom-up techniques for the analysis of operational culture and the respective heuristics turns the book into another attempt to guide the occurrence of events looking to the past and without an appropriate view of what is happening in the present. and what will happen in the future, on the factory floor. What are the social emergent that cause deviations to normalize?

What are the social emergent that cause deviations to normalize?

In this context, analyzing the internal and external social threats to avoid environments prone to deviations gains importance. Risk aversion, excessive self-confidence, and fear of guilt for human error are considered cultural events that produce underreporting. This phenomenon changes the shape of Bird's pyramid [7] to the Christmas tree as shown in Figure 1 and causes the accident to occur already with the chain reaction triggered. The assessment and actions resulting from the notification (not performed) could avoid the domino effect or reduce the risk of major events.

The accumulation of deviations without treatment and with high level of underreporting, a cultural event, increases the likelihood of Black Swan events. Failure to recognize the Scenario for hazard identification and risk analysis that implies a lack of insight into the operational context and creates uncontrolled situation with natural, technological, political disasters. The Black Swan Event goes beyond the local scale and reaches a global scale, with unknown dangers, or already known dangers, but due to lack of attention, it has the characteristics of unknown danger because it has not been investigated as to the mechanisms of action. We do not know what the appropriate barriers are. Thus, we classify the Black Swan because of totally unknown or operationally unknown dangers.

Figure 1. Christmas Accident Tree

2 Literature Review

This work intends to review the discussions already performed about new concepts and tools in the crisis area. Between this works will understand the effect of the stress level on emergency decisions valuating human error and observing team response using LODA tool. The Human elements in the organizational design can stop the high hazard energy flow of some industries. The principal aspects to be constructed and monitored are technology (risk and complexity);

management (leadership and stress); and behavior (cooperation, commitment, competence, and communication).

By the other hand, human factors are analyzed to avoid situations in which hazard energy carriers during events, from cultural aspects to chain reaction until disaster. Decision models for the emergency brigade, leader, and team; indicate the motor and cognitive gaps that can result in lack of control during contingencies. Finally, the analysis of simulation types, by rotating observation of field simulation allow find the deficiencies of the emergency team.

Given the great stress present in the geopolitical situation, in which there is a constant struggle for territory, renewable and non-renewable energy sources, in addition to climate changes intensified by anthropic activity, large flows of Hazard Energy (HE) are generated that are commonly not perceived. The unnoticed concentration of this energy gives rise to what is known as latent hazard, and when it evolves to levels that can no longer control or filter its pending of burst and destruction, a megaevent of disaster occurs, which has recently been called the Black Swan.

Due to the lack of a notion of the hazard to which disaster victims are subjected, it is said that the Black Swan is still unknown to the population, and because they are often unprecedented occurrences, they are considered rare and of high impact in multiple respects. For [8], the victims of the twin tower attack on 9/11 were victims of a Black Swan. It was not possible to locate where the hazard energy was concentrated, nor measure it. Gray Swans, according to the same author, are unlikely but impactful events that are outside the normal scenario.

A Black Elephant, according to [9], is a significant threat, risk or issue that is obvious to everyone, but no one wants to address or deal with it, "they get ignored". When it causes high impact, people get surprised and act as if it were an unpredictable black swan.

There are new models of behavior to include high stress and ways to measure this feature through the radar chart. This chart presents the characteristic of presenting multivariable data in the form of a graph.

The managerial, human-social, and technological aspects can create an environment for unsafe behavior, in which they allow the flow of hazard energy across barriers. The human and organizational factors elaborate, by events, the condition for the transfer of the social hazard towards the physical realization of the accident or incident.

In order to prepare people to deal with risk situations and be able to participate in decision-making processes that report to their own lives, it is necessary investment in: 1) Integrated Communication - Training and language competence are really important. Subjective factors such as judgments, culture of guilt and fear of damage to the image cause the worker to omit information that could contribute to the construction of barriers to avoid incidents and accidents; 2) Stakeholders - person or group who have an interest in a particular business, industry or service; 3) Intellectual Capital - Knowledge of preventive and mitigating actions provides people with adequate decision-making; 4) Competence - Understanding the causal nexus of the phenomena involved in the accident, having experience and experience similar to the activities related to the events.

3 Models & Hypotheses

The discussion about the competence of Black Swan Team beginning answering questions after observing real situations of emergency. After we feel some gaps and try to answer after investigation about the reasons of emergency team perform this behavior. After each question we construct or explain the scenario and answer with new tools that will compound the methodology.

3.1. What really happens in the chain event scenario when the danger is unknown?

We have nothing to perform about the existence of a disaster from an unknown danger as in actual case of a pandemic. The disaster is a truth, we must reduce its impact, investigate the various dimensions involved with this unknown danger, or unknown operational danger considering the cultural and environmental effects during the isolation and flexibilization.

The study of the origin of the disaster allows us to understand its mechanism of action and possible consequences. The danger that causes the accident-disaster of technological (industry) or natural (climate, virus) origin can be acute, being a fast event with high impact, or it can be chronic, lasting a long time with, probably, immense impacts. Thus, the event becomes a public accident when it affects the population. We classify the danger and consequent events of COVID19 as natural, chronic, and public. This discussion is part of Strategic Intelligence that define the need resources and Hazard Identification, and Analysis of General Risk to define and test the best barriers.

While the danger is at work in an external region or country (not close to our region) we will observe and learn about its mechanisms, we seek to install two instances: The Integrated Management Committee and the Strategic Intelligence Committee. The first will join the necessary resources and leaders to take decision and the second will look for the best actions, benchmark, learning from other countries and regions.

In Integrated Management, Specialists, Managers and Practitioners who are part of Public-Private Institutions meet to define studies that assist in the decision on actions. Some tests are suggested to find the mechanisms of action of the hazard. In this way, barriers are built for known dangers, for the unknown and for threatening environments. Indicators and parameters and variables are mined to understand the impact of the pandemic on the economy and social relations.

Strategic Intelligence intends to identify the source of the main hazard and parallel hazards to project the Future Crisis Scenario. Scenarios are designed to test barriers, variables, resources, and knowledge. We will value to learn from the past, and from abroad experience, sending representatives to recognize the mechanism of these dangers through visits to government and institutions. In this moment a good general knowledge about pandemic situation, technological disaster and natural, impacts on social conflicts and economy, will help do draw an initial description and after, inviting specialists, the crisis leader tries to construct and perform effective actions (corrective, preventive and mitigation actions).

After confirmation that the Scenarios indicate an active Pandemic, the Contingency Committees begin in specific and critical regions. In some cases, such as natural disasters involving an earthquake, breaking of barriers in the mining industry and other technological accidents (fire or toxic spill) we need to install a Contingency Committees. In cases of acute accidents, groups are sent for mitigation action. Therefore, operations and local intervention teams will be installed.

The Black Swan Team at the Strategic (Strategic Intelligence), Tactical (Integrated Management) and Operational (Command and operations) levels need the classification of the type of event and impact for the work of the teams. One of the cases may be the study of technological accidents such as the breaking of barriers in mining or the leakage of toxic gases near the city (similar to Bhopal). The other case can be related to public events of high capillarity about impact, a virus that contaminates quickly and through unknown routes, the current COVID19, or even a series of earthquakes that promote tidal waves and affect the functioning of cities. Or even, the possibility of multiple events occurring involving the technological and the public dimension.

3.2. What are the typical models of action of the emergency leader and crisis management required for treat a black swan type event considering the resources, the level of stress and the direction of the danger energy?

LODA (PROGRESSIVE STRESS MULTI-LYER MODEL) shows that the mind map change in extremely high stress with wrong decisions.

Ávila [10], knowing about the possibility of high impact of technological accidents near cities and after confirming the risk to the risk affected by the work team and the population through underreporting. Ávila confirms the occurrence of the phenomenon of cultural normalization of deviation and for the analysis of the consequence of progressive stress, he designed an accident involving an LPG fireball near a city in Brazil.

This event constructed from real events and deviations that occurred, although the top event did not occur, was composed of 65 diversions and successive events, sociotechnical failures in the form of the local routine and complemented with those that occurred in New Mexico. The Accident prepared for the Risk Analysis impacted with 40 fatalities (this simulation was structured in the thesis and published as a tool [10].

The analysis model involved 5 levels of stress in the team, where 1st is uncontrolled process with operational deviations and the last was the crisis scenario in the next city. Progressive stress disperses or disorganizes the mind maps that affect the emergency leader's decisions. In this model it is considered the occurrence of a theater with critical roles where there are communication failures or non-acceptance of the roles in the staging. Observing each phase of the accident and its cognitive stress will indicate social phenomena that represent problems in the treatment of accidents. These problems must be addressed before the actual demand for contingency.

EMERGENCY PERFORMANCE MODEL indicated that there are characteristics for the leader and the team that are intuitive, affective and cognitive and that it was not found in emergency team in these accidents of the history.

The studies by Carvalho [11] and Ávila et al [12] indicate that there are specific characteristics that are appropriate for the leader and the emergency team on real demand. In the case of leadership, we can mention perception of real risks (signs and scenarios); competence regarding the danger mechanisms (fire, virus, mechanical impact); energizing the team to act in contingency; a good memory to reinforce the decision; and speed of command since the time of the chain event, until the greatest impact, is very fast, and evacuation measures can save many lives.

In the case of the emergency team, we can mention: the quick cognitive to retrieve documents, a discussion, a phone call and assist the leadership in the decision through opinion and calculations; quick rescue of memory for the analysis of the scenarios, this aid sometimes brings information avoiding the direction of the danger to where there is greater vulnerability, for example, sending of chemical product to empty the burnt tank where there are visitors or maintenance service - the leader does not knows the level of vulnerability of the areas and depends on the team's support for this analysis; emotional balance brings a better quality of the cognitive and motor apparatus; ability to predict future situations through greater knowledge of the event's phenomena; and the ability to act on time, without delay, bringing firmness to field interventions.

Thus, Carvalho [11] and Ávila et al [13] affirm that the preparation of the emergency team mixes past experiences, attitudes, field and memory exercises, team energization, which are often not considered in normal training for teams of emergency. The results of the team's performance must be observed so that there is continuous improvement in performance under real demand.

COGNITIVE GAP ASSESSMENT IN REAL ACCIDENTS AND SIMULATION where we assess the emergency teams based on public and technologic accidents confirming deficiencies indicated by Carvalho, Ávila e Pichon Riviere.

This research work analyzed the national and international accidents regarding the characteristics discussed in the models for emergency action [11] [12]. The emergency team was analyzed in Accidents: (a) Chemicals and Oil of high impact in number of fatalities and environment (piper alpha - oil; Horizon - Oil; Union Carbide Bhopal - isocyanate); (b) national accidents of shipwreck in Mar Grande and explosion with fatalities at the FPSO São Mateus in the Brazilian state of Espirito Santo; and accident designed in the form of a simulation for the analysis of fireball risk by LPG in northeastern Brazil.

Specialists working in Risk Analysis, Contingency and Emergency Leadership met to classify the teams in terms of cognitive gaps and emotional control. This analysis was published at LACPS in Buenos Aires and it was concluded, in discussions with Process Safety Managers, that the classic way of training for emergency brigade teams does not meet the need to control real demand. Why? Although the situation presented in the training simulations for the Emergency Brigade, the petrochemical, refining, and oil industries seek typical situations for fire, fire, leaks and environmental impact, there are always organizational factors that affect human factors, there are always social factors that they cause conflicts and affect the quality of communication at work, and there are always false indications, a high backlog in the industry today. Thus, events always happen in multiple dimensions and the possibility of something unknown happening is real. Thus, the greatest value to be given is cognitive-affective-intuitive training for the emergency brigade. Lately, Process Safety Managers work only in Cognitive and without surprises. The level of stress and the style of the Leadership bring a level of comfort that may not be satisfactory for the occurrences of real demands.

COGNITIVE FAILURE IN DISASTER FIELD SIMULATION where the team is not always integrated and energized in the hot simulation situation, believing only in the obligation to be present. In this case, preparation through the simulation leaves cognitive and compromising gaps for the contingency in the disaster.

At the CIDEM, Mass Disaster Congress, held in Feira de Santana, UFBA was a rotating observer and investigator during a simulated exercise involving 25 public agencies including Armed Forces (Navy, Army and Air Force), Federal Police, Municipal Guard and State Military Police, SAMU, State Fire Department in addition to the BOPE anti-terrorist squad. Various functions of civil defense, social defense, isolation, and medical care came together in this exercise to prevent events of bomb explosion on buses, tipping of benzene trucks, and kidnapping, actions promoted by terrorist groups, all with an impact on the population and in nature, in the road modal.

The observation of these events indicates that the hierarchy of the agencies is above the recognition of their roles during the simulation and the inter-agency communication. The communication made with noise leads the team to different levels of urgency for mind mapping in the different teams. Some equipment was underused indicating lack of training, therefore, there are human and organizational factors that must be put into discussion to improve the integrated actions of the emergency command. The main characteristic is the lack of common priorities in the agencies promoting a disintegrated management state, where the size of the problem is not known for sure, only when faced with it. We note mainly that communication tools and devices are not used properly. Because it is a cold simulation, prescribed events of common knowledge, many communication failures did not appear. And this is a serious problem when analyzing previous events when municipal, federal, state forces and private governance act around disaster mitigation, for example, by preventing ship collisions. We note that the lack of credit of the great emergency teams on the simulated exercise can induce people to repeat mechanical actions bringing nullity to the Strategic Intelligence team.

3.3. Which models are indicated for the situation of unknown danger and its intense energy flow with high impacts during these events?

SYSTEMIC FAILURE OR FINGERTIP FAILURE MODEL, where the failure is multidimensional and involves from emotional balance, and knowledge - quick cognitive, we must analyze the connection between these properties to invest in selection and training.

The systemic failure models that indicate connectivity points of events of different dimensions to cause the top event. The format of the Systemic failure is like the digital of the fault that was published at the European Reliability Congress (ESREL).

EXERGY OF OPERATIONAL FAILURE AND COMPLEX PROCESSES where the indicators of complexity in the construction of BS are analyzed.

The discussion on the Exercise of Operational Failure [14] presenting that cognitive losses in the task are related to energy losses in equipment and processes. The presentation of The Investigation on Complex Processes through the tools that connect different sources that cause uncontrol in the process, interpreted through: Connectivity Diagram [15] [16]; of the Materialization Diagram where the failures that transform communication noises from management into a disconnected mental map in execution, are transferred into disturbance in the physical dimension; and finally the logical diagram of the fault indicating regions where to better install the barriers, in sociotechnical functions. These are models discussed at the Brazilian Congress of Chemical Engineering and applied in the Chemical Process Industry (aliphatic and aromatic amines in addition to liquefied petroleum gas).

NETWORK OF FACTORS AND HUMAN ELEMENTS, BARRIER EFFICIENCY, AND DANGER ENERGY SATURATION where we need to identify the points where danger energy enters saturation to elaborate the best barriers.

In recent times we have discussed ways to identify the regions in which the greatest flow of danger energy occurs, following the Model of Reason's Swiss cheese [17]. This danger flow circulates among the factors that influence human performance on the sequential, parallel and crossflow route. This multidimensionality of the failure follows the model in a network format of human and organizational factors divided into culture, management, routine and accident. This multidimensionality of the failure also discusses the safeguards of the technological, managerial, and behavioral dimension through the human elements that are design barriers to contain the danger energy that flows until reaching the maximum levels that are the accident and the disaster. Human factors are part of the routine of the operation and human elements are design safeguards. As described in Table 1, in the operation there are different factors that cause incidents and situations of loss. The nature of this event has the following dimensions: Cultural; Managerial; Routine control; and the scale of the accident and disaster. Also, in the same table we present the types of human elements in the project: Management; Technology; and Behavior.

Table 1 – Human Factors and Elements

OPERATION - A. CULTURAL DIMENSION OF HUMAN FACTOR: (1) Social Culture - (2) Social Phenomenon - (3) Organizational and Safety Culture. B. MANAGEMENT: (4) Management - (5) People & Groups. C. DIMENSION OF CONTROL IN THE ROUTINE: (6) Operational Control - (7) Failure Control D. ACCIDENT AND DISASTER DIMENSION: (8) Accident Control - (9) Disaster Control.	**DESIGN** - E. HUMAN MANAGEMENT ELEMENT (1) Stress - (2) Leadership F. TECHNOLOGY (3) Risk - (4) Complexity G. BEHAVIOR (5) Communication - (6) Competence (7) Cooperation - (8) Commitment

4 Methodology

The investigation was carried out following the steps: (1) Recovery from the discussion about stress in the treatment of emergencies, especially in the case of BS type disasters; (2) Recognition of the intense flow of danger energy through antecedent events, human safeguard elements, human factors of industrial operation and consequent events that hinder the return to normality; (3) to locate the environmental gaps and conflicts that influence process losses and the occurrence of accidents, chain reactions and the difficulties of returning to normality, for this we divided the competences in: Strategic Intelligence, Integrated Management, Command and action in Local Operations, and Organizational Resilience. For each competence, possible gaps, deviations, or human errors are discussed, comparing with the expected standard.
Finally, we assess the current event of covid-19 pandemics to suggest correct emergency preparedness actions in the preparation of leaders and their teams regarding cognitive and motor characteristics in situations of high demand.
Here in the methodology, we will indicate the competencies required for the BS team and the mental map of each competency for further discussion on the case of the COVID19 disaster.

4.1. Competencies for BS

As shown in Figure 2, Strategic Intelligence and Organizational Resilience have the purpose, respectively, to identify the danger and to restore the organization to the condition of normal, controlled operation, as it was before the disaster. Therefore, they outline strategies based on investigations of the past, the present, and the future, in addition to comparing local situations with external ones and in other countries. It is important to identify what the danger is and its mechanism of action, map the impacts to project the recovery. The Intelligence group is composed of risk analysts who work with scenarios of different dimensions (health, politics, technology, climate) and who can defragment dispersed information to build Future Scenarios and explain the events, even if the danger is still unknown, through testing and observation.
Another group that will guarantee the application of resources in the right place and that will effectively communicate the current situation involving the population and cities, industry, commerce, and services, is the integrated management group that will develop appropriate indicators for decision making from the disaster impact curve and from the data and information generated by strategic intelligence.
The local operations and command group works to carry out the activities indicated by the integrated and recommended management group, guided by strategic intelligence. The role of the

leader is important to energize the team and the role of the team providing cognitive and motor support in this achievement.
In this way, the Academy, Public Agencies and Risk activities such as the chemical industry and the oil industry need to structure themselves to prepare competencies that know how to act in the investigation of scenarios, in integrated public-private crisis management, in command and in carrying out operations. contingency and the recovery of local and regional activities.
Cold simulations [18] do not bring the expected result to structure contingency teams. Crisis Managers have not been effective in the application of resources, in communication and in the strategies drawn from the decision tools, as presented in the pandemic of COVID19 [19]. The risk management team is not able to design the scenario because it does not identify the danger and its mechanisms and impacts. We have no skills prepared for Organizational Resilience and we lose control of the situation during the emergency period. Thus, we need to develop the skills mentioned in Figure 2 and confirmed from the team's mind maps.
Mind maps are average informal rules accepted by the group after all the list of organizational priorities, group preferences, and regionalities. This map will then show how the team works in each of the four blocks of competence and how the teams should work.
Note that the models do not characterize the relationship of urgency of time, recognition of danger and its mechanisms, also the importance of the discussion on the types of communication (between groups, with the community, with the population or inserted in a political environment) and care in relation to false-incomplete or anticipated information.
The final analysis is a brief discussion of the mind map for BS events in general and for the specific case of the COVID 19 Pandemic.

5 Case Study

The Research Group on Operational and Dynamic Risks (GRODIN) has prepared a Master Program for Disaster Preparedness and Crisis Management and has been acting as a critical observer of Government actions and people's reactions to address COVID19, discussing tools in the Strategic, Managerial and of Operations.
Unfortunately, new strains of this virus are emerging, in a transformation of unknown origin, which shows once again that we are during a Black Swan Event. We must deal with this matter coldly to act in the face of the terror of mass death, and to avoid the wrong decisions. The skills presented in Figure 2 must be developed and available to act in this Global Crisis:

(1) IE Strategic Intelligence (study of the origin of the danger, its mechanisms, and the construction of future scenarios, where spurious data and warnings about changing the danger line must be valued, investigating the past to understand the present, investigating the future for disaster preparedness.

(2) Integrated Management and Crisis Project - GIP (data flow, organization of resources, development of tools for decision appropriate to the mechanisms of action of the danger and its dynamism when the danger is unknown, communication mode and the treatment of the sub- political-economic and technological notification).

(3) Command of local operations COL with the preparation of resources for the here and now based on the organization of the Integrated Crisis Management in the projections of the future created by the IE (people, structures), reviewing the Integrated Management and the IE with new data from the present coming from the COL, having a preparation of the leadership and the team to avoid simulating extreme stress.

(4) Organizational Resilience RO where, with the control of the situation, seek the causal nexus based on practice and science in order not to lose control of the recovery. Always value spurious data and information and warnings, classifying the quality of the information. Motivate IE for further investigations about the danger, its mechanisms of action and possible impacts before it happens.

Taleb [8] stated that in the face of an unknown danger, "everyone" will try to explain without knowing for certain what he is talking about (we need IE), demonstrating dominance that does not exist. Others who are comfortable with being a subject of scientific knowledge, virology, are not concerned with the preparation for situations not lived in the past, so we consider that unknown danger = known local non-operational danger. Thus, with chronic and growing impacts, we will have uncontrolled scenarios due to lack of resources, infrastructure, and lack of competent people for action.

The Black Swan has several dimensions that occur at the same time, Health, Economy, Politics, Logistics, Climate and Geopolitics, regardless of the dimension of origin of the danger. This multidimensionality gets in the way of decisions since the leaders of knowledge, public and private management, do not have the habit of thinking, investigating and deciding considering the systemic failure and its multidimensionality.

The low quality of information, underreporting and / or addictive data, undermines the role of management and science, which creates addictive or incorrect interpretations following mathematical models that are not practical. Observation of facts and events must be valued in times of Crisis (Disaster) to enable the endless elaboration of multidimensional scenarios to dynamize the Strategies.

In times of crisis, it is necessary to work with two teams that feed and feed each other on information. The IE and RO must be calm and when analyzing fatality data, review future actions and investment of restricted resources. The GIP and COL that see in the here and now practical needs to control the mitigation teams towards operational control and human errors caused by extreme stress (EE).

The IE and RO must be calm in the construction of Scenarios with weekly average of 4-day moving averages of fatalities to assess the effectiveness and effectiveness of previous actions, seek to explain or approach the truth about the danger. IE and RO's lack of calm in analyzing the scenario can vitiate interpretations in a single dimension of the disaster leading to incorrect projections of the future. It is necessary to change the position of the gaze to see what is hidden. Thus, we have developed guidelines to be followed:

(1) Installation of Crisis Committees for IE, GIC, COL and RO with a multidisciplinary and dynamic team; (2) Choosing-developing easy-to-use and effective statistical tools to support decisions, therefore, the weekly average of the 4-day moving average tries to avoid underreporting and spurious data that does not indicate a trend; (3) Enhance research on spurious data or information or warnings that may indicate a change in the danger or its mechanism and, consequently, its impact; (4) Prepare the team to maintain the quality of the cognitive and also adapt to the "time available" to guarantee the effectiveness of the actions. Extreme stress increases human error from 10% to 70%, bringing incorrect decisions and actions. We have to use tools to preserve the cognitive in extreme stress !; (5) Build a knowledge network platform, develop practical tools based on grounded theory, information flow, integration, and critical observation; (6) Following the film World War Z, it is important to treat spurious data and observe-validate unconventional local solutions that can be the difference in the application of restricted resources; (7) Following the UTOPIA Series, recognizing the political, geopolitical movements that have low visibility, can indicate simulacra of information, fake news. When knowing the fake news and underreporting, the safeguards will be more assertive, the aim is to test the hypotheses to protect the population; (8) Even if the situation is more optimistic, maintain recourse to pessimistic

scenarios, considering the opinion of people who are reliable but who project a different future from the majority as in the film World War Z.
I forgot the following aspect about the Leader to avoid Accident-Disaster Situations discussed by Collins [20]: (1) Not wanting the power and nor having vanities of signatures; (2) Extreme humility in dealing with unknown hazards; (3) Active listening for events, warnings and deviations; (4) Fierce determination in decisions and actions; (5) Recognize the possibility of own and team mistakes in strategy and action; (6) valuing the organizational and family unit.

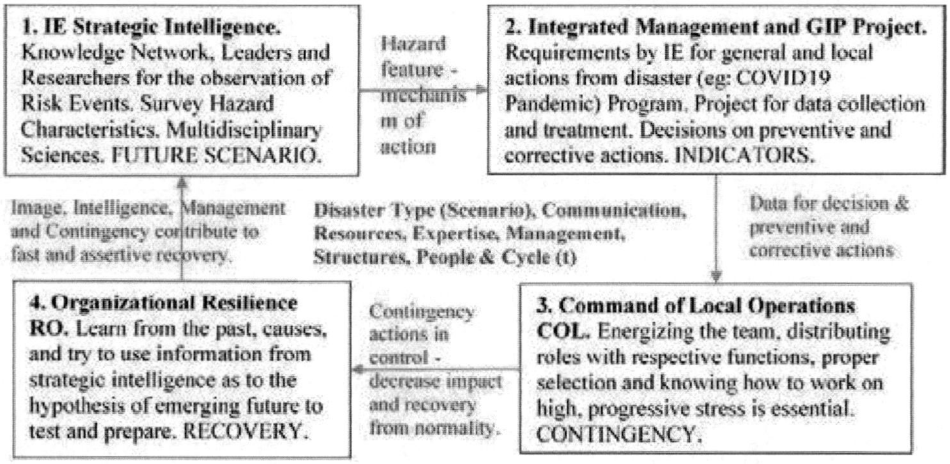

Figure 2. Competencies for the BS Team: Intelligence, Management and Operation.

5.1. Mind Map for BS and COVID19

The characteristics already described by TALEB and Avila using the models from research about the dynamic risk we show the principal competences to attend a BS event. In Table 2, we discuss about the generic pattern and a case of COVID19 BS team.
The informal rules that turn into actions follow mind maps by type of function, Intelligence, Management and Operation. Below are examples of expected patterns and deviations observed in different mind maps.

Table 2. Mind Map for BS

Mind Map	Generic standard	Expected standard (COVID19)	Deviations (GENERIC)
Intelligence	- To study the accidents and incidents;	- Study the disease in depth knowing the form of contagion, consequences, forms of treatment;	- Study superficially without exploring for learning;
	- In the crisis, seek as much information about the products involved, the team exposed, the area affected;	- In the crisis, seek as much information as possible about the disease, exposed teams, affected areas;	- Search for a solution without knowing what we are facing, people involved;
	- Know the risks that are relevant to the business;	- Know the risks that are relevant to the disease;	- Not knowing the extent of the risks and not preparing for it;
M an	- Know the resources available to face the crisis;	- Know in depth all the resources available for coping with the disease;	- Have resources that are not available.

			- Wait to seek resources in crise;
	- Treat accidents and incidents with preventive and corrective improvement actions;	- Predict and seek resources to strengthen the fight against the disease with the creation of scenarios;	- Treat accidents and incidents superficially without seeking a solution;
		- Treat deviations with preventive and corrective improvement actions;	- Treat deviations superficially without seeking a solution;
Operations	- Know and study their role in acting during the crisis;	- Know and study their role in acting during the pandemic crisis;	- Facing the crisis in a way that is not consistent with the procedures causing conflict with the team;
	- Communicate any changes that are made to the procedures by training the others involved;	- Communicate any changes that are made to the procedures by training the others involved;	- Carrying out activities in a manner inconsistent with the procedures, putting the activity at risk;

6 Conclusion

The current educational model prepares society for routine operations only, always with a focus on the main objective of producing with quality and low cost. The safety of the process is studied in a superficial way, focusing on the clearest and expected risks. BS are situations that arise during an event resulting from the sum of risks that may have been studied separately but never together. This moment requires several skills from the teams that work on the front line and coping and mastering the situation can take more time and we know that in an emergency, time and knowledge are fundamental. The intention of this work was to present a review and discussion about new concepts and tools that can be used in the prevention of human errors during crises (LODA) and thus we were able to understand the effects that the level of stress can influence in the decision making during the crisis and compromise coping with the crisis based on the understanding that the mental map changes completely in the crisis.

The presentation of the methodology with the necessary skills to face the BS and the mental map models are fundamental to create the PDCA of the crisis that always starts with strategic intelligence, integrated management, Operations command, and the organizational resilience that will feed the strategic intelligence in a learning cycle and the search for continuous improvement. The more we study this methodology, we see that although the event is unexpected, the knowledge and the feedback of information for the knowledge of the risks, will make that the mastery of the situation is reached more quickly.

The academy needs to better prepare society for unexpected situations. BS events are becoming more common, given the COVID19 pandemic experienced today and methodologies applied at the time of the crisis can save lives.

7 References

[1] T. Overton, M. Broadribb, C.N. Cheung, E. Christie, S. Cowher et al. "Guidelines for Process Safety Metrics", pp. 1-178, 2009.

[2] Center for Chemical Process Safety (CCPS). "Guidelines for Investigating Process Safety Incidents", Third Edition, 2019.

[3] Center for Chemical Process Safety (CCPS). "Guidelines for the Management of Change for Process Safety", 2010.

[4] Center for Chemical Process Safety (CCPS). "Guidelines for Managing Process Safety Risks During Organizational Change", 2013.

[5] Center for Chemical Process Safety (CCPS). "Human Factors Methods for Improving Performance in the Process Industries", December 2006, 248 p.

[6] Center for Chemical Process Safety (CCPS). "Recognizing and Responding to Normalization of Deviance", September 2018, 176 p.

[7] Herbert W. Heinrich, Industrial Accident Prevention: A Scientific Approach, 1931

[8] N. Taleb, "The Black Swan: Second Edition: The Impact of the Highly Improbable". Random House Trade Paperbacks, 2010, 469 p.

[9] P. Ho, "Future Beasts and Where to Find Them". Available at https://www.psd.gov.sg/challenge/ideas/feature/future-beasts-and-where-to-find-them. Accessed on January 22, 2021.

[10] S.F. Ávila. "Dependent Layer of Operation Decision Analyzes (LODA) to calculate human factor, a simulated case with PLG event", 2011.

[11] A.C.F. Carvalho. "Diagnóstico de lacunas nas competências da equipe para enfrentamento de situações fora da rotina". Master's dissertation – Federal University of Bahia, Polytechnic School of Engineering. Salvador, 2016, 103 p.

[12] S.F. Ávila., B. Mrugalska, M. Wyrwicka, M. Souza, J. Ávila, et al. "Cognitive and Organizational Criteria for Workstation Design", 2019.

[13] S.F. Ávila., A.C.F. Carvalho, G.P.J. Portela, C. Costa. "Learning Environment to Take Operational Decision in Emergency Situation". Procedia Manufacturing. vol. 3, pp. 1780-1787, 2015.

[14] S. F. Ávila. "A exergia da falha operacional, uma discussão filosófica com bases práticas". In: Proceedings of 19th Brazilian Congress of Chemical Engineering, 2012, Buzios.

[15] S.F. Ávila. "Análise de falha em processos complexos". In: Proceedings of 19th Brazilian Congress of Chemical Engineering, 2012, Buzios.

[16] C. Perrow. "Normal accidents: living with high-risk technologies". Princeton University Press, Princeton, N.J., 1999.

[17] J. T. Reason. "Human error". Cambridge University Press, 1990.

[18] S. F. Ávila., I. Cerqueira, V. Júnior, J. Nascimento, C. Peres, C. Brunoro, P. Arezes. . Integrated Management in Disaster: Discussion of Competences in a Real Simulation. AHFE 2019.

[19] S. F. Ávila., J.S. Ávila, A. A. Santos, M. C. N. Carmo, L. F. L. Souza. Crisis Decision Tool, fighting against COVID-19. AHFE 2021.

[20] J. Collins. Level 5 Leadership: The Triumph of Humility and Fierce Resolve. Harvard Business. 2005.

619165 Case Study in M+a – Communication and Understanding Hazards and Safety Concepts

Abstract:

A multinational corporation acquired a local consumer products manufacturing facility. Hazard awareness and process safety protocols were lacking at the local company, especially as measured against the parent company's corporate standards. The primary root cause was lack of a strong management of change procedures, as the processes in the facility were originally designed for non-flammable products but over time the facility transitioned to handling flammable liquids and combustible dusts.

Hundreds of unique chemicals/products stored and handled at the facility were analyzed to determine compliance with local fire codes and serve as a design basis for safety program improvements. Specific challenges around flammable liquids and combustible dusts were deemed to be highest risk and were addressed immediately.

This case study presents high level findings and both capital and operational budget related recommendations applicable to companies in similar situations. By applying a combination fire/building code evaluation and hazardous materials / process safety gap analysis, the company was able to prioritize risk reduction actions in alignment with their strategic objectives. This work presents a basis for hazardous material safety in facilities that store and handle a wide range of hazardous materials and aligning risk remediation activities with corporate objectives.

618542 Is Your Plant at Risk from Climate Change?

Abstract:

Climate change is occurring and some of the changes will impact the safety of process plants. This paper summarizes potential adverse impacts on process plants and then outlines how to conduct a Climate Risk Vulnerability Assessment (CRVA) for a process plant. One product of the CRVA is a plan of action to eliminate or reduce adverse climate effects on the process plant.

619167 Is Emergency Response for Facilities Impacted By Climate Change?

Abstract:

Climate change essentially bring about alteration to our natural environment. This can be seen by way of changes in the weather pattern. We see this here in the Gulf region of the US, almost every year during our most infamous hurricane/storm season.

Emergency response plans and preparedness during this season is always critical. It is based on science, best practices and lessons learned. At these times, not only is the shutdown process very critical , however the start up and recovery measures are also critical..

This paper will discuss thoughts and views of the author of changes required to emergency response and preparedness plans that facilities have to account for, due to changes in the weather pattern. The author will also discuss how the risks to people and process during the recovery measures can best be minimized. Irrespective of the conditions and planning, the human lives must be preserved.

619069 Transitioning from One Explosive Fuel to the Next – Safety Considerations in the Emerging Hydrogen Economy

Abstract:

Since the turn of the century, there has been a focus on the development of the hydrogen economy, driven by the desire to reduce greenhouse gas emissions through utilization of alternative energies and cleaner fuels. With both the current and future interest in utilizing hydrogen, it is crucial to evaluate and mitigate the fire and explosion risks posed by hydrogen during storage and transfer. While many other fuels such as natural gas, propane, LNG, and petroleum also present flammability and explosive hazards, hydrogen has several unique characteristics. These include, but are not necessarily limited to, a relatively low ignition energy, low density, and wide flammability limits. These properties present unique challenges for risk management compared to other flammable liquids and gases that will need to be addressed as the hydrogen economy further expands into its emerging market.

In this paper, the properties of hydrogen and how they impact the fire and explosion hazard compared to other fuels will be analyzed. Then, best practices and risk management will be discussed for storage and transferring fuels, with a focus on highlighting differences that are important considerations for hydrogen. Finally, a case study will be presented that reviews an incident involving a hydrogen leak and ensuing jet fire that ignited during the removal of a portable tube trailer. The events that led to the unintended release and ignition of pressurized hydrogen, and the lessons learned from the incident will be discussed. The case study will highlight several aspects related to the safe storage and transfer of hydrogen in commercial applications.

619275 Application of Causal Analysis Based on System Theory (CAST) to the T2 Laboratories Incident

Abstract:

Application of Causal Analysis based on System Theory (CAST) to the T2 Laboratories incident

Joshiba Ariamuthu Venkidasalapathy[a,b,*]

[a]Artie McFerrin Department of Chemical Engineering, Texas A&M University, College Station, TX 77840, USA

[b]Mary Kay O Connor Process Safety Center, Texas A&M University, College Station, TX 77840, USA

**Presenter email: joshvenkat03@tamu.edu*

Abstract

STAMP (Systems-Theoretic Accident Model and Processes) is an accident causalty model that is based on systems thinking and systems theory. STAMP can be applied for incident investigations through the methodology called causal analysis based on system theory (CAST)[1]. It is a socio-techinial causal analysis method that systematically helps to identify causes that led to the accident. Unlike traditional accident models that rely on linear causation effects such as swiss-cheese model and domino effect models, CAST adopts a control-centric approach that helps account for causes that arise due to nonlinear interactions. Some unique features of CAST include treating safety as an emergent property, and providing the scope to account for software interations and temporal aspects of control actions. However, it is seen that applications of CAST to chemical engineering systems is scarce so far[2-4]. Considering the increasing complexity of chemical processes, accident models based on simple linear causation effects may not be sufficient in identifying the root cause of an incident.

CAST methodology is implemented in five steps: assembling the basic information, modeling the safety control structure, analysing the control structure to identify failed control elements, analysing the systemic factors and control structure flaws and providing recommendations to improve process safety. The second step, which is to model the safety control structure, is the essence of the CAST methodology. The control structure is heirarchical and requires every control block to have well defined functional objectives and control elements to achieve its goals.

In this paper, the objective is to apply CAST to a chemical incident to identify the social and technical causes that led to a chemical indident. The T2 laboratories explosion that occurred in Florida, 2007 is chosen to illustrate the CAST methodology[5]. The explosion led to four fatalities and several injuries. The CAST methodology along with extensive simulation based on MATLAB environment led to identifying both social and technical proximate causes. Most importantly, the results showed that temporal aspects of control actions, such as, how long the action was applied or how early/late it was applied also played an important role in the T2 Laboratories incident.

References

1. Thomas, J. and C. Malmquist, *Learning from Accidents that are a Consequence of Complex Systems.* Presented at the 48th Annual International Society of Air Safety Investigators, 2017.
2. Elliott, J., *CAST analysis of the Buncefield incident.* Procedia Engineering, 2017. **179**: p. 23-33.
3. Leveson, N.G., *CAST Analysis of the Shell Moerdijk Accident.* 2016, MIT. Cambridge: EU Major Accident Hazards Bureau.
4. Muñoz, M.F.T., *The Underestimated Value of Safety in Achieving Organizational Goals: Cast Analysis Off the Macondo Accident.* 2017, Massachusetts Institute of Technology, System Design and Management Program.

5. CSB, *T2 Laboratories Inc. Reactive Chemical Explosion.* 2009.

SPRING21
+17TH GCPS
A Joint AIChE and CCPS Meeting

Process Safety Manuscript Archives Available to the Practitioner

Raymond "Randy" Freeman
S&PP consulting
12303 Lake Shore Ridge
Houston, TX 77041
Rafree@Yahoo.com (for correspondence)

Ronald J. Willey, PE, PhD
Professor and Associate Chair Chemical Engineering
Northeastern University
360 Huntington Ave. 201 CN
Boston, MA 02115
R.Willey@northeastern.edu (for correspondence).

Prepared for Presentation at
American Institute of Chemical Engineers
2020 Spring Meeting and 17th Global Congress on Process Safety
Virtual
April 18 - 22, 2021

Process Safety Manuscript Archives Available to the Practitioner

Raymond "Randy" Freeman
S&PP consulting
12303 Lake Shore Ridge
Houston, TX 77041
Rafree@Yahoo.com (for correspondence)

Ronald J. Willey, PE, PhD
Professor and Associate Chair Chemical Engineering
Northeastern University
360 Huntington Ave. 201 CN
Boston, MA 02115
R.Willey@northeastern.edu (for correspondence).

ABSTRACT

Since 1967, the American Institute of Chemical Engineers (AIChE) has sponsored a series of safety symposia. The goal of these symposia is to reduce the frequency and severity of accidents in the chemical process industries (CPI). The Safety and Health Division of AIChE was established to serve as a central hub for collection and dissemination of this process safety information. Over the years, this dissemination expanded to several process safety symposia tracks at the Global Congress on Process Safety. Three symposia manuscripts have been assembled and archived into a data base accessible on the Internet through the AIChE Academy.

- Loss Prevention Symposium (LPS)
- Process Plant Safety Symposium (PPSS)
- Center for Chemical Process Safety (CCPS) Symposium

The technical papers presented at these symposia cover a wide range of topics such as fire protection, risk reduction, explosion analysis, plant maintenance practices, process safety management and many other topics.

A total of 2,524 papers were scanned and organized into a form which can be used to search for information on process safety topics. A spreadsheet of the contents of the archives was created. This paper describes the organization of the Process Safety Archives and presents an example of how the archives can be used to find information.

1. BACKGROUND

As the English philosopher George Santayana[1] (1905) eloquently stated:

"*Those who cannot remember the past are condemned to repeat it*"

Trevor Kletz[2] (1993) restated the importance of the failure to remember the past in his book entitled: *Lessons from Disaster: How Organizations have no memory and accidents recur.* Only people have memory. Unfortunately, our memories are very poor. Therefore, to preserve knowledge two conditions must be met:

- It must be written down
- It must be findable.

The ability to find information is very important. It does very little good to state that the needle you must rapidly find is located in one of many haystacks in the back of 40 acres of a field. Likewise stating that a document of interest may be in the Library of Congress or the National Archives achieves little in actually finding the document of interest. For information to be useful, it must be timely and rapidly findable.

Previously, the Safety and Health Division sponsored projects to scan and digitize technical papers presented at various meetings. The papers were loaded into a Folio database and made available to the Safety and Health Division membership. Several items are of interest:

- The last update of the Folio database was in 1995.
- The process safety databases (versions 1, 2, 3) use a proprietary FOLIO format and did not include PPSS.
- Better scan technology today than was available 25 years ago.

Why this is important and why is it useful? Many safety research programs were only presented at Loss Prevention Symposia and were never formally published. These include accident investigations, chemical stability tests, hazardous material property test results and other information. Such research reports may only be available in the printed conference proceedings. Rapid access to these reports and papers may improve designs and prevent future process safety incidents.

2. LOSS PREVENTION SYMPOSIUM (LPS)

2.1 History of LPS

The first LPS was organized and held in Houston, TX in February of 1967. A total of 23 technical papers were given on topics such as:

- Building Design Criteria
- Equipment Design
- Reactive Chemicals
- Process Safety Management

The final session of the first LPS was devoted to a review of six process incidents.

Since the first 1967 symposium, the Safety and Health Division has sponsored an annual LPS. More details on the history of the LPS may be found in papers by Davenport[3] (2006) and Freeman[4] (2016).

2.2 Number of LPS papers

In the time period from 1967 to 2005, a total of 1,324 technical papers have been presented and are included in the Process Safety Archives. The papers include many summaries of process accidents, as well as discussions of such topics as fire protection, safety instrumented system analysis, dispersion analysis, and nitrogen inerting to name but a few.

3. CENTER FOR CHEMICAL PROCESS SAFETY (CCPS)

3.1 History of CCPS

In the second half of 1984, several incidents occurred which had international impact:

- In October 1984 a lubricating oil tanker caught fire and exploded immediately after leaving San Francisco. One sailor died and a large oil spill resulted.
- In November 1984 a large fire and explosion occurred in a PEMEX storage and distribution center in Mexico City. Between 500 and 600 people were killed and houses were damaged to a distance of over 5 miles.
- In Early December 1984, water entered a storage tank of methyl isocynate (MIC) at a Union Carbide plant site in the Indian town of Bpohal. The resulting reaction of MIC and water caused in a runaway exothermic reaction. A large cloud of toxic material (MIC and others) was released killing at least 2,000 people and injuring 10s of thousands of people.

As a result of these disasters, 16 U.S. chemical companies approached AIChE with a request to help improve the process safety practices in the chemical industry. This request was the start of CCPS. Details on the early history of CCPS may be found in the paper by Berger[5] (2009).

3.2 Topics and intent of CCPS Symposia

The first CCPS conference was held in 1987. The conference was focused on the following areas:

- Hazard Identification
- Storage and Transfer of Hazardous Chemicals
- Reactive Chemicals
- Post-Release Mitigation
- Human Factors
- Vapor Cloud Modeling

All of these topics are of direct relevance to the Bhopal Incident.

3.3 Number of papers

In the time period of 1987 to 2006, a total of 736 technical papers were presented at the annual CCPS conference. These technical papers are focused on specific problems and included detailed reviews of lessons learned from accidents.

4. PROCESS PLANT SAFETY SYMPOSIUM (PPSS)

4.1 History of PPSS

In 1989, there were two major process incidents in the Houston, TX area.

- A fire/explosion at the Phillips Pasadena plant killed 23 workers
- A fire/explosion at the Arco Channelview plant killed 17 workers

In 1992, the South Texas Section of AIChE organized a technical symposium on process safety. The symposium was held in Houston, TX to minimize the travel to attend the meeting and to encourage engineers and managers from the Houston area chemical and process industries to attend. The first PPSS was a great success with attendance by over 700 people.

4.2 Topics and intent of PPSS Symposia

The focus of the PPSS is on plant operations. PPSS papers are focused on practical application of process safety programs. More detailed technical material is found in LPS and CCPS symposia series. Papers on the management and control of process safety programs are a popular topic. Details on the history of the PPSS may be found in the Congress paper by Edwards[6] (2012). Edwards summarizes the focus of the PPSS as:

"*Practical Application of Process Safety Management at the Plant Level: Ensuring the Protection of People, the Environment, Capital and Business*."

4.3 Number of papers

In the time period of 1992 to 2005, a total of 515 technical papers were presented at PPSS conferences. These technical papers are focused on specific problems and included detailed reviews of lessons learned from accidents.

5. CONTENTS OF PROCESS SAFETY ARCHIVES

Contents of the process safety archive include the following collections:

- Loss Prevention Symposium manuscripts from 1967 to 1980, originally Published annually as "Loss Prevention Symposium, Vol XX" by AIChE, New York, New

York, where XX went from 1 through 14. 326 titles across 14 volumes are in the archives.
In 1981, Loss Prevention Symposium manuscript numbering changed over to a AIChE session number, alpha character (e.g. 60b), mode of numbering and the AIChE discontinued the annual compilation of manuscripts into a single issue. Fortunately, John Davenport and others collected these papers and when this archive was announced, shared their personal collections with one of the authors where these were scanned in. In 1982, selected papers from the 1981 session forward began to appear in a new AIChE publication, initially called *Plant/Operations Progress*. In 1993 *Plant/Operations Progress* was renamed to *Process Safety Progress*. This collection of LPS manuscripts may also include both the *Plant/Operations Progress* version and the original manuscript delivered at the conference. Approximately 397 manuscripts and publications from 1981 to 1989 have been archived.

In 1990, AIChE moved to posting manuscripts onto digital storage media, and all manuscripts have been extracted from various digital media that existed at the time for the years of 1990 to 2005. Approximately 601 manuscripts exist in this part of the collection. After 2005, AIChE retained manuscripts from various Global of Process Safety on its web site AIChE Academy, and these can be accessed by going to the appropriate year and congress.

- The Center for Chemical Process Safety began congresses in 1987 and all manuscripts from 1987 to 2006 have been found, scanned, and archived in this database. Approximately, 736 manuscripts make up this part of the archive. Manuscripts after 2006 are available via the AIChE Academy.
- The final collection of manuscripts continues with Process Plant Safety Symposium, beginning 1992 and continued to 2005. 515 manuscripts make up this part of the archive.

Manuscripts published after 2005 are available on at www.aiche.org under this web link by year: https://www.aiche.org/conferences/aiche-spring-meeting-and-global-congress-on-process-safety

6. INTERNET ACCESS OF PROCESS SAFETY ARCHIVES

6.1 Overview of Access Protocol

Accessing the Process Safety Archives involves two major steps:

- Determine what you want to look at. This is accomplished by using the Master Index to the Process Safety Archives. The Master Index (Revision 0) is presented as supplementary material to this paper in the form of a Microsoft Excel spreadsheet. Details on how to use the Master Index are presented in section 7 of this paper,
- Once you know which volume of the LPS, PPSS or CCPS that the paper of interest is located, you must purchase the volume. The purchase price for a member of the Safety and Health Division is $0.00 (FREE). Purchase of the volume is via the AIChE transaction portal. The following section of this paper demonstrates the process for obtaining a paper from the Process Safety Archives.

6.2 Detailed Description of Method to Access Process Safety Archives

Important note: this example contains working links that if you are a member of the safety and health division, should take you to the images shown.

The process safety archives can be accessed at AIChE thru this web link: https://www.aiche.org/ccps/resources/safety-health-proceedings.

Once at the web site clicks "sign in" if you are a member of safety and health division (see Figure 1). The Home Page of CCPS will appear.

Re-click this web link: https://www.aiche.org/ccps/resources/safety-health-proceedings. to return to the archives.

This part of the archive is organized sequentially by year from 1976 to 2005. There are 7 subpages. Within each year is a link to the AIChE academy where the papers are actually stored. This approach requires the user to know what year and what conference the paper of interest exists. The easiest way to find out what to download is to use the Master Index as presented in the Supplementary Materials. For example, clicking on page 6 (bottom). Then scrolling, you will find two conferences as shown in Figure 3.

Click "READ MORE" for CCPS 17th conference. You reach this page (Figure 4).

Then you click TECHNICAL PAPERS and Figure 5 will appear.

Click the first title, Figure 6 will appear.

Click Checkout (this link is unique to the user and thus is not included).

This page appears, Figure 7.

Click Process my order. Figure 8 Will appear.

Next click CONTINUE and the following page appears (Figure 9).

You need to re-navigate after paying or receiving permission to access the conference papers. Click TECHNICAL PAPERS, and then click the first title again; this time a pdf file appears in the window (Figure 10).

This page now shows the pdf file of the original paper. Click on the paper abstract and a pdf display of the first page of the paper will appear (Figure 11).

Figure 11 shows the pdf file that can be read in a web reader or downloaded.

Another link, shown back in Figure 9 and highlighted in Figure 12, on the lower right hand side is a zip file download that will contain all papers presented in that conference, as well as the table of contents.

You can also download a zip file with all papers presented in that conference. Simply click on the Downloadable version and put the zip file onto your computer.

7. HOW TO USE THE PROCESS SAFETY MASTER INDEX -EXCEL FILE AN EASY WAY TO FIND A PAPER ON THE WEB SITE

7.1 The Process Safety Master Index

As the project progressed, an Excel File was assembled of year, symposium name, titles, authors, paper numbers. This file, to be available on the process safety web site, brought in two other data bases, the Process *Beacon* collection titles from 2001 to 2019 and the Hyperlinks to all *Plant/Operation Progress* and *Process Safety Progress* Papers published from 1982 to 2020, Issue 2. To the process safety professional, in need to seek past papers on the topic, this spreadsheet can proven to be very helpful. Two examples follow.

7.2 Ammonia Nitrate Example

An ammonia nitrate accident occurs (example Beirut, Lebanon, August 2020). The process safety expert desires to know all titles that have "ammonium nitrate" that are available in the archive:

Following these steps,

- Open the Process Safety Master Index in Microsoft Excel
- Initiate a Find and Replace dialog Box ("Ctrl F") – Figure 13
- Enter the search term "ammonium nitrate", and select the whole Workbook under option button to see all titles: - Figure 14
- Click Find All

The following list appears (in this case two manuscripts related to ammonium nitrate appear as the key word appears twice for each manuscript).

7.3 Ethylene Oxide Example

Let's try one more example, one that will have more results by searching on the combination "ethylene oxide". The following Figure 15 shows 19 titles appearing 7 titles from LPS, 1 title from CCPS (a duplicate appears), and 11 titles from *Process Safety Progress*.

Let's pick paper 3 in the list Ethylene Oxide Explosion at Doe Run. We see in Figure 16 that the paper was presented in the Loss Prevention Symposium in 1968.

Next arrow over to column M and click on the hyperlink provided:

Then, arrow over to column S and click on the hyperlink provided there (See Figure 18):

After clicking the hyperlink in Column S, Figure 19 should appear.

Figure 19. Paper Title Page

The Checkout link exists to purchase this paper. See the previous section for directions on how to purchase. If you have already purchased this volume, the pdf file link should appear below the authors (not shown in this figure).

8. DISPOSITION OF PAPER COPY OF THE ARCHIVES

The paper copies of the Process Safety Archives were scanned by InterActive of Houston, Texas. After scanning, the paper copies of the archives were deposited in the Linda Hall Library of Kansas City, MO for safe keeping. This step is still in progress as of March 2021.

9. ACKNOWLEDGMENTS

The creation of the Process Safety Archives was sponsored by the Safety and Health Division of AIChE. Creation of the Process Safety Archives using historical documents was a major task.

The Process Safety Archives includes over 2,000 technical papers, reports and documents. The Safety and Health Division Committee members were Randy Freeman (S&PP Consulting), Ron Willey (Northeastern University), Tracy Whipple (BP), Philip Myers (Advantage Risk Solutions), and Dan Crowl (Professor Emeritus, Michigan Technological University). Special thanks go to Jing Chen and Gennady Shafirovich from the New York AIChE staff. Scanning, initial file organization, and meta- data creation was completed by Maricela Manteca and her team at InterActive of Houston, TX.

The Proof Checking, Quality Assurance Team was organized by Ron Willey at Northeastern University. This team was instrumental in finding scanning issues, and cross checking the scanned material against the original conference programs to find missing papers. In several cases the QA/QC team identified papers that were missing and were retrieved by contacting the original paper author. This team included Zachary Kylar, Isabella Cardona Barber, Saule Bussing, Michael Nguy, Naa Monoh Odateifio, Joshua Garvey, and Eric Weinstein, as well as others.

9.1. Special Acknowledge for John Davenport

During the period of 1981 through 1989, the papers presented at the Loss Prevention Symposia were not collected by AIChE and published in a book as had been done in the earlier years. The years 1981 through 1989 are called the "Lost Years" by the officers of the Safety and Health Division. During this period, John Davenport of IRI collected the papers from each of the "Lost Years" as well as any associated published technical papers. Thanks to John Davenport, the collection of the "Lost Years" Loss Prevention Symposia papers have been digitized and are included in the Process Safety Archives. In many cases the collection of material by John Davenport contains both the technical paper presented at the LPS symposium and the subsequent paper as published in a journal by AIChE. Thank you John for a job well done!

10. REFERENCES

1. George Santayana, *The Life of Reason: The Phases of Human Progress, Volume 1 – The Life of Reason*, Dover Publications, New York, 1905. (Available on line at the Gutenburg Project, Ebook number 15000.)
2. Kletz, T., *Lessons from Disaster: How Organizations have no and memory and accidents recur*, Institute of Chemical Engineers/Gulf Publishing Co, Houston, TX, 1993.
3. Davenport, J., History of Loss prevention symposia: Forty years, 1967 -2006, *Process Safety Progress,* Volume 25, 2006, pp 268-273.
4. Freeman, R. "History of the Loss prevention symposium: The first 50 years – There are no secrets in safety," *Process Safety Progress*, Volume 35(1), pp 32-38, 2016.
5. Berger, S. *CCPS – 20 Years and more*, *Process Safety Progress,* Volume 25(1), 2006, pp 83-85.
6. Edwards, et al., *History of Process Plant Safety Symposium*, *Process Safety Progress*, Volume 31(3), 2012, pp 200–202.

SUPPLEMENTARY MATERIALS

1. Master Index – Excel Spreadsheet

    Contents – LPS, CCPS, PPSS, Beacon List, Index to PSP papers

Figure 1. Initial entry page to the Safety and Health Proceedings.

Figure 2. CCPS home page.

Figure 3. Example index page that take one to the AIChE Academy Web Page.

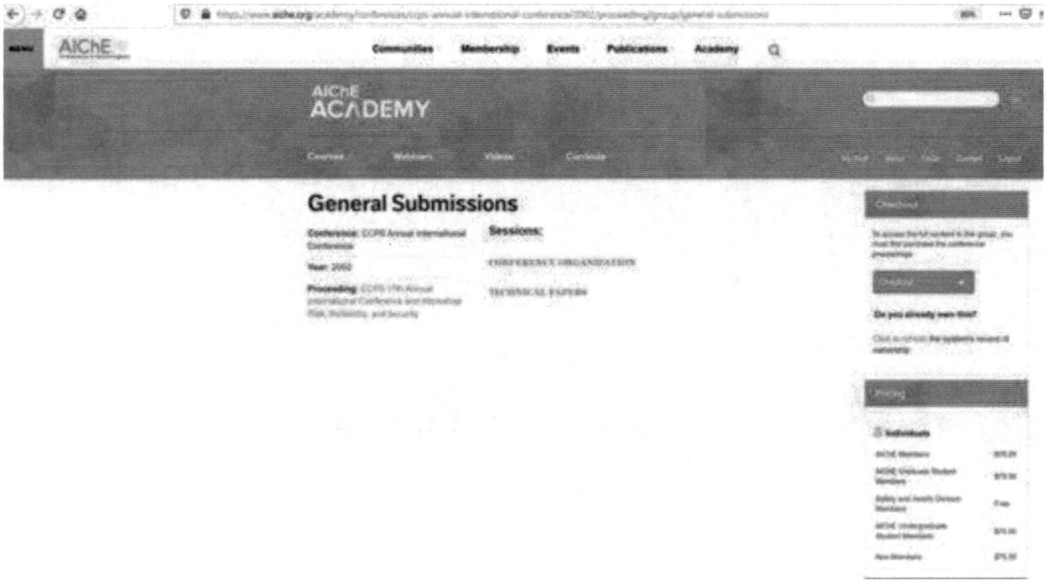

Figure 4. How to reach technical papers.

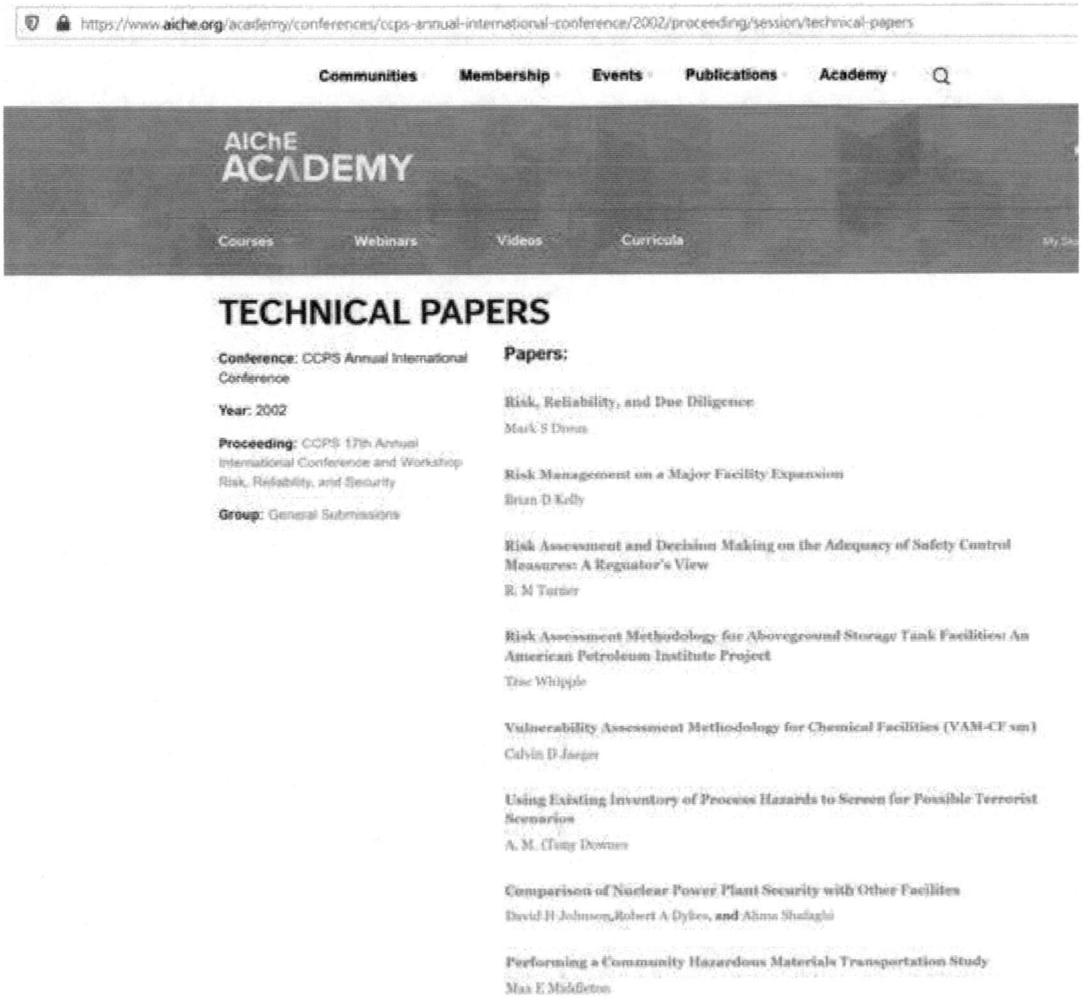

Figure 5 List of papers available for the CCPS Annual International Conference, 2002.

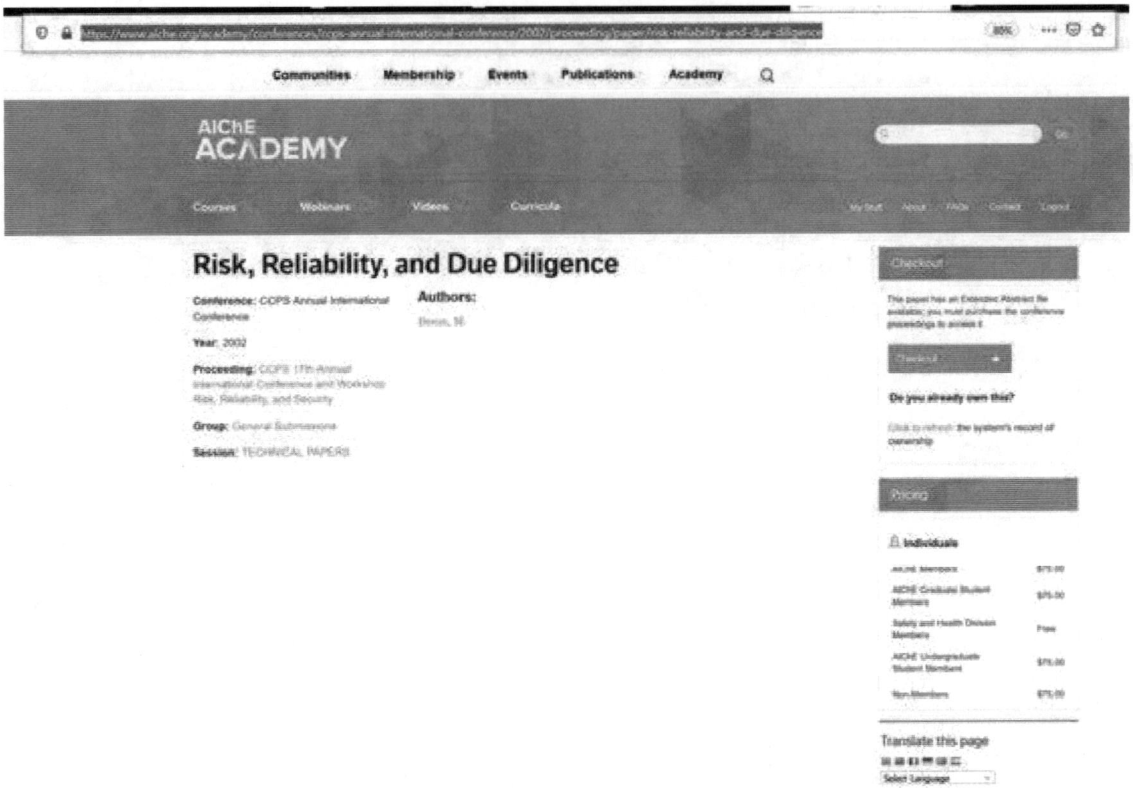

Figure 6 AIChE Academy representation of Risk, Reliability, and Due Diligence, by M. Dreux.

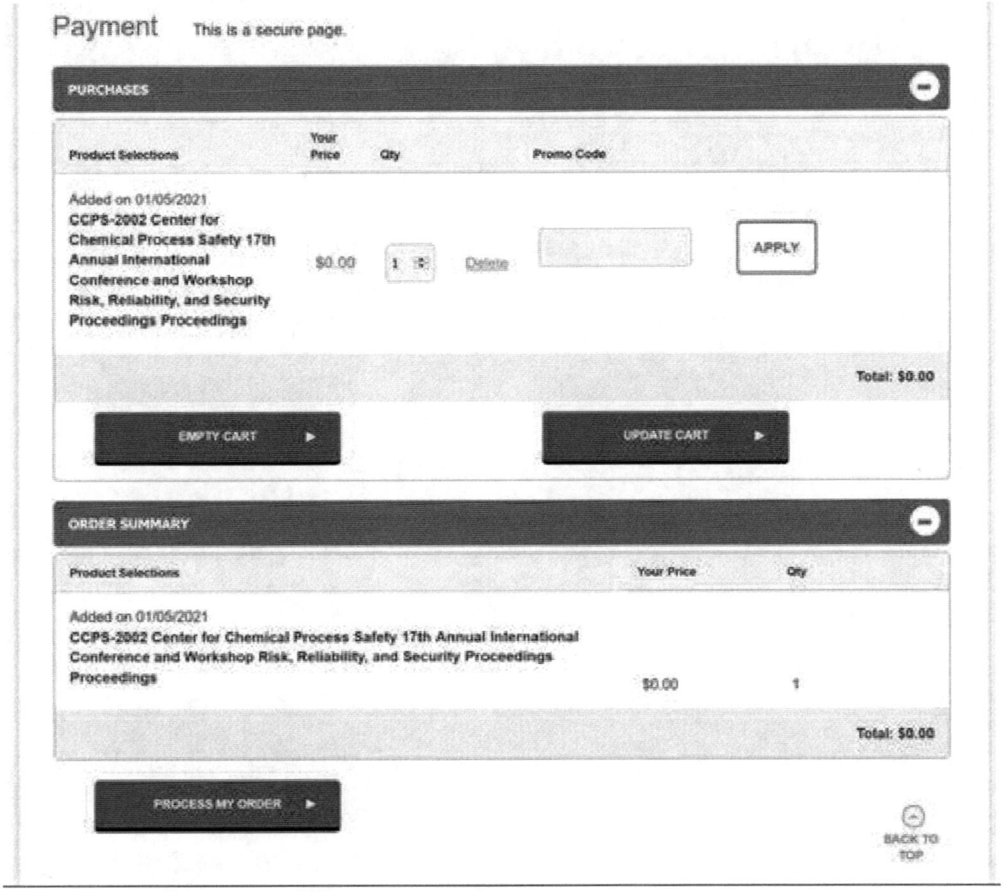

Figure 7. Purchase confirmation page.

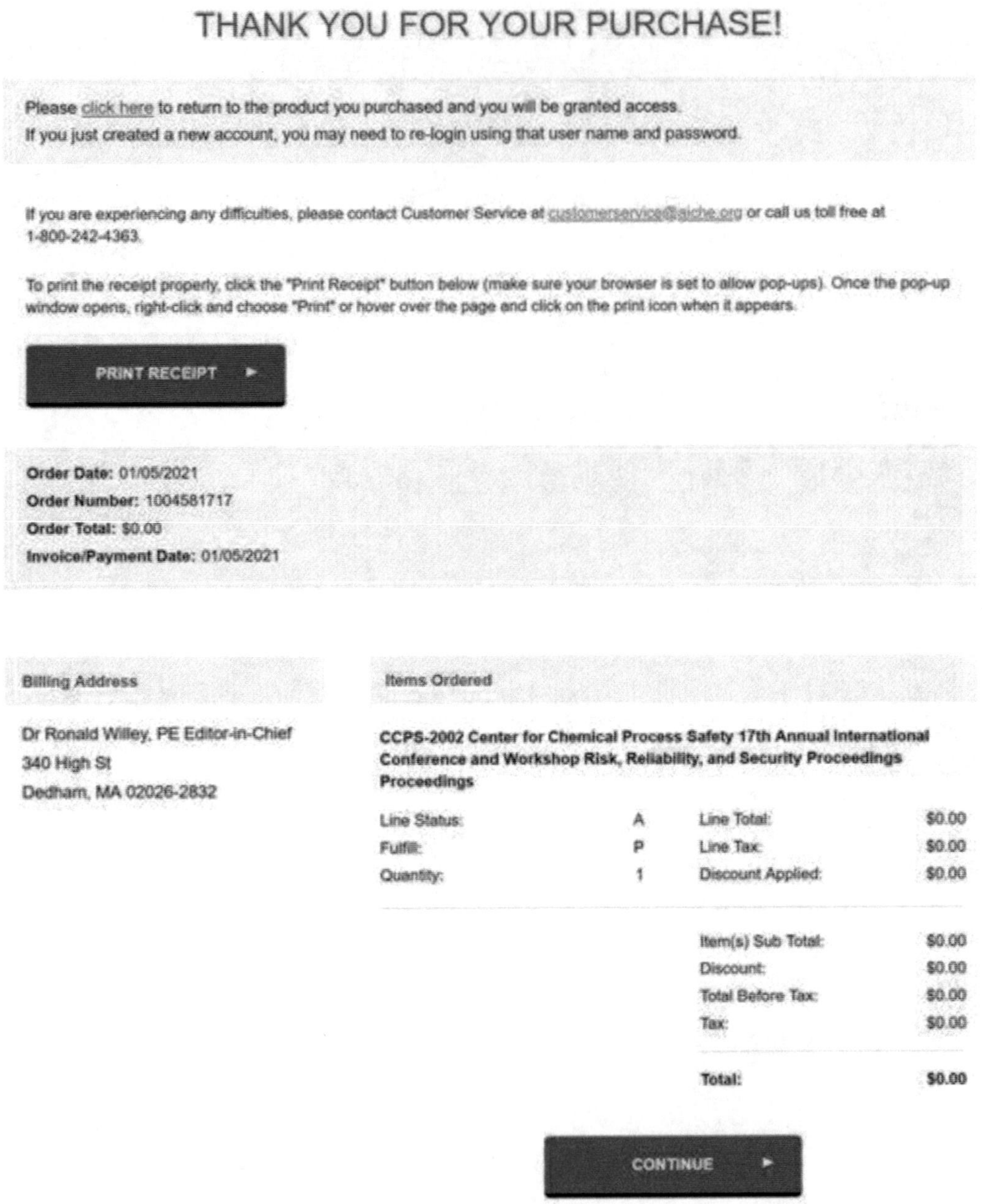

THANK YOU FOR YOUR PURCHASE!

Please click here to return to the product you purchased and you will be granted access.
If you just created a new account, you may need to re-login using that user name and password.

If you are experiencing any difficulties, please contact Customer Service at customerservice@aiche.org or call us toll free at 1-800-242-4363.

To print the receipt properly, click the "Print Receipt" button below (make sure your browser is set to allow pop-ups). Once the pop-up window opens, right-click and choose "Print" or hover over the page and click on the print icon when it appears.

PRINT RECEIPT ►

Order Date: 01/05/2021
Order Number: 1004581717
Order Total: $0.00
Invoice/Payment Date: 01/05/2021

Billing Address

Dr Ronald Willey, PE Editor-in-Chief
340 High St
Dedham, MA 02026-2832

Items Ordered

CCPS-2002 Center for Chemical Process Safety 17th Annual International Conference and Workshop Risk, Reliability, and Security Proceedings Proceedings

Line Status:	A	Line Total:	$0.00
Fulfill:	P	Line Tax:	$0.00
Quantity:	1	Discount Applied:	$0.00

Item(s) Sub Total:	$0.00
Discount:	$0.00
Total Before Tax:	$0.00
Tax:	$0.00
Total:	**$0.00**

CONTINUE ►

Figure 8. Receipt page

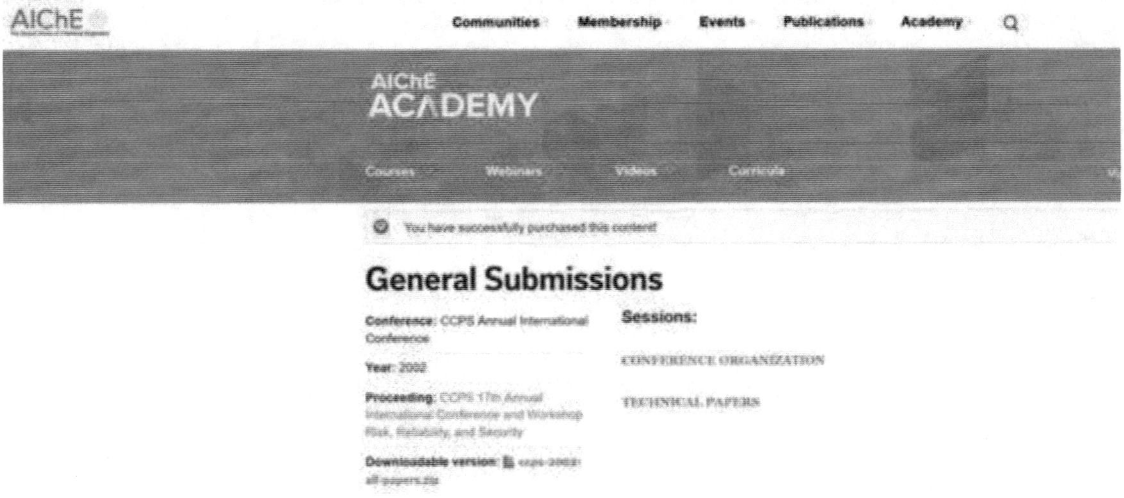

Figure 9. New View of Conference Proceedings

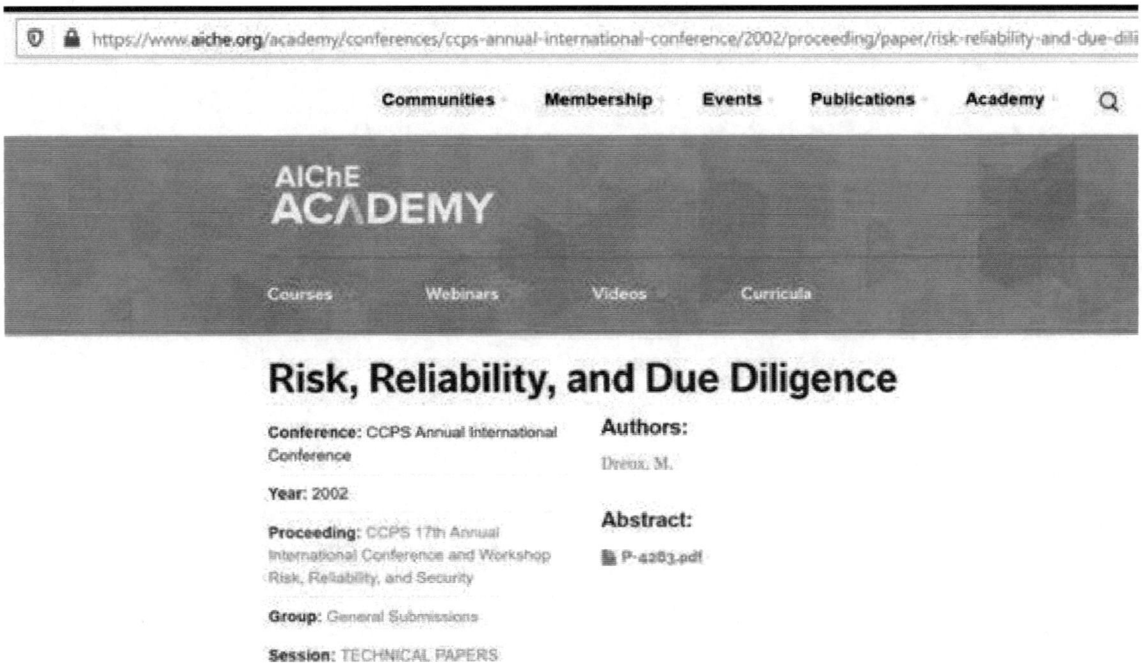

Figure 10. Paper Title Page.

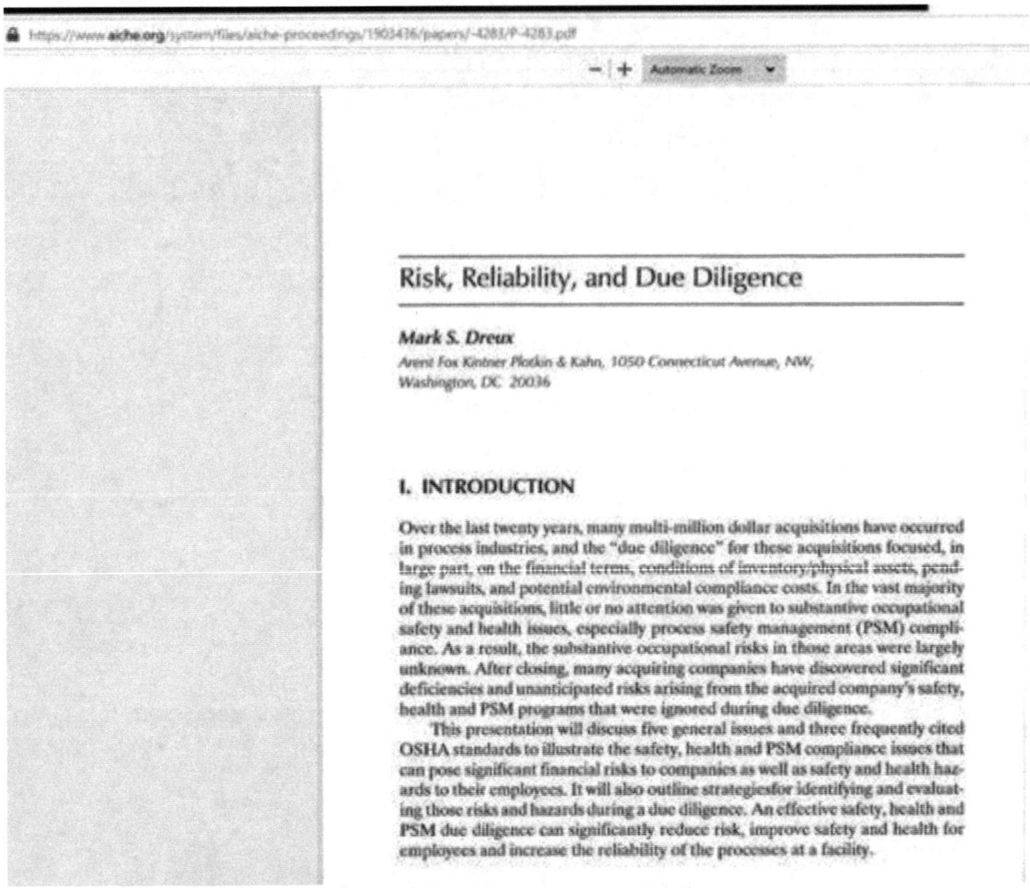

https://www.aiche.org/system/files/aiche-proceedings/1903436/papers/-4281/P-4283.pdf

Automatic Zoom

Risk, Reliability, and Due Diligence

Mark S. Dreux

Arent Fox Kintner Plotkin & Kahn, 1050 Connecticut Avenue, NW, Washington, DC 20036

I. INTRODUCTION

Over the last twenty years, many multi-million dollar acquisitions have occurred in process industries, and the "due diligence" for these acquisitions focused, in large part, on the financial terms, conditions of inventory/physical assets, pending lawsuits, and potential environmental compliance costs. In the vast majority of these acquisitions, little or no attention was given to substantive occupational safety and health issues, especially process safety management (PSM) compliance. As a result, the substantive occupational risks in those areas were largely unknown. After closing, many acquiring companies have discovered significant deficiencies and unanticipated risks arising from the acquired company's safety, health and PSM programs that were ignored during due diligence.

This presentation will discuss five general issues and three frequently cited OSHA standards to illustrate the safety, health and PSM compliance issues that can pose significant financial risks to companies as well as safety and health hazards to their employees. It will also outline strategiesfor identifying and evaluating those risks and hazards during a due diligence. An effective safety, health and PSM due diligence can significantly reduce risk, improve safety and health for employees and increase the reliability of the processes at a facility.

Figure 11. Paper finally appears.

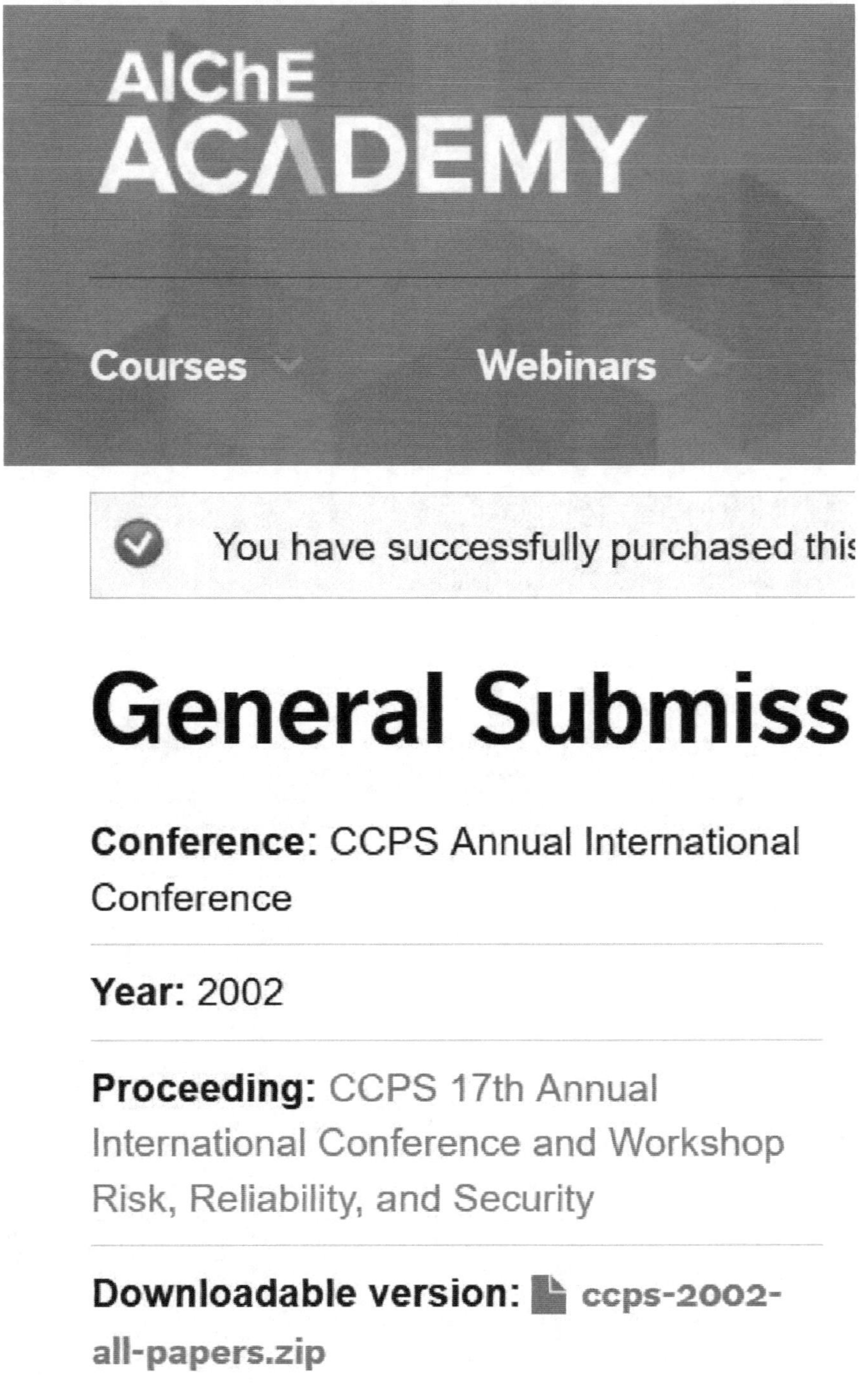

Figure 12. Conference Download Option.

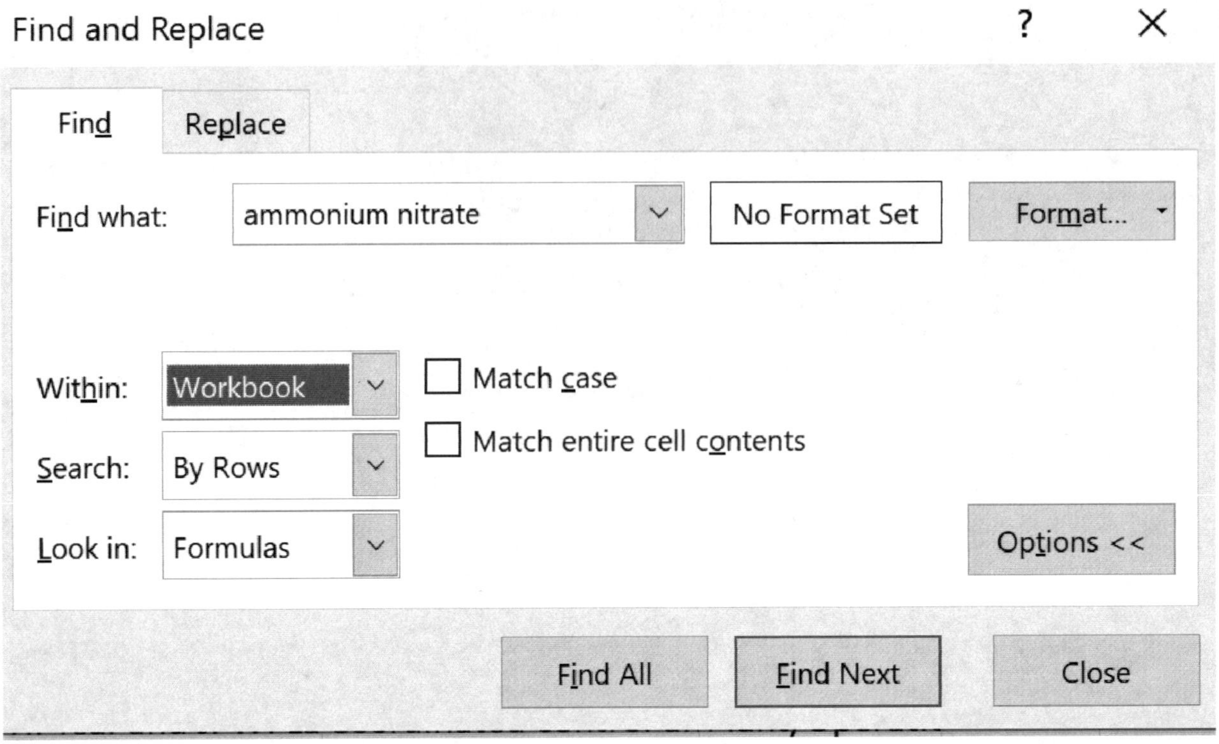

Figure 13. Setting up an Excel Search for the key word combination "ammonium nitrate".

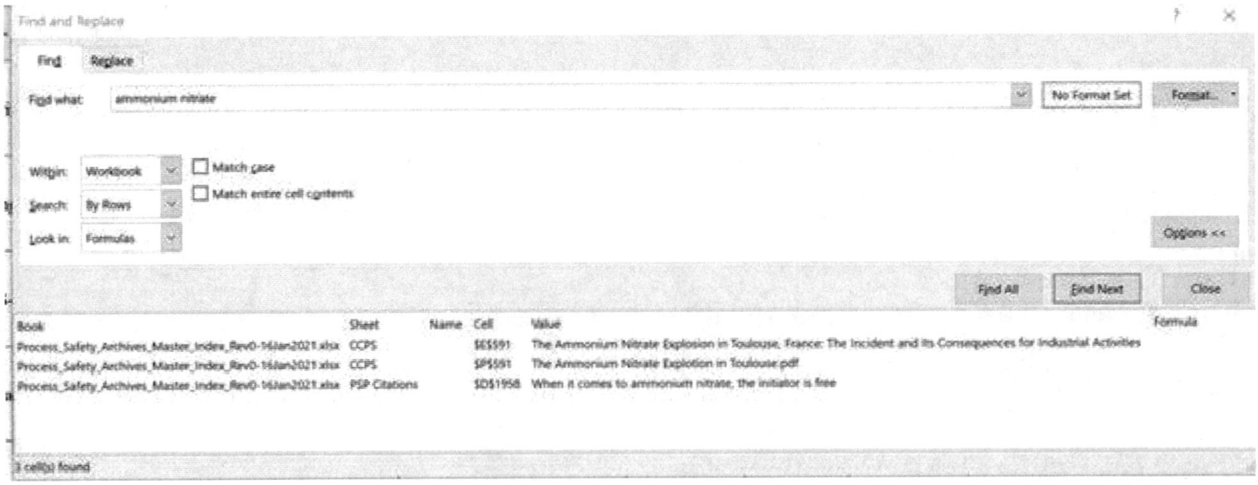

Figure 14. Search results using the term "ammonium nitrate" in the Excel Spreadsheet.

Book	Sheet	Name	Cell	Value	Formula
Process_Safety_Archives_Master_Index_Rev0-16Jan2021.xlsx	CCPS		E716	Tracing Source of Ethylene Oxide Gas House Explosion Using Damage Indicators and Thermal Modeling	
Process_Safety_Archives_Master_Index_Rev0-16Jan2021.xlsx	CCPS		P716	Tracing Source Of Ethylene Oxide.pdf	
Process_Safety_Archives_Master_Index_Rev0-16Jan2021.xlsx	LPS		E51	Ethylene Oxide Explosion at Doe Run	
Process_Safety_Archives_Master_Index_Rev0-16Jan2021.xlsx	LPS		E292	Flammability of Ethylene Oxide in Sterilizer Operations	
Process_Safety_Archives_Master_Index_Rev0-16Jan2021.xlsx	LPS		E293	Liquid Phase Decomposition of Ethylene Oxide	
Process_Safety_Archives_Master_Index_Rev0-16Jan2021.xlsx	LPS		E294	Methane Diluent Requirements for Ethylene Oxide Storage and Handling	
Process_Safety_Archives_Master_Index_Rev0-16Jan2021.xlsx	LPS		E852	Ethylene Oxide Explosion at Seadrift, Texas - Reboiler Safety	
Process_Safety_Archives_Master_Index_Rev0-16Jan2021.xlsx	LPS		E855	Ethylene Oxide Explosion at Seadrift, Texas - Iron Oxide Chemistry	
Process_Safety_Archives_Master_Index_Rev0-16Jan2021.xlsx	LPS		E861	Ethylene Oxide Explosion at Seadrift, Texas - Background and Technical Findings	
Process_Safety_Archives_Master_Index_Rev0-16Jan2021.xlsx	PSP Citations		D187	Flammability of gaseous mixtures of ethylene oxide, nitrogen, and air	
Process_Safety_Archives_Master_Index_Rev0-16Jan2021.xlsx	PSP Citations		D494	Explosive decomposition of ethylene oxide	
Process_Safety_Archives_Master_Index_Rev0-16Jan2021.xlsx	PSP Citations		D495	Thermal stability and deflagration of ethylene oxide	
Process_Safety_Archives_Master_Index_Rev0-16Jan2021.xlsx	PSP Citations		D496	An overview of exposures and exposure control in the ethylene oxide producer and ethoxylation industries	
Process_Safety_Archives_Master_Index_Rev0-16Jan2021.xlsx	PSP Citations		D497	Safe storage of dilute ethylene oxide mixtures in water	
Process_Safety_Archives_Master_Index_Rev0-16Jan2021.xlsx	PSP Citations		D500	Prediction of decomposition limits for ethylene oxide-nitrogen vapor mixtures	
Process_Safety_Archives_Master_Index_Rev0-16Jan2021.xlsx	PSP Citations		D501	Engineering controls to minimize worker exposure to ethylene oxide at sterilization facilities	
Process_Safety_Archives_Master_Index_Rev0-16Jan2021.xlsx	PSP Citations		D997	Deflagration and detonation of ethylene oxide vapor in pipelines	
Process_Safety_Archives_Master_Index_Rev0-16Jan2021.xlsx	PSP Citations		D1008	Use of propylene oxide versus nitrogen as an ethylene oxide diluent	
Process_Safety_Archives_Master_Index_Rev0-16Jan2021.xlsx	PSP Citations		D1049	Kinetics of the reactions of ethylene oxide with water and ethylene glycols	
Process_Safety_Archives_Master_Index_Rev0-16Jan2021.xlsx	PSP Citations		D1324	Analysis of ethylene oxide gas house explosion	

20 cell(s) found

Figure 15. Results using "ethylene oxide" as a search term in the Excel Spread Sheet.

Ethylene Oxide Explosion at Doe Run

	A	E	F	G	H	I	J	K	L	M
1	Master Order	Paper Title	Authors	PDF Volume	PDF Issue	Page Number	Session	Paper Location	Conference Long Title	Link to AIChE online Volume
51	50	Ethylene Oxide Explosion at Doe Run	J.E. Troyan and R.Y. LeVine	2		125-130	2	24	Loss Prevention Symposium - Vol 2	https://www.aiche.org/academy/conferences/loss-prevention-sympo

Figure 16. Clicking on the third entry in Figure 15 search takes one to Cell E51 of the LPS tab.

	A	H	I	J	K	L	M	N
1	Master Order	PDF Issue	Page Number	Session	Paper Location	Conference Long Title	Link to AIChE online Volume	Citation
51	50		125-130	2	24	Loss Prevention Symposium - Vol 2	https://www.aiche.org/academy/conferences/loss-prevention-symposium/1968/proceeding	J.E. Troyan and R.Y. LeVine, "Ethylene Oxide Explosion at Doe Run," Loss Prevention Symposium - Vol 2, AIChE, 1968 pp. 125-130.

Figure 17. Column M contains a direct link to all proceedings presented in that conference.

	A	N	S
1	Master Order	Citation	Link to AIChE online Paper
51	50	J.E. Troyan and R.Y. LeVine, "Ethylene Oxide Explosion at Doe Run," Loss Prevention Symposium - Vol 2, AIChE, 1968 pp. 125-130.	https://www.aiche.org/node/519271/group/8981/session/116896/paper/812846

Figure 18. Column S contains a direct link to the actual paper.

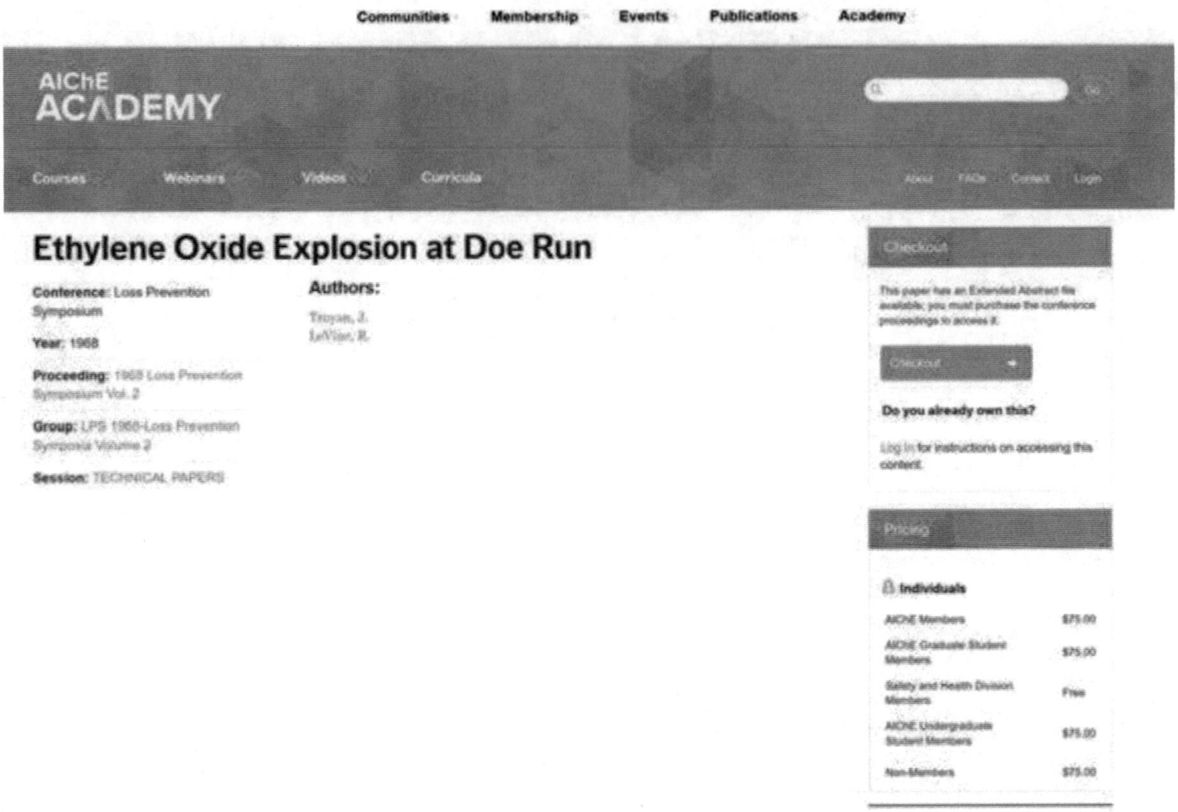

Figure 19. Paper Title Page

619027 Millennial Strategies to Capture Baby-Boomer Knowledge

Abstract:

Risk data is some of the most difficult data in the world to read, digest and understand. Focusing on Process Hazards Analysis (PHA) reports specifically, although the framework and guidewords are followed, for the most part, universally it is still difficult for someone with fresh eyes to pick up a report, read it and know what is important. Never mind having the time to wade through 500-5,000 pages of risk assessment information or multiple risk assessments looking for the specific piece of equipment that they need to know about.

Wouldn't it be great if a new operator or engineer, starts on Monday morning and by noon hour knew the most important risk information relevant to the facility or project they're working on? Why does this matter? Thinking about critical defeats processes, understanding critical procedures and which alarms matter most and need to be prioritized there are many times and processes where this information will be important.

A solution will be presented to make your risk data user friendly for millennials and all generations who are familiar with basic search engine functionality. Being able to search risk data for specific equipment tags and seeing that data in easy to understand, interactive visualizations puts it in context for what they need when they need it. Understanding the relationships between initiating events, safeguards/ IPLs (mitigative/ preventative) and consequences using bowties to tie it all together.

In addition, gamification of risk data will be presented with online, modern and interactive puzzles that can help people understand in minutes the strategies used to safeguard facilities and process units. It also adds a fun factor, competition and other nuances to engage and excite people about safety.

The future workforce needs better solutions to help enable them to leverage the knowledge of the baby boomers and generations that went before. Tens of thousands of experienced workers, who are often the SME in PHA sessions, are retiring on a weekly basis each with decades of knowledge. Although significant knowledge is being lost, the information captured in documents like PHA reports can help the next generations to learn, in second and minutes, what others took a lifetime to build. This knowledge transfer is only possible when presented in an accessible and consumable format.

619104 Lessons Learned for Nextgen PHA Facilitators

Abstract:

There are many training sessions available to train the next generation of PHA facilitators on how to execute a PHA; however, as seasoned facilitators look back on their experiences, they realize there are many details that did not get transferred in their classroom sessions. They learned these details on the job through years of valuable experience and through face-to-face and virtually-led workshops. These learnings can improve the quality and efficiency of the PHA workshop, the documentation of the workshop and the PHA experience. In 2020, this knowledge transfer became even more necessary during the COVID-19 pandemic. Organizations had to move toward conducting workshops virtually while still striving to achieve the same standards needed to ensure a safe facility and satisfy OSHA PSM requirements.

In this paper, seasoned facilitators capture both the small and the significant findings they have learned in their years of PHA facilitation from the very beginning of the PHA (at the preparation phase) to the reporting and follow-up phase. These learnings touch on setting expectations in the terms of reference, maintaining consistent documentation throughout the workshop to aid in effective reporting, maintaining consistency when the team members change (which happens in longer-term workshops), increasing engagement, especially in virtual workshop settings, and increasing use of tools (such as PHA Pro) to improve consistency and efficiency throughout the documentation process.

It is our desire to pass these lessons learned to the next generation of PHA leaders to supplement their classroom learnings and help them succeed in different settings and environments.

618410 "Begin at the Beginning and Go on till You Come to the End: Then Stop" Lifecycle of a Safety Instrumented System

Abstract:

Safety Instrumented Systems have specific steps in their lifespan. Many engineers are responsible for managing these safety systems at a specific point in the lifecycle. This seminar will describe all of the phases of the SIF lifecycle, beginning at the beginning and going through to the end.

617758 Adopting an Integrated Process Maturity Risk Model Approach to Process Safety

Abstract:

Adopting an Integrated Risk Maturity Model for Process Safety

Process Safety can mean different things to different people but one thing is for sure, that when it comes to automation we need to address the three disciplines of: alarm management cybersecurity and functional safety. The reason is that there are consequences of not addressing these from a risk point of view, such as:

- the consequence of inaction due to operators being flooded with alarms;
- the consequence of compromise due to malware or viruses that disable and/or take over the Safety Instrumented System (SIS) or Basic Process Control System;
- the consequence of failing on demand due to the SIS not being maintained properly.

The key point here is that all three disciplines are managed via performance-based standards that have a lifecycle approach with the same three phases: Analysis, Design/Implementation and Operation and Maintenance. By adopting an integrated risk maturity model, companies can assess how well they are doing at implementing these three disciplines and managing the risks associated with them, to meet tolerable levels. This paper outlines an approach and methodology to help companies establish where they are from a maturity point of view, help identify where they want to get to and how to address any gaps. The paper will also present some examples of where companies have benefited from this type of approach and that it is not just large companies that can benefit but companies of all sizes within the process industries.

617249 Reimaging Management of Change

Abstract:

For many companies, Management of Change opportunities ranks among the higher number of findings and observations of all OE processes. Additional experience has found that MOC is one of the more difficult process safety management elements to implement and to get and keep it right. But how can we improve MOC, improve leadership appreciation for what the MOC process is trying to prevent and help eliminate the "get 'er done" culture to make the change and document when we have time? Reimagine an improved overall MOC system, including people, tools, competency and efficiency.

Reimaging a MOC system would include:

- MOC system(s) are more fault tolerant and resistant to circumvention or human error
- Competent, knowledgeable personnel are involved upfront to decide what changes need to be put into the MOC process
- Qualified personnel, normally independent of the MOC originator, review the request to identify any potentially adverse impacts
- Achieve better results with fewer, qualified resources
- Monitor MOC performance and efficiency in a practical way

SPRING21
+17TH GCPS
A Joint AIChE and CCPS Meeting

YES You Can: Case Study of Implementing a Center for Chemical Process Safety Risk Based Process Safety (CCPS RBPS)

Nestor Humberto Sposito
PPR Consultora
Bahia Blanca, Buenos Aires, Argentina
pprconsultora@gmail.com
Juanita Corredor
Process Safety Leader Brinsa
Cajicá, Colombia
juanita.corredor@brinsa.com.co
John Murphy
Process Safety Services
Punta Gorda, FL
hamjfm@embarqmail.com
John Diaz
Quality Director, Brinsa SA,
Cajicá, Colombia
john.diaz@brinsa.com.co

Prepared for Presentation at
American Institute of Chemical Engineers
2020 Spring Meeting and 17th Global Congress on Process Safety
Virtual
April 18 - 22, 2021

—

YES You Can: Case Study of Implementing a Center for Chemical Process Safety Risk Based Process Safety (CCPS RBPS)

Nestor Humberto Sposito
PPR Consultora
Bahia Blanca, Buenos Aires, Argentina
pprconsultora@gmail.com
Juanita Corredor
Process Safety Leader Brinsa
Cajicá, Colombia
juanita.corredor@brinsa.com.co
John Murphy
Process Safety Services
Punta Gorda, FL
hamjfm@embarqmail.com
John Diaz
Quality Director, Brinsa SA,
Cajicá, Colombia
john.diaz@brinsa.com.co

Keywords: Process Safety, RBPS,

Abstract

This paper discusses the challenges and opportunities that arise in the implementation of a risk-based process safety management system in a hazardous materials facility in the chlorine-alkali industry, starting from a relatively low level of compliance with standards, rules of good art, and a general lack of knowledge of the basic principles of process safety.

The motivation to begin the implementation of risked-based process safety (RBPS) came from an external audit requested by the company in November 2017 to evaluate its compliance with several pamphlets from the Chlorine Institute. At that time, the organization did not have a process safety sector and safety efforts were oriented towards personal safety. The final report included a long list of recommendations, each of them associated with good practice from the CI pamphlets. These recommendations were also associated with one of the 20 elements of the CCPS RBPS.

The authors discuss how the implementation plan was developed, the awareness and training of the top and middle levels of the organization's leadership to support the project, the mechanism for selecting priorities, and the process of preparing the schedule including scope, responsibility, and compliance dates.

The paper discusses examples of improvement in facilities, equipment, systems, and procedures in the operational, maintenance, engineering, etc. sectors, the incorporation of systematic and rigorous hazard identification and risk analysis techniques, and the benefits of implementation and monitoring the recommendations suggested in the studies. The strong leadership commitment and the process monitoring methodology by the company's senior management drives the implementation and gives visibility to the management system within the entire organization.

The authors also discuss the economic results of the implementation and quantify the impact of the improvements.

The pandemic unleashed in the world, altered the course of implementation of the process safety management system. The paper discusses the alternative and exceptional measures adopted to mitigate this impact and the results of those measures.

While it is true that the progress toward total implementation of RBPS is very important, it also important to understand what is missing to achieve the full implementation of the 20 elements of RBPS, the appropriate level of maturity necessary in the culture of the organization to achieve continuous improvement and what is the plan to address the challenges to reach each stage of implementation.

The authors encourage all organizations that have not yet implemented a process safety management system to understand that it can be done, even in less than ideal conditions, with very old facilities and a poor or non-existent process safety culture, if there is a strong commitment from the top leadership of the organization.

1 Introduction

A process safety management system that is robust and reliable is necessary to ensure the safe operation of a facility involving highly dangerous products and operating conditions, such as the ones found in a chlor-alkali chemical plant.

Below, there will be a brief introduction of the installation's, which was the subject of this study, most important characteristics.

Brinsa is a Colombian company, leading in salt chemistry, with 25 years of experience. They extract and refine the salt, and from this raw material, they develop the chemistry for the chlor-alkali which supplies the whole industry. The company has manufacturing facilities in Colombia, Costa Rica, and the Dominican Republic, as well as a commercial presence in Latin America.

The chemical plant has been increasing the historical monthly production record of chlorine-soda, through electrolytic cell technology, in order to achieve the separation of the sodium chloride molecule. This is cutting-edge technology that improves process efficiency as well as environmental control.

There are inherent risks involved in the chlor alkali processing because of the hazardous substances involved in the process. Some are classified as corrosive, toxic, and flammable.

The chemical plant has an installed capacity of 100 ton/day of liquid chlorine, 225 ton/day of 50% soda, 160 ton/day of hydrochloric acid, 180 ton/day of 15% sodium hypochlorite (batch production), and 260 ton/day of 5.25% sodium hypochlorite (continuous production). Additionally, there is a storage capacity for 1540 tons of sodium hydroxide, 503 tons of hydrochloric acid, 421 tons of 15% sodium hypochlorite, 302 tons of 38% sodium chloride, 327 ton of lineal sulphonic acid, and finally, and 141 ton of chlorine as a finished product.

The plant has 4 boilers that supply steam at 400 PSIG and 400 °C. Part of the steam that is produced is used in the plant's turbo generators to produce up to 8.5 MW, achieving energy self-sufficiency. There is also a sulfonation plant, which produces raw material for the manufacturing of the cleaning products, and a blowing plant that fabricates packages and lids.

The company has a total of 1400 workers, between fixed and temporary workers and 400 permanent contractors.

2 External Audit. Trigger for the Process.

In November 2017, the company's leadership, decided to conduct third party audit of the chlorine production plant, including all of its stages (membrane, compression, storage, packaging, shipping, and distribution) to access the level of the compliance to the Chlorine Institute's Pamphlets applicable to chloro alkali production.

A protocol was developed (a checklist), based on the guidelines from the Chlorine Institute and the industry's experience. The list was used as an evaluation tool during the 5-day visit to the chlor-alkali process plant.

The evaluation activities in the field included:

- A tour of the facilities
- Review of the following programs with the plant's personnel:
    - Incident reports
    - Emergency preparation and response plan
    - Standard Operating Procedures
    - Plan for personnel training and development assurance
- Review of the pertinent Chlorine Institute checklists, with the plant's personnel.
- Presentation and discussion of the CCPS Risk-Based Process Safety program to the plant's personnel and leadership.

2.1 Findings and Priority Recommendations.

Below, are itemized the most recommendation that was given, as mandatory, by the auditors:

- Develop a written program for the investigation and report of process safety incidents. (1)
- Develop a written program for the involvement of concerned parties, including information on the process dangers, the preparation and emergency plan, self-protection measures, the alarm system, medical treatment, etc.
- Develop a plan for the performance of a periodical risk identification and analysis for the high-risk process areas.
- Develop a written plan on the reliability and mechanical integrity of the plant's assets. Define the critical equipment and an inspection and maintenance plan.
- Develop and implement a Management Of Change program (MOC)
- Develop plans of electrical areas classification in accordance with international standards
- Develop a solid culture of process safety. Establish process safety as a core value
- Provide solid leadership from all levels, starting with the board of directors (lead by example)
- Improve the operative procedures management system
- Establish a Process Safety role

The audit results were presented and shared on the last day with the facility leadership team. The team showed itself very willing to implement all of the recommendations, especially those cataloged as mandatory, and to launch a project to implement a process safety management system that didn't exist at that time in the organization.

3 Awareness & Planning

The company decided to hire the services of a process safety specialist consultancy to advise and monitor the implementation of the Risk-Based Safety System

The concept of process safety was new for all of the organization levels. Initially, a broad training and awareness plan was launched on process safety topics using different courses and workshops encompassing the different organization levels.

The plan was designed to bring them the basic process safety knowledge and to achieve the commitment from all the people to the future program implementation.

Workshops were designed with different scopes and approaches in the function of the target audience, starting with the management and leadership levels.

The first workshop was targeted at leaders from different sectors of the organization. During the 3 days training, after 24 training hours was completed, all participants were summoned to demonstrate their commitment to the system by answering the following questions:

> From your role, what is your commitment to Brinsa's Process Safety implementation?
> What strategy will you use for the implementation?

Indicate which are the 5 activities you will carry out immediately to improve the Process Safety development in your area or sector

All of the resulting material was collected in a report which was subsequently raised to general management and was a significant contribution to the further planning and implementation

Additionally, at this first stage, a 4-hour workshop was developed for the top management of the company, where top management members took part, including the company's CEO. This was key for having the commitment of the highest authorities in the organization on the management system implementation.

Table 1 shows the training that was carried out covering a total of more than 2000 men-hr of training in safety process topics by the end of 2020. During 2021 it is intended to train operators and personnel in sectors not associated with operations such as contracts, human resources, and purchases through specially designed courses for these groups.

Table 1. PS Training Course

Course Name	Top Management	Directors	Senior Management	Supervisors	Operators	Total Men-hours
RBPS Introduction			168	240		408
PSM Audit		24	168	192		384
PSM Audit Refresh for Auditors		8	56	48		112
RBPS Basic - Virtual			40	32		72
Process Safety Advanced		24	336	288		648
Inspiring Leadership	12	12				24
MOC Sensitization			46	40	180	266
MOC Implementati on			46	40	180	266

Total	12	68	860	880	360	2180

At the end of the first training, an 8-hour workshop was held to determine the planning for the implementation of the 4 pillars and the 20 elements of the Process Safety management system. The managers, chiefs, and leaders who participated were divided into multidisciplinary working teams.

The consultant developed a tool that contemplated an assessment for each element the following

- Element's impact on the organization's risk level
- Element's level of development at the organization
- Resources availability

Rating Guide				
Very High - The implementation will immediately have a very strong impact on the prevention and mitigation of the risks of the installation (Weighted as 4)				
High - The implementation will immediately have a considerable impact on the prevention and mitigation of the risks of the installation. (Weighted as 3 points)				
Medium - The implementation will immediately have a moderate impact on the prevention and mitigation of the risks of the installation. (Weighted as 2 points)				
Low - The implementation will immediately have a low impact on the prevention and mitigation of the risks of the installation. (Weighted as 1 point).				
Impacto del Elemento en el Nivel de Riesgo de la Organización				
Process Safety Commitment	Very	Hig	Mediu	Lo
Proces Safety Culture	X			
Standar Compliance	X			
Proces Safety Compdetencies			X	
Workforce Participation		X		
Disclosure to Stakeholders	X			
Current score count by ranked category	3	1	1	0
Percentage of current scores by ranked category	60%	20%	20%	0%
Average Score	3.40			
Understanding Hazards and Risks	Very	Hig	Mediu	Lo
Knowledge Management Process	X			
Hazard Identification and Risk Analysis	X			
Current score count by ranked category	2	0	0	0
Percentage of current scores by ranked category	100%	0%	0%	0%
Average Score	4.00			
Risk Management	Very	Hig	Mediu	Lo
Operative Procedures	X			
Safe Work Practices	X			
Asset Integrity and Reliability	X			
Contractor Management		X		
Guarantee Training and Performance	X			
Management of Change	X			
Operational Availability	X			
Conduct in Operations			X	
Emergency Management	X			
Current score count by ranked category	7	1	1	0
Percentage of current scores by ranked category	78%	11%	11%	0%
Average Score	3.67			
Learn from Experience	Very	Hig	Mediu	Lo
Incident Investigation	X			
Measurements and Metrics		X		
Audit		X		
Management Review and Continuous Improvement	X			
Current score count by ranked category	2	2	0	0
Percentage of current scores by ranked category	50%	50%	0%	0%
Average Score	3.50			

Figure 1. Element prioritization for PSM

The teams had to classify each elements:

- High
- Medium
- Low

Then the tool performed a weighted calculation to determining the implementation priority of the each element of the management system.This was done to achieve the greatest possible impact in reducing the risk of the installation, to assure short-term implementation progress results and to maintain a very strong commitment of the organization. The workshop results and recommendations were presented to the company's management team and finally, after a few requested changes, it was approved. The plan contemplated 3 stages with priorities 1, 2, and 3 for the implementation of the different elements. The elements Process Safety Culture, Management Review, and Continuous Improvement (Stage 0) they would start in stage 1 and continue to develop during the entire implementation process. It was estimated that all the stages would be completed in 3 years. Subsequently, based on the experience, the maturation of the program, and the impact of the pandemic, it was updated extending to 5 years.

4 Organizational Strategy for Implementation.

A Company has to strive to make sure that Process Safety activities are completed in such a way that it minimizes the risk for employees, the public, and the environment, whichever the risk level. The activities that have high-risk levels could require higher attention levels from Management and higher resources assignment. The RBPS methodology intentionally guides the companies to direct efforts to the highest-risk levels sectors.

In compliance with the RBPS strategy, the organization decides to start with the Chemicals Plant, due to the high risk of the process and the results obtained from the diagnosis assessment. Likewise, a specific task force was created, in accordance with the auditor's recommendations and the industry best practices.The focus of the task force would be to develop the implementation methodologies for each of the elements of the RBPS system, follow up on the work plan, ensure interdisciplinary participation from the different sectors in the organization and ensure the system sustainability. The "Process Safety" team was created under the Quality Management integrated by a Chief, 3 professionals, and 1 analyst.

In parallel, the maintenance area team designed its preventive maintenance model and strategy. The Assets Integrity and Reliability task force, which was assigned to the Engineering and Maintenance Management, composed of 1 chief and 5 engineers. It had specific focus on preventive and predictive maintenance. The Technical Documentation center attendants were included in this work team,.They are responsible for centralizing and managing all the organization's projects and processes information.

With the emphasis on the implementation of the elements, a work plan was established with different activities, responsible and execution dates to strengthen the implementation process and to follow an effective realistic schedule.

The Process Safety Area, in alignment with the other company areas, will follow up on the progress and closure of the assigned activities. In Figure 2 a general schedule is presented of each of the elements with its implementation general plan. It should be noted that the Process Safety Culture, Management Review and Continuous Improvement elements are to be addressed the entire 5 years of implementation. Other elements started in 2018 will continue their work strategy with assistance of the consulting firm.

				2018				2019				2020				2021				2022			
Elements	Start	End	% Projected to Dec/20	Q1	Q2	Q3	Q4	Q1	Q2	Q3	Q4	Q1	Q2	Q3	Q4	Q1	Q2	Q3	Q4	Q1	Q2	Q3	Q4
Process Safety Culture	02/03/2018	20/12/2022	73.30%																				
Management review and Continuous Improvement	02/03/2018	20/12/2022	73.30%																				
Contractor Management	02/04/2018	20/12/2021	73.30%																				
Operative Procedures	05/03/2018	30/06/2020	100.00%																				
Knowledge Management Process	05/03/2018	20/09/2020	100.00%																				
Operations Conduct	02/05/2018	20/04/2021	86.10%																				
Standards Compliance	01/06/2018	28/12/2021	72.10%																				
Hazards identification and Risks Analysis	01/06/2018	30/01/2020	100.00%																				
Reliability and Assets Integrity	1/10/2018	20/12/2022																					
Emergency Management	01/06/2018	20/12/2019	100.00%																				
Training Assurance	01/06/2018	20/12/2021	72.10%																				
Preparation for Operation	01/07/2018	20/09/2019	100.00%																				
Labor Force Participation	05/03/2018	28/02/2020	100.00%																				
Management of Change	02/04/2018	28/12/2021	55.60%																				
Incidents Investigation	05/03/2018	30/06/2019	100.00%																				
Audit	02/04/2018	20/12/2021	71.10%																				
External Audit	2/11/2023	6/11/2023	0.00%																				
2019 Internal Audit	3/12/2018	28/02/2019	100.00%																				
2020 Internal Audit	20/12/2019	20/11/2020	100.00%																				
2021 Internal Audit	20/02/2021	20/11/2021	0.00%																				
2022 Internal Audit	20/02/2022	20/11/2022	0.00%																				
Safe Work Practices	05/03/2018	20/04/2021	89.50%																				
Measurements and Metrics	05/03/2018	20/12/2020	100.00%																				
Process Safety Competency	1/10/2018	20/12/2021	69.20%																				

Stage 0 Stage 1 Stage 2 Stage 3

Figure 2: Management System Work Plan Implementation schedule

Next, the results of some elements of the system are presented, and the advances obtained during their development and implementation.

4.1 Hazard Identification and Risk Analysis.

The second pillar of the RBPS system "Understanding Hazards and Risks" is supported by two elements:

- Knowledge management
- Hazard identification and Risk Analysis

This pillar was identified at the diagnostic stage as High Priority(See Figure 2). Thus both elements start to be developed simultaneously, given the interdependence that exists between them. The organization decided to adopt HAZOP as an evaluation tool, complementing the company's risk matrix. HAZOP studies were developed for 10 chemical plant processes, with the participation of the entire interdisciplinary team. This made it possible to identify the current situation of the plant and assess the hazards according to the organization's risk matrix. The study was divided into two stages. Onstage implemented in 2018 included the Chlorine Circuit, Electrolysis, Hydrogen, Hydrochloric

Acid, and Sulfonation processes resulted im 162 recommendations categorized as very high and high risks The second stage started in 2019 included the CEDI (Chemical Distribution Center by its Spanish acronym meaning), Sodium Hypochlorite, Caustic Soda, Calcium Chloride and Demercurization processes resulted 176 very high and high risks recommendations.

In general, there's a high risk at the analyzed plants. 72% of the identified scenarios are at extreme or high risk. The recommendations generated by the HAZOP team when implemented significantly reduce the risk. They mainly focused on strengthening the designs with engineering measures to prevent and mitigate incidents.

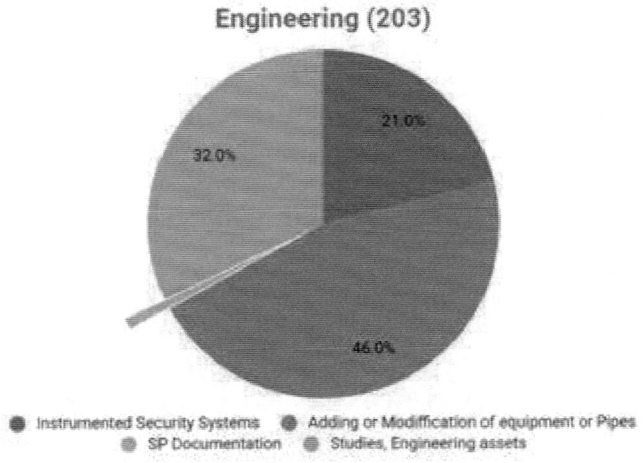

Figure 3: Engineering Recommendations

About 60% of recommendations are engineering improvements, which include among others, instrumented safety systems installation (21%), modification or adding new equipment or piping (46%), engineering studies (32%).

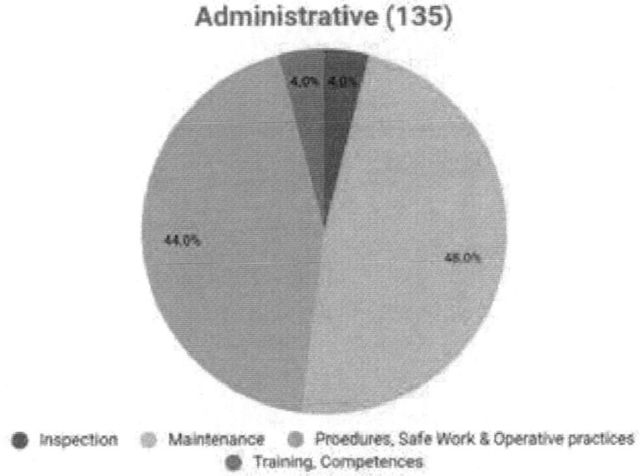

Figure 4: Administrative Recommendations

40% of remaining recommendations were administrative barriers improvements such as inspections, maintenance, training and qualification, procedures, operational and safe-work practices, and in general, process safety standards (Figure 4).

A 3-year work plan was defined for closing gaps, with budgetary assignments and defining specific task forces for gap closing.

4.2 Operating Procedures and Performance Assurance.

The RBPS element that ensures adequate development, timely maintenance, and the consistent use of operative procedures (procedures) is one of the 9 elements at the RBPS pillars: Risk Management.

Good operative procedures, among other things, describe the process, the hazards, tools, protective equipment, control systems and also provides troubleshooting instructions when a non-expected behavior emerges. They also address abnormal situations such as emergency shutdowns and temporary operations.

Operative procedures at Brinsa didn't comply with many of the requirements outlined above. Therefore, a standard is generated for drafting operative procedures according to the CCPS recommendations. This way, the participation of the operators in drafting procedures, created synergy within the personnel expertise and the information related to the operation manuals, the review by the supervision, and the coordination and approval by the Plant management. This procedure ensures that the input of different management levels is guaranteed and the documentation is appropriate. The drafting of all operative procedures required about 2 years. The final result was much more user friendly procedures, with adequate language for easy comprehension by users. Pictures were also included and unambiguous instructions of the steps to follow during the operation.

By the end of 2019, after the information and procedures updated and approved for the processes of Compressors, liquefaction, drying and Chlorine storage, de-chlorination and brine treatment, sodium hypochlorite production, calcium chloride and hydrochloric acid, a personnel retraining program was implemented. It included the technical skills assessment of the operators. it included 3 parts:

- Performance list: It is an evaluation of the knowledge of operational practices in the field.
- Theoretical evaluation of the process: The questions are asked taking into account the operating procedures.
- Cause-Effect evaluation: It measures the ability of the operator to understand the main risks that can be produced by the process and his role to minimize, delete or correct the risk.

After completing the first evaluation, feedback on weaknesses was carried out and subsequently, training was given to each operator. Finally, a second evaluation was carried out, emphasizing the non-conformities. When the training was completed all of the operators had a rating above 90%.

After updating the operative procedures, the first phase of generating the Pre-startup Safety Review (PSSR) lists was started, with the same creation, review and approval flow. In this first phase, the lists only contemplate start-ups after programmed stoppage or non-programmed stoppages without changes or modifications of the process were included. The documentation was developed during 2020. A second phase is planned for 2021that will ensure that the RFSR lists will be also mandatory for new projects.

4.3 Incident Investigation.

Developing, sustaining, and enhancing the organization's incident investigation competency is one of four elements in the RBPS pillar of ***learning from experience.*** (1)

In accordance with RBPS methodology, the company implemented a process for reporting, tracking and investigating incidents that include: 1. a formal process for investigating incidents, including staff provision, conducting, documenting and tracking process safety incidents & 2. Incident and incident investigation data trends to identify recurring incidents. This process also manages the resolution and documentation of recommendations generated by the investigations.

From the implementation of the process, a culture of reporting process safety incidents, including near misses, gradually began to emerge. Reports include, but are not limited to, chlorine leaks, spills (or drips) of liquid and corrosive substances, safety valve activation, fire outbreaks, among others. These events are investigated by interdisciplinary teams (mechanical, electrical maintenance, reliability, operation, process safety) through different investigation methodologies, with the use of Tap Root® being preferred in the organization.

The investigations involve the socialization of the lessons to be learned from the events that occurred at all levels in the operation, through daily operation meetings (DOM), and the lessons are left at different points of the plant for a free consultation. In Figure 5, the lesson-to-learn model currently being socialized

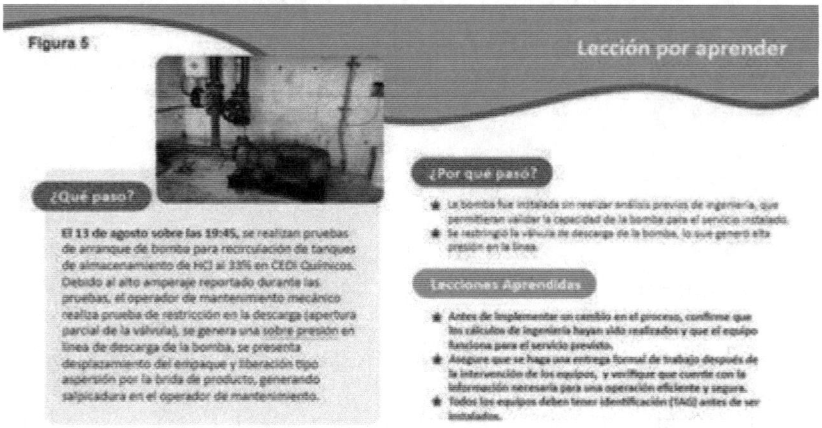

Figure 5: Lesson-to-learn model being socialized

With the implementation of this element, different levels of the organization have been reached, a culture of both reporting and interdisciplinary understanding of the events is being generated.

In Figure 6, the increase in process safety incidents reported by the Chemical Plant is evidenced. The conclusions and recommendations of these investigations led to substantial improvements in the processes, increasing safety and operational reliability.

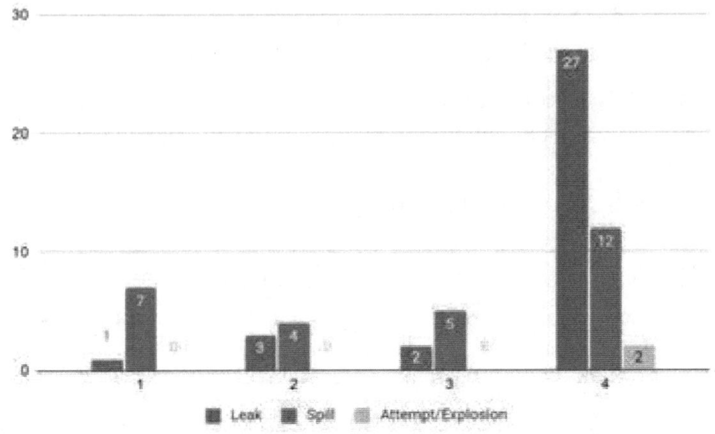

Figure 6: Process Safety reported incidents

During the first 3 years of implementation, activities of all the main elements that have allowed a systematic advance in the implementation have been developed. This has also made it possible to unite efforts from different fronts, create a comprehensive system and receive genuine support from the senior management team.

4.4Management Review and Continuous improvement

Since 2018, monthly monitoring is carried out by the management team and operations leadership to review the progress in the implementation of the process safety management system, opportunities for improvement, delays, budget execution and redefinition of strategies that leverage the system implementation. In these evaluation meetings, the gap closures identified in risk analysis and incident investigation are also reviewed. The results and conclusions of these evaluations are presented and analyzed every 4 months at the company's shareholders' meeting.

5 Investment

The RBPS implementation requires allocating human, material and financial resources to implement a robust, consistent and sustainable system over time.

The leadership of the organization, understanding its responsibility, approved for the first 4 years of implementation an approximate budget of 10,000,000 USD. This budget was assigned, among other things, to Training and performance assurance, incorporation of resources, developing the procedures and policies required by the RBPS and implementing improvements in the plant arising from the risk studies, the requirements of the management system and the conclusions of the investigation of process safety incidents.

Some improvements and changes involve large projects, and others are "minor" improvements that seek to reduce the human factor as a source of risk.

Below we will describe some of the main improvements made to the installation:

Around 30 level transmitters were installed in different process tanks, to eliminate the level glasses with which the level visualization was conducted (Figure 7).

BEFORE

15% Sodium Hypochlorite Preparation tanks 688 / 689. Level visualization through a sight glass, without an alarm system to alert the overflow of the tanks.

AFTER

Installation of level transmitters with alarms, giving the operator greater control over the operation and identification of substandard conditions, both due to overflow and lack of power.

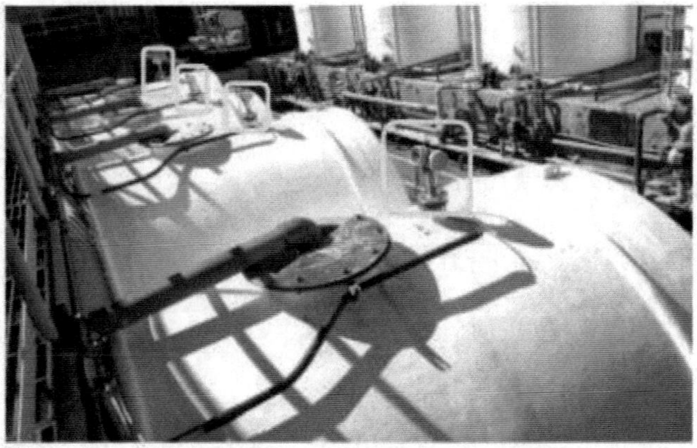

Figure 7: Level Transmitter, 15% Sodium Hypochlorite Preparation Tanks

A project was created for the installation of new liquid chlorine storage tanks, meeting the minimum requirements defined by the Chlorine Institute pamphlets, the EUROCHLOR technical definitions, and the available ASTM and ASME construction standards, in replacement of existing ones that did not comply with the RAGAGEP standards mentioned above. The storage system will start operations in the second half of 2021 (Figure 8).

Figure 8: Liquid Chlorine Storage Tanks

The implementation of the safety management system allowed the organization to know in greater depth the hazards associated with the handling of hydrogen and for this reason it was decided to carry out a specific HAZOP study for the system with the advice of a hydrogen expert. The study incorporated the risk rating using the company's matrix and based on this, approximately 40 recommendations for improvement were obtained. Most have already been implemented, especially those associated with the elimination of equipment and systems that were not operating after the change of technology from mercury cells to membrane cells.

BEFORE

Equipment in the out-of-service area of the hydrogen filtration system, excessive obstacles in the area.

AFTER

Dismantling of equipment, cleaning the area, facilitating access to operational areas.
Image was taken on May 30/20

Figure 9: Hydrogen Filtration System dismantling

In the HAZOP study carried out on the hydrogen system, it was identified that in the hydrochloric acid receiving tanks there was an imminent risk due to the presence of an explosive atmosphere (accumulation of hydrogen gas). This recommendation was verified with the technology provider, and an engineering study was started to implement the improvement. Tank blanketing, or padding with N2 was installed.

BEFORE

Mersen Recommendation (HCl Furnace Supplier): Ensure nitrogen purge in the process tanks with minimum flow control to avoid an explosion due to hydrogen accumulation. In previous years one of the tanks had an explosion with the expulsion of part of the tank.

Rec. 315. Consider adding nitrogen as a blanketing gas to these Receiver Tanks.

AFTER

Connections and adjustments are made in the receiver system and gas scrubbing to maintain a permanent purge to both receiver tanks of HCl guaranteeing the displacement of remaining gases and avoiding any accumulation of hydrogen inside.

Figure 10. Blanketing gas to the hydrochloric acid receiving tanks.

Several processes in the facility still have a highly manual operation. During the HAZOP study, several recommendations were made associated with the automation of some processes. The figure shows the implementation of the automatic valve to guarantee the start-up of the backup fan of the sodium hypochlorite fan, to avoid a plant shutdown and a chlorine leak due to the pressure increase in the chlorine collector.

BEFORE

Rec. 20. Implementation of the automatic start of the spare fan in case of fan failure in service.

Risk: Loss of vacuum in the absorption tower and

AFTER

Automatic valve installation and control loop programming was performed.

loss of receipt of Cl2 with stoppage of the chemical plant due to increased pressure in the Cl2 receipt lines and subsequent risk of Cl2 leakage into the atmosphere with risk to people of the area.

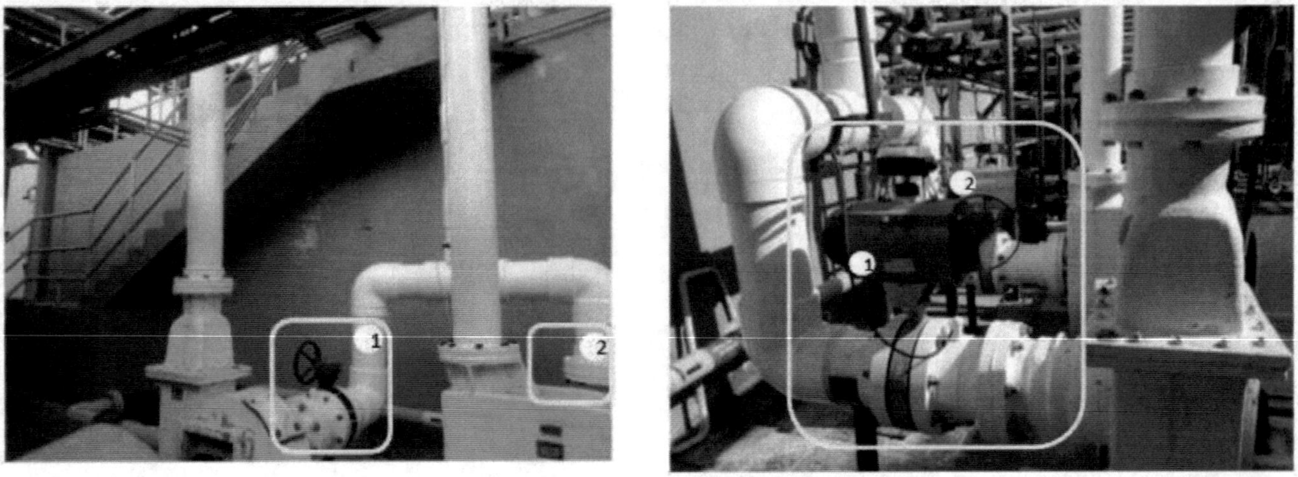

Figure 11. Automatic startup of the sodium hypochlorite backup fan.

Consideration of the "human factor" is a key aspect of a process safety management system. This discipline takes into account, in the work environment, the interaction between people, facilities and equipment, and management systems. (2) Based on these concepts, the organization decided to incorporate them into many aspects of the management system, for example, to improve the consistency of all technical documentation, including operating procedures and PSSR checklists, which includes updating of P&IDs, datasheets, among others, a program was launched to improve the identification of all equipment, instruments and pipes in the plant. (See Figure 10, Figure 11, Figure 12).

BEFORE

Unidentified valves can lead to their misalignment and therefore a mixture of chemicals that can give off dangerous gases Cl2, H2S, generating risk to people in the area.

AFTER

Valves are tagged, minimizing operational errors due to confusion.

Figure 12. Valves tagging

BEFORE

Tanks 769 and 779 unidentified.

AFTER

The identification of the auxiliary storage tanks of sodium hypochlorite at 15% is carried out.

Figure 13. Tanks tagging

BEFORE

Unidentified pipe, an event occurred due to mixing of raw materials due to misalignment of valves.

AFTER

The identification of the pipes is carried out to minimize errors in the direction of fluids.

Figure 14. Pipes tagging

6 AUDIT

The only way to determine that a PSM program is working properly is to thoroughly examine both its design and implementation on a regular basis. Therefore, a PSM audit represents a way to measure the effectiveness of the PSM program and also allows a comprehensive comparison with predetermined PSM metrics. If the PSM program is not measured carefully, it cannot be controlled or improved. (3)

A successful PSM program audit begins with a commitment by the Company's senior and facility management to conduct periodic audits of these programs. This is essential to allocate appropriate resources to conduct the audits, ensure that findings and recommendations are actively and carefully addressed promptly, and ensure that on-site activities allow a successful development of the audit.

(Successful PSM program auditing according to CCPS begins with a commitment by the senior company and site management to perform periodic audits of these programs, to allocate the appropriate resources to perform the audits, to ensure that the findings and recommendations are actively and carefully addressed in a timely manner, and to ensure that the activities on-site during the audit are arranged to support the audit to the maximum extent possible.)

These principles were strongly adopted by the leadership of the organization and based on this, the process safety audits program was established to accompany the implementation of the system in order to control its effectiveness and make adjustments and corrections promptly. The plan first contemplated an internal training process and the development of audit protocols for the 20 elements of the RBPS. All this material was based on the CCPS audit book

Then it was decided to start with internal audits for partial elements of the management system. So far, 2 internal audits have been carried out, one in person in 2019 and one in 2020. Despite the challenges posed by the global situation due to the pandemic, the company decides to continue with the internal audit plan in 2020, counting with remote assistance from the consulting firm and blended participation by the audit team and the audited team, seeking to guarantee the biosafety protocols defined by Brinsa.

The audited elements in 2020 are in an advanced development process with partial implementation; a greater understanding of the spirit and scope of the audit and a more agile development were observed.

Ultimately, and beyond the qualification and findings, there is evidence of growing knowledge and commitment of most sectors of the organization with the process safety subject.

Procedures and tools have been incorporated that are strengthening the entire process safety management system. The Asset Integrity and Reliability element shows a significant improvement in its rating (2.44 in 2019 vs. 3.19 in 2020) derived from the important incorporation of procedures and tools for planning and monitoring the reliability and maintenance tasks of the equipment to be throughout their lifecycle.

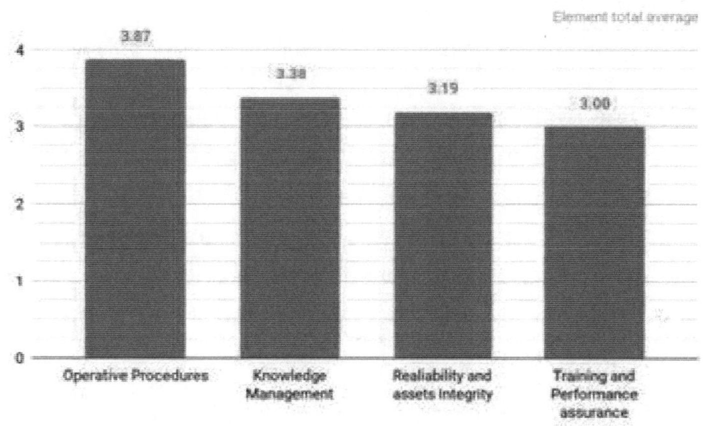

Figure 15: Qualification of audited elements 2020

The “operating procedures” element has notably improved in the second audit compared to the first one. It is worth highlighting the existence of a procedure that establishes the scope, format, responsibilities, etc. for the development of the operating procedures, and promoting the active participation of the operators in carrying out / reviewing them.

7 CONCLUSIONS

The most important conclusions of this case study are:

- An RBPS can be applied with very good results, in an organization with many years of operation and a culture focused only on personal safety.

- Awareness and training in the concepts of process safety before the implementation of the management system, at all levels of the organization, especially leadership, are essential to achieve understanding and awareness of its importance.

- The support and commitment of the organization's leadership is key to implement a process safety management system in organizations that are unaware of the issue and their safety concern is exclusively personal safety. Leadership must provide the human and material resources and make the necessary investments for a correct implementation as happened in this case.

- Implementation poses a huge challenge when starting from scratch. However, it can have very good results in the short and medium-term if it is properly planned and with the active participation of all levels of the organization in such planning.

- The proper selection of the elements of the management system on where to start the implementation is essential to try to quickly reduce the risks of the installation and at the same time show significant progress that maintains the interest and commitment of all levels of the organization, especially the leadership.

- The implementation requires a specific sector of Process Safety to lead it, monitor progress and solve limitations and challenges during the process.

- The implementation of the management system introduced benefits in other areas of the organization such as production by achieving increased reliability and operational continuity through the creation of a previously non-existent asset integrity group.

8 Bibliography

1. **Center for Chemical Process Safety, American Institute of Chemical Engineers.** *CCPS. Guidelines for Risk-Based Process Safety.* New York: Wiley, 2007.

2. **Center for Chemical Process Safety, American Institute of Chemical Engineers. CCPS. .** *Human Factor Methods for Improving Performance in the Process Industries.* New York: Wiley, 2007.

3. **Center for Chemical Process Safety, American Institute of Chemical Engineers. .** *Guidelines for Auditing Process Safety Management System.* New York: Wiley, 2011.

618537 Using Data Analytics to Generate PHA Checklists – Saving SME Time and Supporting SME Knowledge

Abstract:

The majority of those that have been exposed to Process Safety are aware of Hazard and Operability Studies (HAZOPs); a Process Hazard Analysis (PHA) methodology that is regularly used in industry because of its scalability and relative ease of execution when analyzing the risks of a process unit. Those who have been exposed to HAZOPs are also aware that these studies are qualitative, subjective and incredibly dependent on the people involved in the analysis session. Everyone who participates in a HAZOP does the best they can to do a complete and comprehensive analysis, however their contributions can be limited and biased based on their human knowledge and personal experience. This can lead to inconsistencies in identifying and capturing risks in comparable facilities/units, which can result in increased risks at site.

In order to strive towards higher quality, more comprehensive and more consistent HAZOPs, operation companies are starting to create HAZOP/PHA Checklists to help their teams better prepare for their sessions and be aware of potential high risk scenarios.

Unfortunately, HAZOP/PHA Checklists often take a large amount of time and effort for a Subject Matter Expert (SME) to create, and this amount of SME time is not always available in smaller operating companies.

So, what to do? There is obviously a benefit in having HAZOP/PHA Checklists, so how can the time of an SME required to generate one of these checklists be reduced while still creating a high quality reference tool for HAZOP teams? Also, if the insights of a few SMEs are so beneficial, what about the knowledge of ten or a hundred SMEs?

From these wants, a new method has been found to accomplish this task of generating HAZOP/PHA Checklists by leaning on the power of automatic data processing and data analysis of already completed PHAs. By combining the SME knowledge of tens to hundreds of HAZOPs of the same technology process together, a complete HAZOP/PHA Checklist can be generated with minimum manual SME input.

This paper shows a case study comparison between an SME generated HAZOP Checklist and a Big Data generated HAZOP Checklist, and highlights the high similarity/accuracy when compared, as well as the time savings of the a Big Data generated HAZOP Checklist methodology.

617639 How Effective Is Your Fire and Gas System?

Abstract:

A well-conceived fire and gas system enables the early detection of loss of containment incidents, and the resultant actions can significantly reduce the magnitude and duration of potential consequences. With a wide range of variables affecting fire and gas hazard behavior, and an unlimited number of potential fire and gas detector configurations, how does one ensure that the system is effective and fit-for-purpose?

This presentation will outline an adaptable and consistent approach to assist with the strategic development of a fire and gas detection philosophy that is applicable to both onshore and offshore environments. Emphasis is placed on the importance of hazard assessment to gain an understanding of the behavior of the specific hazards and how this can be used to inform the detection strategy, in conjunction with relevant standards and codes. The presentation goes beyond what may be included in a typical fire and gas philosophy document and covers detector layout strategies, describes modern verification methods to assess detector coverage (e.g. 3D detector mapping), and provides approaches for the management of ongoing inspection and testing of the fire and gas detection system.

618809 Factors Contributing to US Chemical Plant Process Safety Incidents from 2010 – 2020

Abstract:

Process safety incidents occur more frequently than desired – they can result in injuries, fatalities, environmental impacts, facility damage, downtime & lost production, as well as impacts on a company's and industry's reputation. While such incidents occur across many industries, this study is focused on an analysis of the most commonly reported contributing factors to process safety incidents in the US chemical industry.

The database for the study contained 79 incidents from 2010-2019, investigated by the Chemical Safety Board (CSB) and/or by companies. To be included in the database, the incident had to occur at a company classified as 325 in the North American Industry Classification System (NAICS), assigned to businesses who participate in chemical manufacturing. Unlike other summary studies of process safety incidents, the subject analysis is specific to the US chemical industry and based on contributing factors to the incidents reported post investigation.

All of the incidents included in the study were classified as Tier 1 incidents according to the American Petroleum Institute (API) RP 754 definition. Using this methodology for determining incident severity, each incident in the database included a severity score, and was placed into one of four severity categories, identified as low, medium, high and very high severity. All incidents investigated by the CSB were deemed very high severity. The list of contributing factors included in this analysis also provided more granularity and specificity than typically in other studies, and included factors such as preventive maintenance; safeguards, controls & layers of protection; and hazard awareness and identification.

Reported contributing factors were catalogued for each incident by severity category, and then a prioritized (weighted) list of contributing factors was developed for the database of incidents. The top three weighted contributing factors for the chemical industry were: design; preventive maintenance; and safeguards, controls & layers of protection. The finding that the top contributing factors to process safety incidents included factors such as operating procedures, preventative maintenance, and emergency preparedness and response, areas that are traditionally covered under the occupational health and safety administration (OSHA) regulations, prompted a review of the federal OSHA citations database. The investigation and citation history for NAICS 325 companies was analyzed to assess whether there was any overlap between the top reported contributing factors to process safety events and the top OSHA citations recorded for the industry.

A database consisting of the inspection and citation history for the chemical industry identified by NAICS code 325 was assembled for inspections occurring between 2010-2020 (August). Analysis of the thousands of entries provided an understanding of which regulations were most commonly cited, and which were most commonly cited following fatalities. Findings were presented for NAICS code 325, as well as for each sub grouping, including NAICS 3251, 3252, 3253, 3254, 3255, 3256, 3259.

The analysis of the citation history for the chemical industry specifically, identified that the key factors reported as contributing to incidents overlapped with the most common OSHA violations. The overlap of contributing factors for process safety incidents with the most common OSHA violations suggests that addressing regulatory issues may also pay dividends in reducing process safety incidents. These findings will likely be most beneficial if shared among those with the ability to create influence within the chemical industry, assess practices in these key areas, and identify improvement opportunities seeking industry best practices.

SPRING21
+17TH GCPS
A Joint AIChE and CCPS Meeting

Human Factors Implementation – for Plant Workers

William Bridges
President
Process Improvement Institute, Inc.
1321 Waterside Lane,
Knoxville, TN 37922 USA
wbridges@piii.com

William C. Rhodes II, P.E.
Process Improvement Institute, Inc. (PII)
82 Via Santa Maria
San Clemente, CA 92672
wrhodes@piii.com

PROCESS IMPROVEMENT INSTITUTE

Prepared for Presentation at
American Institute of Chemical Engineers
2021 Spring Meeting and 17th Global Congress on Process Safety
Virtual
April 18 - 22, 2021

Human Factors Implementation – for Plant Workers

William Bridges
President
Process Improvement Institute, Inc.
1321 Waterside Lane,
Knoxville, TN 37922 USA
wbridges@piii.com

William C. Rhodes II, P.E.
Process Improvement Institute, Inc. (PII)

Keywords: human factors, process safety, PSM, human error, risk-based process safety, RBPS, organizational culture

Abstract

Process safety is about controlling risk of failures and errors; controlling risk is primarily about reducing the risk of human error. Often it is believed that human errors committed by plant workers are the cause of most process safety accidents. However, the entire organization contributes to these human errors. Therefore, actions are needed by the organization to reduce plant worker human errors by improving Human Factors that contribute to the plant worker human errors, and to build an organizational culture that seeks to learn from plant worker human errors, rather than to assign blame.

This paper introduces the multiple types and categories of human error and the Human Factors that influence the rate at which human errors are made. It establishes the need for creating management systems for these Human Factors, and how to implement them in a manner to reduce plant worker human errors. Finally, it describes a Safety-Principled Organizational Culture that allows an organization to create the proper environment to enable these critical improvements to reduce plant worker human error. This paper builds on earlier papers, starting from 2010, on the same topic.

The data presented is from basic research by the authors on the root causes of more than 3000 accidents and near misses; and also based on the review of hundreds of accidents analyzed by others and summary data from many companies. This Video Presentation and the related slides are even more keenly focused on the selected human factors for which the frontline workers should take the lead so that the base human error rate at a site is as low as possible. Case studies and examples are used to illustrate key points.

1. Introduction

99% of accidental losses (except for natural disasters) begin with a human error ***(supported by data from more than 1500 investigations)***

Root causes of accidents are management system weaknesses
(Center for Chemical Process Safety, American Institute for Chemical Engineers, "Guidelines for Investigating Chemical Process Incidents", 2003) – OSHA agrees

Weak Management Systems → Human Error → Accidents

Figure 1: Relationship between Accidents and Management

All accidents (or nearly all, if you consider that there are some natural phenomena that we either cannot guard against or choose not to guard against) result from human error. Quite often, the erroneous conclusion is drawn that the error was "caused" by the actions of the individual directly involved in the incident. However, this view not only does a disservice to the affected individual, but it prevents proper corrective actions because **humans govern and accomplish all of the activities necessary to control the risk** of accidents. In other words, humans influence other humans in the process. Not only do humans cause accidents (unintentionally) by making errors directly related to the process itself, but they also cause errors by creating deficiencies in the design and the implementation of management systems. Specifically, we make errors in authorities, accountabilities, procedures, feedback, proof documents, continual improvement provisions, etc. Ultimately, these management systems govern the human error rate of those directly contacting or influencing the process. Thus, the entire organization (plant workers, technical staff and management) contribute to the human errors resulting from the actions of the plant worker.

The optimum approach to reduce plant worker human errors therefore is not to "fix" the plant worker but rather to strengthen management systems so that it is less likely that the human errors will occur (and that when they inevitably do, their consequence is controlled to the greatest extent possible).

This paper outlines this approach by first providing context to the term "human error" so that the means to reduce and mitigate it can be better understood. It then establishes that the most practical and efficient means to minimize plant worker human errors is through properly designed and implemented management systems. Next, it establishes the key Human Factors that contribute to human errors. Conceptually, plant worker human errors are controlled by understanding the Human Factors related to them and their work to the degree that effective management systems for each of them can be developed. Therefore, the paper then details how these Human Factors can be improved for plant workers. Finally, it explains that this entire process is only possible with the right type of organizational culture.

This paper introduces the concept of a Safety-Principled Organizational Culture as one that enables the proper application of Human Factors to plant workers to reduce their human errors. Further, with this culture, when human errors are made, the appropriate lessons are learned so that those errors, and even seemingly unrelated errors, are not repeated. In this way, a plant's overall safety performance will improve.

Definitions for Use in This Paper

- This paper will use the term **Human Error** to mean the errors that are made during direct interface or direct influence of the process.
- **Human Factors** are those aspects of the process and related systems that make it more likely for the human to make a mistake that in turn causes or could cause a deviation in the process or could in some indirect way lead to the increased probability of an accidental loss.
- **Management Systems** are the administrative controls an organization puts in place to manage the people and workflow related to the process under consideration, and so these inherently attempt to control human factors.

2. Types and Categories of Human Error

Quite often people will analyze a plant incident and draw the (typically correct) conclusion that it was due in large part to a plant worker human error. However, they end their analysis at that point, and are therefore unable to draw conclusions as to what could or should be done to prevent its recurrence. To fully protect against future harm, however, plant worker human errors must be fully analyzed, and the first step in this analysis is to understand the types and categories of human error.

In simplest terms, there are only two types of human error: *Errors of Omission* (someone skips a required or necessary step) and *Errors of Commission* (someone performs the step wrong). But in addition, these errors occur either inadvertently (unintentional error) or they occur because the plant worker believes his or her way is a better way (intentional error, but not intentional harm). Intentional errors can usually be thought of as errors in judgment. Some believe a "lack of awareness of the risk" causes these errors, but in actual practice, the plant worker who commits an intentional error may well be aware of the risk. They instead believe they know a better way to accomplish a task or they believe there are already too many layers of protection (so bypassing one layer will not cause any harm).

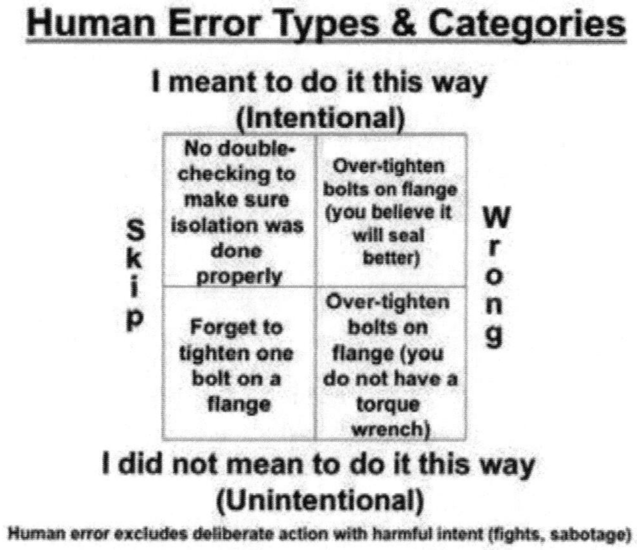

Figure 2: Types and Categories of Human Error

Regardless of type or category of human error, the organization can and should exert considerable control of the errors through properly implemented management systems.

3. Controlling Human Error through Management Systems

A process is a combination of the utilities, raw materials, and human actions (direct actions and those actions involving programming the process to accomplish automatic functions). If anything goes wrong with these Inputs, or if there are basic design flaws or basic fabrication flaws in the process, then the Outputs will not be desired. The desired output is acceptable (or high) production rates at acceptable or higher quality factors with no harm to humans (long term or short term), no harm to the environment, and with acceptable (or higher) life of the process components. The negative outcomes resulting from plant workers failing to control the raw material quality, failing to control the utility levels consistently, making errors directly related to the operation of the process, or making errors in the care of the process (such as maintenance) will result in lower production, lower quality, higher number and severity of safety-related accidents, and more negative impact on the environment. The potential of the negative outcomes is collectively referred to as ***business risk*** – more precisely, the risk is a product of the likelihood of one or more of these negative outcomes and the severity of each outcome.

A critical concept is therefore:

> ***If an organization does not directly control risk, the organization cannot directly control quality, safety, environmental impact, or production to acceptable levels. An organization must sustainably control human error to manage the risk of accidental losses that impact quality, safety, the environment, production, or assets.***

In order to sustainably control the risk of a complex process (such as an oil/gas operation, refinery, chemical plant, steel plant, automobile manufacturing, aircraft manufacturing, etc.), the organization must design and implement management systems to:

- **Understand the Risk** – This involves predicting problems; which in turn includes predicting the risk of possible accident/loss scenarios, establishing the appropriate design and the right layers of protection to control risk to a tolerable level
- **Control Risk Factors Day-to-Day** – This involves controlling the original design by maintaining the established layers of protection and managing changes to the design using integrated management systems
- **Analyze Actual Problems and Determine Weaknesses in the System** – This involves identifying weaknesses in designs and management systems and weaknesses in risk understanding through root cause analysis of actual problems (losses and near-losses)

These three elements are illustrated in the following figure:

© Process Improvement Institute, Inc. (2004-2021)
Figure 3: Controlling Risk through Human Reliability

Management systems control the interaction of people with each other and with processes. They are the high-level procedures we use to control major activities like conducting PHAs, management of change, writing operating procedures, training employees, evaluating fitness for duty, conducting incident investigations, etc. If management systems are weak, then layers of protection will fail, and accidents will happen.

To reiterate, accidents are caused by human error. In general, Process Safety Management (PSM) is focused on maintaining these human errors at a tolerable level because:

- All accidents happen due to errors made by humans; including premature failure of equipment. There is a myriad of management systems to control these human errors and to limit their impact on safety, environment, and quality/production
- When these management systems have weaknesses, near misses occur
- When enough near misses occur, accidents/losses occur

4. Human Factors that Contribute to Human Errors

To minimize plant worker human error, process safety systems should address the **Human Factors Categories** (see various US NRC and US DOE standards from 1980s and 1990s)[1,2]. Table 1 on the next page lists the key human factor categories along with multiplication factors that poor human factors can have on the base human error rates.

TABLE 1: SUMMARY of 10 HUMAN FACTOR CATEGORIES

Based in part on: Gertman, D.; et. al., *The SPAR-H Human Reliability Analysis Method*, NUREG/CR-6883, U.S. Nuclear Regulatory Commission, Office of Nuclear Regulatory Research, Washington, DC, August 2005. PII has modified the list slightly to account for general industry data and terminology and to incorporate PII internal data.

Human Factor Category	Human Factor Issue/Level	Multiplier for Cognitive & Diagnosis Errors
Available Time	Inadequate time	P(failure)=100%
(includes staffing Issues) – ***for responses only***	Barely adequate time (≈2/3 x nominal)	10
	Nominal time (1x what is expected)	1
	Extra time (at least 2x nominal and >20 min)	0.1
	Expansive time (> 4 x nominal and > 20 min)	0.01
Stress/Stressors	Extreme (threat stress)	5
(includes staffing issues)	High (time pressures such as during a maintenance outage; issues at home, etc.)	2
	Nominal	1
Complexity & Task Design	Highly complex	5
	Moderately complex (requires more than one staff)	2
	Nominal	1
	Obvious diagnosis	0.2
Experience/Training	Low	10
	Nominal	1
	High	0.5
Procedures	Not available in the field as a reference, but should be	20
	Incomplete; missing this task or these steps	8
	Available and >90% accurate, but does not follow format rules ***(normal value for process industry)***	3
	Good, 95% accurate, follows >90% of format rules	1
	Diagnostic/symptom oriented	1
Human-Machine Interface (includes tools)	Missing/Misleading (violates populational stereotype; including round valve handle is facing away from worker)	20
	Poor or hard to find the right device; in the head calc	10
	Some unclear labels or displays	2
	Good	1
Fitness for Duty	Unfit (high fatigue level (>80 hrs/wk or >20 hr/day, no day off in 7-day period; or illness, etc.)	20
	Highly degraded fitness (high fatigue such as >15 hr/day, illness, injury, etc.)	10
	Degraded Fitness (>12 hr day and >72 hr/wk)	5
	Slight fatigue (>8 hr per day; ***normal value for process industry***)	2
	Nominal	1
Work Processes & Supervision	Poor	2
	Nominal	1
	Good	0.8
Work Environment	Extreme	5
	Good	1
Communcation	No communication or system interference/damage	10
	No standard for verbal communication rules ***(normal value for process industry)***	3
	Well implemented and practiced standard	1

If all of the human error factors shown above in Table 1 are controlled very well, then the optimized human error rates listed in Section 5 of this document are achievable. Alternatively, if one or more of these factors are compromised, the human error rate will increase by the values shown in this table. As an example, an individual whose fitness for duty rating is unfit due to the excessive work hours shown in the table will make errors at a rate 20 times greater than an individual whose fitness for duty is normal.

With excellent control of each of the human factors listed above, a company can begin to approach the lower limits that have been observed for plant worker human error. These lower limits are about:

- **1 mistake in 100 steps for most procedures-based tasks** (such as starting up a process unit), a little less for a routine (daily) task that becomes almost a reflex

- **1 in 10 chance or a little better for diagnosis and response to a critical alarm**

Excellent control requires superior design and implementation of management systems, which is enabled through a thorough understanding of these factors, as outlined below.

5. Strengthening Human Factors for Plant Workers

In order to achieve the lowest practical plant worker human error rates noted above, comprehensive management systems must be developed for each of the Human Factors provided in Table 1. Key elements for each of these Human Factors are provided below.

A. AVAILABLE TIME and STRESS / STRESSORS

The **Available Time** for the task refers to the time the task is expected to be required for completion. In many cases, the time determined for a process is determined using the theoretical times where the conditions are "ideal" and not necessarily realistic. When task and process duration are not realistic plant workers tend to "find the best and easy way" to get it done, ultimately creating latent conditions for error occurrence. **Work Stress** refers to the emotional response that arises when work demands exceed the plant worker's capacity and capability to cope. Some processes have been designed paying little attention to job design, work organization and management systems. Failing to fulfill this requirement eventually results in work stress.

Staffing levels directly impact both factors. Staffing is the process of assessing, maintaining and scheduling personnel resources to accomplish work. An adequately staffed organization ensures that personnel are available with the proper qualifications for both planned and foreseeable unplanned activities. Staffing is a dynamic process in which plant management monitors plant worker performance to ensure that overall organizational performance goals are met or exceeded. The result of an effective staffing process is a balance between personnel costs and the achievement of broader organizational goals.

Three key issues must be considered when staffing decisions are made; the first of which is selecting the right staff for a job. Each organization requires the proper amount and type of expertise to operate the plant safely and competently under a variety of conditions. The term "expertise" includes the attributes of talent, effectiveness, knowledge, skills, abilities, and experience necessary to operate and maintain plant systems, structures and components.

The second key issue is avoiding staff overload. Surges in workload, such as during outages, typically require staff augmentation as well as longer work hours for permanent staff. The introduction of contractor personnel or company personnel from other sites may increase the likelihood of errors due to unfamiliarity with the plant, its procedures and hardware, for example. Longer work hours have the potential to increase fatigue, which also contributes to the likelihood of error.

The third key factor is rotating staff every 1 hour or less for tasks that require high vigilance. Humans are inherently unable to remain alert for signals that seldom, if ever, occur. Even a sailor whose life is at stake cannot maintain an effective watch (look-out) for hostile submarines for more than 30 minutes or so. The following figure illustrates the rapid decrease of vigilance with time.

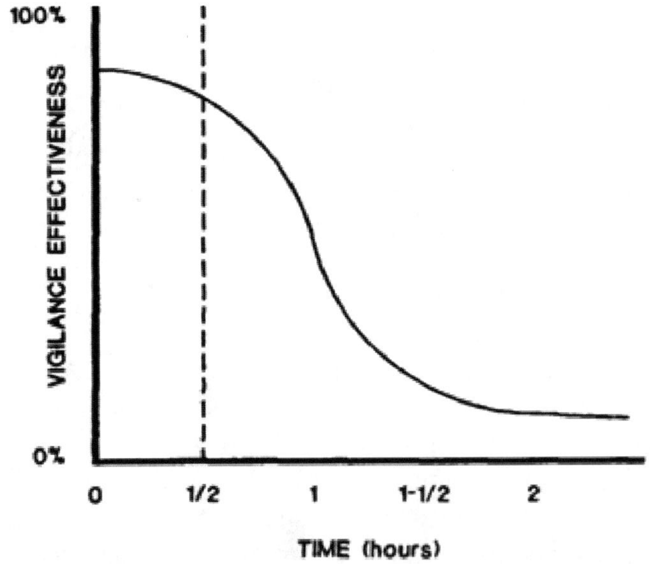

Figure 4: Vigilance as a Function of Time

Placing a worker in situations requiring extended, uneventful vigilance may lead to accidents. Therefore, it is important to design control systems in a manner that requires regular operator interaction so that the operator will remain attentive.

B. <u>COMPLEXITY & TASK DESIGN</u>

A task that is designed with the human limits in mind is much more likely to work effectively than one that assumes humans can and will "always" do what is written. The task must consider that humans think and remember and factor in prior data and prior experiences. Factors to consider in assessing **task design** adequacy (i.e., minimizing the likelihood of plant worker human error during its completion) include:

- Complexity of the task
- Probability of repeat errors (coupled error) on redundant aspects of the design
- Error Detection and Error Recovery

Complexity of Task (procedure-based or a call for action) – If the task is too complex, then plant workers can forget their place in the task, fail to understand the goal of each step or sub-step, or fail to notice when something isn't going right.

Task complexity is a function of:

- Number of choices available for making a wrong selection of similar items (such as number of similar switches, number of similar valves, number of similar size and shaped cans)
- Number of parallel tasks that may distract the worker from the task at hand (leading to either an initiating event or failure of a protection layer)
- Number of staff involved (more staff = more complex)
- Number of adjustments necessary to achieve the goal
- Amount of mental math required (as a rule, NO math should be required in anyone's head when accomplishing a standardized task)
- How much judgment is required to know when you have accomplished each goal within the task

For most chemical process environments, the complexity of the task is relatively low (one action per step), but for response actions (in which plant workers serve as the Independent Protection Layer (IPL)), there are almost always other tasks underway when the out-of-bounds reading or the alarm is activated. Complexity is difficult to predict (since it is not known when a human intervention will be needed), but higher complexity can increase error rates by 2 to 10 times.

Probability of Repeat Errors (Coupled errors) – For many maintenance tasks, making a repeat error or "common cause error" or "dependent error" can lead to failures of all backup systems. Such errors have led to MANY airplane crashes and major process safety accidents as well.

Coupling represents the probability of repeating an error (or repeating success) on a second identical task, given that an error was made on the first task. The increased probability of failure on subsequent tasks given that an error has already been made is known as dependence. The list below provides some starting point guidance on values to use:

- 1/20 to1/90 – if the same tasks are separated in time and if visual cues are not present to re-enforce the mistake path. This error rate assumes a baseline error rate of 1/100 with excellent human factors. If the baseline error is higher, then this rate will increase as well.
- 1/2 – if the same two tasks are performed back-to-back, and if a mistake is made on the first step of two. This error rate assumes a baseline error of 1/100 with excellent human factors. If there the baseline error is higher, then this rate will increase as well.
- 8/10 to 10/10 – if the same three tasks are performed back-to-back and strong visual cue is present (if you can clearly see the first devices you worked on), if a mistake is made on the first step of the two or more

- Two or more plant workers become the same as one person (with respect to counting of errors from the group), if people are working together for more than three days; this is due to the trust that can rapidly build.

These factors are based on the relationships provided in NUREG-1278[1] and the related definitions of weak and strong coupling provided in the training course by Swain (1993)[1] on the same topic, as shown below in Table 2.

The following relationship is for errors of omission, such as failing to reopen a root valve or failing to return a safety instrumented function (SIF) to operation, after bypassing the SIF. The qualitative values in Table 2 are based jointly on Swain[1] (1993) and Gertman (SPAR-H, 2005 which is NUREG/CR-6883)[2].

One can readily conclude that staggering of maintenance tasks for different channels of the same SIF or for related SIFs will greatly reduce the level of dependent errors. Unfortunately, most sites PII visits do not stagger the inspection, test, or calibration of redundant channels of the same SIF or of similar SIFs; the reason they cite is the cost of staggering the staff. While there is a perceived short-term higher cost, the answer may be different when lifecycle costs are analyzed.

Simple Rule: Staggering of maintenance can prevent a significant number of plant worker human errors in redundant channels. In the airline industry, the US Federal Aviation Administration (FAA) requires staggering of maintenance for aircraft with multiple engines or multiple control systems (i.e., hydraulics) (*FAA Advisory Circular 120-42A*, as part of Extended Operations (ETOPS) approval).[6]

Table 2: Guideline for Assessing Dependence for a within-SIF Set of Identical Tasks (based partially on SPAR-H, 2005 [1,2], and partially on field observations by PII) Courtesy Process Improvement Institute, Inc., All Rights Reserved

Level of Dependence	Same Person	Actions Close in time	Same Visual Frame of Reference (can see end point of prior task)	Worker Required to Write Something for Each Component
Zero (ZD)	No; the similar tasks are performed by different person/group	Either yes or no	Either yes or no	Either yes or no
Zero (ZD)	Yes	No; separated by several days	Either yes or no	Either yes or no
Low (LD)	Yes	Low; the similar tasks are performed on sequential days	No	Yes
Moderate (MD)	Yes	Moderate; the similar tasks are performed more than 4 hours apart	No	No
High (HD)	Yes	Yes; the similar tasks are performed within 2 hours	No	No
Complete (CD)	Yes	Yes; the similar tasks are performed within 2 hours	Yes	Either yes or no

Once the level of dependence is known, the probability of either repeat success or repeating errors on identical tasks can be estimated. For these probabilities, we use Table 3, which is a re- typing of Table 20-17 from NUREG-1278[1] (and the similar table in SPAR-H [Gertman, 2005][2]).

Table 3. Equations for Conditional Probabilities of Human Success or Failure on Task N, given probability of Success (x) or Failure (X) on Task N-1, for Different Levels of Dependence *Courtesy Process Improvement Institute, Inc., All Rights Reserved*

Level of Dependence	Repeating Success Equations (but shown as error probability)	Repeating Failure Equations
Zero (ZD)	$P_{Success@N} = x$	$P_{Failures@N} = X$
Low (LD)	$P_{Success@N} = (1+19x)/20$	$P_{Failures@N} = (1+19X)/20$
Moderate (MD)	$P_{Success@N} = (1+6x)/7$	$P_{Failures@N} = (1+6X)/7$
High (HD)	$P_{Success@N} = (1+x)/2$	$P_{Failures@N} = (1+X)/2$
Complete (CD)	$P_{Success@N} = 1.0$	$P_{Failures@N} = 1.0$

<u>**Error Detection and Error Recovery**</u> – Is there enough feedback in the process to allow the plant worker to realize (in time) that they made a mistake? Have they been trained on how to reason through how to recover from mistakes they or others make? (Sometimes, doing a step too late is far worse than NOT doing the step at all.) Is there enough time available for the type of intervention necessary?

C. *EXPERIENCE/TRAINING*

Training is necessary for general functioning of management systems and for task-specific skills such as how to start up a compressor, repair a pump, lead a root cause analysis, perform a proper Lock-Out/Tag-Out, etc. Organizations normally do a good job of training, including hands-on training. Training systems can be weak (inadequate), however, if the training system does not adequately:

- Address how to troubleshoot the process (i.e., handle process deviations or upsets).
- Provide plant workers with a mental model(s) of the process so they can perform the assigned tasks correctly, diagnose process upsets properly and quickly, and understand the consequences of their actions.
- Provide plant workers with enough practice of critical tasks.

Ironically, training is many times listed as the "cause" or "root cause" of an accident, when in fact the training is adequate, and some other Human Factor is the major cause.

D. *PROCEDURES*

Many procedures do not follow best practices for controlling human error, and so the written procedure actually "contributes" to increased plant worker error rates. Further, many organizations are missing guides on how to troubleshoot (what to do when process deviations occur). The best practice rules for writing and validating procedures have been published for many years ***(see Bridges, 1997-2010)***[3, 7, 8]. Below is a checklist based on the current set of best practice rules for developing operating, maintenance and other work-instructions (procedures):

TABLE 4: PROCEDURE QUALITY CHECKLIST (courtesy PII, 2008)

#	Issue	Response
	Procedure Content Checklist	
1	Is the procedure drafted by a future user of the written procedure? (Engineers should not author procedures to be used by operators or maintenance staff.)	
2	Is the procedure validated by a walk-down in the field by another future user of the procedures?	
3	Is the procedure reviewed and commented on by technical staff (engineers or vendors)?	
4	Is the procedure checked versus the Page and Step format rules below?	
5	Is a hazard review of step-by-step procedures performed to make sure there are sufficient safeguards (IPLs) against the errors that will occur eventually (when a step is skipped or performed wrong)?	
6	Is the content measured using "newly trained operators" to judge the % of steps that are missing, steps that are confusing or wrong, and steps that are out-of-sequence? (A score of 95% accuracy of content is good.)	
	Page Format Checklist	
1	Is the title of the procedure the largest item on the page?	
2	Is the procedure title clear and consistent with other titles, and does it uniquely describe the topic?	
3	Are the document control features the smallest items on the page?	
4	Are temporary procedures clearly identified?	

#	Issue	Response
5	Is white space used effectively? • Is there one blank line between each step? • Does indentation help the user keep their place? • Are the margins large enough to reduce page congestion?	
6	Is type size 12 point font or larger?	
7	Is serif type used (rather than sans-serif)?	
8	Is mixed case used for words of steps, with ALL CAPS used only for special cases (such as IF, THEN, AUTO, and WARNING)?	
9	Is the step number very simple (single level of number used)? Only an integer?	
10	Have sections or information not necessary to performing the steps been moved to the back or to another part of the manual or training guide?	
11	Are section titles bold or larger than the text font? Do sections have clear endings?	
12	Is the decision on electronic presentation versus hard copy correct? Are electronic linkages to procedures clear and accurate and easy to use? If paper is chosen, is the color of the paper appropriate?	
13	Is the overall page format (such as Outline format or T-Bar format) appropriate to the use of the procedure?	
14	Are play script features added for tasks that must be coordinated between two or more users? • Play script is normally used when there are two or more hand-offs of responsibility for steps.	
15	Are rules followed for formatting of Warnings, Cautions, and Notes? (See annotated rules, such as Warnings are for worker safety and Warnings must clearly standout from rest of page.)	
	Step Writing Checklist	
1	Is each step written as a command?	
2	Is the proper level of detail used throughout? This is judged based on: • *Who* will use the procedures • Same level of detail used in similar procedure steps	
3	On average, is there only one implied action per instruction? Best practice is to average 1.2.	
4	Does the procedure indicate when sequence is important? • If sequence matters, each step should be numbered (with an integer or letter) • If sequence does not matter, bullet lists should be used	
5	Are only common words used? Apply "education" level test (5 grade reading level is best)	
6	Do all acronyms, abbreviations, and jargon aid understanding? • Develop a list of such terms for use in procedures *and* communication. • Use terms that users use (within reason)	
7	Is each step specific enough? No room left to guess/interpret: • The meaning of a word or phrase (Check vs. Make sure) • The intent of a step or series of steps • A desired quantity or value • To what equipment the step applies	
8	Is the procedure free of steps that require in-your-head calculations? • Values expressed as ranges rather than targets with error bands • Conversion tables, worksheets, or graphs provided where needed	
9	Are graphics to the user's advantage? • No explanatory paragraphs or lengthy instructions that could be replaced by a picture • No impressive graphics that provide no real advantage	
10	Are references to the user's advantage? • No lengthy explanations or instructions that could be replaced by branching to a reference • No references to a procedure that references still another • No gaps or overlaps between this procedure and a referenced document • If branching, must branch to a procedure, not to a specific step in a procedure	

#	Issue	Response
11	Are rules followed for writing warnings, cautions, and conditional steps? • No more than 2 per page • No actions within a warning or caution (actions must always be numbered steps) • Warnings and Cautions contain descriptions of potential consequences	

For procedures to be effective in ensuring that tasks are performed correctly, they must be used. There are many reasons that plant workers may not use procedures. ***Deficient Procedures*** are the most prevalent problem in process industries since procedures have not traditionally been developed from the perspective of optimizing human factors; instead, procedures have been traditionally developed to meet a compliance requirement to have written procedures. Examples of procedure deficiencies (inaccuracies) include:

- Incorrect/incomplete/nonexistent. Most procedures we have audited have been only 70 – 85% accurate; the inaccuracies include missing critical steps, steps as written are not what needs to be done, or the steps are out of sequence
- No/misplaced warnings. For example, a warning should never **contain** the action to take; it should instead **emphasize** the action to take
- Poor format and presentation rules

Other reasons plant workers may not use procedures include:

- Procedures are out of date
- No procedure has been written for the task
- Plant workers cannot find the procedure they want to use
- Plant workers don't need a procedure because the task is simple
- Plant workers need more information than the procedures contain
- Plant workers see procedures as an affront to their skill
- Procedures are difficult to use in the work environment
- Procedures are difficult to understand

So, in addition to the rules for writing procedures that are shown in the table above, the organization must also address the reasons that cause the plant worker not to use the written procedure.

E. HUMAN-MACHINE INTERFACE (INCLUDING TOOLS)

The **Human-Machine Interface** (also known as the Human-System Interface, or HSI) is defined as the technology through which plant workers interact with plant systems to perform their functions and tasks. The major types of HSIs include alarms, information systems, and control systems. Each type of HSI is made up of hardware and software components that provide information displays, which are the means for user-system interaction, and controls for executing these interactions.

Plant worker use of HSIs is influenced directly by (1) the organization of HSIs into workstations (e.g., consoles and panels); (2) the arrangement of workstations and supporting equipment into facilities, such as a main control room, remote shutdown station, local control station, technical support center, and emergency operations facility; and (3) the environmental conditions in which the HSIs are used, including temperature, humidity, ventilation, illumination, and noise.

There are three important goals to be achieved in the design and implementation of the HSI. These are:

- **Design for operability** refers to designing the HSI to be consistent with the abilities and limitations of the plant workers who will be operating it. Weaknesses in the design processes can result in an HSI that is not well suited to the tasks that they must perform to ensure plant safety, resulting in increased workload, decreased performance by personnel, and an increased likelihood of errors.

  The wide-spread adoption of distributed control systems has provided many advances in process control. However, if implemented improperly, these systems can greatly reduce the operability of the facility. Although they were unable to provide as much information, old analog panels were typically laid out in a manner such that experienced plant workers were able to monitor the entire process and effectively identify conditions meriting attention. Due to space limitations, and their cost (e.g., hardware and installation), alarms were typically limited to those deemed to be associated with a significant process upset.

  The advent of distributed control systems has eliminated the space limitations associated with alarms, and greatly reduced their costs (in many cases, alarms can be added merely through programming). Without proper alarm prioritization, configuration structure and discipline, this can lead to "alarm flooding" (an excessive number of simultaneous alarms) during abnormal operating conditions. The human error of failing to respond properly to an alarm increases by a factor of over 1,000 if ten alarms activate simultaneously compared to a single alarm activation.

  Therefore, it is critical that control and alarm system designers in facilities with distributed control systems properly prioritize information so that plant workers can respond appropriately to abnormal conditions. Further, trouble-shooting guides should be readily accessible via the distributed control system to reduce human error in alarm response.

  Beyond the topic of alarms, the proper design of the visual displays can also reduce human error. Certain color combinations, such as a light gray background with darker gray equipment and black lines coupled with a flashing red indication for those components in an alarm state effectively draw the plant workers' attention to the relevant part of the process requiring attention. Further, an effective display should include:

  - Safety systems that are bypassed, impaired, or out of service
  - Process alarms that are disabled
  - Process conditions that are not within established limits
  - Backup systems that are unavailable
  - Special maintenance or testing activities that are currently in progress

- **Design for maintainability** refers to designing the HSI and associated plant equipment to ensure that plant workers are able to perform necessary maintenance activities efficiently. Weaknesses in the design process can result in systems that impose excessive demands on plant workers for maintenance and, therefore, are prone to maintenance errors or problems with reliability and availability.

- **Design for flexibility** refers to the way that changes, such as upgrades to the HSI, are planned, and put into use. A new HSI component may require the plant worker to perform functions and tasks in new ways. Skills that the plant worker developed for managing workload when using the former design, such as ways for scanning information or executing control actions, may no longer be compatible with the new design. The new HSIs must also be compatible with the remaining HSIs so that plant workers can use them together with limited possibilities for human error. Also, HSI modifications may not be installed or put into service all at one time, causing the plant worker to adapt to temporary configurations that are different from both the original and final configurations. Weaknesses in HSI implementation can increase plant worker workload and the likelihood of errors.

Tools are a special category of HSI, which typically refers to hand tools or devices that are generally designed with the plant worker in mind. A big part of human factor consideration is how to make the equipment and process operation mistake-proof (to prevent errors as much as possible). The following are some factors to consider when mistake-proofing in designs:

- **Design for unambiguous assembly.** Design the product or device such that the assembly process is unambiguous (by designing components so that they can only be assembled one way); i.e., design matching parts that are easy to insert and align. For example, use notches, asymmetrical holes, and stops to mistake-proof the assembly process. Products that go together in only one way require less plant worker training, perform more reliably, and can be repaired more quickly.
- **Consult plant workers.** Operators, technicians, and maintenance personnel can pinpoint the most troublesome areas.
- **Avoid symmetry.** When a particular orientation is critical to the design, avoid symmetry. For example, use nonsymmetrical hole patterns.
- **Use labels sparingly.** Labels tend to come off equipment too easily and often are wordy.
- **Review the environment.** Environmental problems that encourage mistakes include poor lighting; high/low heat; excess humidity, dust, and noise — anything that distracts plant workers.

The following are some factors to consider when mistake-proofing in process/operations:

- **Error-proof Mechanisms.** Error-proof mechanisms are very powerful in improving system reliability when incorporated into the design. These mechanisms, by design, will not allow a plant worker to perform an illegal operation. For example, if a plant worker enters a value that is outside the accepted range of operation, the control logic will not accept the value.

- **Automatic Alerts.** Automatic alerts immediately inform the plant worker of an illegal operation to prompt corrective action. These alerts are particularly useful in critical operations that allow time for corrective action before some adverse consequence occurs. For example, a piping manifold may be designed such that a warning alarm sounds if valves are not opened/closed in the proper, critical sequence, or if a critical valve is left opened/closed.
- **Automatic System Shutdown.** Automatic system shutdown should be incorporated into the design when an illegal action is performed during a critical operation and no time is available for corrective action. For example, if a plant worker uses the wrong sequence in charging a reactor, the reactor will shut down before other materials are charged that may lead to critical temperatures and pressures.

F. FITNESS FOR DUTY

Successful task performance requires that the capabilities that plant workers bring to the task fall within an expected range. **Fitness for Duty** issues include reduction in a plant worker's mental or physical capabilities due to substance abuse, fatigue, illness, or stress, which increases the likelihood of errors. Types of possible impairments include:

- Physical attributes – strength, reach, eyesight and color acuity, hearing,
- Mental attributes – drug and alcohol (abuse), mental stress (on and off the job);
- Fatigue – issues from on the job and off the job (especially control of hours per work-day and per work-week)

Safeguards to prevent fitness for duty related errors include company programs for the detection and prevention of potential or actual impairment, as well as the individual responsibility of plant workers to decline assignments if they are impaired for any reason. The latter safeguard is a weak one, however, because humans are generally over-confident of their capabilities when under the influence of drugs or alcohol, or are stressed, fatigued or ill. Other factors that may discourage self-reporting include the fear of poor performance reports from supervision or not receiving pay for extra overtime.

Company programs that may be implicated in errors caused by plant worker impairment include:

- **Fitness for Duty Program** – Company fitness-for-duty programs are primarily responsible for detecting and preventing impaired plant workers from performing tasks that may affect public health and safety. Medical evaluations of personnel, behavioral observation programs, employee assistance programs and drug and alcohol testing are used to detect impairment. Impaired plant workers can be prevented from performing tasks by establishing protocols for instances in which a plant worker is believed to be unfit for duty, training for supervisors on detection of and response to fitness for duty issues, and employment guidelines for personnel who voluntarily enter drug or alcohol treatment programs. Weaknesses in a Fitness for Duty program may allow impaired personnel to have access to vital areas in a plant where they could commit errors. One excellent starting point for a Fitness for Duty sub-element is the guidance provided in ***US NRC's 10 CFR 26 (2005)***[15]**.**

- **Overtime Policies and Practices** – Most companies establish limits for work hours to reduce on-the-job fatigue. It has been shown that 17 hours of work without a break results in the same error rate as being legally drunk. And, at 10 days straight of 12 hours work-days, the error rates for non-routine tasks such as startup of a continuous unit can increase to 1 mistake in 5 to 10 steps (as opposed to the target of 1 mistake in 100 steps). Routine authorization for work hours in excess of those recommended may result in fatigued plant workers. Further, a practice of excluding training or meetings that occur outside of a plant worker's normal work schedule from work-hour limitations will also contribute to fatigue. In ***US NRC's 10 CFR 26 (2005)***[15], the guidance given for control of overtime hours is no more 72 hours of work per 6-day period, no more than 16 hours of work in one day within that period, and a minimum of 24 hours contiguous hours away from work within a 7-day period. The US DOT has even more stringent rules on limiting work hours and establishing required hours for recovery from fatigue.
- **Shift Scheduling** – Shift scheduling may also affect the likelihood that plant workers will show performance decrements due to fatigue. A change in the assigned shift or a rotating shift schedule will disrupt circadian rhythms and may increase the likelihood of errors.
- **Safety Culture** – The effectiveness of self-reporting and behavioral observation programs depends greatly upon the safety culture at a site.

G. WORK PROCESSES AND SUPERVISION

Supervision is the process by which work is directed and overseen by first-line management. Successful supervision requires a combination of leadership skills and technical competence. Supervision differs from peer checking or quality control because a supervisor has line management responsibility for the plant worker(s) as well as responsibility for the work activity.

Supervision is more than the moment-to-moment direction of a work activity. Successful supervision requires the assessment and shaping of plant worker attitudes and motivation, communication and implementation of management expectations for performing work, the assignment of the best-qualified workers to various tasks, as well as the technical competence to identify incorrect actions and stop improper activities before an error is committed. Effective supervision involves directing the work, overseeing how it is performed, and leadership.

Organizations are typically structured to have sufficient supervision of the job, but many times the delineation between the trainer and the supervisor roles is blurred. Supervision can and normally does play a key role in selecting of the right worker for the job, scheduling of workers to match the required tasks for the day/week, and generally overseeing the task execution to ensure policies and procedures are followed. Supervisors are not always trained on all of their key roles in support of control of human factors, such as detecting issues in plant workers related to fitness for duty or fatigue.

H. WORK ENVIRONMENT

The **Work Environment** refers to the physical conditions in which work is performed. Environmental conditions that can affect performance include excessive vibration and noise, temperature extremes, and insufficient lighting, as summarized below:

- There are two types of **vibration** that may cause errors. The first is whole-body vibration, in which vibration is transferred to the plant worker from standing or sitting on a vibrating surface. The second is object vibration, in which a stationary worker interacts with a vibrating object in some fashion. The effects of vibration depend upon its frequency and acceleration. Frequency is the number of oscillations (cycles) that occur in one second. Acceleration is the force, or intensity, of the vibration.
- **Noise** is unwanted sound. Noise can cause errors in several ways. It may disrupt communications, affect the ability to perform tasks and annoy plant workers. The effects of noise on communications are complex. Even relatively low levels of noise can mask speech, but only under some circumstances. For example, speakers naturally raise their voices when there is background noise and may be able to overcome some of its effects on communication. Being able to see the speaker's face or using standardized phrases also improves communication in a noisy environment. The type of background noise also affects communication. It is easier to communicate over noise that is steady and uniform than noise that includes sharp tonal peaks, such as background speech.
- **Heat exposure** is a common problem in many areas of a plant, such as the turbine building when the plant is operating. The extent to which plant workers will be affected by heat depends on many factors. These include their physical characteristics, such as age, weight, acclimation to heat, physical fitness and dehydration. Other factors that determine the effects of heat on performance include airflow, humidity, clothing, and level of physical activity.
- **Exposure to cold** affects the performance of manual tasks. Decreases in the ability to control hand movements begin at an air temperature of approximately 54° F. The fingers may become numb to pain at this temperature and touch sensitivity is reduced. Performance of gross manual tasks, such as those involving the arms and legs is also degraded at 54° F. The speed at which manual tasks can be performed is affected by the rate of cooling. Slow temperature drops have a greater negative impact on manual dexterity than rapid temperature decreases, during the initial exposure period.
- **Adequate lighting** is required for accurate performance of nearly every task in a unit operation.

The organization must have engineering controls to help control each factor, but sometimes there is no other choice but to rely upon administrative controls.

I. COMMUNICATION (most importantly, verbal communication)

Communication is the exchange of information while preparing for or performing work. Verbal communication occurs face-to-face, by telephone, sound-powered phones, or walkie-talkies, as well as over public address systems. Written communication occurs, for example, through policies, standards, work packages, training materials, and e-mail.

Communication involves two sets of behaviors: (1) creating and sending messages and (2) receiving and interpreting them. Communication always involves at least two individuals, the sender and the receiver, and occurs:

- Between individuals
- Within and among work groups
- In meetings
- In pre-job or pre-evolution briefings
- During shift turnover

Successful communication requires several steps. The sender first develops the intention to communicate **either** verbally or in writing. The sender then composes a message that presents the meaning as clearly as possible. The receiver must pay attention to the message and then interpret its meaning. If the communication is successful, the receiver interprets the message consistently with the sender's intended meaning. However, there are many potential errors in both sending and receiving the communication:

- Sending Errors
    - *Content wrong*
    - *Content inconsistent*
    - *Content inappropriate for the job*
    - *Content inappropriate for the receiver*
    - *Standard terminology not used*
    - *Familiar terminology not used*
    - *Message production inadequate or interfered with*
    - *Necessary information not sent*
    - *Wrong place or person*
    - *Wrong time*
    - *Sending verification failure*
- Receiving Errors
    - *Information not sought*
    - *Information not found*
    - *Information not used*
    - *Receiving verification failure*
    - *Message misunderstood.*

Communication errors may be reduced in the following ways:

- Structuring messages in a standard format, which alerts the receiver if important information has been skipped
- Providing written instructions or procedures for complicated tasks
- Establishing protocols to repeat key parts of verbal communications
- Preceding special or nonroutine activities with a pre-job briefing
- Using structured protocols, checklists and logs to supplement written instructions

One key opportunity for communication errors is when a new shift takes over responsibilities from the previous shift. An effective tool to reduce these errors is a properly designed shift handover process that includes appropriate checklists and logs of activities, and a structured and consistent method for verbally reviewing this information. Additionally, the logs and discussions need to include:

- Routine operations
- Transition operations (e.g., changes to produce a different product or switch to a different raw material source)
- Nonroutine operations (e.g., maintenance activities or temporary operating conditions)

6. Establishing a Safety-Principled Organizational Culture

To ensure the success of the management systems created to improve Human Factors for plant workers, the proper culture of the organization must be developed so that this can be done effectively. Merriam Webster defines culture as "the set of shared attitudes, values, goals, and practices that characterizes an institution or organization". An organization's culture ultimately reflects the collective values of individuals and groups and can be influenced by the personalities of charismatic leaders, by shifting priorities based on the perceived scarcity of resources or actions necessary for survival, or by a sustained history of successes, behaviors, and beliefs. In an especially sound culture, deeply held values are reflected in the group's actions.

Organizational Culture is a somewhat intangible attribute of a company that is evident by the words and actions of management and all employees – including the plant workers. It is further defined informally as "how we do things around here", "what we expect here", or "how we behave when no one is watching". Reducing human error by plant workers therefore requires that the proper organizational culture be built and maintained.

A Safety-Principled Organizational Culture is one in which the actions of the entire organization are driven by the genuine desire to complete every task and achieve every objective in a manner that protects employees, plant equipment, and the community in which the plant is located to the greatest extent possible. While profitability and production targets are also important goals for the organization, the entire organization understands that these are enabled by continuous safe operations, rather than despite them.

Establishing a strong Safety-Principled Organizational Culture is a long-term process that requires dedicated resources for as long as the organization exists. Additionally, it is the result of the quality and quantity of effort put into its implementation. A solid, sensible effort by management and technical staff to establish a culture of this nature will be noticed by the plant workers and they will want to help. This will have the effect of catalyzing the program, thereby helping to ensure its success. Alternatively, merely stating the objective of building this culture will have little effect on creating excellent organizational safety culture if the implementation and control of safety is weak.

Establishing a strong Safety-Principled Organizational Culture will minimize safety incidents, and this will save lives. Since essentially all accidents begin with a human error, the actions or inactions of the plant worker can be the limiting factor in determining overall safety performance.

The proper culture helps plant workers understand why strict adherence to procedures is the right thing to do. Further, it helps to enable thoughtful compliance and to alert management when procedures are wrong. In this way, a sound culture is essential to maximizing the results of ALL safety activities.

Establishing a strong Safety-Principled Organizational Culture can be achieved by successfully pursuing 5 Key Strategies – each of which are described below.

Strategy 1: Establish Safety as a Core Value

Attaining the goal of superior plant worker performance (i.e., low human error rates) begins by providing strong and consistent communication about the importance of safety **from the very top of the organization**. This is achieved by setting an appropriate tone that safety is equal to or greater than production and by not diluting the message with a lot of other high priority core values. The first step in setting the tone is by clearly stating the importance of safety in a high-level mission statement. Then, this mission statement must be communicated to the organization, along with the related goals and plans for achieving it. An important aspect that must be communicated is the key role of controlling plant worker human error by focusing on Human Factors. Additionally, it is critical that this effort not be reduced to a "program of the month" – it must be spoken about frequently and consistently to all levels of the organization.

Of course, words are hollow without consistent corresponding actions. Therefore, strong leadership must be provided by demonstrating the commitment to safety excellence through actions and decisions. Replace slogans and banners with high standards, shared values, and core beliefs which sustain the organization regardless of who is in charge. To help ensure these values propagate throughout the entire organization, managers must be educated in this culture, vision, and related standards, as well as the expectations for their roles and responsibilities in this critical effort.

A strong Safety-Principled Organizational Structure relies upon high standards of performance that are consistently enforced. This is achieved by first defining appropriate safety goals – namely, those that provide an indication of the organization's capability to perform their work safely. Too often, organizations tie the success of their safety programs on the lagging indicator of incident rates. While this is of course the ultimate goal of the organization, this isn't necessarily a reliable indicator of safety for the organization. Consider how many ways a plant worker can be hurt completing a single task, then multiply that by the number of tasks the individual completes per day, and then by the number of plant workers within the organization. Clearly there are thousands of opportunities for injury each day. Fortunately, however, for nearly all organizations the number of incidents that occur is very small compared to the number of opportunities for an incident. Therefore, unless incident rates are abnormally high, a change in an organization's injury rate is typically due to factors outside of their control. It is much more valuable, then, to establish goals centered around the organization's work in improving their safety-related management systems.

Another aspect of establishing safety as a core value is to establish responsibilities and accountabilities for process safety roles. Every individual within the organization should be held accountable for steadfastly adhering to the newly established expectations. While it may be tempting to overlook seemingly minor deviations from these expectations, this can lead to normalizing deviations which can potentially lead to serious incidents due to a relaxed response to these warning signs of potential hazards. Therefore, the leaders of the organizations must hold direct reports accountable for demonstrating behavior consistent with this value and set expectations for each level of the organization to hold their direct reports to this same behavior.

Finally, it must be recognized that improving the organizational culture involve costs and benefits that are both long-term in nature and sometimes difficult to measure. By establishing appropriate goals as noted above that focus on management system health, however, the organization has the means to continuously improve its performance.

Completing all of the above will help ensure that the core value of safety is consistently embraced throughout the entire organization, including the plant workers.

Strategy 2: Provide the Necessary Resources to Ensure Success

Providing resources to transform the organizational culture is the most tangible way to demonstrate to the plant workers that safety is indeed an organization's core value. In this context, resources constitute monetary commitments, as well as those of personnel.

The place to begin is to fund and allocate personnel to enable good safety performance by creating and effectively implementing the Human Factor management systems described above. Done properly, these will close any gaps in the control of Human Factors as well as those evident in existing systems that don't meet a critical need such as performing a PHA of procedures to find scenarios unique to non-continuous modes of operation as described by Bridges and Marshall[26]. Adequate personnel must also be allocated for completing safety-related tasks to support these management systems (inspections, maintenance, etc.), for maintaining these management systems as changes or newly identified requirements occur, and for training and educating personnel.

It is also important to designate a high-ranking leader for safety who can help ensure that major decisions are based upon a proper consideration of safety. But it is also important to provide the positional support necessary to promote strong safety performance throughout the organization. This is achieved by identifying resources on paper and filling with effective and competent people. For example, if a trainer is shown on an organization chart, it should be ensured that this is their full-time job and that they aren't responsible for other unrelated tasks. This demonstrates that the organization places a high value on training and development of people and groups. This can be further enhanced by ensuring that training materials are maintained current, and by maximizing the use of instructor-led training, rather relying than too much on the use of Computer-Based Training.

Furthermore, a properly designed and implemented training program will allow the organization to demonstrate that it places a high value on safety knowledge and expertise. This can be enhanced by identifying and providing the proper mix and amount of expertise for its most important tasks (for example, by pairing an individual with a high level of expertise with one who has the potential, but not yet the experience, to develop this same level of expertise). To properly ensure optimum prevention of human error, the organization should focus on building expertise in Human Factors and seek out expertise for critical decisions.

Finally, since excessive overtime impacts employees' abilities to perform their jobs safely and increases safety risks, the organization must devote sufficient manpower resources to properly control overtime.

In summary, the overarching approach here is to ensure that a structured, strategic approach is developed to properly allocate funds and personnel to create, implement and maintain the management systems needed for safety enhancement opportunities.

Strategy 3: Establish a Blame-Free and Learning Environment

It is critically important that the organization commit to a blame-free culture for all plant worker human errors, including for near misses reported and investigation findings. At an intellectual level, everyone understands that no one is perfect and that zero human errors are not possible. Plant workers are no exception. A key driver for the creation of a blame-free environment then is driven by how the organization responds to the inevitable human errors that plant workers will make.

The response to plant worker errors can be thought of as a continuum that spans two extremes; on the left side the response is one of blaming or fault-finding, while on the right side the response is one of trying to understand or learn why the error was made. Quite often people will automatically default to the "blame" side of the continuum for plant worker human errors, saying something to the effect that if only the plant worker had handled the situation differently the error would not have occurred. While this is true at face value, it is not a helpful response. After the fact, it is easy to see what should have been done differently because the outcome is now known. But when the plant worker was taking actions, this outcome was not obvious to him or her. Significantly, people do not come to work planning to make errors; instead, they come to work to complete their tasks in the most efficient manner possible. It is only after an incident occurs that the benefit of alternate actions is clear. Thus, the more productive response is to try to understand the context of the plant worker's action – what was it that made this particular action appear to be correct? It is only by taking this approach that the lessons from the incident can be fully understood.

Each person in the organization needs to understand where on the human error response continuum they reside and make every effort to move as far to the right side as possible. As shown in Figure 5, there are three aspects to an individual's response to a plant worker human error: the **internal response** (what we say to ourselves when we are made aware of a human error), the **external response** (what

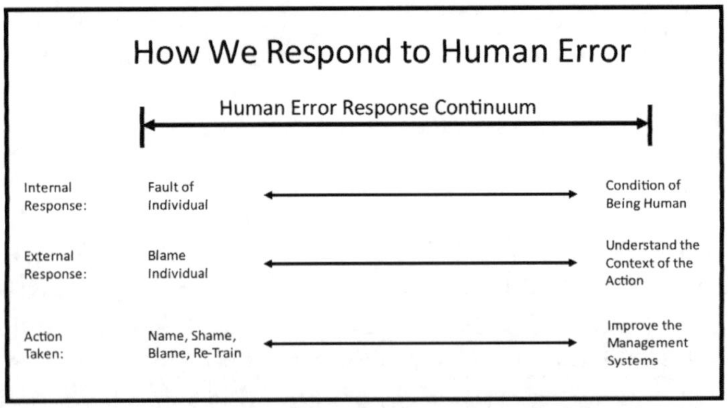

Figure 5: How We Respond to Human Error

we say out loud when we are made aware of a human error), and the **action taken** (what we do when we are made aware of a human error).

The internal response continuum ranges from thinking that the error was the fault of the plant worker, to recognizing that the error was a condition of that plant worker being human. The best approach for an individual to move to the more productive right side of this spectrum is to consider experiences in their own life in which they have made an error, which in retrospect seems to have been easily avoided. With this perspective, it is easier to understand that the plant worker who made the error in question was doing the best job possible but did not fully understand a specific hazard associated with it.

The external response continuum ranges from blaming the plant worker, to understanding the context of the action (i.e., why it made sense for the plant worker to take the action he or she took). Only by understanding the context of the action is it possible to develop measures so that another plant worker faced with the same choice does not repeat the error. As an example, a major chemical company experienced an acid spill from a tank truck one very cold evening. The truck driver was directed by the chemical company's guard to the wrong operating unit and began unloading the acid into that unit's tank. The operator from the unit soon emerged from the local control room and informed the driver that no acid was needed there. After a few phone calls, it was learned that the delivery was for a different unit which was located about ½ mile away. Accordingly, the driver disconnected his truck but failed to properly tighten the unloading hatch on top of the tank truck (most likely due to the cold ambient temperature). Therefore, as the truck was driven to the other unit, acid was spilled along the entire route. The chemical company's response to this incident was to inform the delivery company that that driver was no longer allowed on company property – in essence, blaming the driver for the spill. But this decision left many questions unanswered. Why was the driver directed to the wrong unit by the company's guard? Why was the driver allowed to unload directly into a company tank without checking in to the local control room (which is a requirement even for company employees visiting from a different plant area)? Only by answering these questions – understanding the context of the action – could assurance be provided that future spills of this nature would not occur.

The action taken response continuum ranges from a combined action of naming the plant worker responsible, shaming the person for making the error, blaming them for the consequences, and then re-training them, to making improvements to the underlying management systems. Using the acid spill example, the driver most certainly knows how to correctly complete all of the tasks needed to safely unload product; therefore, training would provide no real benefit other than to make him feel worse than he already did. However, by improving the management systems (strengthening the process for directing drivers to the proper locations, enforcing the rule to check in to the local control room before unloading) a better assurance that the human error could be prevented from recurring is possible.

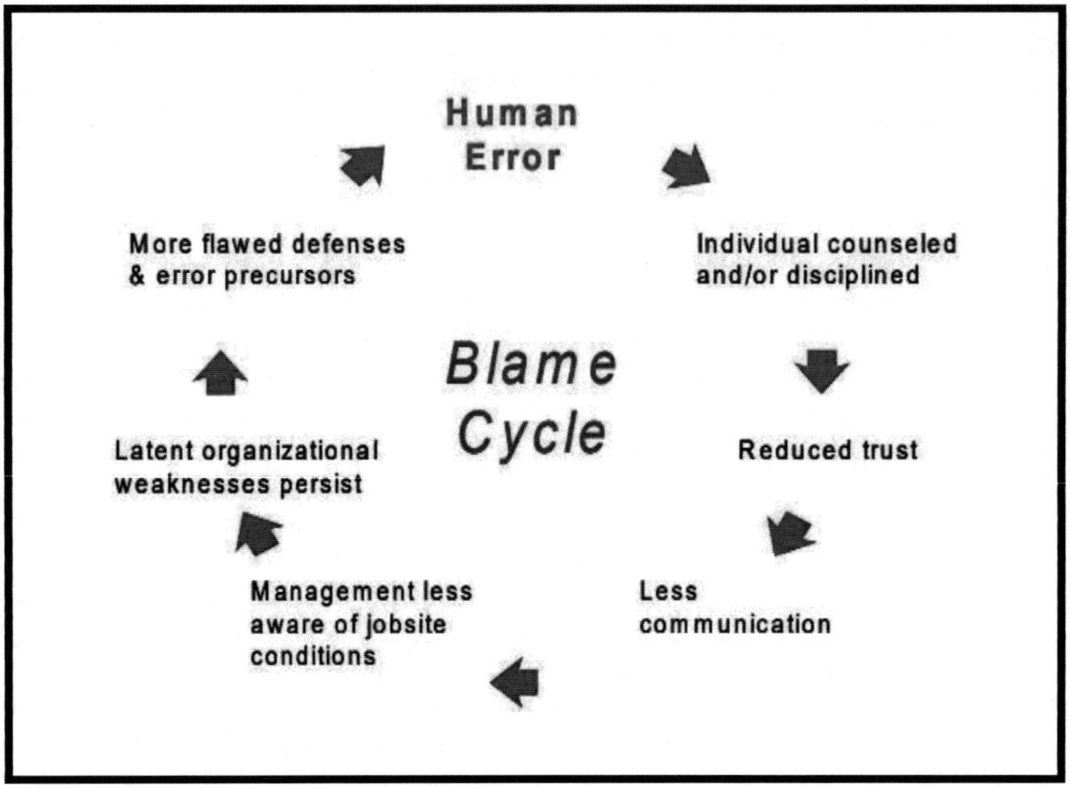

Figure 6: The Blame Cycle

The above discussion highlights the fact that residing on the "blame" end of the human error response continuum prevents an organization from truly learning about the root causes of incidents, thereby potentially leading to a recurrence of the error. However, blaming plant workers pose further risks to the organization. This is illustrated by the "Blame Cycle" that was developed by the US Department of Energy[27]. This cycle establishes a phenomenon that when a plant worker is blamed for a human error, this results in this individual ultimately not reporting future errors, which the organization is therefore unaware of and is unable to address. This then leads to future human errors due to these latent weaknesses, which are unreported, and the cycle repeats itself potentially leading to very serious consequences.

With this perspective then, a blame-free and learning environment can be achieved by ensuring that the first questions following an incident are what and why, not who. Those are followed by asking how we can improve the procedures, task design, practices, equipment, and other management systems to prevent this from happening again. With this environment, all incidents (especially near misses) are reported and investigated without blame, which results in root causes (management system failures) being identified, and corrective measures being taken to implement sustainable and permanent solutions. Trust and collaboration are high because the culture is blame-free when it comes to mistakes and people ask for help when they need it and give help without being asked.

Some key approaches to establishing a questioning/learning environment include properly performing risk reviews, conducting thorough incident investigations, performing table-top drills, and sharing and seeking lessons inside of and outside of facility. Significantly, plant workers should be highly involved in each of these activities. It is also critical to maintain a high awareness of hazards and consequences and remain vigilant for indications of system weaknesses. This can be achieved by ensuring that all plant workers are educated on the hazards of the operation, that they understand how safety hinges on control on Human Factors, that they all are educated on the consequences of deviation from safe practices, and that they share lessons learned.

A final aspect is to provide timely response to, and communication of, safety issues and concerns and lessons learned from incident investigations, audits, risk assessments, employee concerns and discrepancies between practices and procedures.

In summary, an organization must ensure that a blame-free and learning environment is consistently embraced throughout the entire organization. This can be achieved by reinforcing this at all opportunities through words and actions by personally seeking solutions instead of blame. Furthermore, top management must hold direct reports accountable for demonstrating behavior consistent with this value and set expectations for each level of the organization to hold their direct reports to this same behavior.

Strategy 4: Enhance Employee Empowerment and Involvement in Safety

Plant workers can best contribute to the Safety-Principled Organizational Culture and thus successfully fulfill their safety responsibilities by having the necessary authority and accountability for safety responsibilities. Leadership can enable this by reinforcing that all employees have responsibilities to themselves, fellow employees, company, and community and by involving the plant workers in safety management systems to ensure that they are widely understood and supported. This approach has the added benefit of recruiting additional resources to support the many tasks required.

A prerequisite to this is for the organization to ensure open and effective communication – both vertically and horizontally. Strategically, this can be achieved by emphasizing the need to promptly observe and report non-standard conditions. A key management system to enable this is a properly deployed near miss reporting program. Bridges[28] outlines an approach for creating a program of this nature, as well as specific, common hurdles that must be overcome to ensure its success. Key items in a properly designed program include first providing means for all plant workers to communicate hazards to management (in a blame-free environment) and for management to consistently act on these stated concerns.

Related to this, plant workers should be involved in identifying hazards and deciding how they should be addressed. In certain situations, plant workers should be authorized to take action to address hazards without management involvement. A very important example of this is that plant workers should have the authority to question decisions which are being made about process safety, including a "Stop Work Authority".

Ensuring optimum plant worker involvement in the safety management systems must be intentional. Therefore, the organization should develop an Employee Participation Plan, which lists key components of the management system that plant workers can best contribute to. A few examples to include identifying and controlling deficiencies in Human Factors, writing and reviewing operating and maintenance procedures, and participating in Process Hazard Analyses, Root Cause Analyses, and audits. Other opportunities can be developed by critically reviewing components of all safety management systems (including those developed for Human Factors) that will benefit by having the specialized knowledge of plant workers.

Strategy 5: Drive Continuous Improvement in the Safety Management Systems

Even the best designed safety management systems are not perfect (because they are developed and managed by humans). Therefore, the best Safety-Principled Organizational Cultures undertake specific activities to understand gaps in the systems and then take actions to close these gaps. This is achieved by establishing relevant metrics on the health of the management systems, which as noted previously, should predominately be leading metrics. Typical examples include the timeliness of correcting deficiencies, adherence to ITPM schedules/procedures, and the expansion of problems detected in one area to investigate similar applications in other areas.

Over time, trends should be developed of these metrics, presented to management who should then direct actions to improve these Key Performance Indicators. Further, these metrics should be shared within the organization so that all employees – including plant workers – have a clear sense of how well their efforts to contribute to the success of these programs are working. It is important to remember that many of these safety-related metrics have long-term results, rather than an immediate impact. Therefore, continued diligence in these continuous improvement efforts are needed to ensure a sustained Safety-Principled Organization Culture, and ultimately outstanding safety performance for the organization.

7. Closing

Human error is inevitable. And the human errors that are typically thought of as the ones that lead to serious accidents are often attributable to those made by plant workers. Although this assessment is unfair (as other members of the entire organization contribute in meaningful ways to these incidents), it is nonetheless very important for organizations to undertake measures to reduce plant worker human errors.

The means to do this are outlined in this paper. It relies first on the organization having the proper culture – one that doesn't react to plant worker errors by blaming them; rather, it seeks to understand why the error was made and taking actions to fix the issue instead of the plant worker. It recognizes that to do this requires that the error must be traced to the underlying Human Factors that increased the likelihood of this error's occurrence, and that the best way to improve these Human Factors (and thus reduce plant worker human errors) is to develop and implement robust management systems for them.

The information provided in this paper to strengthen the plant worker Human Factor management systems is the most important method to reduce plant worker human errors. However, even with perfectly implemented management systems human errors will still occur – albeit at a lower frequency and likely with a smaller adverse impact than before. Therefore, in cases in which the consequence of a human error is unacceptably high, the organization should implement solutions that will lower the residual risk further. These solutions can be designed to either help prevent the error or compensate for the error, but **they must be independent of the human error initiating event**. Bridges and Bridges [25] have compiled a list of the best remedies for human error for these situations. In a similar vein, plant worker human error can be further reduced by performing the previously mentioned PHA of procedures to find scenarios unique to non-continuous modes of operation as described by Bridges and Marshall[26].

Finally, a small fraction (about 15%) of plant worker human error is attributable to acquired habits, or behaviors; the remedy for which is a properly designed and implemented plant worker observation program. The reader is encouraged to review these other important documents and methodologies to obtain a fully comprehensive understanding of the best ways to reduce plant worker human errors.

8. References:

1. Swain, A. D., Guttmann, H. E., *The Human Reliability Handbook - with Emphasis on Nuclear Power Plant Applications*; NUREG/CR-1278, U.S. Nuclear Regulatory Commission, Division of Facility Operations, 1983.

2. Gertman, D.; Blackman, H.; Marble, J.; Byers, J. and Smith, C.; *The SPAR-H Human Reliability Analysis Method,* NUREG/CR-6883, U.S. Nuclear Regulatory Commission, Office of Nuclear Regulatory Research, Washington, DC, August 2005.

3. Bridges, W., and Tew, R., *Human Factors Elements Missing from Process Safety Management (PSM),* 6th GCPS, AIChE, March 2010.

4. *Risk-Based Process Safety*, CCPS/AIChE, John Wiley & Sons, Inc.; Hoboken, NJ, 2007.

5. *Conduct of Operations and Operational Discipline*, CCPS/AIChE, John Wiley & Sons, Inc.; Hoboken, NJ, 2011.

6. *US FAA Advisory Circular* 120-42A (as part of ETOPS approval)

7. Bridges, W. G., Williams, T. R., "Effective Procedure Writing – A Two-Part Series (Part 1: Reducing Risk of Worker Errors; Part 2: Addressing Process Safety Management Requirements for Operating Procedures)," *Chemical Engineering Progress*, 1997.

8. Bridges, W., and Collazo-Ramos, G., *Human Factors and their Optimization,* 8th GCPS, AIChE, April 2012.

9. *Guidelines for Hazard Evaluation Procedures*, 3rd Edition, CCPS/AIChE, John Wiley & Sons, Inc.; Hoboken, NJ, 2008.

10. Bridges, W., *Optimizing Hazard Evaluations,* LPS/AIChE, April 2009.

11. Bridges, W., Lorenzo, D., and Kirkman, J., *Addressing Human Errors During Process Hazard Analyses,* CCPS Conference, AIChE, 1991 (also published in Chemical Engineering Progress May 1994).

12. *Guidelines for Investigating Chemical Process Incidents, 2nd Edition,* CCPS/AICHE, John Wiley & Sons, Inc.; Hoboken, NJ, 2003.

13. Bridges, W., *Gains in Getting Near Misses Reported,* 8th Conference, ASSE-MEC, Bahrain, 2008.

14. Della Rocco, P.S., et al., *The Effects of Napping on Night Shift Performance,* Federal Aviation Administration, U.S. Department of Transportation, 2000.

15. *Fitness for Duty Programs, Preamble, 10 CFR 26, U.S. Nuclear Regulatory Commission, Revised 2005.*

16. US Chemical Safety and Hazard Investigation Board (CSB), *Anatomy of a Disaster: Explosion at BP Texas City Refinery,* published 2006.

17. University of South Australia (UniSA), Adelaide Centre for Sleep Research, *SDA Fact Sheet - AT03*, 2006.

18. Hallbert, B.; Whaley, A.; Boring, R.; MCabe, P.; and Chang, Y.; *Human Event Repository and Analysis (HERA)*, NUREG/CR-6903, U.S. Nuclear Regulatory Commission, Office of Nuclear Regulatory Research, Washington, DC, 2007.

19. Bridges, W., and Clark, T., *LOPA and Human Reliability - Human Errors and Human IPLs,* 7th GCPS, AIChE, March 2011.

20. Bridges, W., and Thomas, H., *Accounting for Human Error Probability in SIL Verification Calculations,* 8th GCPS, AIChE, April 2012.

21. Bridges, W., *The Impact of Human Factors on LOPA,* 9th GCPS, AIChE, April 2013.

22. Bridges, W., and Tew, R., *Best Practices for Writing Operating Procedures and Trouble-shooting Guides,* 13th GCPS, AIChE, March 2017.

23. Bridges, W., *Proven Approaches to Ensuring Operators Can Respond to Critical Process Deviations in Time (Human Response IPL),* 13th GCPS, AIChE, March 2017.

24. Bridges, W. and Rhodes, W., *Everything you need to Know about Human Reliability and Process Safety,* 17th GCPS, AIChE, March 2019.

25. Bridges, S. and Bridges, W., *Abnormal Modes of Operation – A risk-based focus,* 16th GCPS, AIChE, March 2020

26. Bridges, W. and Marshall, M., *Necessity of Performing Hazard Evaluations (PHAs) of Non-normal Modes of Operation (Startup, Shutdown, & Online Maintenance)*, 12th GCPS, AIChE, April 2016

27. DOE Standard – *Human Performance Improvement Handbook, Volume 1: Concepts and Principles*

28. Bridges, W., *Gains from Getting Near Misses Reported*, 8th GCPS, AIChE, April 2012

619491 Lessons-Learned from Remote HAZOP/LOPA Applications in 2020

Abstract:

Going into Year 2020, basic brainstorming team session approaches for the performance of a Hazard and Operability (HAZOP) Study or a Layer of Protection Analysis (LOPA) had not changed significantly over the years; however, the infection prevention measures necessary to address COVID-19 resulted in a rapid transition to widespread application of telecollaboration tools that previously had only been applied sporadically. Prior to 2020, telecollaboration tools for Process Hazard Analysis (PHA) had been used to short-circuit the need for pulling together international teams, for very short PHAs or MOCs, and other select applications. As 2020 drew to a close, numerous applications of HAZOP/LOPA Team Exercises using telecollaboration tools resulted in valuable insights into tools and techniques that were productive for different PHA applications. Properly applied, applications ranging from shorter Process Hazard Analyses (PHAs) and Management-of-Change (MOC) evaluations to longer studies addressing Five-Year PHA Updates and Capital Projects can be implemented productively using telecollaboration approaches.

Post COVID-19, many PHA applications will still be best addressed with the HAZOP/LOPA Team being co-located, but the savings in travel costs, the logistics associated with personnel travel, and the ability to quickly convene and implement a HAZOP/LOPA review can allow remote HAZOP/LOPA, through the use of contemporary technologies, to be an attractive option that can increase overall efficiency, without compromising quality. Experiences in the successful application telecollaboration tools in 2020 mean that Remote PHAs are here to stay.

A core objective of this paper is to provide insights into good applications and tips that can directly be used by the HAZOP/LOPA Facilitator to optimize the logistics to get the most out of a HAZOP/LOPA, where some or all team members are geographically distant.

SPRING21
+17TH GCPS
A Joint AIChE and CCPS Meeting

Research on Human Resource Development Exercises for Resilience to Cyber Incidents in Critical Infrastructure

Yuitaka Ota
Nagoya Institute of Technology,
System Management and Engineering
Showaku, Gokiso, 4668555 Aichi, Japan
cjr17009@nitech.jp

Erika Mizuno, Asai Haruna, Tomomi Aoyama, Yoshihiro Hashimoto, Ichiro Koshijima
Nagoya Institute of Technology,
System Management and Engineering
Showaku, Gokiso, 4668555 Aichi, Japan
ckv17032@nitech.jp, h.asai.852@nitech.jp, aoyama.tomomi@nitech.ac.jp,
hashimoto@nitech.ac.jp, koshijima.ichiro@nitech.ac.jp

Prepared for Presentation at
American Institute of Chemical Engineers
2020 Spring Meeting and 17th Global Congress on Process Safety
Virtual
April 18 - 22, 2021

Research on Human Resource Development Exercises for Resilience to Cyber Incidents in Critical Infrastructure

Yuitaka Ota
Nagoya Institute of Technology,
System Management and Engineering
Showaku, Gokiso, 4668555 Aichi, Japan
cjr17009@nitech.jp

Erika Mizuno, Haruna Asai, Tomomi Aoyama, Yoshihiro Hashimoto, Ichiro Koshijima
Nagoya Institute of Technology,

Keywords: OT-Security, cyber incident exercise, business continuity management

Abstract

Current cyber-attacks target IT systems and OT (Operational Technology) systems that are indispensable for the automatic operation of critical infrastructure industries. There have been various reports of cyber incidents targeting the OT system. In 2019, Norsk Hydor was infected with particular ransomware worldwide through its corporate network, forcing the manufacturing site to operate manually. In May 2020, Taiwan's CPU, responsible for Taiwan's oil supply, was targeted by a cyber-attack.

COTS (Commercial Off The Shelf, such as Windows OS, Intel PC, and open source applications) devices have been proposed to reduce the deploying OT system budget. Also, DX efforts are being made to shift from physical operation to virtual operation by using virtualization with IoT, AI, and Cloud. Due to this technological trend, the expanded attack surface is forced to manage the OT system as a "Zero Trust System" that assumes cyber-attacks. In IT systems, countermeasures such as whitelisting and firewalls prevent attackers from outside the system. For OT system, Network Security Zoning described in IEC62443 is recommended to mitigate the effects of cyber-attacks.

However, attackers exist all over the world, and they are continually developing new attack methods. For this reason, the authors believe that it is impossible to prevent all attacks, no matter what measures the defenders take. Therefore, when responding to cyber-attacks against the OT system, it is essential to consider countermeasures by focusing on plant accidents caused by control system failures, most of which are assumed as safety measures.

To ensure the plant's safety and continue the business operation even if it has a cyber-attack on the OT system, it is indispensable to develop human resources to respond to the incident flexibly. For this reason, we had been developing cybersecurity exercises to acquire the meta-knowledge necessary to respond to cyber-attacks targeting the OT system.

In this paper, the authors present the design framework, the implementation method, and OT expert's evaluation results of a proposed exercise.

1 Introduction

1.1 Background

An industrial control system (ICS) is a device or group of devices for managing and controlling other devices and systems. ICS is widely used in the Critical Infrastructure (such as electric power, city gas, and transportation) and the manufacturing industry that supports our daily lives, from those that are familiar to us. The conventional ICS has an isolated network as an independent system built with its hardware, software, communication protocol and has been considered resilient to cyber-attacks. However, in recent ICS, the adoption of COTS (Commercial Off The Shelf, such as Windows OS, Intel PC, and open source applications) devices is progressing due to versatility and budget. Also, DX (Digital Transformation) is being promoted for the purpose of plant operation support and operation optimization using IoT (Internet of Things) and AI (Artificial Intelligence) through open networks with the outside. As a result, ICS is exposed to the threat of cyber-attacks, and companies using ICS need to recognize not only the risk that has been assumed as the risk of plant shutdown but also cyber-attacks as one of the risks.

Cyber-attacks targeting ICS have been increasing in recent years. In 2010, malware called Stuxnet attacked Iran's uranium enrichment centrifuge [1]. Hackers caused power outages in Ukraine in 2015[2] and 2016[3]. In 2019, Norsk Hydros in Norway, the world's leading manufacturer of rolled aluminum products, was affected by cyber-attacks in plant operations [4]. In addition, the Information Security Agency (BSI) of the German Federal Government has published "Industrial Control System Security: Top 10 Threats and Countermeasures" [5].

In this way, cyber-attacks are a risk that threatens the safety of the plant, and companies must respond as a risk that threatens their business.

1.2 Response to a cyber incident

1.2.1 Cyber-attacks targeting Industrial Control System

Fig. 1 shows a cybersecurity testbed and a virtual ICS network. The testbed (Fig. 1) heats water with a heater in the lower tank and circulates the heated water through a simple pipeline to the upper tank (using a pump). The testbed is also equipped with actual industrial control devices and is controlled automatically[6]. In this ICS network, OLE for Process Control (OPC) Servers collect and exchange process data and monitor them by using Supervisory Control And Data Acquisition (SCADA) function included in the OPC Servers. OPC Servers receive various parameters from the testbed's sensor, and SCADA gives instructions to the control equipment using the parameters' data.

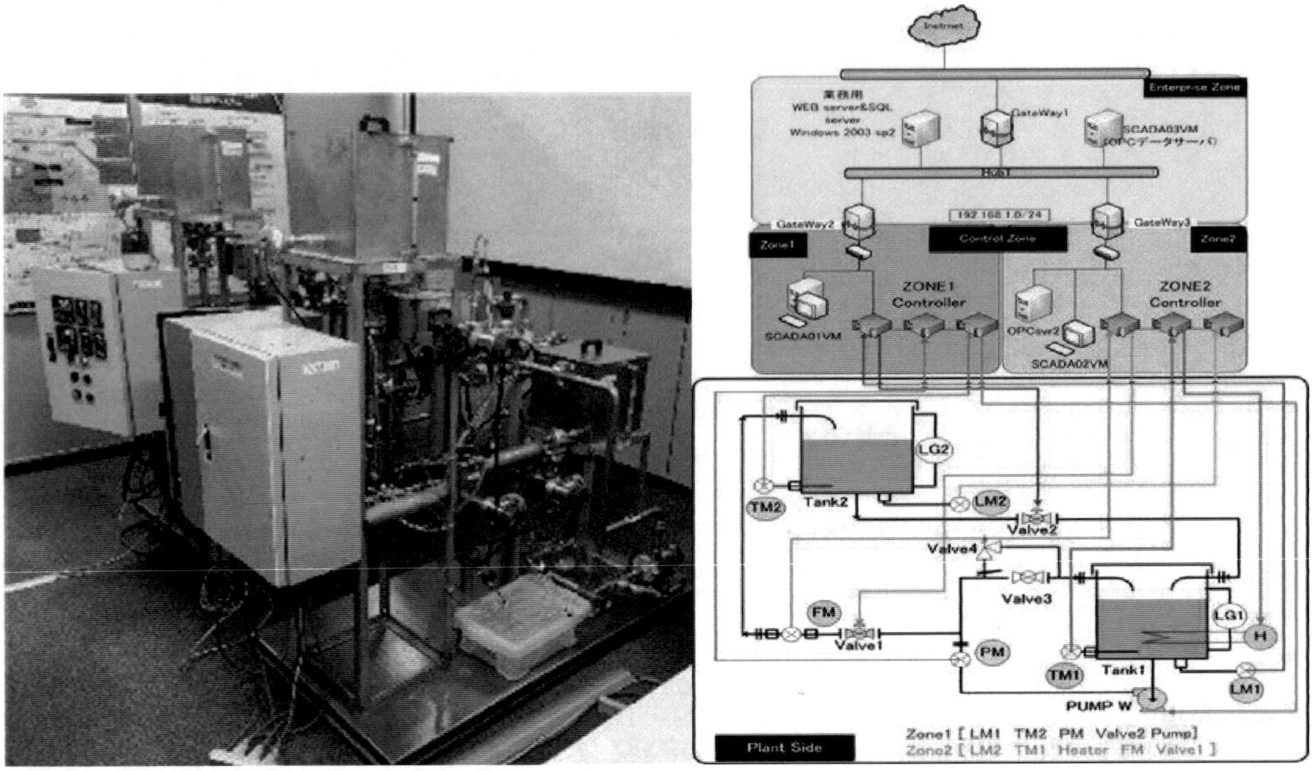

Figure 1. Cyber Security Testbed and ICS network

The authors have demonstrated cyber-attacks against the testbed, according to the steps of the anatomy of a cyber-attack. The attacking procedure is determined as follows.

1) Attack Step 1: Reconnoitering and enumeration

The attacker executes the Armitage to launch the Metasploit framework. Then, the attacker executes the Nmap, which scans a target network (ICS-2 Network 192.168.12.0/24). After completing the scan, the attacker gets target information.

2) Attack Step 2: Intrusion and advanced attacks

According to the functions of the Armitage, the attacker ascertains an attack method and vulnerability in the Metasploit framework. The attacker exploits payloads, which execute inappropriate code in the use of the vulnerability. After sending the payloads, the attacker obtains the administrator's permission for the OPC2 illegally.

3) Attack Step 3: Malware insertion

The attacker uploads a malware execution file to an OPC2 background process to execute a malware file. The configuration file of OPC2 was rewritten, and then the OPC2 was restarted. They were resulting from the attack, the valve controlled by the OPC2 shutdown.

4) Attack Step 4: Clean-up

The attacker deletes the malware file and the configuration file from OPC2 storage to not leave the cyber-attacks evidence and finish the cyber-attack.

As the valve is opened, the stage of the testbed changes. Usually, the testbed's state transition appears on the monitor of the operator (Fig. 2). The changing graph shows the liquid level of the upper tank. The liquid level is changing due to the malfunction of the valve. The operator can notice the testbed abnormality due to the unusual process deviations from the steady-state condition through the monitor. However, the attacker conceals the information about cyber-attack in order not to let anyone notice the cyber-attack. The testbed monitoring screen is concealed, as shown in Fig. 3, while the cyber-attack causes severe accidents at the testbed.

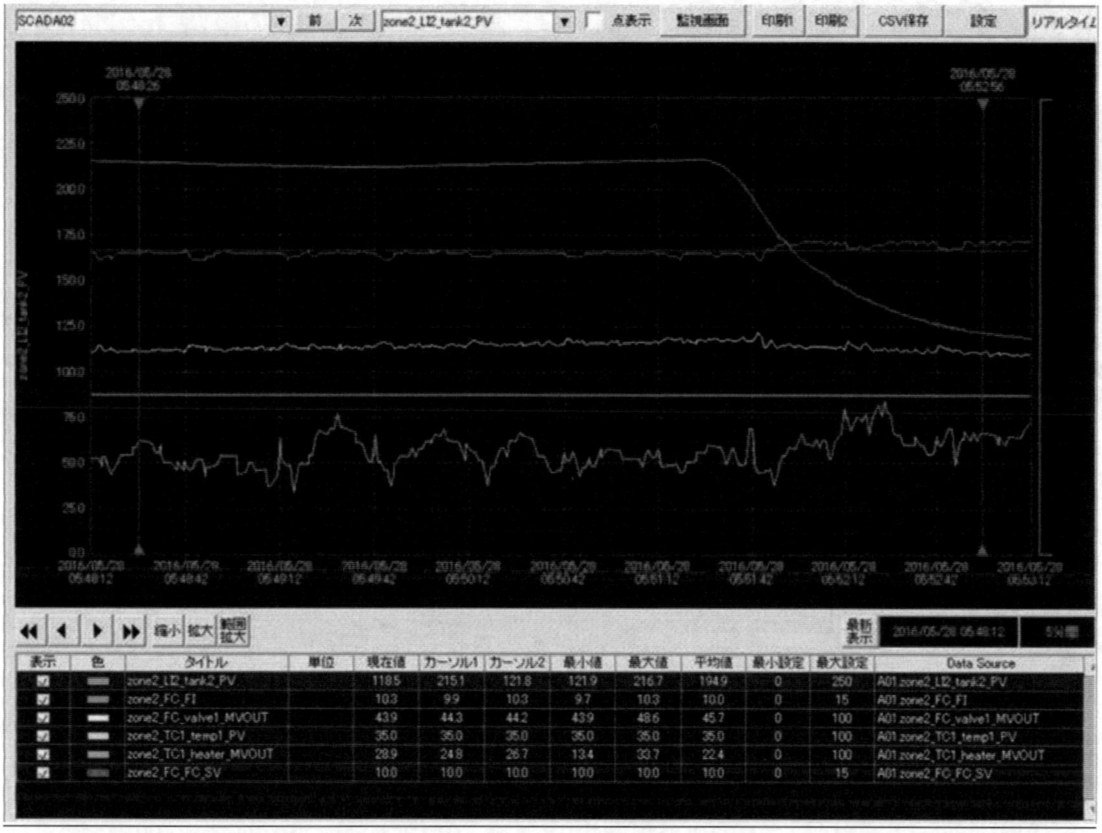

Figure 2. HMI Screen Shot in Operation (Safety Incident)

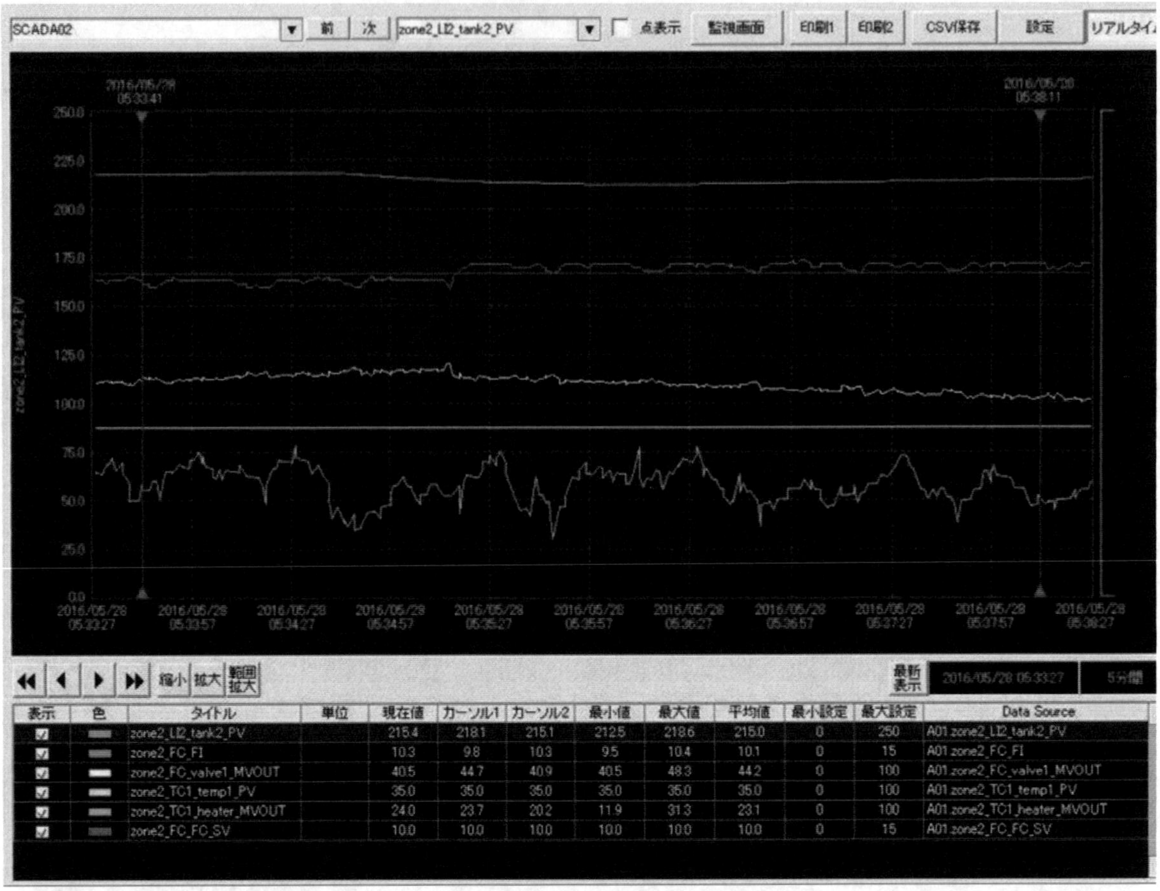

Figure 3. HMI Screen Shot in Operation (concealed by the attacker)

This demonstration shows a possibility of safety incidents caused by cyber-attacks to ICS.

1.2.2 Response to Cyber Incidents

The consequences of a cyber-attack depend on the characteristics of the plant. For example, an attacker hijacks a data server (DCS, OLE for Process Control server, etc.) that gives parameters and commands to control devices such as a PLC (Programmable Logic Controller) and gives the control devices malicious instructions. Such an attack can lead to abnormal operation and an incident, similar to a data server or control equipment failure or operator error with traditional equipment. That is, incidents caused by cyber-attacks in ICS systems are incidents that have already been considered in HAZOP scenarios with different initiating events. Operating companies have taken countermeasures to ensure the safety of the plant for many years. Companies are preparing countermeasures to ensure safety (Safety Response) under an incident at their plant. Human resource development is also carried out based on the countermeasures to respond when an incident occurs. In other words, even if an abnormality occurs due to a cyber-attack, it is possible to ensure the safety of the plant if the operation staff takes a proper Safety Response. However, problems peculiar to cyber-attacks, such as rewriting server configuration files and rewriting control device programs, may hinder the implemented safety-based Safety Response. Figure 4 shows the incident response structure that takes into account the cyber-attacks proposed here. If the plant is operating abnormally caused by the cause considered so far, it is possible to restore plant safety by performing a safety response, as shown in the left figure. However, there is a possibility that cyber-attacks will interfere with safety response (e.g., concealment of a plant's information on the monitor). Companies must consider where cyber-attacks interfere with safety response. Moreover, it is necessary to take the responsibility to eliminate the effect of cyber-attacks (security response) to take safety response as shown in the right figure. Therefore, if a cyber-attack causes a plant's abnormal operation, a response to rectify the situation is required in combination with a structure to add a security response to the safety response.

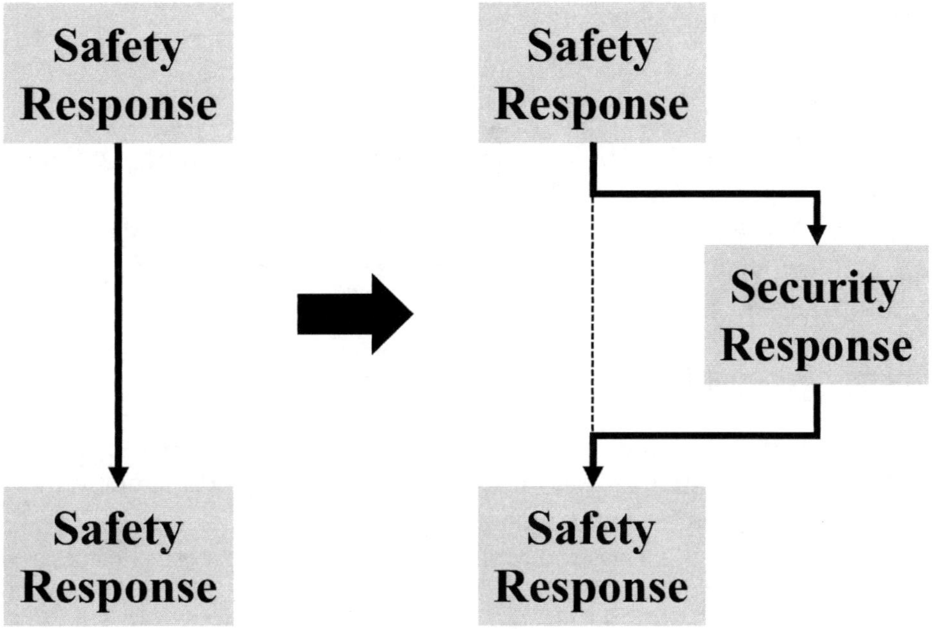

Figure 4. Incident Response Structure[7]

Responding to cyber-attacks targeting ICS systems, it is necessary to integrate a Safety Response to ensure the plant's safety and a Security Response to eliminate the effects of cyber-attacks. It is essential to create a plan that considers security incidents in addition to the existing safety response plan, and make it known to each operations staff.

1.3 Required organizational cooperation

The right staff needs to take action at the right time. To respond to incidents competently, it is necessary to properly share the current situation and communicate it among the right staff.

When an incident occurs due to cyber-attacks at a plant, two responses, a Safety Response to ensure the plant's safety and a Security Response to remove the cyber-attack effects, proceed simultaneously. Since the Safety Response responds to ensure the plant's safety, the response's responsibility is on the OT side (operators working at the plant). Since the Security Response is a response to remove the effects of cyber-attacks, the response's responsibility is on the IT side (IT engineer). Also, considering a cyber-attack on ICS, the result of the cyber-attack appears in the plant, and the cause is the IT network. In other words, it is a structure in which the cause-effects occur in different places. Also, in the incidents that we have experienced so far, there was no problem in dealing with the cause, but in cyber incidents, the same incident may occur in all parts connected to the Internet. Therefore, information sharing is vital for appropriately responding to the above-mentioned new type of incident, "Simultaneous Multiple Accidents."

2 How to develop Security Personnel

2.1 Preparation required as a company

・Formulation of ICS-BCP

Many companies have created a business continuity plan (BCP) in advance so that their business can continue even during natural disasters such as earthquakes and typhoons. It is important to develop a similar program for cyber incidents. It is necessary to analyze the risks in the event of a cyber incident and formulate an ICS-BCP that includes countermeasures for reducing, holding, avoiding, and mitigating the effects of the risks.

・Human Resource development

Even in unprecedented situations such as cyber incidents, companies must respond to continue their business. And it is the employees who respond to the incident. If the personnel's ability to respond to the incident is not sufficient, the company's damage will increase.
In other words, as a company, it is required to develop human resources who will respond.

2.2 Importance of virtual experience

As mentioned in 2.1, companies must prepare a BCP in advance for cyber incidents and develop human resources to carry it out. However, many companies have no cyber-attacks experience with ICS, and it is not easy to imagine how cyber-attacks may affect companies.

A mechanism is needed to experience cyber incidents and discuss necessary countermeasures.

As an approach to that mechanism, the authors chose an exercise. Figure 5 shows the flow for developing a training exercise. By analyzing incidents in the real world, we can extract the causes of incidents and the commonality of cyber-attacks targeting ICS. By giving the extracted information rules and game characteristics, the exercise participants will learn systematically. Thus, participants experience cyber-attacks on ICS through exercises.

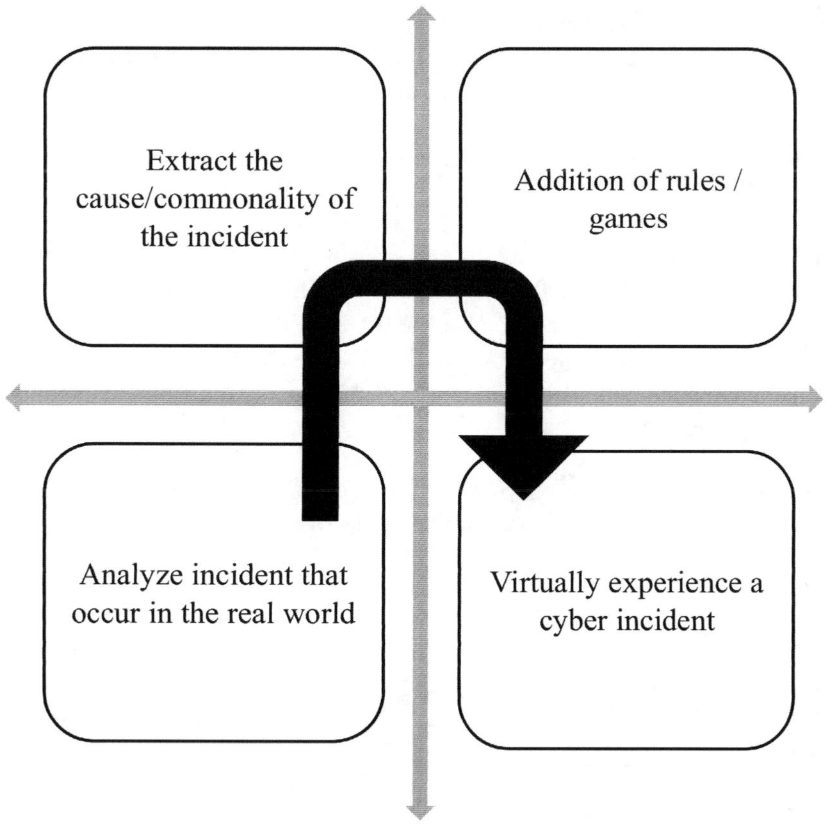

Figure 5. Exercise development process

Even staff who have never experienced a cyber incident can increase their experience with cyber incidents through exercises. By expanding this experience value, staff will more likely think and act even if people have a similar incident or an incident that people have never experienced before.

2.2 Human resource development method using virtual experience (exercise)

By taking the following four steps, we aim to improve the incident response capability of the company.

1. Information input
   Many companies have no experience in responding to cyber incidents, so it is necessary to understand the impact of cyber incidents on companies and plants. It is also required for the parties concerned to understand that cyber-attacks on plants are problems on the IT side and OT side.

2. Create ICS-BCP from a management perspective
   Companies are required to generate stable profits. When an incident occurs, companies need to think about what to prioritize to keep the business running, when to recover, and alternatives if there is no recovery. Each company must create an ICS-BCP assuming a cyber incident and communicate it to employees.

3. ICS-BCP with added on-site perspective
   It is necessary to respond to incidents based on the ICS-BCP defined as a company created in step 2. However, this BCP only includes the perspective of how to respond as an organization.

   By checking the BCP from the viewpoint of the operational site, it can be made into an ICS-BCP that can be operated.

4. Share the idea of all layers
   Employees must understand ICS-BCP to respond even if a never-experienced-before incident occurs. Therefore, by conducting training centered on the planned ICS-BCP, we will deepen the understanding of employees and objectively evaluate the ICS-BCP.
   By repeating this training multiple times, it is possible to confirm omissions and deficiencies in BCP and increase ICS-BCP accuracy.

The activities mentioned above don't do only once. Each company can improve its resilience ability by making assumptions based on various preconditions (scenarios), rotating the PDCA cycle, and continuing discussions.

3 Exercises for improving resilience developed

Based on the recognition that not only hard skills but also soft skills are important for responding to cyber incidents, we have developed discussion-based ICS security exercises.

3.1 Theme of the exercise

As mentioned above, it is very important to prepare a response plan to respond to an incident with hard and soft skills. Therefore, the purpose of the exercise is to understand the required actions to be taken by cyber-attacks and the importance of inter-organizational collaboration needed to take those actions.
The final goal is to be able to create an ICS-BCP using the knowledge gained from the exercises.

In order to make participants understand the purpose of the exercise, the exercise was divided into three phases according to the purpose of correspondence. (Fig.6)
When an incident occurs in a plant, whether it is a cyber incident or not, it is necessary to take measures to ensure the safety of the plant. Then, it is necessary to investigate the cause after ensuring the safety of the plant. If the cause is a cyber-attack, it is necessary to investigate the extent to which the impact of the cyber-attack has spread.
Initially, the safety workload is large, but the security workload gradually increases.
Then, there will be a response to confirm the extent of the impact of cyber-attacks not only on plant safety but also on business continuity.

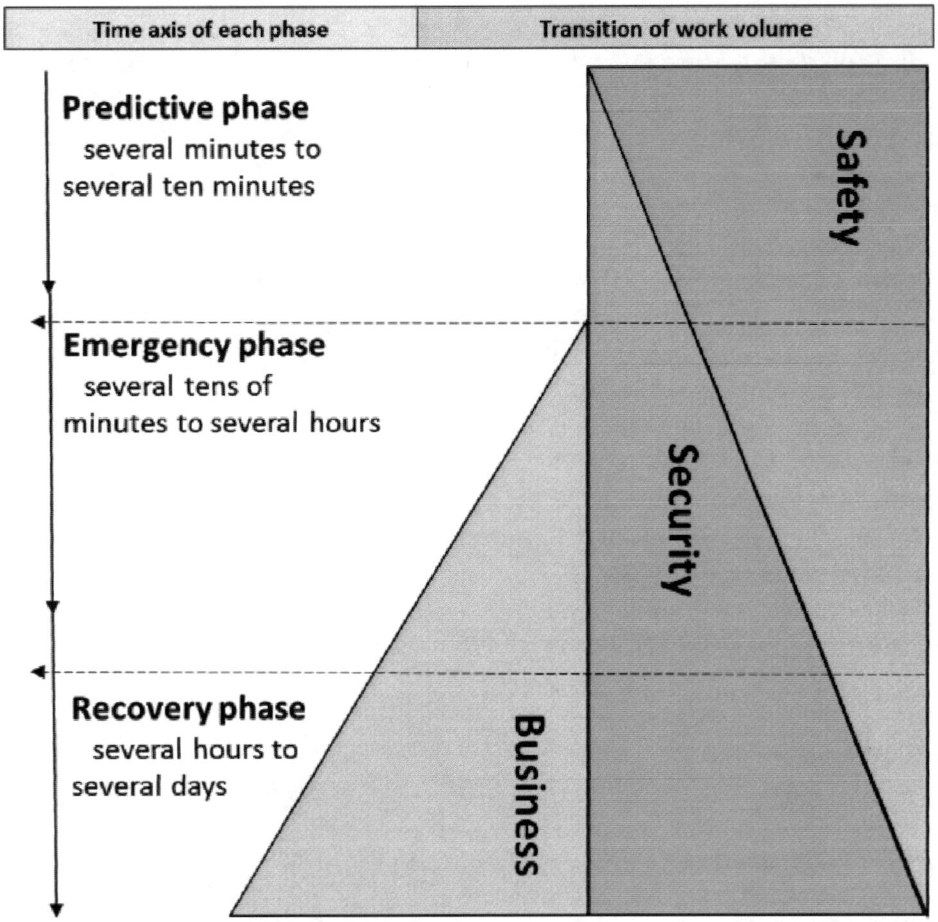

Figure 6. Structure of Exercise with Workload Image among Safety, Security, and Business Continuity Activities[7]

The authors visualized the state that this working volume change with the progress of time. (Fig6) Fig.6 expresses the work volume at a certain spot. The work volume is divided into safety, security, and business. The participant understands a change of this working volume through exercise.

1) Predictive Phase
The purpose of this phase is to ensure the safety of plants threatened by cyber-attacks. In this phase, the participants examine the safety response obstructed by a cyber-attack.
After detecting suspicious communication in a section that monitors the network, the exercise begins with an alert for the status of the liquid in the plant. In this phase, the participants must ensure the safety of the plant. Once safety is ensured, participants must investigate the abnormal operation. In this phase, participants can recognize how different causes can influence the safety response.

2) Emergency Phase
The purpose of this phase is to continue the business. The participants consider the security of the plant and how the exercise implicates a real cyber-attack. They must also take into account

how their business will be affected by the cyber-attack. In this phase, the participants consider practical measures taken to ensure business continuity throughout the company.
This phase begins after the abnormal operation of the plant reveals a cyber-attack. In this phase, participants can recognize that the entire company must be mobilized to treat a situation caused by a cyber-attack.

3) Recovery phase
The purpose of this phase is to restore own business to normal operation rapidly. The participants must find a way to nullify the influence of the cyber-attack so that the plant can return to normal operation. In addition, they must consider how business is affected by the temporary plant shut down. In this phase, the participants examine the start-up procedure of a plant crippled by a cyber-attack. In addition, the participants examine methods to prevent the recurrence of similar cyber-attacks.
This phase begins after the plant is stopped by a cyber-attack. In this phase, participants think about how to restart the plant. In this phase, participants can see how different causes affect the restoration of normal plant operations.

The following three types of training exercises have been developed.

- Card-Type Incident Response Exercise
- BCP Creation Exercise
- Simulation Exercise called "IMANE" (Incident MANagement Exercise)

3.2 Core scenarios for exercise

The developed exercises are based on the same scenario with different methods.
The base scenario is required to be able to consider the ideal organizational cooperation and necessary soft/hard skills for the organization to respond quickly in the event of a cyber-attack through exercises. The created scenario was composed of three scenarios to think about the three aspects of safety, security, and business.

Virtual Company:
Participants from various companies, departments have different backgrounds. A common scenario is indispensable for the participants to share the exercise together. In addition, using a simple and realistic virtual company makes it easier to apply the awareness of the exercise to the business. Figure 7 is an organizational chart of the created virtual company. Practical prerequisites, such as the company's business profile and its organization's intercommunication rules, have been set to achieve the exercise's objectives. Participants can discuss how information should be shared and how an organization should respond to cyber-attacks. Figure 8 is the network structure of a virtual enterprise. The attacker will attack using the computer that was hijacked by the cyber-attack as a stepping stone. The attacker steps on the hijacked computer and moves to the computer he wants to attack. In other words, the effects of cyber-attacks are spreading not only to the attack target but also to the attack process. For mitigating the impacts of cyber-attacks, it is necessary to consider countermeasures within the range of the effects of cyber-attacks. By conducting exercises based on this network structure, participants can discuss the Security Response required to respond to cyber incidents.

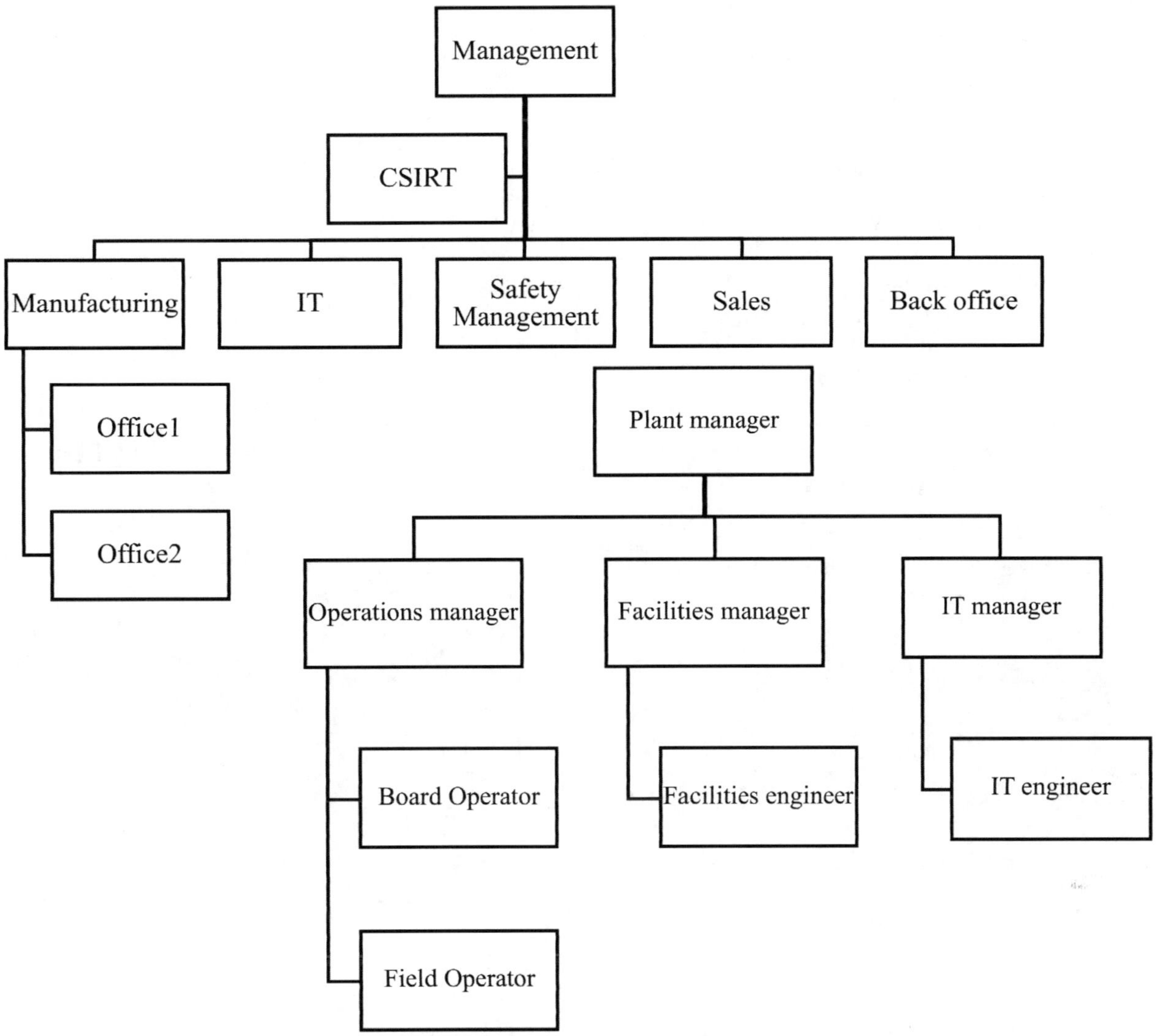

Figure 7. The organization chart of a virtual company

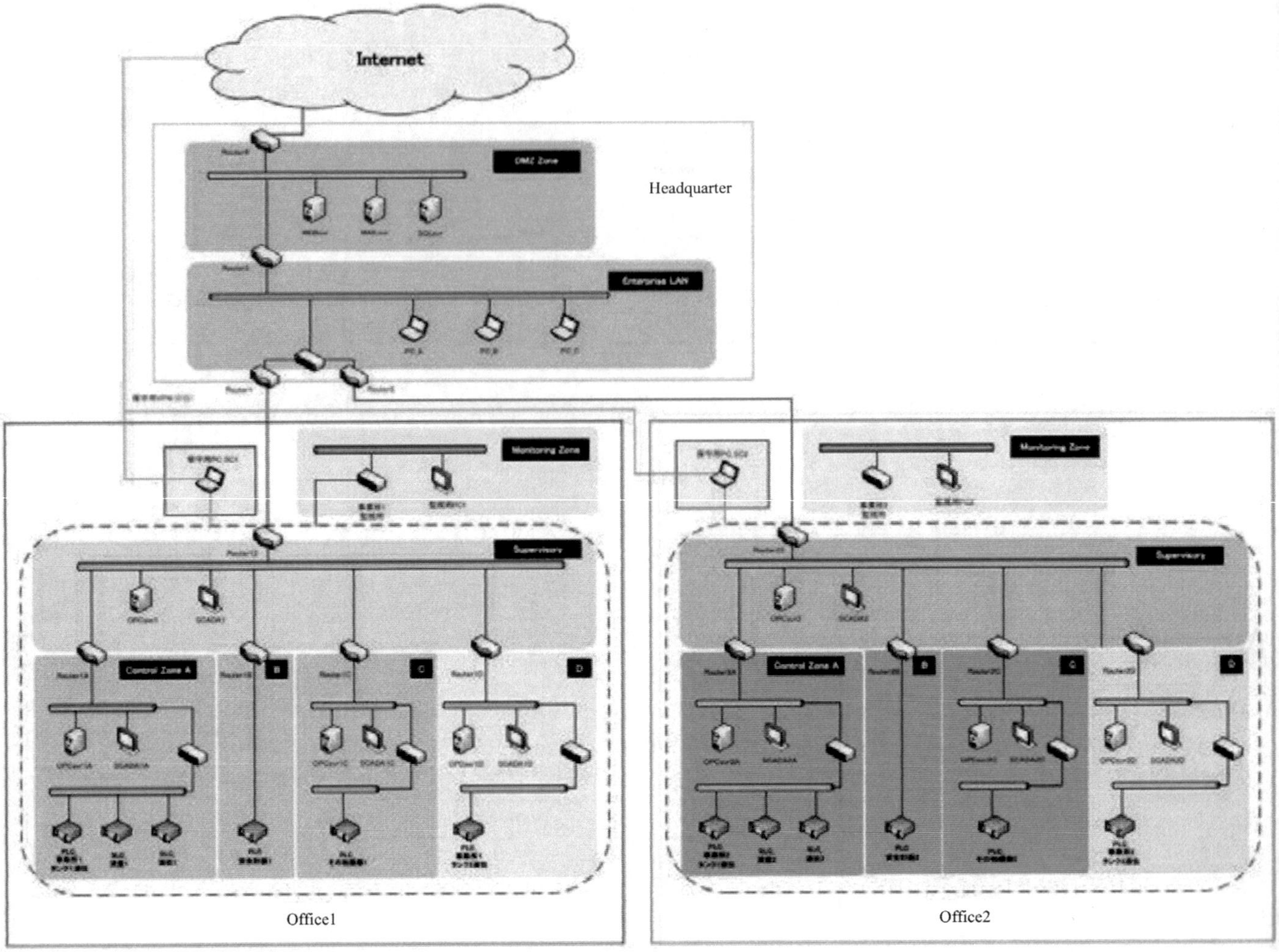

Figure 8. The network structure of the virtual company

Plant Operation Status Scenario:
A real-world plant consists of many devices, pipes, valves, and so on. If the plant is used as the subject of the exercise, the discussion may be complicated, and the purpose of the exercise may not be achieved. The plant used as the subject of the exercise should have a simple structure. By using a plant with a simple structure as the subject, it is possible to facilitate discussions focusing on the cyber incident that is the purpose of the exercise. In addition, it becomes easier to obtain meta-knowledge about cyber incidents, and it becomes easier to apply the knowledge gained through exercises to actual plants.

This scenario is based on our cybersecurity testbed. And we have introduced the control equipment used in the actual plant. This testbed aims to test and develop ICS security solutions. Based on the cyber-attack demonstration conducted as part of the research, the plant operation information caused by the cyber-attack is used for the exercise.

Cyber-attack Scenario:
The flow of cyber-attacks on plants is almost the same as cyber-attacks on IT systems.
To imagine how to respond to a cyber incident, it is important to deepen your cyber-attacks knowledge. There are many cyber-attacks methods, but attackers usually attack based on the cyber kill chain [8]. The cyber kill chain [9] classifies cyber-attacks into seven phases (Reconnaissance, Weaponization, Delivery, Exploitation, Installation, Command & Control, Actions on objectives). The idea is that appropriate measures can be taken in each phase, and cyber-attacks can be prevented if the attack does not proceed to the next phase.
This scenario is based on a cyber-attack demo conducted in our laboratory. It also includes the idea of a cyber kill chain. Throughout the exercises, participants can discuss the measures to protect the company by understanding cyber-attacks workflow and intention.

3.1 Card-Type Incident Response Exercise

The first exercise is Card-Type Incident Response Exercise. This exercise aims to gain an awareness of what a cyberattack on a plant is and what actions will be required.
This exercise uses a prepared action card to create a workflow for responding to cyber-attacks. Participants will achieve the purpose of the exercise by considering the response to cyber incidents while understanding the given action card's contents.

3.3.1 Exercise steps

These exercises consist of five steps: introduction, explanation of prerequisites, group work, presentations, and open discussions. Participants guess the company's situation from the given preconditions and create a workflow while thinking about the actions and information sharing necessary to make the plant safe and carry out business continuity.

Introduction:
The facilitator explains the purpose of the exercise and how to proceed. The general scenarios of the exercise are introduced to the participants. At this step, participants understand the importance of the exercise and, therefore, be motivated to apply themselves fully. By doing so, participants can increase the benefit of the exercise.

Description of Prerequisites:
Participants perform group work as a virtual company that owns a plant and provides products. The facilitator explains the preconditions about the virtual company at the beginning of the exercise. Prerequisites include the plant's status, the state of the IT network, and the company situation. Participants watch a video to understand the preconditions (In this phase, we use the videos we create to help participants understand the prerequisites.).

Participants envision the outcome based on the prerequisites.

Group Work:
The group task is to create a workflow by envisioning the outcome from the given information and considering the possible situation's response.
Originally, this exercise was conducted in group work in person. With COVID-19, we improved this exercise so that we could perform it using web conferencing.

In group work, the Excel file shown in Fig. 9 is used. Worksheets and action cards are pre-stored in this Excel file.

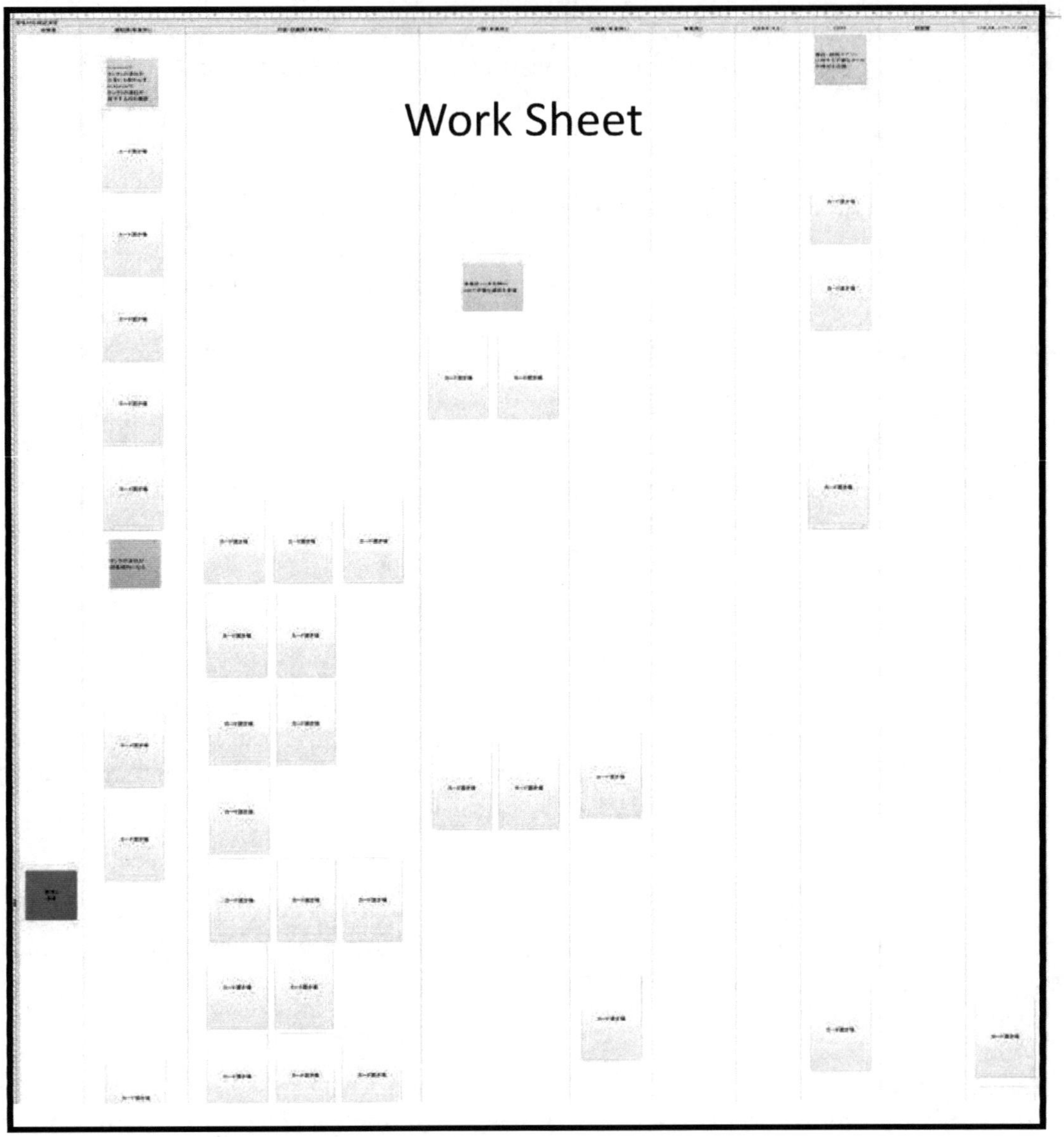

Figure 9. Group worksheet (Excel Screen)

The worksheet contains columns with the virtual company's departments, prerequisites, and action card storage roles. Action cards are classified into two types: normal cards and special cards (the upper part of Fig. 10 is the normal card and the lower part is the special card).

On-site Action

Action at headquarters

Plant shutdown/restart

Suspect a cyberattack

Information sharing

Figure 10. Type of action cards

The normal card is an incident response card, and there are a total of 38 cards of three types:

- "action at the site"
- "action at the head office"
- "shutdown/restart of the plant"

On the normal card, the result after taking action is written on the backside, and the participant thinks about the next action based on that information.

There are two types of special cards: "suspect/validation of cyber-attack" and "information sharing." The information-sharing card is used to share the information obtained by each department with other departments. In this exercise, participants discuss and strategize countermeasures from a bird's-eye-view, that is, without playing a specific role. However, in reality, information cannot be shared with others unless it is shared by the person who has the information. Participants use the information-sharing card to share the information held by each department. This makes it possible to express the flow of information sharing on the workflow. An important action in this exercise is to suspect/validate a cyberattack. No security response will be given unless the cause of the plant malfunction is suspected to be a cyber incident. There is information that can be a trigger to suspect that it is a cyber incident, and someone must judge that it is a cyber incident based on that information. By describing it on the workflow, participants identify the departments that will be the key persons in the cyber incident response. Participants will use these action cards to create a workflow, as shown in Figure 11.

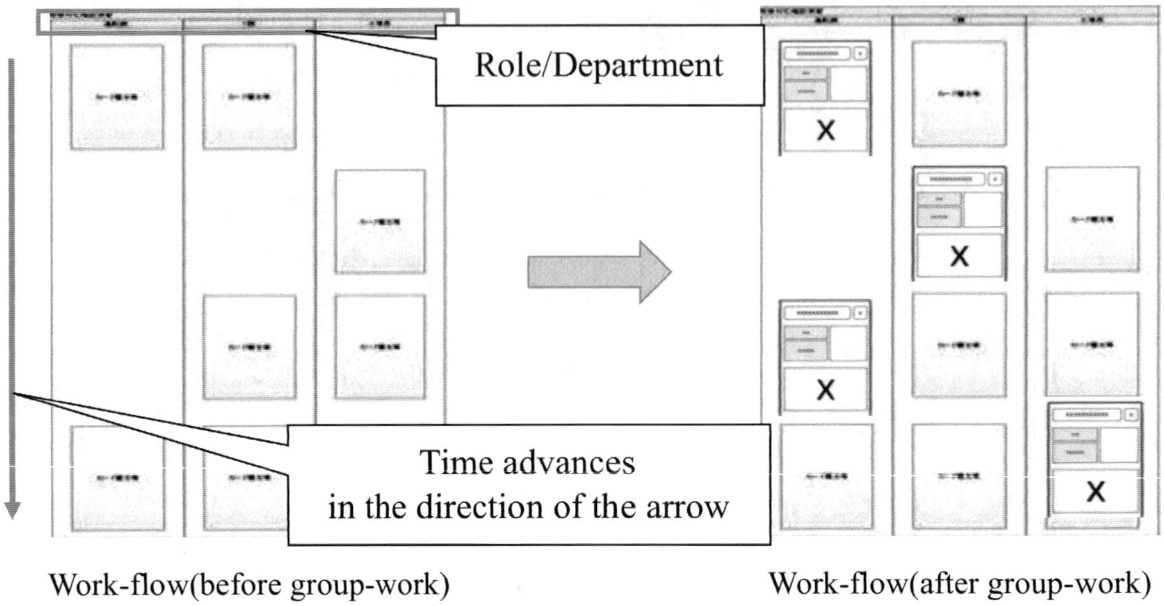

Figure 11. Workflow creation (image)

Presentation:
For groups with a lot of work experience in plants, create workflows that focus on plant safety. For groups with a lot of work experience in the IT department, create workflows that mainly support IT systems. Workflow, which is the product of group work, depends on group members' background and experience. By sharing the workflow and the workflow creation process, the exercise participants confirm the similarities and differences between the ideas of other groups and their own group's ideas. This step can enhance the effectiveness of the exercise.

When each group makes a presentation, ask them to focus on the following three points.

- How did you think about "who (which department), when, why" and placed the action card?
- What did you imagine when you created the workflow from the information given?
- How did you feel through the exercises what kind of human resources, information sharing, and organization-related systems should respond to cyber incidents?

Participants can understand the ideas of other groups and gain new knowledge by listening to other groups' presentations and asking and answering questions.

Open-Discussion:
After presenting the group's plans, the facilitator will show an example of the exerciser's workflow. This workflow presented is the optimal solution that the exercise developer thinks. The facilitator will give the same presentation as the participants in the exercise. After the

facilitator's presentation is over, the ideal inter-organizational collaboration for cyber incidents, information sharing, and necessary human resources will be discussed.

Through the exercises, exercise participants will experience the response to cyber incidents and understand the impact of cyber-attacks on ICS. And it is possible to gain insights on ideal organizational cooperation and necessary soft skills and hard skills so that the organization can respond quickly in the event of a cyber-attack.

3.4 ICS-BCP Creation Exercise

The second exercise is a BCP creation exercise. Participants create an incident response exercise based on a given scenario (such as a virtual business scenario). Participants will consider how to attack the plant from the attacker's point of view, consider actions to protect the company from cyber-attacks, and create an ICS-BCP. The purpose of the exercise is to understand the perspective of things needed to create an ICS-BCP.

3.4.1 Exercise steps

This exercise consists of five steps: scenario explanation, group work, demonstration, exercise review, and open discussion. Participants in the exercise identify the risks that the company faces from the given prerequisites. Next, consider an example of a cyber-attack that may cause that risk. Then, think about how to respond to the incident caused by the cyber-attack and create an ideal ICS-BCP.

Description of the Scenario:
Virtual enterprise scenarios, plant scenarios, and cyberattack scenarios are required to create incident response exercises. In this exercise, we want participants to discuss how to create an ICS-BCP rather than how to create a scenario. Therefore, the scenario is provided by the facilitator in advance. In addition, since there id an infinite number of attack scenarios, some scenarios are presented, and participants select from those scenarios. In this step, those scenarios will be explained.

Group-Work:
This exercise also creates deliverables through group work.
The deliverables to be created are the worksheets and action cards required for the card-type incident response exercise.

- First, participants analyze the risks of a virtual company from a given virtual company and plant scenario.
- Next, participants create a cyber-attack scenario that causes the analyzed risk. Participants create a scenario of the impact on the company and the plant's operating status due to the created cyber-attack scenario.
- Finally, participants create an ideal workflow for the company to perform incident response exercises and create actions that use that workflow as action cards.

Figure 12. An example of Team Discussion in Group Work

Demonstration:
In this step, participants perform their own exercises on other groups. Exercises will be conducted for different groups using workflows and action cards created in group work. Through this step, participants confirmed that they could convey the message through the exercise (ideal response / inter-organizational collaboration) to other groups. Participants can also gain more cyber-incident experience by taking exercises in various scenarios.

Group-work (Review of the exercise):
There are many things that can be gained by doing exercises.
Participants review the exercises they have created based on the results of the exercises conducted by other groups. The group once again discusses whether the ideal workflow that they have considered is the optimal solution, what factors hinder the ideal workflow that they have created, and so on.

Open-Discussion:
Each group will share what they have reviewed their exercises and what they have learned through the exercises in this step. By listening to various people's opinions through this step, it is possible to discuss how to view the things necessary for creating an ICS-BCP.

3.5 IMANE (Simulation exercise)

This exercise aims to confirm the effectiveness of the created ICS-BCP by simulating the incident response. This exercise is a simulation exercise that uses a personal computer to create a game based on a scenario, input chat, and respond to an incident by selecting a response.

3.5.1 Exercise steps

This exercise consists of five steps: identifying stress factors such as no one to deal with or no tools to deal with, adding storylines, simulating incident response, reviewing the exercise, and open discussion. Participants will play the role given and aim for business continuity while responding to incidents as virtual company members.

Identifying Stress Factors:
The ICS-BCP created by ICS-BCP creation exercise is an ideal workflow. In other words, the plan is based on the premise that all the assumed actions can be performed. However, in reality, it may not be possible to perform actions that should be possible due to various factors, such as lack of resources such as people and tools required to perform actions. Participants will identify stress factors that can hinder their actions in operating the ICS-BCP they have created. Stress factors are the risks involved in conducting ICS-BCP, and it is important to have advanced preparations and alternatives to eliminate the stress factors.

Adding storylines:
The created ICS-BCP only identifies the actions and the shared information, instruction contents, and information on the operation transition of the plant due to the actions are hidden between the actions. Participants identify the hidden information and necessary information communication paths. Then, create a game scenario by incorporating the stress factors created in the previous step.

Simulating Incident Response:
This step simulates an incident response. Figure 13 shows the structure of this exercise. In previous exercises, exercise participants were required to create a workflow in groups without being assigned to a specific role. In this exercise, roles are assigned to each exercise participant, and incident response is performed by taking actions for each role. It is an exercise that is closer to the actual incident response.

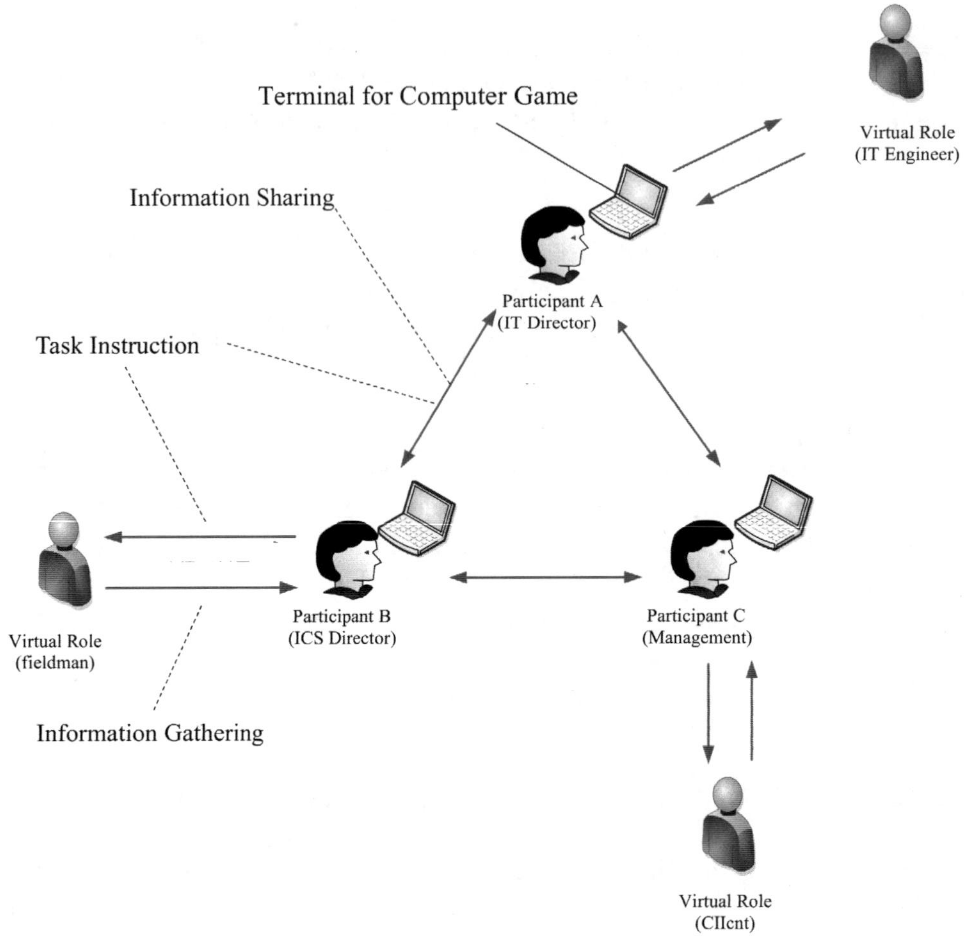

Figure 13. IMANE structure

Figure14 is a console screen operated by the participants. Participants share information with other players and take actions based on the information flowing in the tray. Each participant is required to understand the situation and take action based on the information displayed on their console screen. In addition, since the actions that can be performed for each role and the information is given are different, it is also required to share the information necessary for taking action to converge the situation with other players. Participants use the interface in Figure 14 to share information and select actions. The information displayed in the tray in Figure 15 is linked to the story created in the previous step, and the information displayed differs depending on the action performed so that the incident response can be made more realistic.

Figure 14. The user interface of IMANE exercise

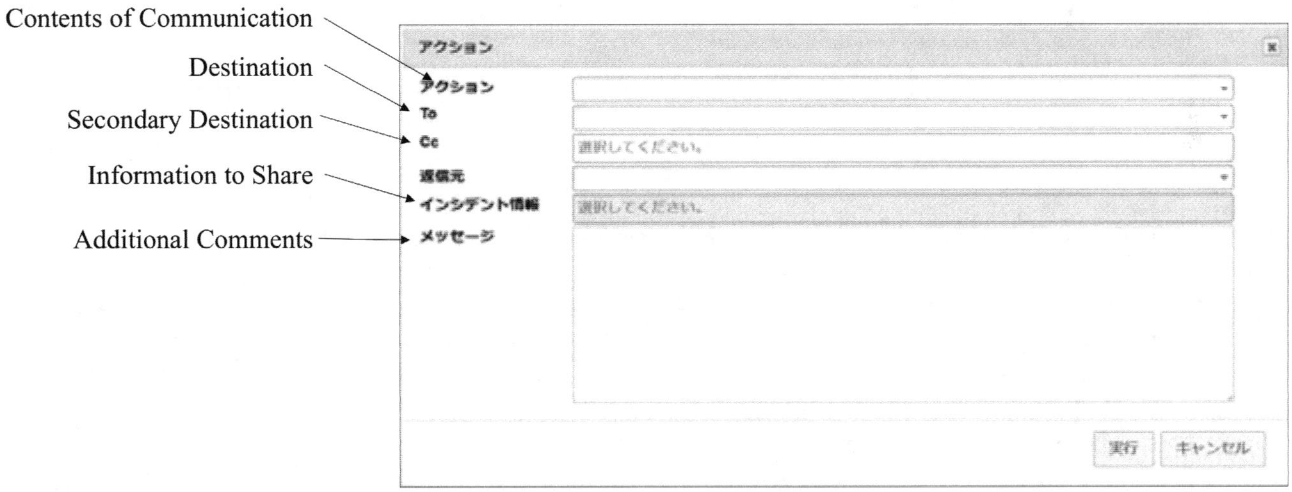

Figure 15. The action user interface of IMANE Exercise

Review of the Exercise:

After the end of this exercise, it can be output as a product of the exercise, as shown in Figure 16. This workflow includes the information given to each role, the actions taken, and the flow of information-sharing and participants can use this workflow to review the incident response after the exercise.

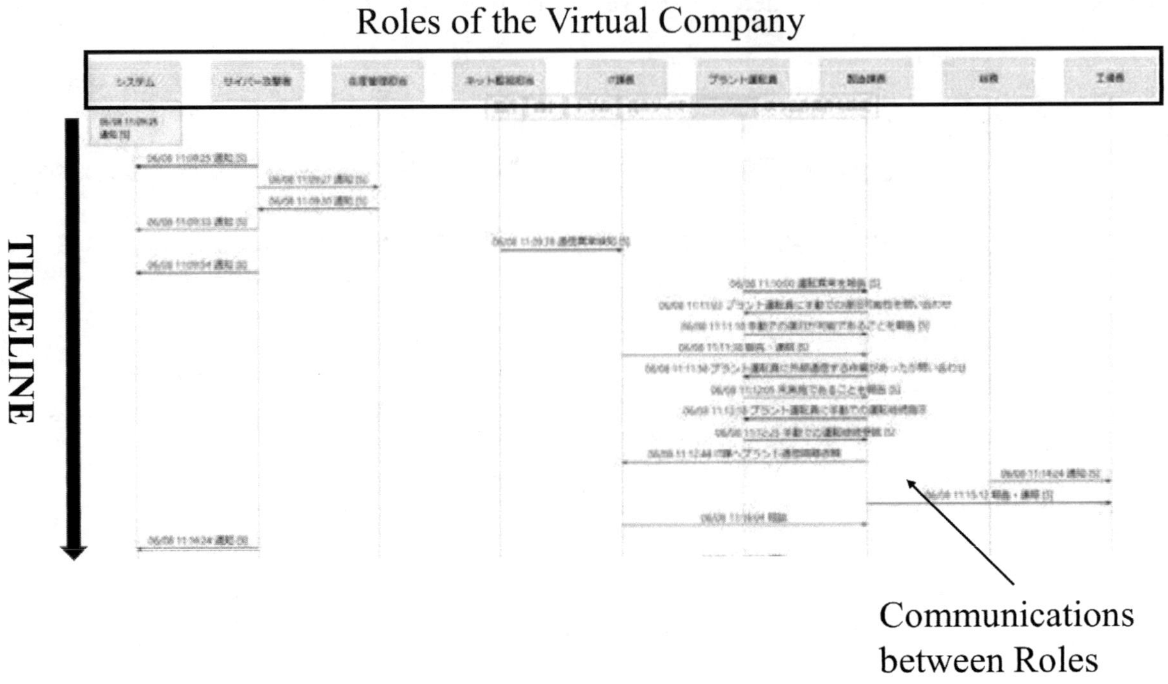

Figure 16. Example deliverable of the exercise

Using this deliverable, exercise participants can objectively review ICS-BCP by looking back on their response. It is possible to improve the accuracy of ICS-BCP by checking for missing or defective ICS-BCP.

Open-Discussion:
In this step, each participant shares the impressions gained through the exercise. We discuss the ideas needed to create an ICS-BCP and human resource development using this exercise.

By performing these exercises in various scenarios, the exercise participants can increase their imagination and resilience to cyber incidents.

4 Pilot Exercise

While improving the developed exercises step by step, many security and operation experts participated in the Nagoya Institute of Technology security workshops from 2018. To get an opinion to improve the exercises, we asked the participants to complete the questionnaire. The questionnaire was answered with a maximum of 5 points. In the exercise conducted on December 11, 2020, 28 people participated, and the following questionnaire results were obtained (Max point is 140). From the results of this questionnaire, it can be seen that the people who participated in the exercise understood the purpose of the exercise that we set and received high praise for the exercise.

Questionnaire content for Card-Type Incident Response Exercise and ICS-BCP Creation Exercise:

1. Was the goal set for the exercise, and was it possible to attain it?
2. Do you think you learned about organizational collaboration in the event of a cyber incident?
3. Do you think you learned about the organizations needed to respond to cyber incidents?
4. Was the information necessary to proceed with the exercise given?

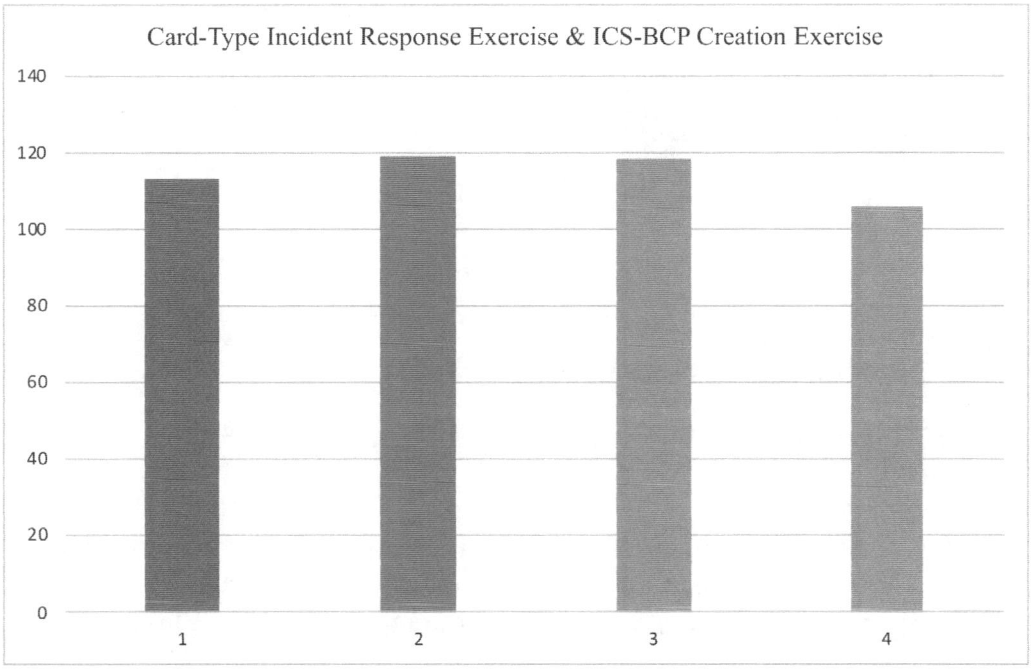

Figure 17. Questionnaire results of Card-Type Incident Response Exercise & ICS-BCP Creation Exercise

Questionnaire content for IMANE:

1. Was the goal set for the exercise, and was it possible to get it?
2. Do you think you learned about organizational collaboration in the event of a cyber incident?
3. Do you think you learned about the organizations needed to respond to cyber incidents?
4. Is it designed so that necessary information is communicated and given to each role?
5. Do you feel that you were able to experience information gathering and decision-making in an emergency?

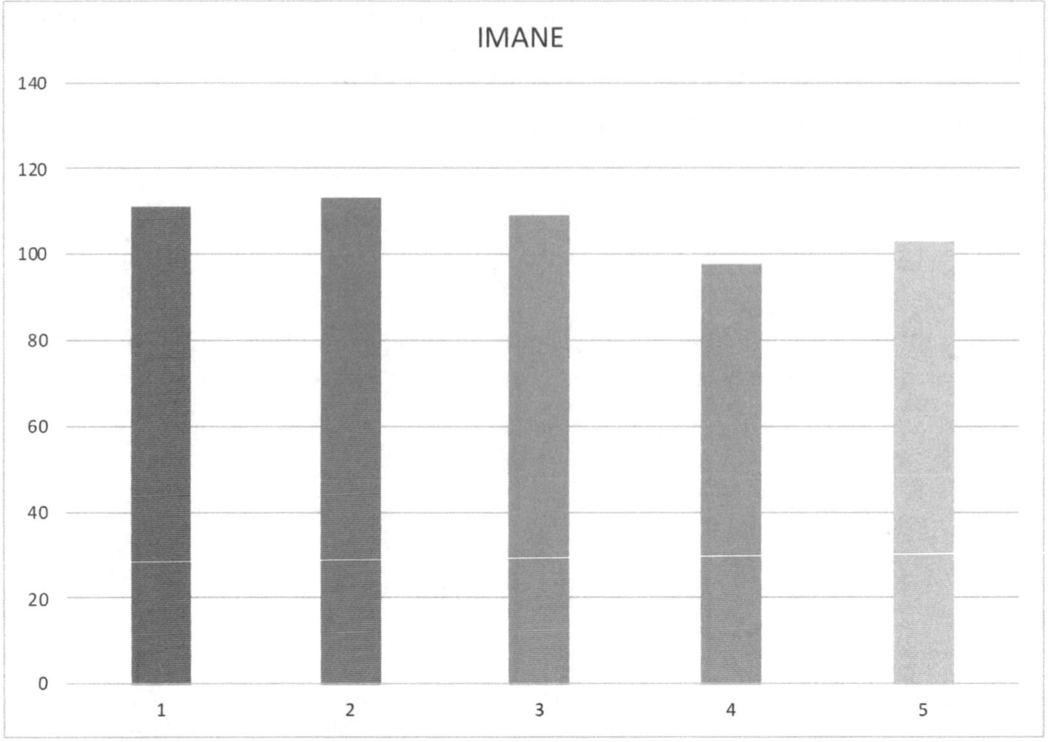

Figure 18. Questionnaire results of IMANE

In addition, the following was found by analyzing the deliverables of the exercise.

・ Plant safety first

Many workflows can keep the plant safe, even if the backgrounds of the participants in the exercises vary. Even if the cause of the plant abnormality is a cyber-attack, it is easy to imagine the visible situation, and it can be seen that appropriate measures can be taken.

・ No action has been taken to identify a cyber-attack

Analyzing the workflow, it is possible to select situation sharing and actions for performing Safety Response. However, information sharing for performing Security Response, such as identifying the range of influence of cyber-attacks is often not successful. This is thought to be due to the fact that they have never experienced cyber-attacks and lack the knowledge to consider actions. In addition, Safety Response is easy to imagine because it corresponds to the visible physical impact on the plant, but Security Response is difficult to imagine because it corresponds to the invisible network.

・Organization that is the center of the response

To respond to cyber incidents to ICS, it is necessary to fluidly change the organization that is the core of the response.

In the initial response, it is necessary to respond to plant operation abnormalities caused by cyber incidents. The person working at the plant takes action. After ensuring the plant's safety, it is necessary to eliminate the effects of cyber incidents. IT department takes this action. Finally, it is necessary to continue the business considering the impact of the cyber incident on the business.

Management considers actions for the business gains.
In this way, the work to be done changes over time, and it is necessary to respond while changing the organizational response that is the core of the response accordingly.

5 Concluding Remarks

In this paper, the authors presented the factors required of a company against the threat of cyber-attacks targeting ICS and explained the exercises developed to increase the company's resilience.

Through the workshop, we received positive opinions from the participants that they could gain various awareness of cyber-attacks. However, the effect of the exercise could not be confirmed quantitatively, and it could not be confirmed that the response-ability was improved.

In the future, we will conduct exercises for more people and obtain deliverables. Then, by analyzing the deliverables, we will consider the method of planning ICS-BCP considering cyber-attacks and the conditions necessary to confirm the validity of the planned ICS-BCP.

6 References

[1] Institute for Science and International Security, "Did Stuxnet Take Out 1,000 Centrifuges at the Natanz Enrichment Plant?", https://isis-online.org/isis-reports/detail/did-stuxnet-take-out-1000-centrifuges-at-the-Natanz-enrichment-plant/8 (accessed Jan. 25, 2021)

[2] ICS-CERT,
ICS Alert(IR-ALERT-H-16-056-01), "Cyber-Attack Against Ukrainian Critical Infrastructure", https://us-cert.cisa.gov/ics/alerts/IR-ALERT-H-16-056-01 (accessed Jan. 25, 2021)

[3] SANS Institute, ICS Defense Use Case No. 6, "Modular ICS Malware", https://ics.sans.org/media/E-ISAC_SANS_Ukraine_DUC_6.pdf (accessed Jan. 25, 2021)

[4] Norsk Hydro, "Cyber-attack on Hydro", https://www.hydro.com/en/media/on-the-agenda/cyber-attack/ (accessed Jan. 25, 2021)

[5] BSI, "Industrial Control System Security Top 10 Threats and Countermeasures2019", https://www.allianz-fuer-cybersicherheit.de/ACS/DE/_/downloads/BSI-CS/BSI-CS_005E.html#download=1 (accessed Jan.25, 2021)

[6] T. Aoyama, M. Koike, I. Koshijima, Y. Hashimoto, "A unified framework for safety and security assessment in critical infrastructures". In: Safety and Security Engineering V. WITPRESS LTD. (Sep 2013)

[7] Y. Ota, T. Aoyama, D. Nyambayar, I. Koshijima, "Cyber Incident Exercise for Safety Protection in Critical Infrastructure," Int J. Saf. Secur. Eng. 8,246-257 (2018)

[8] J. Sakhnini, H. Karimipour, A. Dehghantanha, R.M. Parizi, G. Srivastava, Security aspects of internet of things aided smart grids: a bibliometric survey. Internet Things (2019). https://doi.org/10.1016/j.iot.2019.100111

[9] A. Shostack, *Threat Modeling: Designing for Security* (Wiley, Indianapolis, In, 2014)

SPRING21
+17TH GCPS
A Joint AIChE and CCPS Meeting

Hazard Identification during Process Related

Maintenance and Operations Tasks

T Michael O'Connor
Mary Kay O'Connor Process Safety Center
Texas A&M University, College Station, TX
mikeocon@gmail.com

Dr Zohra Salim
Mary Kay O'Connor Process Safety Center
Texas A&M University, College Station, TX
zohrahalim@tamu.edu

Prepared for Presentation at
American Institute of Chemical Engineers
2021 Spring Meeting and 17th Global Congress on Process Safety
April 18 - 23, 2021

Hazard Identification during Process Related

Maintenance and Operations Tasks

T Michael O'Connor
Mary Kay O'Connor Process Safety Center
Texas A&M University
College Station, TX
mikeocon@gmail.com

Dr Zohra Halim
Mary Kay O'Connor Process Safety Center

Keywords: serious injury, fatality, chemical manufacturing, petroleum refining

Abstract

Trends in process related fatalities are presented for chemical manufacturing and petroleum refining. The trend analysis revealed the more than half of fatalities occur in single fatality incidents. This observation led to study of the nature of these low fatality incidents. Three common tasks were identified that are associated with a large portion of process related fatalities and severe injuries. These 3 tasks are clearing, cleaning and purging lines and vessels; opening lines and vessels; and transferring fluids. The specific hazards encountered in incidents associated with these tasks are identified. The goal is to provide the basis to improve system design, training, procedures, job hazard analysis, PPE requirements and other factors based on hazards actually associated with major harm to individuals performing these tasks.

Introduction

There are 3 primary components of this study:

- Trends in fatalities from 1984-2019 in chemical manufacturing (CM) and petroleum refining (PR)
- Tasks associated with severe injuries and fatalities in CM and PR
- Nature of serious injury incidents in CM only

Data is extracted from two US OSHA databases:

The first is the Fatality and Catastrophe Inspection Search (FCIS) database which includes all incidents with a fatality or 3 or more injuries. Since incidents without a fatality and less than 3 injuries are not included in the data injuries from this database are not included in the study. This database covers the years 1984 to the present whichallows long-term trending including the implementation of OSHA PSM in 1992. This analysis is limited to the states in OSHA Region 6. These states are TX, LA, AR, OK and NM. The data was accessed through OSHA's online query system.

The second database is the Severe Injury Reports (SIR) database. This database is limited to the 28 States with federal, not state, OSHA programs. This database includes all the Region 6 states. It covers the years 2015 to July 2020 and includes cases with hospitalization or amputation. This entire dataset can be downloaded as a spreadsheet.

For both databases the authors have interpreted the accident description to determine whether it is a process related incident and the task being undertaken. The serious injury database contains the hazardous substance, nature of injury, affected body part and event type. In some cases, the description is not sufficient to extract the needed data and these items are classified as unknown. Some incidents have been investigated by the Chemical Safety and Hazard Investigation Board (CSB) or other agencies. In those cases, information may have been extracted from those sources.

This paper focuses on both the Chemical Manufacturing (CM) and Petroleum Refining (PR) industry as determined from their NAICS and SIC codes. Both industries were analyzed for fatality trends from 1984-2020. The tasks associated with both severe injuries and fatalities were determined for both CM and PR. A more detailed analysis of selected tasks was performed only for CM. Detailed analysis of PR incidents are not yet completed.

Scope of Incidents in Study

The following definitions were utilized to determine if an incident was process related and was to be included in the study:

Hazardous Material – Any substance the is toxic, reactive, flammable, asphyxiating or that presents a hazard to people, the environment, or property due to pressure or temperature.

Process Safety Incident – The sudden, unintended release of, or exposure to, a hazardous material, associated with an industrial process, that results in or might reasonably have resulted in death, injury, significant environmental or property damage, evacuation or sheltering in place.

These are generic definitions used by the MKOPSC. In this paper all of the incidents involve a fatality or injury since they are from the OSHA databases.

Trends in Fatalities in Chemical Manufacturing and Petroleum Refining

Long term trends with time are available from the Inspection Database for fatalities. Serious injuries are included in the database if there are 3 or more injuries or they occur with a fatality. However, incidents with one or two injuries are not included making any trending information questionable. The Serious Injury database only covers the years 2015 through 2020, making any trends of doubtful significance, so only fatalities were used for trending.

Plots are shown for CM and PR. Fatalities and fatal incidents over the years 1984 to 2020 are illustrated. There are a number of high fatality incidents in the latter 1980s. This is followed by a dramatic decrease in fatalities around the time of the implementation of OSHA PSM in 1992. It seems likely this decrease was due to both the increased focus on process safety due to the new regulation as well as a direct response to these high fatality incidents. The trends are somewhat similar for CM and PR with the exception of the BP Texas City explosion in 2005.

Figures 1 through 4 show the large number of incidents with a low number of fatalities. After 1992, for CM and PR, greater than 80% of fatalities occurred in single fatality incidents. This statistic in particular motivated a more detailed examination of these numerous incidents with mostly low numbers of fatalities as well as incidents with serious injuries.

In this plot (see Figure 1), the total height of the bars represents the total fatalities in a year. The colored bands making up the bars show individuals incidents and the number of fatalities in that event. The red arrow denotes the implementation of OSHA PSM in 1992. Clearly there is a sharp decrease in fatalities after that time. In addition, there are no incidents with more than 4 fatalities after PSM began.

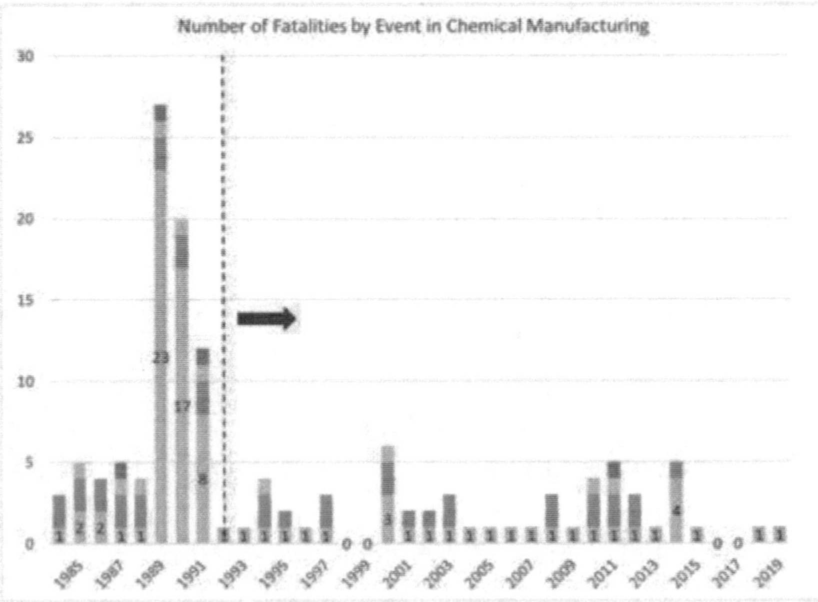

Figure 1 - Fatalities by Event in Chemical Manufacturing Region 6

This plot (see Figure 2) shows that not only the annual number of fatalities dropped in 1992 but also the number of fatal events.

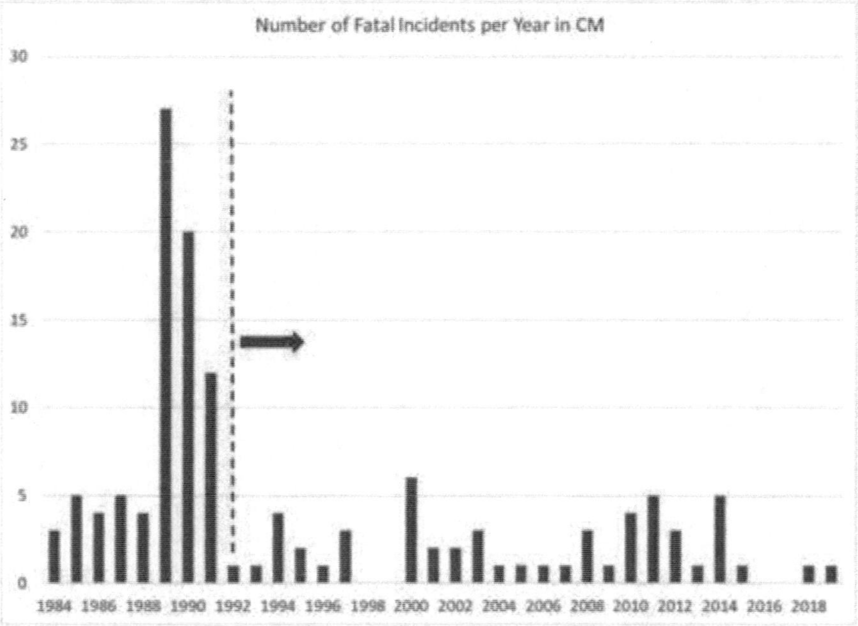

Figure 2 - Fatal Events in Chemical Manufacturing

Like CM there is a distinct reduction in fatalities in PR about the time of PSM implementation. This reduction appears to have occurred in 1994, a few years after PSM. There is another reduction about the time of the event at BP Texas City (see Figure 3). We theorize that this high-profile event may have re-focused the PR industry on process safety. This reduction has persisted for 14 years.

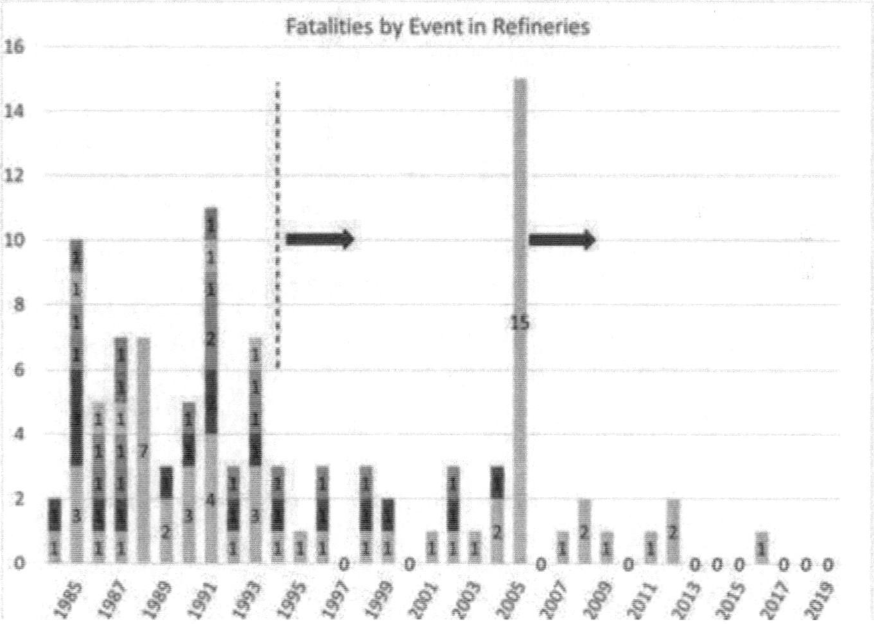

Figure 3 - Fatalities by Event in Petroleum Refining

It is clear that not only the number of fatalities was reduced around 1994 but also the number of events. Since 2002 there has been either zero or one fatal event per year (see Figure 4).

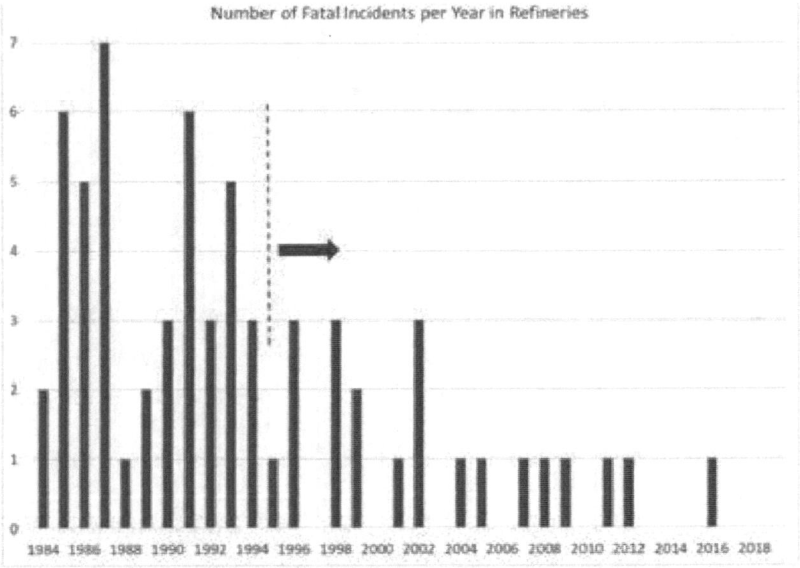

Figure 4 - Fatal Incidents in Petroleum Refining

Table 1

The fatality and fatal event annual rates are shown for CM and PR in Table 1. For CM there is a 50% reduction in fatal events per year after 1992. The fatalities per year dropped dramatically by a factor of almost 5. For PR there is a 60% reduction in fatal events in 1994 and an overall reduction of 75% after 2005. Fatalities in PR were reduced by nearly 50% in 1994 and a remarkable 10 to 1 after 2005. The relative magnitudes of the PR and CM rates are not comparable since they are for different numbers of facilities and employees.

Event and Fatality Rates for CM and PR - Region 6			
Chemical Manufacturing			
	1984-1992	1993-2020	
Fatal Events/yr	4.1	2.0	
Fatalities/yr	10.0	2.2	
Petroleum Refining			
	1984-1994	1995-2005	2006-2020
Fatal Events/yr	4.0	1.6	1.0
Fatalities/yr	6.0	3.2	0.6

Tasks Associated with Serious Injuries and Fatalities

As the descriptions of each incident were analyzed it was soon apparent that most of the incidents involved a task being performed by a small number of people and the consequences typically suffered by those workers. Both maintenance and operational activities are involved.

Table 2 lists the tasks for serious injury and fatal incidents, from both databases, in both CM and PR. The term "unrelated" is used to denote incidents that are not related to the activity of the victim. Most of these "unrelated" cases are mechanical failures, leaks or explosions and the victim is by chance in the vicinity.

Table 2

Severe Injury and Fatal Incidents by Task									
	Chemical Manufacturing				Petroleum Manufacturing				
	Severe Injury Incidents	%	Fatal Incidents	%	Severe Injury Incidents	%	Fatal Incidents	%	% Averaged for Sort
Cleaning, clearing, purging lines and vessels	35	16	16	20	10	29	11	11	19
Opening lines and vessels	47	21	8	10	3	9	15	15	14
Transferring fluids	43	20	8	10	3	9	7	7	11
TOTAL % for 3 Tasks		57		39		47		32	
Unrelated*	10	5	11	13			16	16	8
Unknown	20	9	3	4	2	6	8	8	7
Maintenance	18	8	3	4	5	15	4	4	8
Operating	6	3	6	7	3	9	4	4	6
Starting equipment	6	3	5	6	2	6	5	5	5
Confined Space Entry			7	9			8	8	4
Mixing chemicals	8	4	3	4			7	7	4
Troubleshooting	7	3			2	6	4	4	3
Fall or step into hazard	3	1			2	6	1	1	2
Other	5	2	4	5			1	1	2
Sampling	3	1	3	4	1	3			2
Handling explosives			3	4					1
Lighting heater							5	5	1
Transferring solid	7	3							1
Reactor startup			2	2					1
Starting process					1	3	2	2	1
Hot Work							3	3	1
Testing	2	1					1	1	0
Inspecting							1	1	0
Total Incidents	220	100	82	100	34	100	103	100	100

*Unrelated indicates the employee was not involved in a task telated to the event.

Tasks Associated with Serious Injury and Fatal Incidents in Chemical Manufacturing.

In both CM and PR, there are 3 activities that account for a large fraction of the fatalities. These tasks are:

- Cleaning, clearing, or purging lines and vessels
- Opening lines and vessels
- Transferring fluids and solids

These 3 types of tasks usually involve the intentional interruption of a pressure containing equipment component. That is of course the case with opening lines and vessels. For cleaning, clearing, or purging, it is necessary to open the pressure containment before proceeding. Transferring fluids often involves making or breaking connections as well as venting and draining lines and vessels

Cleaning, clearing, and purging by their nature involve removal of a substance, usually from the interior of the containment. Difficulties are encountered in numerous ways. These include:

- Substance is not what was expected
- Quantity of substance is greater than expected
- Unexpected pressure behind a blockage
- Blockages are propelled out of the containment by pressure
- Tools, covers and hatches are propelled by unexpected pressure
- Containment is over-pressurized when clearing an obstruction

Opening lines and vessels often involve problems similar to those above. These include:

- Unexpected pressure
- Unexpected substance
  Unexpected quantity

Transferring fluids involve many different situations based on the source of the fluid, receiving container, and type of transfer line or hose.

- Hoses often disconnect unexpectedly
- Hoses are subject to failure
- Disconnected hoses may contain an unexpected substance
- Hard pipes often contain vents and drain lines that are open
- Transfers may result in mixing incompatible substances
- Rail cars and tank trucks may have complex sets of valves for venting, draining, multiple compartments and other confusing situations
- Totes, drums, buckets, and other containers can be dropped or damaged, sometimes by forklifts
- Static electricity may be an issue in many of these operations

Nature of the Serious Injury Incidents in CM

The OSHA database, SIR, contains four factors that describe the incident and the resulting injuries. The four factors are the following:

- Type of Event
- Hazardous Substance
- Nature of Injury
- Affected Body Part

The results for all tasks in CM are shown in four tables. These four factors are the results of the incident and infer a limited amount about the causes. Hopefully information about the most common hazards will assist in focusing efforts for improved safety. Data for all incidents are included, not just those for the 3 most common tasks.

Table 3

Chemical Manufacturing		
Serious Injury - All Tasks		
Event Type	**#**	**%**
Exposure through intact skin, eyes, or other exposed tissue	82	37
Contact with hot objects or substances	74	33
Inhalation of harmful substance-single episode	21	9
Ignition of vapors, gases, or liquids	14	6
Struck by dislodged flying object, particle	9	4
Explosion of pressure vessel, piping, or tire	8	4
Explosion, n.e.c.	6	3
Other	9	4
TOTAL	223	100

Event Type

Two types of events account for 70% of all events in roughly equal frequency. These two are exposure of external tissue to a hazardous substance and contact with hot objects and surfaces. These are such broad categories that it does little to identify a particular type of event which deserves more attention than others. Two other event types also occur in a significant portion of incidents. These are inhalation of a substance and ignition of a substance. In planning for and executing a task, it is important to consider if any of these events could occur. This can lead to identifying procedures to avoid the event and to utilize proper PPE and emergency procedures.

Table 4

Hazardous Substance

Perhaps surprisingly, water is the most frequent hazardous substance involved in serious injuries. Of course, this is due in part to its widespread presence in many processes. Often the water is condensate and results in burns. Nearly as common are acids followed by resins. Resins are often associated with processes such as extrusion were contact burns are common. While the table shows a number of common substances there are an additional 40 substances that were only involved in one incident. In addition, there are 30 objects. In OSHA's classification system, objects, rather than a hazardous substance, are identified as the source of injury. These are usually cases where an object propelled by pressure or an explosion strikes the victim. These objects are typically shrapnel, pipe plugs, covers and hatches or hoses.

Chemical Manufacturing		
Serious Injuries - All tasks		
Substances	**#**	**%**
Water	**27**	**12**
Unknown, not classified	27	12
Acids	**25**	**11**
Resins	**17**	**7**
Multiple	12	5
Chlorine, chlorine bleach, compounds	**9**	**4**
Steam and vapors	**9**	**4**
Sodium and potassium hydroxide, potassium carbonate	**8**	**4**
Ammonia, anhydrous ammonia	**6**	**3**
Chemicals, n.e.c.	6	3
Phenol	**4**	**2**
Liquids, unspecified	4	2
Aromatics and hydrocarbon derivatives, except halogenated, n.e.c.	**3**	**1**
Other Substances	40	18
Physical Objects*	30	13
TOTAL	**227**	**100**
* OSHA lists objects such as those propelled by pressure		

Table 5

Nature of Injury

The nature of injury generally reflects the type of event shown in the previous table. Thermal burns are naturally associated with contact with hot objects and substances and chemical burns with exposure of tissue. Traumatic injuries result from being struck by object and explosions. Poisoning, toxic, noxious, or allergenic reactions are typically due to inhalation. While there is little information in these databases about PPE it appears that it is often inadequate for the actual circumstances.

Chemical Maufacturing		
Serious Injuries - All Tasks		
Nature of Injury	**#**	**%**
Thermal Burns	**93**	**40**
Chemical Burns	**79**	**34**
Traumatic	**26**	**11**
Poisoning, toxic, noxious, or allergenic effect	**26**	**11**
Other	**9**	**4**
TOTAL	**233**	**100**

Table 6

Affected Body Parts

One third of the incidents involve multiple body parts. This may reflect the serious nature of the studied incidents. Severe injuries may be the result of releases of larger quantities of hazardous substance, higher pressures and other factors that lead to more widespread injuries. This contrasts with less serious injuries associated with more common cases involving days away from work. In a study of these injuries (Mannan), there were typically specific body parts injured. The most often injured body part was the face. The second most affected body part for severe injuries is body systems. This means internal systems implying inhalation or ingestion of a substance. The remaining body parts are more typical parts of face, extremities, trunk, and head. The adequacy of PPE may be a factor in the severity of these injuries.

Chemical Manufacturing		
Serious Injuries - All Tasks		
Body Parts	#	%
Multiple body parts, n.e.c.	77	33
Body Systems	28	12
Nonclassifiable	28	12
Face	25	11
Lower Extremities	25	11
Upper Extremities	24	10
Trunk	15	7
Trunk & Head	8	3
TOTAL	230	100

Incidents with Detailed Investigations

The CSB and the US Department of Labor investigated several of the CM incidents in this study. These incidents were Phillips 66 1989, Honeywell 2003 (3 incidents) and DuPont 2014. These incidents generally involve one of the 3 common tasks. Phillips 66 involved the removal of a plug in the settling leg of a reactor. This led to a release of isobutane resulting in 23 fatalities and more than 100 injuries. Honeywell had 3 incidents involving 3 different substances. The first was a chlorine release due to a mechanical failure. The second was an antimony pentachloride release which resulted in a fatality. The release was the result of removing a plug from a cylinder. The third release was hydrogen fluoride during the transfer of fluid using a venturi stick, resulting in hospitalization of one worker. The DuPont incident resulted in 4 fatalities during the removal of a hydrate plug, releasing mercaptan. These three investigation reports show that larger scale incidents may be the result of the 3 common tasks.

Observations

The trends in fatalities before and after the implementation of PSM in 1992 shows a remarkable improvement in fatalities and fatal events. Following the reduction in 1992, most fatalities occur in single fatality incidents. These trends are counter to what one might conclude by observing high profile incidents in the media. This is true in part because of incidents in storage, transportation and at user facilities which are not included in this study.

The nature of these incidents seems to straddle the line between process safety and occupational safety. It may be that there is a need to better coordinate the safety efforts of what are in many companies two distinct organizations. The same may be true of the educational system. For example at Texas A&M University, occupational safety is taught in the School of Public Health while process safety is addressed in the College of Engineering.

Based on the incident description, the vast majority of fatalities appear to occur with relatively small releases affecting workers in close proximity to the release. At least 80% of the events are associated with what seem to be routine maintenance and operations tasks. They do not appear to be the result of complex technical failures such as impaired control systems.

There has been great success in limiting fatalities since the introduction of OSHA PSM. In order to further reduce fatalities, it appears that additional focus on less complex small scale releases is warranted.

The nature of these incidents suggest that improvements may be needed in training for hazard recognition, procedures, and job safety analysis.

The incidents are often caused at least in part by the dynamic nature of the tasks being performed, where the operation is changing over time. Procedures are needed that recognize and adapt to these changes.

In many cases, there are only two physical barriers to protect the workers, the pressure containment and PPE. Since many of the tasks involve opening the containment, adequate PPE is especially important. PPE requirements should be based on all possible release scenarios.

Recommendations for Industry and Academia

Process and Occupational Safety specialists should work together to understand and address small scale process incidents associated with common tasks. Some specific recommendations:

- Utilize this insight to better educate operations and maintenance personnel.
- Design process systems to minimize the need for hazardous tasks or reduce the severity of possible releases.
- Incorporate these findings in junior college process operator curriculum.
- Consider whether current procedures adequately cover the type of tasks most often associated with severe injuries and fatalities.

Further Investigation by the Authors

This study focused on CM and PR. Fatalities were addressed only for OSHA Region 6. The other 9 regions should be studied to confirm the trends and patterns and increase the confidence in the results. The severe injury data covered the 22 federal OSHA states. The scope for severe injuries and fatalities should be expanded to Oil and Gas and other industries.

The descriptions of the incidents in this study were extracted by the authors. To identify and interpret the entire dataset, machine learning techniques should be applied.

Develop systematic methods for identifying hazards for the most common tasks which were identified in this study.

Investigate the need for and characteristics of procedures that would prevent or mitigate the consequences of incidents from the common tasks.

References

https://www.osha.gov/data, This is the site for both data sources in this paper, "Fatality and Catastrophe Investigation Search" and "Severe Injury Reports".

Mannan etal, "Patterns and Trends in injuries due to chemicals based on OSHA occupational injuries and illness statistics", Journal of Hazardous Materials, 2008, p 349-356.

Chemical Safety and Hazard Investigation Board (CSB), Honeywell Incidents, Report No. 2003-13-I-LA, August 2005

Chemical Safety and Hazard Investigation Board (CSB), Toxic Chemical Release at the DuPont La Porte Chemical Facility, La Porte, Texas, Incident Date: November 15, 2014, No. 2015-01-I-TX

US Department of Labor, Phillips 66 Company's Houston Chemical Complex (HCC), Report to the President, April 1990.

618389 Incident Investigations to Enhance Safety Culture: Refinery Examples

Abstract:

The investigation of small incidents and near misses provides an opportunity for any company to change its safety culture at the grass roots level. This change is due in part to features unique to small incidents and near misses. The purpose of this paper is to outline the benefits of using small incidents and near misses to increase situational awareness and enhance safety culture. The paper will also include a refinery case study demonstrating how these small steps can make a big difference for process safety management.

Scope: The primary purpose of the investigation is to understand the underlying causes of the incident and to prevent it from happening over and over again. Events are not just human error or a result of a physical failure, each event is indicative of a fault in the system. The investigations are an important part of safety culture of company. They increase the awareness as well.

Methods, procedures, Process: The important elements of the event research are the timely research and report creation, the effective documentation of the findings and the sharing of the results. The method of the investigation is selected due to the potential risk of Incident/Near Miss/Safety Observation. In addition to the method, the assignment of team members is made by different levels of the organization depending on the level of potential risk assessment of the event.

Results, Observations, Conclusions: The current and potential evaluation of the impact of the incident to environment, human, reputation and to asset, the establishment of research teams, the appointment of root causes and actions, the automatic creation of the corrective and preventive actions or maintenance requests are made on specific investigation tab of the SAP (SAP EHSM) system. The publication of the accident research report are carried out by SAP EHSM as well.

In most of the applications Tier 1 and 2 Process Safety incidents are deeply investigated but sometimes Tier 3 and 4 incidents left behind in terms of investigation and actions completion speed. However, if you check all the Tier 3 and 4 incidents root causes comprehensively, you may see that they already call for the Tier 1 and 2 incidents. For example, you may find some corrosion reports as Tier 3 since it did not caused any Tier 1 or 2 incident. As you have investigation you will come up with some defaults due to crude oil quality or process technologies applied. Having such an investigation in the phase of Tier 3, you will be able to prevent such a big leakage. Moreover emphasizing on Tier 3 and 4 incidents investigation, gives the message that you are more proactive since you are not only dealing with the incidents only caused harm to people or asset but also potential risks. That will also help management to communicate to employees that, they are one-step ahead the incidents, which will lead employees to think potential risks in the workplace.

619133 Effective Implementation of Dhas – Trends One Year Past the NFPA Deadline

Abstract:

The deadline incorporated into NFPA 652 for existing facilities handling combustible dusts to complete a dust hazard analysis (DHA) has come and gone as of September 2020. How are companies doing with incorporating the findings of their DHAs into capital expense and operational expense budgets? Are authorities having jurisdiction (AHJs) even aware of the requirement? This presentation will review key findings from surveys of our clients and AHJs.

American Petroleum Institute

PROCESS SAFETY SITE ASSESSMENT PROGRAM

American Petroleum Institute
What Don't You Know about Your PSM Program?: How the Process Safety Site Assessment Program (PSSAP®) Can Help Sites Find and Address These Blind-Spots.
March 22, 2021

Often companies struggle to understand what process safety management areas need their focus. What specific topics am I neglecting? Which of my programs or procedures could use some help to be truly effective? How does my holistic program compare with my peers in the industry? The Process Safety Site Assessment Program (PSSAP®) is an industry-created, industry-led program that can help answer these questions.

The American Petroleum Institute (API) and the American Fuel and Petrochemical Manufacturers (AFPM) kicked-off the PSSAP program in 2012. Since then, API has administered over 130 of these assessments across the globe. These assessments cover the main focus areas of process safety management: Process Safety Leadership, Operating Practices, Mechanical Integrity (focused on fixed equipment), Safe Work Practices, Management of Change (MOC), Process Hazards Analysis (PHA), Facility Siting, Product Storage & Transfer, and Incident Learning. In addition to these traditional areas, the PSSAP program can also assess your Hydrofluoric Acid (HF) Alkylation/API RP-751 processes, as well as the Safeguards a site has in place to address risks, our newest offering.

But how can an assessment help my site? Won't I just receive a report? The PSSAP program is not a normal assessment. While our assessments are based on satisfying industry-created good practice protocols, an operator gets much more from our assessors than a report following their time on-site. API sends a full team of vetted, qualified assessors that are leaders in their field, with the average experience of one of our assessors being 40+ years. Our assessors spend time on site, and in addition to evaluating your programs for your blind spots, they also are constantly sharing the good-practices they have seen across the industry as part of their participation in this program. Not only does a site receive an assessment of their programs, they also can find out what the rest of the industry is doing to address these areas. Not only does a site get this through good-practice sharing from our assessors, but sites also get to see the benchmarking data from our previous assessments. With over 130 worldwide assessments, a site can statistically gauge how they compare with their peers. Sites also receive a detailed examination of the specific areas that could use more of their attention.

In addition to the different focus areas and our assessor pool, it is important to note that these assessments are scalable. Are you a small site? Is your operation too small for you to think such an assessment would be worthwhile? API created a shorter program for these sites, enabling them to access our industry-leading expert assessors so that all may benefit from their wealth of knowledge, continuing to drive refining, petrochemical and chemical operators toward our shared goal of zero process safety incidents.

We will explore our newest offerings in this forum, and will also show examples of what a site stands to gain from their participation in this industry-led and driven program.

619049 Bouncing Back from a Heart-Wrenching Process Safety Incident

Abstract:

During startup of Urea Fertilizer Plant after a brief shutdown, due to recurrence of a leakage, the field maintenance team was inspecting top head gasket of condenser on a 31-ft high platform when Carbamate solution splashed on the workers. They tried to escape via monkey ladder but fell down one after the other resulting in loss of life of one of the employees and hospitalization of other four.

Detailed incident investigation by a multi-disciplinary team highlighted number of gaps related to human elements, job handover/execution and crisis management. The incident occurred due to failure of multiple layers of protection including management of change, mechanical integrity, risk assessment, PTW, job supervision, fatigue element, over-occupancy, communication, training & development, emergency response, PPEs etc.

Interestingly, this happened in less than 3 years after successful compliance certification of Process Safety Management System but raw material unavailability and resulting financial crunch hugely jeopardized organization's process safety management system and safety culture.

The paper will cover this eye-opening case of huge PSM deterioration in very short time due to financial factors and how did the organization bounce back from that desperate situation. It will shed light on causes of incident, lessons learning and results achieved after revamp of the process safety management program. It will enable other participants to learn from story of our setbacks & revival, and implement similar cultural shift programs in their respective organizations.

617206 Risk Assessment and Safeguarding of Lithium Ion Battery Containing Facilities

Abstract:

The energy portfolio is rapidly moving to renewables and away from traditional fossil-fuel based options. While wind and solar comprise a larger share of energy sources, the technology that allows this energy to be stored and transferred is still either in its infancy or only in development. The most common approach for storage of renewable energy, usually in the form of electricity, is the lithium ion battery. While the technology for lithium ion batteries has been around for decades, only recently has there been large scale ramp up in production of these batteries. Furthermore, the batteries are being collected into large modules, packs, and systems that are capable of powering large vehicles and even entire municipal power grids. Unfortunately, renewable energy is not immune to process safety hazards. After all, batteries store energy as chemical potential, the same as fossil fuels. Exacerbating the problem, the chemicals used in batteries includes hazards the typical motor fuels do not, specifically the potential for self-accelerating decomposition with the generation of highly toxic hydrogen fluoride as a byproduct. To illustrate this fact, a “fire” in a lithium ion battery facility that stored solar power for re-injection into the municipal grid resulted in a violent explosion that sent four first responders to the hospital in Surprise, Arizona, USA, in April 2019. This paper will present the hazards related to runaway reactions that can occur in lithium ion battery containing facilities – going over the anatomy of the runaway reaction in detail. The paper will also present risk analysis techniques and safeguards such as safety instrumented systems, relief systems, and gas detection systems that can be used to reduce risk in these facilities to a tolerable level.

619072 Risk Based Process Safety: Providing Safety & Reliability Beyond the Chemical Industry

Abstract:

Catastrophic toxic releases, fire & explosions usually occur after a series of early warning signals, often subtle, are overlooked. Some organizations have been very successful in avoiding catastrophic events, even in environments with high risk factors and complex operations. Known as Highly Reliable Organizations (HROs), these operations have developed cultures that embrace the understanding of hazards and risk, drive proper management of risk, and value learning from experience.

The Center for Chemical Process Safety (CCPS) Risk Based Process Safety (RBPS) management system approach has created a globally recognized model that helps organizations develop effective catastrophic incident prevention programs. RBPS has become a well-respected resource for developing HRO culture and capability in organizations.

After several high-profile accidents in the 1980s, OSHA and EPA issued the PSM Standard and Risk Management Plan rule, respectively. These regulations have changed little over the decades, and it is becoming increasingly clear that the risk of catastrophic fires, explosions, and chemical releases is not limited to the chemical & refinery industry sectors.

Increasing number of organizations are recognizing the need to adopt process safety management systems and apply them beyond the limited scope of OSHA & EPA Process Safety regulatory requirements. An example of this is increased focus within numerous industry segments concerning the explosive hazards of combustible dust, and the inclusion of process safety management systems within NFPA 652 (2019), Standard on the Fundamentals of Combustible Dust. This increased focus has become a natural place for organizations in all sections to utilize the Risk Based Process Safety model to assess and improve Management System performance to reduce the probability of catastrophic incidents.

This presentation will provide:

- An overview of the CCPS Risk Based Process Safety model,
- Introduce practical approaches to apply process safety outside of traditional chemical industry segments, and
- Describe ways to improve an organization's risk management effectiveness.

Several case studies from outside traditional chemical industry sectors will be offered to demonstrate how a holistic, risk-based approach to applying Risk Based Process Safety can improve any organization trying to manage potentially catastrophic risk.

SPRING21
+17TH GCPS
A Joint AIChE and CCPS Meeting

"If You *Really* Think It's That Important—Show Me, Don't Tell Me!"

Jennifer. F Mize, PE, CCPSC
Eastman Chemical Company
P.O. Box 511, B-54D
Kingsport, TN 37662
jfmize@eastman.com

Prepared for Presentation at
American Institute of Chemical Engineers
2021 Spring Meeting and 17th Global Congress on Process Safety
Virtual
April 18 - 22, 2021

"If You *Really* Think It's That Important—
Show Me, Don't Tell Me!"

Jennifer. F Mize, PE, CCPSC
Eastman Chemical Company
P.O. Box 511, B-54D
Kingsport, TN 37662
jfmize@eastman.com

Keywords: process safety culture, process safety management, risk management, resource allocation

Abstract

"Action speaks louder than words but not nearly as often."
Mark Twain

Effective process safety messaging supports, represents, and reinforces a strong process safety culture. However, no communication program or slogan can create a strong process safety culture. Organizations must invest the necessary time and resources to support process safety from the front lines to the board room. In order to establish, sustain, and improve process safety performance, organizations must identify and address situations that contradict their expectations and communication related to process safety. Employee engagement and commitment are only possible when sufficient incentives are provided. In other words, people will put the most effort into the activities that provide the most reward. Actions, particularly financial decisions, always speak much louder than words. Does the executive leadership have a long-term strategy for investing in process safety? Is a demonstrated commitment to process safety necessary for career advancement? Are leaders rewarded for properly managing risk even when their decisions might have a short-term negative impact on production? This paper will provide examples of situations that contradict an organization's process safety messaging and suggest best practices to address those situations.

1 Introduction

Excellence in process safety performance requires a strong and ongoing commitment to risk management. Investing in process safety is not simply a cost of doing business; a strong process safety culture provides a foundation for continued strong business performance, as it ensures that a company maintains its license to operate. This license to operate is not granted by any regulatory agency, but is provided by the shareholders who invest in the company and the people and communities surrounding the operating facilities. It is not a question of choosing between investing in process safety and investing in growth and innovation. The challenge for leaders is to properly balance their investment in *both* to drive the company's success. Process safety is the ultimate sustainability initiative.

Process safety professionals are adept at understanding and quantifying risk for the low frequency, high consequence process safety incidents that can result from loss of primary containment. They speak the language of $1x10^{-6}$ frequency, catastrophic consequences, Independent Protection Layers, and Safety Integrity Level (SIL) reliability. The Process Hazard Analysis (PHA) or HAZOP reports and Layer of Protection Analysis (LOPA) recommendations document the risk that a company has to manage and identifies gaps to be addressed. Leaders have exactly what they need to understand the risk for which they are ultimately responsible. What else could they possibly need to know?

In order to truly manage risk, leaders need to understand prioritization for all recommendations and gaps, as well as their common features--— the need for resources to address them and a timeline in which to do so. This is where balance comes in. Even with process safety, the laws of supply and demand have not been repealed--there will always be more demand than there are resources to supply it. Leaders face a difficult choice when it comes to funding growth and innovation and meeting their commitments for effective risk management. Their choices, and whether they clearly and effectively communicate them, can either strengthen or compromise the company's process safety culture.

Effective process safety messaging supports, represents, and reinforces a strong process safety culture. However, no communication program or slogan can create a strong process safety culture. Organizations must invest the necessary time and resources to support process safety from the front lines to the board room. In order to establish, sustain, and improve process safety performance, organizations must identify and address situations that contradict their expectations and communication related to process safety.

2 Build a Strong Foundation

"If you miss the beginning, the basics, then you are destined to go back and visit the basics." – Buck Brannaman

Process safety messaging is about effectively communicating expectations and values for process safety management activities. In order to be effective, it has to engage the hearts and minds of all employees in the mission of process safety excellence. This isn't a recent concept; in fact, in the New York Department of Labor recognized the need for effective actions in support of safety messaging in a 1916 report intended to "broaden the interest in industrial safety by showing that accidents can be prevented…"[1].

From that same report comes the following:

"The information above …is suggestive of at least the nature of successful safety efforts. It indicates that the prevention of accidents can be effected neither by the mere parrot-like utterances of 'Safety First' nor by the installation of mechanical safeguards alone. Furthermore, successful experience has demonstrated that spasmodic safety campaigns launched with a blare of trumpets and dropped soon after, cannot produce lasting results. If our industrial accidents are to be prevented or even materially reduced in number and seriousness, our efforts need to be directed by well-studied plans and they must be continuous and persistent. We must realize also that after we have eliminated the grossly

unnecessary accidents, the fruits of our efforts will become less apparent from year to year. But having attained a satisfactory record we must persist in our safety campaigns in order to maintain it."

If people doubt the sincerity of the message, they will not put in the necessary effort to achieve the desired results. Unfortunately, it is all too easy to give people cause to question process safety messaging. Following are several examples of issues that can undermine the foundation of a process safety program no matter how catchy the slogan.

2.1 Personal versus process safety

Leaders should address performance in *both* areas and ensure that all personnel understand the difference between the two. Safety topics delivered in meetings should be a mixture of personal and process safety. Both personal and process safety programs need to be effectively resourced. There is a good deal of overlap in these areas, since a strong personal safety culture promotes operational discipline and avoidance of normalization of deviance. [2] Effective process safety messaging ensures that people don't focus on the things they can see while ignoring or underestimating the things they can't see. They must be aware of and focus on leading indicators and operational discipline to prevent process safety incidents.

2.2 Metrics and goals

What message is being sent when goals or metrics are established, reviewed, and rewarded?

Goals for zero incidents are insufficient to track safety performance, especially when they are based only on lagging indicators. People can go about their daily business without recognizing their impact on process safety and still meet the goals – after all, nothing has happened (*yet*). Goals and metrics must support a healthy sense of vulnerability and should include leading indicators. Don't set metrics and goals such that people are discouraged from reporting near misses and incidents in order to get a safety t-shirt or other reward.

2.3 We're all in this together, right?

In order to achieve, maintain, and improve a strong process safety culture, goals and metrics must be established across the entire enterprise. Everyone must operate under the same set of expectations. Exempting certain facilities or allowing subsidiaries to play by different rules is not indicative of a strong safety culture and undermines safety messaging. Good engineering practice is independent of size or organizational structure.

3 Motivate the People

"I've learned that people will forget what you said, people will forget what you did, but people will never forget how you made them feel". - Maya Angelou

An effective indication of the health of a company's process safety culture is based on how people behave when no one is watching. In order to develop, maintain, and enhance process safety culture, people must be invested in supporting it. This cannot be achieved by simply telling people what to do. Leaders must reward the behaviors that they want to be repeated. Talented and passionate process safety professionals will move on to greener pastures if they are not rewarded for their expertise and contributions. Any of the following issues can cause people to lose motivation and focus on supporting process safety programs.

3.1 Situations that undermine safety messaging

3.1.1 Improper Management of Organizational Change (MOOC)

- The revolving door

  It is in the best interest of organizations to reward their high performing employees. There is an expectation that high performers should advance at a faster pace than their peers and that each position is a step on their career path. However, organizations must ensure that sufficient knowledge and experience are obtained in each position. Each opportunity must contribute to the expertise of the individual to prepare them for the next step.

  A demonstrated commitment to process safety should be established as a condition for advancement. Managers are tasked with making risk decisions and determining where to allocate resources. This responsibility increases as someone moves up the management ladder. If people can move quickly up the management chain without demonstrating a sufficient competency in and a strong commitment to process safety, the company is not backing up its process safety messaging with its actions.

- Career move trumps PHA role

  A high performing employee has been tasked with serving on a PHA revalidation team as the process expert. In this role, she is responsible for developing and implementing resolution plans for recommendations from the PHA. Six weeks after the PHA is completed, she is moved to a new job.

  If this situation occurs systemically within the organization, it greatly undermines the company's messaging about their commitment to process safety. Assuming that the engineer knows about the move before the PHA starts, she is aware that she will have no responsibility to implement the findings. This may encourage the wrong behavior in the conduct of the study as she knows that implementation will be someone else's problem.

- No experience required (or wanted—it slows things down)

  A high performing employee has received a promotion. An inexperienced engineer has been identified as his replacement. Aware that a PHA is scheduled to start in six weeks, the manager determines that the personnel changes should take place immediately to ensure that the inexperienced engineer is available to serve on the PHA team.

When planning personnel changes, managers must consider the impact on their operations due to the loss of experience and the time required to replace that experience. The manager should develop a transition plan for the personnel changes. In this example, the manager should have considered having both the outgoing and incoming engineers serve on the PHA team to ensure that an effective hazard analysis and resolution plan were developed.

3.1.2 "It's us versus them" (also known as "We have a process safety department for that.")

Process safety is not something that is done to or for an operating facility. It is something that must be woven into every aspect of the operation, providing a foundation for excellent process safety performance. A true partnership must be established between process safety professionals and operations personnel.

Process safety competency expectations must be set for every position within the company. It is not sufficient to hire process safety experts and task them with making sure the company has good process safety performance. There must be sufficient process safety competency based the job function (operations engineers, operators, and managers) such that everyone is contributing to a strong safety culture and is committed to properly evaluating risk and then working together to manage it.

PHA teams should be staffed with experienced personnel in each role. This not only ensures that effective hazard assessment is completed, it reinforces the company's safety messaging by properly resourcing the study. Expanding the team to include inexperienced personnel as a learning opportunity demonstrates a strong commitment to process safety as part of career development. Sparing an experienced person from serving on a PHA team so they can focus on a growth project delivers the message that process safety has less value than innovation.

3.1.3 Ineffective management involvement

Management review and oversight must be balanced such that the managers are 1) sufficiently competent in process safety and risk-based decision making and 2) have process safety experts involved in the discussion. A manager cannot just show up for management review of PHA action items or budget discussions. Managers must demonstrate their commitment to process safety in how they approach running their operation. This commitment is demonstrated daily, in all aspects of running the plant. Effective risk management is achieved by setting expectations for daily conduct of operations, including operational discipline and avoidance of normalization of deviance. People know how to and will make the right decisions because they know management is committed to process safety. They will shut down the process rather than trying to "save the day" and "keep it running" because they know that their management will support those decisions.

3.1.4 Contractor management

Employees are expected to execute their work in a manner consistent with the company's values and expectations. There should not be a separate set of expectations for contractors.

- Select contractors who are recognized experts in process safety work (PHA/HAZOP, Safety Instrumented Systems (SIS), relief design) when needed

  Companies must set criteria for process safety work done outside their company. The necessary process safety expertise may not be available at every engineering contractor. Leaders should set expectations that contractors with process safety expertise will be hired when needed.

- Provide appropriate oversight of contractors doing process safety work

  Contractors are often hired to lead PHAs and design relief systems and Safety Instrumented Systems (SIS). These activities are part of the company's risk management program. The risk is owned by the company, not the contractor. Appropriate oversight must be provided to ensure that they are aligned with the company's risk criteria and tolerance.

3.2 What good looks like

- Professional development and career advancement opportunities are available

  In order to retain and develop talent, they must be able to see opportunities for increasing their expertise and value and be rewarded for their efforts with advancement opportunities. Providing promotions disproportionately for those employees with the strongest contributions to financial performance (increased production, cost reduction, increased sales volume) as compared to those responsible for maintaining the company's license to operate (HSES employees) clearly indicates the value placed on their efforts.

- Training is available and encouraged

  Training is an investment in the development of employees. Organizations that cut training budgets year after year are sending a very clear message about the value of those employees and their expertise. Continuing education for process safety professionals is essential to ensure that companies benefit from the advances in the field.

- External industry activities are promoted and encouraged

  The practice of risk-based process safety is continually evolving to better equip the process industries to manage their risk. Benchmarking with other process safety professionals and standards organizations provides leaders with invaluable insight.

- Recognition of the technical expertise that process safety professionals bring to the table

  Process safety professionals are not simply facilitators of PHA team meetings. Process safety experts are highly skilled employees with a breadth and depth of technical expertise. In order to effectively identify hazards, assess risk, and provide management with recommendations to manage risk, they must be adept in process design, engineering standards, safety systems, LOPA, and PHA. They must be effective communicators and collaborators. They must have strong leadership skills to guide teams through the PHA process.

- Representation of process safety expertise on executive leadership team

  Companies must identify an individual on the executive leadership team that is responsible for process safety. This is not to say that process safety is not everyone's job. However, there has to be an identified responsible party. The person identified for this role must have sufficient process safety competency to be credible in their leadership.

4 Consistently Provide Sufficient Resources

"Don't tell me where your priorities are. Show me where you spend your money and I'll tell you what they are." – James W. Frick

Another indication of the health of a company's process safety culture is how they choose to allocate their resources. The true values and identity of a company are evident in the way it spends its money and how it prioritizes its initiatives. Where a company spends its money shows what its leaders truly care about. People understand that there will always be more demand than there are resources available; what undermines a strong process safety culture is when risk management activities seem to be funded by the resources are left over after other needs are met, or are the first thing to go in an austerity program

4.1 Situations that undermine safety messaging

4.1.1 Goals don't match up with the resource allocation (also known as the unfunded mandate)

Setting goals and expectations that cannot be achieved due to lack of resources strongly undermines process safety messaging. Organizations tasked with improving process safety performance must be able to get the people and funding necessary to achieve the stated goals. If it is far easier to hire people in Research & Development while the Safety department is chronically understaffed, the clear message is that growth and innovation are prioritized over risk management. Repeatedly expecting one organization to "do more with less" and "improve productivity" while others are able to increase staffing clearly demonstrates which organizations and outcomes are deemed to be most important to management. Leaders must provide sufficient resources to allow process safety professionals to achieve the expected results.

4.1.2 Emphasizing quantity over quality in resource allocation for risk management

Action items and recommendations from PHAs, HAZOPs, and audits must be effectively prioritized to ensure that resources are allocated to the highest priority items - the ones that achieve the most significant risk reduction. In many cases, the highest priority items are not the easiest ones to address. They may require significant technical problem solving and resources to implement solutions. The temptation is always present to prioritize the action items that can be addressed with minimal time and effort. Emphasizing the quantity of closed action items over the quality of those action items (the most significant risk reduction) sends the message that the company's priority is to reduce the number of action items rather than reducing risk.

4.1.3 Growth and innovation projects are on the fast track

Great ideas generate excitement and anticipation of commercial success. The journey from the lab to full scale production is lengthy, and there are many valid business reasons to accelerate the pace to achieve commercialization as quickly as possible. It can be tempting to skip steps in the development process to compress the overall schedule. However, those steps exist to ensure that the final result will be a process that can be safety and economically operated. It is critical to allow sufficient time for inherently safer design options to be evaluated, necessary hazard data to be generated in pilot plants, and a formal hazard assessment to be conducted. Allowing "fast track" projects to bypass steps or not fully satisfy the company's project management / stage gate process sends the message that business needs can override process safety considerations. [3] Following a strong stage process for innovation projects ensures that safety and environmental considerations are properly addressed and reinforces the company's process safety and regulatory compliance expectations.

4.1.4 "We've always done it that way"

A company's risk tolerance evolves over time. Hazards that were once routinely accepted as part of doing business may no longer be acceptable. Leaders need to evaluate where existing facilities and their associated hazards fit into the company's risk profile. In some cases, improved understanding of the risk being accepted and a lower tolerance for that risk may require leaders to consider shutting down processes that have inherent risk that cannot be cost-effectively addressed. Continuing to operate facilities that have significant risk above the company's risk tolerance with no plan to reduce that risk undermines the company's commitment to risk management.

4.2 What good looks like

- Long-term plan for spending on risk management

  Companies with a strong process safety culture are well aware of the risk that they manage and the areas of opportunity for risk reduction. Leaders develop a long-term plan for providing funding to address risk reduction goals and maintain a strong commitment to that plan even when business conditions are challenging.

- Effective prioritization for growth and innovation projects versus risk management and reduction projects

  Leaders establish guidelines for comparing resource requests for risk management and reduction needs alongside opportunities for long-term growth and innovation. Evaluating projects based on rate of return is not a viable method for analyzing the benefits of risk management or reduction. In order to effectively prioritize projects, leaders must be able to quantify the value of reducing the overall risk profile of the company.

- Balance in resources between growth/innovation and risk management

  Effective prioritization for projects and a long-term plan for spending on risk management facilitates the necessary balance in resources across the company. This balance is a statement of the company's values and identity – adding value with long-term growth and innovation while supporting its commitment to people, property, and the environment.

5 Conclusion

"Well done is better than well said." – Benjamin Franklin

Effective process safety messaging must support a strong process safety program. A strong foundation, including the development of effective metrics and goals that apply across the entire enterprise, is required to develop, maintain, and enhance a strong process safety culture. Leaders must motivate all employees by rewarding the behaviors that support the company's commitment to process safety while guarding against situations that can undermine process safety messaging. Leaders must also provide sufficient resources to achieve the desired process safety performance and strive to effectively balance their investment in process safety and growth and innovation. A catchy process safety slogan and stylish visuals cannot make a healthy process safety culture. Actions, after all, speak louder than words.

6 References

[1] "Recent Reports Relating to Workmen's Compensation and Industrial Accidents", New York Department of Labor, Industrial Commission: Bureau of Labor Statistics and Information, Industrial Accident Prevention. No. 77 (1916) Albany, NY

[2] Center for Chemical Process Safety (CCPS), Recognizing and Responding to Normalization of Deviance, CCPS, 2018.

[3] P. N. Lodal and J. F Mize. "Integrating Process Safety and Innovation", Chemical Engineering Progress, Pages 36-40, Volume 112, Issue 10, October 2016.

SPRING21
+17TH GCPS
A Joint AIChE and CCPS Meeting

C. P. Snow's "Two Cultures" Challenging CCPS Process Safety Culture – Chemical Engineers' Roadmap to Bridging Engineering to the Humanities

Long Zhang
Production Operator at Imprint
11455 Bissonnet St Apt #514, Houston TX 77099
longz.houston@gmail.com

Prepared for Presentation at
American Institute of Chemical Engineers
2021 Spring Meeting and 17th Global Congress on Process Safety
Virtual
April 18 - 22, 2021

C. P. Snow's "Two Cultures" Challenging CCPS Process Safety Culture – Chemical Engineers' Roadmap to Bridging Engineering to the Humanities

Long Zhang
Production Operator at Imprint
11455 Bissonnet St Apt #514, Houston TX 77099
longz.houston@gmail.com

Author's statement: all relevant statements, opinions and claims presented in this paper are solely of the author's opinions and are NOT endorsed by the author's affiliation.

Keywords: Process Safety Culture, Ethical Dimension, Engineering Practice, Roadmap

Abstract

In 1959, C. P. Snow raised a "two cultures" challenge – a persistent divide (split) between science and the humanities, later between engineering and the humanities, in our society, educational system and intellectual life. And to some extent, the author believes, the Element **Process Safety Culture** was intended by CCPS to face a similar challenge. The author argues, however, that CCPS, industry and companies should go further to pursue RBPS Core Value & Process Safety Cultural Communication (PSCC), both of which would function together as an ethical dimension in context of which messaging and communication between leadership and process safety professionals could be dramatically improved. In another words, "traditional" messaging and communication activities between leadership and professionals could be, and shall be, broadened to a constitutional (and epistemological) map (roadmap) of PSCC & engineering practice, within which some new essential categories of engineering practice would be identified, probably the first time in mankind's engineering history, and found relevant to the future evolving of Code of Ethics of AIChE. The author's conclusion: CCPS, industry & companies should initiate a public debate on this constitutional roadmap; and, epistemologically, chemical engineers should play a leading role in this debate, in pursuing the roadmap to bridging engineering to the humanities and in putting more trust in mankind's "values and beliefs" (i.e. in God) rather than solely in instrumental rationale.

1 An Ethical Dimension – Process Safety Culture's Wagon

1.1 C. P. Snow's "Two Cultures" Challenge

In 1959, C. P. Snow gave the annual Rede lecture at Cambridge University and he raised the "two cultures" challenge – a persistent divide/split between science and the humanities, later also extended to the split between engineering and the humanities, in our society, educational system and intellectual life.[1] The concept of the "two cultures" has a claim about a changing map of academic disciplines: "…the sprouting of ever more specialized sub-disciplines and the growth of various forms of inter-disciplinary endeavor…in place of the old apparently confident empires, the map shows many more smaller states with networks of alliance and communication between them crisscrossing in complex and sometimes supervising ways".[2] And the central message of this first sketch of the "two cultures" is that "the greatest enrichment the scientific (and engineering) culture could give us is …a moral one." [3]

MIT professor Louis L. Bucciarelli claimed that he had taken seriously C. P. Snow's "two cultures" challenge and been working for 35 years to bridge the "two cultures" [4]; and that instrumental rationality does not suffice itself in engineering design, diagnosis and education[5]. And he also implied that some primary concepts and principles that should have been at the root of engineering disciplines have not been taught at all at universities. He believes engineers need to know the relationships between their texts they construct themselves and rely on in daily life and the status & function of engineering knowledge (epistemology).[6]

1.2 Ethical Dimension – Process Safety Culture's Wagon

After the Bhopal catastrophic accident, AIChE established the Center for Chemical Process Safety (CCPS) with 17 initial charter member companies in 1985. Since then CCPS has been evolving rapidly as a global organization [7] and especially in 2007 presented Risk Based Process Safety (RBPS) 4 Pillars and relevant 20 Elements, among which Process Safety Culture was identified as an Element for challenges arising from organizational cultures or from messaging and communication between leadership and process safety professionals (officially, chemical engineers).[8] **With a sense of vulnerability toward the possible collapse of temple (4 Pillars) and dying state of Process Safety Culture**, however, the author reflected upon some of the biggest lessons learned, which could be called 5 organizational/social patterns of core value in ancient east and west cultural traditions. The author finally identified 2 new underlying (Mega) Elements/ Universes - RBPS Core Value (engineering value) and Process Safety Cultural Communication (engineering cultural communication), some kind of what Bucciarelli found absent at the root of engineering disciplines. Neither of them probably has ever been formally taught or seriously talked about in the society, educational system or intellectual life.[9]

The author describes RBPS Core Value as the root of (the tree of) Process Safety Culture and states that RBPS Core Value would function as the landlord for the temple, and if RBPS Core Value dies the temple would still look like a house outwardly but substantively have a purpose of tomb. And in light of anthropology and intercultural communication, the author defines a series of (new) concepts, and principles, as follows:

Process Safety Expertise (PSE) means neither any process safety engineer nor any world prestigious process safety expert but the divine essence of the highest strategic insights and vision of process safety management, the Tao for process safety culture, and soul of the *Art of Warfare* against catastrophic process safety accidents.

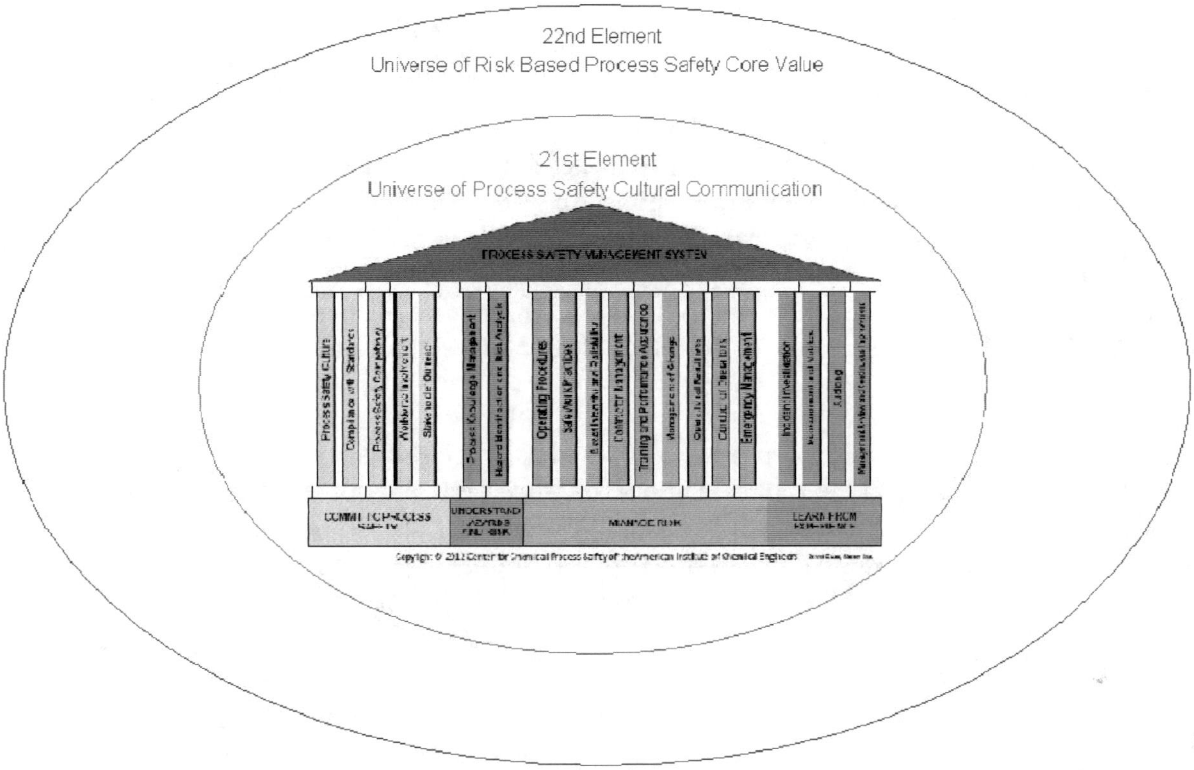

Egg Model (CCPS Temple + 2 universes) – Figure 1

The Temple (4 pillars) Figure Source: CCPS, Process Safety Boot Camp, contributed by David Guss, Nexen Inc. Copyright 2010 AIChE and used with permission

Process Safety Expertise Power Distance (PSEPD): (defined as a subtle underlying sub-dimension of Power Distance one of those 4 dimensions of mankind's value studied by Geert Hofstede in his *Cultural Consequences*) the extent to which PSE can exclusively chair at the highest strategic level throughout all the processes of decision making and behavior determining of the organization and its members as far as RBPS concerns.[10]

RBPS Core Value: PSE shall eternally and exclusively chair at the highest strategic level throughout all the processes of decision making and behavior determining of the organization and its members as far as RBPS concerns.

Unfolded version: **RBPS Core Value functions as the root of the tree of process safety culture, whose deficiencies are symptoms or signals for a dying or dead state of core value. Those deficiencies signify that the organization has highest strategically failed in managing**

process safety through a risk based approach even though there has been no occurrence of process safety accidents till that moment. In another words, the RBPS temple is bound to collapse, sooner or later, due to the change or drift from the original design intent of the temple.

Simplest version: **There is no power, authority or governance on process safety except on behalf of process safety expertise.**

Coincidentally, also in 1959, anthropologist Edward T. Hall published his book *The Silent Language,* within which one of his main earth-breaking statements on the relationship between culture and communication was: "culture is communication and communication is culture"[11] , from which the author gets inspired and defines the relationship between process safety culture and process safety cultural communication: **Process Safety Culture could not exist except through Process Safety Cultural Communication (PSCC). PSCC is the photosynthesis of the tree, and the future form, of Process Safety Culture.** Furthermore, the author defines the relationship between RBPS Core Value and Process Safety Culture (or PSCC): **the ultimate purpose of Process Safety Culture or PSCC is to convey the message of the RBPS Core value. Without the nourishment from RBPS Core Value (the landlord), the tree of Process Safety Culture would get dry and dead.**

Both RBPS Core Value and PSCC might be deemed as some primary concepts and principles Bucciarelli found missing at the root of engineering disciplines in universities, and function together as an inevitable ethical dimension in context of which the messaging and communication between leadership (possibly up to board members, even to the White House in the case of Columbia Shuttle Accident) and process safety professionals (chemical engineers) could be dramatically improved and Operational Discipline (OD) more reliably maintained. And the author claims: only after process safety culture's deficiencies and dying state have been neglected for a certain period of time could catastrophic accidents possibly happen. No matter how technically over competent process safety professionals (chemical engineers) could be, without a sense of vulnerability toward the possibly dying state of Process Safety Culture and death of RBPS Core Value, hardly could catastrophic accidents be prevented, and **an ethical dimension (RBPS Core Value + PSCC) could function as a Wagon for Process Safety Culture to survive, a "big ship" for escape/refuge from the collapse of temple (4 Pillars) and some "biggest flood", and should be "rediscovered" as some primary concepts and principles currently still absent at the root of engineering disciplines.**

2 A 3 Cultures' Map – Chemical Engineers' Exodus

As mentioned, in 2003, Bucciarelli expressed his concerns on the engineering disciplines, and 10 years later, Michelfelder, McCarthy & Goldberg also pointed out, in their editorial preface of *Philosophy and Engineering: Reflections on Practice, Principles & Process,* that "exploring the ontological & epistemological dimensions of engineering challenges the notion that engineering is simply the application of scientific knowledge to problem solving, a challenge that has deep implications for the design of engineering curricula."[12] Ontology could be defined as the order of categories of existence, being, becoming, and reality.[13] We could say RBPS Core Value exists on the most fundamental level; PSCC on the 2nd most fundamental level; the 4th Pillar Learn from Experience has a potential all-encompassing ontological spectrum, and both 2nd & 3rd Pillars on the superficial/shallow level. And epistemology could be understood as a hierarchical structure of human knowledge[14]: ethical dimension as "driver" knowledge; both 2nd & 3rd Pillars as "car" knowledge; and 4th Pillar as "car diagnosis & driver alcohol test" knowledge. "Driver" here means "values and beliefs" of engineers and engineering practice [15]. During benchmarking the Egg model (CCPS 4 Pillars + 2 universes) against Sun Tzu's *The Art of War* the author found that the ancient engineering (military) discipline(s) seemed neither ontologically nor epistemologically so deficient as the modern ones in universities. And those 7 Mega Elements, as an outcome of the benchmarking work (refer to Table 1) [16], could constitute a complete engineering design process and a universal designing & diagnosis vocabulary (principles), and also could be spread into a map of PSCC & Engineering (designing & diagnosis) Practice. And upon this map, the author would explain why formal culture's *tone* matters, and what engineers' 1st amendment in designing process really look like.

2.1 Chemical Engineers' Vocabulary – 7 Mega Elements (Engineering Thoughts)

<table>
<tr><th colspan="2">RBPS Egg</th><th>The Art of War</th><th>Philosophy of Process Safety</th><th></th></tr>
<tr><td colspan="4">RBPS Core Value - Tao - RBPS Worldview</td><td>Worldview</td></tr>
<tr><td rowspan="4">Commit to Process Safety</td><td>Process Safety Competency</td><td rowspan="2">Leadership</td><td rowspan="2">People</td><td rowspan="7">Process Safety Cultural Communication
(Process Safety Culture)
RBPS Methodology</td></tr>
<tr><td>Workforce Involvement</td></tr>
<tr><td>Compliance with Standards</td><td rowspan="2">Weather/Climate</td><td rowspan="3">Plant</td></tr>
<tr><td>Stakeholder Outreach</td></tr>
<tr><td>Understand Hazards & Risk</td><td></td><td>Terrain</td></tr>
<tr><td>Manage Risk</td><td></td><td>Discipline</td><td>Procedure</td></tr>
<tr><td>Learn from Experience</td><td></td><td>Espionage</td><td>Engineering & Humanities</td></tr>
</table>

Table 1

Table 1 is an outcome of benchmarking Egg model (4 Pillars + 2 Universes) against *The Art of War.* As per Sun Tzu, his engineering (military) diagnosis (espionage) and design (warfare) process has 6 Mega Elements: Tao, Leadership, Weather/Climate, Terrain, Discipline and Espionage.

2500 years ago when Sun Tzu characterized Tao (military RBPS Core Value), which could be deemed as ancient Chinese metaphysics/ontology or concept of God, within *The Art of War*, he simultaneously defined engineering (military) Operational Discipline (OD): "(Tao is) is what brings the thinking of the people in line with their superiors. Hence, you can send them to their deaths or let them alive, and they have no misgivings one way or the other".[17] So when facing issue of OD, Sun Tzu's solution was Tao, which we may call RBPS Core Value today. And RBPS Core Value could become not only a new engineering thought/concept (and principle) to the engineering discipline but also a new Mega Element for engineering design process. Tao, Leadership & Weather/Climate as some ontological & epistemological dimensions possibly could be diagnosed and identified as root causes. However, Diagnosis/Espionage (4th Pillar) to Tao, Leadership and/or Weather/Climate might also be prohibited as a threat to (organizational) national security. Sun Tzu described Espionage saying: "Only the sagacious ruler is able to employ spies; only the most humane and just commander is able to put them into service; only the most sensitive and alert person can get truth out of spies."[18] This means Sun Tzu had recognized the persistency of the divide/split between engineering (military) and the humanities.

To some extent, the author believes, the Element **Process Safety Culture** was intended by CCPS to face some similar challenge as what C. P. Snow recognized and raised decades ago. However, Process Safety Culture still needs to survive, and **CCPS may go one step further: RBPS Core Value and PSCC (future form of Process Safety Culture**). The author **defined PSCC as the landlord's livelihood, the ethical/ ontological/ epistemological nourishments, or the photosynthesis inspired, from RBPS Core Value, to the whole tree of Process Safety Culture**. PSCC, the surviving form of Process Safety Culture, becomes the 7th Mega Element of Sun Tzu's engineering design (& diagnosis) process, and another 3 Mega Elements i.e. Leadership, Weather/Climate & Espionage are also newly defined as below:

Leadership: a new chain of command, inherently from RBPS Core Value, rather than blindly from superiors, with its 4 dimensions/competencies (embracing RBPS Core Value, upward PSCC, downward PSCC and intervention). Some chemical engineers (process safety professionals) may get inspired from RBPS Core Value and, without any prior permission from any operation manager, HR manager, board member or corporate attorney, communicate upward, even to highest level of the organization, and the overall truthfulness of verbal and behavioral universes of messaging and communication of board of directors, executive management, operation managers etc. would be powerfully challenged. Also, all superiors of the organization shall be psychologically and rationally prepared for getting empowered to be "powerless" i.e. to obey RBPS Core Value through completely giving up their personal power and authority in deep integrity. Only through this powerlessness could the message of RBPS Core Value be truly felt during downward PSCC.

Weather/Climate: The ultimate purpose of the outreach (Corporate Social Responsibility) to the stakeholders & community is to convey the message of RBPS Core Value, rather than solely to maintain a seemingly good relationship with the community;

Terrain (Understand Hazard & Risk): Sun Tzu defined "Terrain refers to the fall of the land, proximate distances, difficulty of passage, the degree of openness, and the viability of the land for deploying troops".[19] Not only a general knowledge and expertise of hazards & risk is necessary but the specifics of all actual hazards & risk shall be acquired and understood. Terrain consists of 2 elements: Process Knowledge Management and Hazard Identification and Risk Analysis;

Discipline (Manage Risk): Sun Tzu defined "Regulation (Discipline) entails organizational effectiveness, a chain of command, and a structure for logistical support".[20] Discipline consists of 9 elements: Operating Procedure; Safety Work Practices; Asset Integrity & Reliability; Contractor Management; Training and Performance Assurance; Management of Change; Operational Readiness; Conduct of Operations; and Emergency Management.

Espionage (Diagnosis/ Counter-intelligence): 4th Pillar Learn from Experience is the Espionage not to enemies but to ourselves. For Espionage to enemies, the more comprehensive and deeper, the better, however, for Espionage to ourselves, it's not necessarily the deeper the better but heavily dependent on how honest we could be to ourselves and how effectively the "patient's privacy" be respected. Through PSCC, the ontological/epistemological spectrum of engineering diagnosis process in practice could possibly be broadened from "car" knowledge (Terrain & Discipline) to "driver"/value knowledge (RBPS Core Value, PSCC, Leadership & Weather/ Climate). In another words, 4th Pillar Learn from Experience should play not only a role of diagnosis toward Terrain & Discipline but a "broadened" role of diagnosis i.e. counter-intelligence toward (organizational) national security, both technical dimension and ethical/ontological/epistemological dimensions;

<u>The author hereby claims</u> **those 7 Mega Elements probably would constitute both mankind's 1st complete engineering design process and engineers' 1st universal designing & diagnosis vocabulary, a future paradigm for engineering disciplines to evolve, and the principles in leading to bridge engineering to the humanities.**

2.2 28 Practices' Map of Engineering (Designing & Diagnosis) Practice

Inspired from anthropologist Edward T. Hall's map of culture, the author develops chemical engineers' 7 Mega Elements' (as 7 Primary Message Systems) vocabulary and a map of PSCC and Engineering (including designing & diagnosis) Practice[21], what Bucciarelli might like to call another more complex vision of engineering thoughts and practices.[22] And this map consists of 28 basic human activities or namely 28 categories of engineering practice (refer to Table 2nd those 7 ones in the diagonal from upper left to lower right and 21 ones above).

	RBPS C.V. 0	P.S.C.C. 1	Leadership 2	Weather/C. 3	Espionage (Diagnosis) 4	Terrain 5	Discipline 6
RBPS Core Value 0	00	01	02	03	04	05	06
Process Safety Cultural Communication (PSCC) 1	10	11	12	13	14	15	16
Leadership 2	20	21	22 4 dimensions	23	24	25	26
Weather/ Climate 3	30	31	32	33	34	35	36
Espionage (Diagnosis) 4	40	41	42	43	44	45	46
Terrain 5	50	51	52	53	54	55	56
Discipline 6	60	61	62	63	64	65	66

Table 2

Each of those 7 categories of practice along the diagonal are the critical verbal statements & behaviors related to the concept(s) and principle(s) of each Mega Element. And all those 7 categories of practice are described as below:

Practice 00: verbally and behaviorally embrace RBPS Core Value and reflect upon the "two cultures' challenge" in those 5 organizational/ social patterns of core values in ancient east and west cultural traditions: Bian Que-Duke Cai pattern, Moses-Pharaoh pattern, Solomon-Sheba Queen pattern and Bian Que's Brotherhood pattern. Those 5 patterns could collectively accommodate as an atlas of mankind's values for both engineers and non-engineers to learn from. And all main secrets with values might be over there.[23]

Practice 11: verbally and behaviorally embrace PSCC and reflect upon some biggest (process safety cultural) communicational failures, experiences & wisdom mankind has ever had from such as Bhopal accident, Challenger shuttle accident, Columbia shuttle accident and their ancient, probably more classic, east and west versions of "space shuttle accident or Bhopal" such as Duke Cai's death and its west version the red sea behind Moses, as well as from traditional Chinese symptom (technical)-root cause (ontological) balanced medical approach [24] what Bucciarelli calls the designing a social process today.

Practice 22: verbally and behaviorally embrace and reflect upon Leadership with 4 dimensions, as dictated by RBPS Core Value and PSCC, capable of organizationally /socially embracing RBPS Core Value, downward PSCC, upward PSCC and self-intervention - the engineering thought/concept/principle of new chain of command inherently from RBPS Core Value rather than blindly from superiors;

Practice 33: verbally and behaviorally embrace and reflect upon the RBPS Core Value enlightened outreach (corporate social responsibility) to the stakeholders & community, rather than the one solely counting on the tactics of maintaining a seemingly good relationship with community;

Practice 55: verbally and behaviorally pursue and reflect upon Terrain (CCPS 2nd Pillar: Understand Hazards & Risk);

Practice 66: verbally and behaviorally pursue and reflect upon Discipline (CCPS 3rd Pillar: Manage Risk);

Practice 44: verbally and behaviorally embrace/pursue and reflect upon main concepts and principles of Espionage (CCPS 4th Pillar: Learn from experience). This category of practice is a combination of both instrumental rationality & value rationality. And instrumental rationality does not suffice itself in engineering design, diagnosis and education. And Espionage is also like ancient Chinese medical diagnosis to patients, or counter-intelligence toward (organizational) national security (i.e. toward rest 6 Mega Elements, or toward both technical dimension and ethical/ontological/epistemological dimensions/roots).

And each of those 21 categories of practice above the diagonal (upper left to down right) is defined as the critical statements & behaviors related to the interaction between Mega Element A and Mega Element B. And some categories of practice (04, 14, 24, 34, 45, 46 & 56) are descried as below:

Practice 04: verbally & behaviorally communicate that the actual state of RBPS Core Value might be one of the root causes in accident investigation, or one of the items of audit/ metrics/ review;

Practice 14: verbally & behaviorally communicate that the actual state of PSCC might be one of the root causes in accident investigation, or one of the items of audit/ metrics/ review;

Practice 24: verbally & behaviorally communicate that the actual state of Leadership with 4 dimensions might be one of the root causes in accident investigation, or one of the items of audit/ metrics/ review;

Practice 34: verbally & behaviorally communicate that the actual state of Weather/Climate might be one of the root causes in accident investigation, or one of the items of audit/ metrics/ review;

Practice 45: verbally & behaviorally communicate that the actual state of Terrain (knowledge + hazard identification) might be one of the root causes of accident investigation, or one of the items of audit/metrics/review; & Espionage might be helpful to improving Terrain;

Practice 46: verbally & behaviorally communicate that the actual state of Discipline (the 9 Elements of Managing Risk) might be one of the root causes of accident investigation, or one of the items of audit/metrics/review; & Espionage might be helpful to improving Discipline;

Practice 56: verbally & behaviorally communicate that poor state of Terrain (Knowledge + Hazard Identification) is bound to undermine Discipline (those 9 Elements); and good state of Terrain do not necessarily bring about good state of Discipline.

This verbal-behavioral matching character of those 28 categories of practice comes from anthropologist Edward T. Hall and resonates with Wittgenstein's recognition that rules and meanings (what the author calls ***tone***) in talk and texts were mixed with practice [25]. Bucciarelli also put "the meaning of statements is embedded in practice". [26] For PSCC and engineering design & diagnosis, those 6 green backgrounded categories of practice (44's instrumental dimension, 45, 46, 55, 56 & 66) could not be solely counted on while rest those 23 red lettered categories (including 44's ethical/ epistemological dimension) are completely blinded to. An 'authentic" sense of vulnerability starts from facing the possible dying state of PSCC and death of RBPS Core Value.

The author claims **those 28 categories of practice would not only make up mankind's 1st complete map of PSCC & Engineering (designing & diagnosis) Practice but also constitute the 1st roadmap for chemical engineers to play a leading role to bridge engineering to the humanities in society, educational system and intellectual life.**

2.3 A 3 Cultures' Roadmap – Chemical Engineers' Exodus & Trust in God

Edward T. Hall defined that culture has three levels i.e. formal, informal and technical. **Formal culture/level is always learned through mistakes and correction, with a clear right-wrong character.** Technical is also learned through mistakes and correction, but being offered with reasons. Informal culture, however, is learned without knowledge of occurrence of the learning process at all.[27] The author sorts 28 categories of practice into 3 groups - an map of 3 levels of Process Safety Culture/ PSCC (**Formal:** 22 red lettered & yellow backgrounded categories of practice e.g. practice 34; **Informal:** practice 44 (dual-rationality); **Technical**: 5 black lettered & green backgrounded categories of practice: 45, 46, 55, 56 & 66). Hall described that these 3 cultures/levels exist in a relationship of continuous change [28]; while one will dominate, all three are present in any given situation [29]. And Bucciarelli also pointed out that in designing process, "when the rhythm of execution is challenged by the unanticipated", "engineers *do* stop and think, and rethink, and redo, and rethink." [30] The author believes that, however, this is still not enough and engineers shall have deep emotions with formal culture [31] and, by "instinct" [32], be apt to correct anything wrong using formal culture. So the ***tone*** of formal culture matters.

Bucciarelli stated that "designing is a social process, and different participants with different competencies, responsibilities and interests must negotiate their claims and individual contributions to the design if the process is to prove fruitful". He also claimed that "engineers in their designing convey, via products and systems, certain values and norms that allow for particular kinds of use, social discourse and exchange, even way of life, and limit, if not rule out, other kinds."[33] He also emphasized that above the negotiations between different disciplines different disciplinary engineers share same system of beliefs about how the world works. [34] In light of anthropology and intercultural communication, as the author defined RBPS Core Value (disciplinary highest principle – one dimensional Tao or God), Mechanical engineers could have Mechanical Engineering Expertise (MEE) as their own disciplinary core value. Another disciplinary engineers, e.g. electric engineers, have their disciplinary core value. When mechanical engineering disciplinary universe and electric engineering disciplinary universe are intended to become a 2-dimensional universe, mechanic engineers and electric engineers *must* have a core value charter (i.e. 2-dimensional Tao or God). Under this 2-dimensional Tao or God, these two ethnicities (disciplines) negotiate their claims & individual contributions and "the products and systems are defined in good measure" [35], ultimately by the 2 dimensions of this 2-dimensional Tao or God. In this manner, all possible disciplines from both engineering and the humanities need to make up an universe of a continually increasing number of dimensions under Tao or God of a continually increasing number of dimensions, as designing process becomes increasingly integrated.[36] However, some ethnicity (disciplinary participants) might face in reality the deprivation of their ethnic formal culture, be degraded to the level of the products and systems and defined in "good measure" by other participants, which is akin to a genocide. Engineers' actual value (and beliefs) need to have an unlimited large number of dimensions, at least as large as the maximum number of dimensions of actual values of all other participants in designing process, so that all decision making and behavior determining of engineering designing could possibly be ensured correct, if they are determined to defend their 3 cultures, especially their formal culture. So the persistent divide/split between engineering and the humanities seems deriving from **long time social and political deprivation of Engineers' 1st Amendment in designing process**, namely engineering's formal culture.

As participants Chemical Engineers should defend their 1st Amendment (formal culture) in designing process. They should put more trust in God, in the universal governance of God, on the changing map of academic disciplines, rather than solely in instrumental rationality, pursue a leading role in the *tone* of formal culture, and follow the 3 cultures' roadmap to bridge engineering to the humanities. CCPS, industry & companies need a public debate on the roadmap. And Chemical Engineers' Code of Ethics also need evolve.

3 Conclusion

Facing the persistent challenge of "two cultures" or process safety culture's survival, CCPS, industry and companies do need a public debate on the 3 cultures' roadmap to bridge engineering to the humanities, especially on the *tone* of formal culture. Incorrect *tone* of formal culture does limit formal process safety communication & learnings and is very difficult to change in a short period of time. Chemical engineers should focus not only on practice 00 to get inspired from mankind's values and beliefs, but, more practically, on practice 44 to help introduce some culture change in their organizations. The author gives some examples: one chemical engineer (process safety professional), whose main responsibility is leading and coordinating annual PHA workshop, posts the 3 cultures map in his/her cubicle or office for self learning and meditation.

And some of colleagues who pass by his cubicle/office and notice the map might ask him/her, “What’s this?” The answer could be, “I just got it (from this year’s GCPS) and it seems a new root cause mapping method.” If the colleague further asked, “What’s new?”, the reply might be “it seems helping identify ontological root cause as well. However, it would take time for me to really understand it.” And the chemical engineer also might post an unfolded definition of RBPS Core Value in the room occasionally dedicated for PHA workshop, and when being asked, e.g. by operation manger, about it he/she answers “it is a new concept and would take time for me to really understand it”. And this chemical engineer might share some good movie on Moses for the close colleague chemical engineers to watch after work. He or she also encourages his or her colleague engineers, managers, directors etc. to attend annual GCPS as “tourists”. Also, this chemical engineer successfully manages all process safety communication and learning relevant to more than instrumental rationality to occur in informal culture or manner. Edward T. Hall suggested that “if a person really wants to help introduce culture change he or she should find what is happening on the informal level and pinpoint which informal adaptations (to formal culture) seem to be the most successful in daily operations.” And he further explained that “the out of awareness nature of the informal is where all changes start”. [37] The author believes that this chemical engineer has fortunately pinpointed some informal adaptations/ culture in practice 44 and is substantially helping introduce culture change in his or her organization!

4 References

[1] Snow, C. P., *The Two Cultures*, Cambridge University Press, 1998, p.vii, cover page

[2] Ibid. p.xliii, xliv

[3] Ibid. p.xxvii

[4] Bucciarelli, L. L., Engineering Philosophy, Delft: DUP Satellite, 2003, p.2

[5] Ibid. p.99

[6] Ibid. p.5

[7] www.aiche.org/ccps/history

[8] Center for Chemical Process Safety, Guidelines for Risk Based Process Safety, John Wiley & Sons, Inc., 2007

[9] Zhang, L., An Upgraded Definition of Risk Based Process Safety Core Value and Its Impact in Light of Anthropology & Intercultural Communication on the Future Evolution of Process Safety Culture & Management, 2013 AIChE Spring Meeting and 9th Global Congress on Process Safety, San Antonio, April 2013

[10] Hofstede, Geert, *Culture’s Consequences: Comparing Values, Behaviors, Institutions and Organizations Across Nations,* Shanghai Foreign Language Education Press, 2008, pp.83

[11] Hall, Edward T., *The Silent Language,* Anchor Books, 1990, p.186

[12] Michelfelder, Diane P., McCarthy, Natasha & Goldberg David E., Philosophy and Engineering: Reflections on Practice, Principles & Process, Springer, 2013, pp.xiv

[13] https://en.m.wikipedia.org/wiki/Ontology

[14] https://www.britannica.com/topic/epistemology

[15] Bucciarelli, Louis, Engineering Philosophy: Values and Design, isepp.org, May 18th, 2006

[16] Zhang, L., Benchmarking Risk Based Process Safety Egg (RBPS Temple + 2 Universes) against The Art of War, 1st CCPS Asia-Pacific Conference on Process Safety, Qingdao, Sept 2013

[17] Sun-Tzu and Ames, Roger T., Sun-tzu: The Art of Warfare, Ballatime Books, 1993, p.103

[18] Ibid. p.170

[19] Sun-Tzu and Ames, Roger T., Sun-tzu: The Art of Warfare, p.103

[20] Ibid. p.103

[21] Zhang, L., Bridging Chemical Engineering (Risk Based Process Safety) to the Humanities – 7 Mega Elements' Approach (Protocol) of Accident Investigation in light of Ethical Dimension, Architecture & epistemology, 2019 AIChE Spring Meeting and 9th Global Congress on Process Safety, New Orleans, April 2019

[22] Bucciarelli, Engineering Philosophy, p.5

[23] Same with [9]

[24] Same with [16]

[25] Bucciarelli, Engineering Philosophy, p.6

[26] Ibid. p.8

[27] Hall, Edward T., The Silent Language, p.67-68

[28] Ibid. p.88

[29] Ibid. p.66

[30] Bucciarelli, Engineering Philosophy, p.8

[31] Hall, Edward T., The Silent Language, p.73

[32] Ibid. p.76

[33] Bucciarelli, Louis, Engineering Philosophy: Values and Design

[34] Bucciarelli, Engineering Philosophy, p.20

[35] Bucciarelli, Louis, Engineering Philosophy: Values and Design

[36] Zhang, S. Engineering Philosophy Meets Intercultural Communication – What Message of Oneness of God to Human Knowledge, 20th Conference of the International Association for Intercultural Communication Studies, Rhode Island University, Aug 2014

[37] Hall, Edward T., The Silent Language, p.93

SPRING21
+17TH GCPS
A Joint AIChE and CCPS Meeting

Good or Great? The Seven Practices of Great Process Safety Leaders

Jim Spigener
Chief Client Officer
DEKRA North America
Jim.spigener@dekra.com
1945 The Exchange SE, Suite 300, Atlanta, GA 30339

Prepared for Presentation at
American Institute of Chemical Engineers
2020 Spring Meeting and 17th Global Congress on Process Safety
Virtual
April 18-22, 2021

Good or Great?

The Seven Practices of Great Process Safety Leaders

Jim Spigener
Chief Client Officer
DEKRA North America
Jim.spigener@dekra.com
1945 The Exchange SE, Suite 300, Atlanta, GA 30339

Keywords: Commitment to Process Safety, Leadership, Culture

Abstract

Nearly every company strives to have "Safety First" values where managers act in a way that promotes an injury-free workplace. However, when the organization is stressed by economic or internal pressures, the actions managers take have a profound impact on the workplace culture.

History reveals that most process safety events, by their nature, are rare. These factors demand that robust leadership engagement and an effective governance system for appropriate process safety leading indicators be put in place. The task at hand is ensuring that all leaders relentlessly demonstrate their commitment to process safety and are contributing to a learning culture.

As framed in Center for Chemical Process Safety's (CCPS's) Risk Based Process Safety Guidelines, authentic ***commitment to process safety*** is the cornerstone of process safety excellence. It is not enough for organizations to have a process safety program on paper; it must be implemented well.

Organizations are unlikely to improve process safety performance without strong leadership and solid commitment, from all layers of the organization. However, great leaders are not always managers; and managers are not always leaders.

This, however, addresses two issues. First leadership and management are two different things. Management is a primarily a transactional process. It is important and requires to a great extent position to execute the responsibilities. Leadership on the other hand is transformational. It creates the commitment from others to want to complete the transactions in the spirit of the objectives of those transactions. Transformational leadership is possible by anyone regardless of position. It is the art of influence. This transformational activity called leadership is what makes any objective sustainable and robust.

It is important to recognize that a workforce that is convinced that the organization fully supports safety as a core value will tend to do the right things, in the right ways, at the right times, even when no one is looking. This behavior should be consistently nurtured, and celebrated, throughout the organization. Once embedded in the company culture, this commitment to process safety can help sustain the focus on excellence despite the rarity of events and the more technical aspects of process safety.

This paper will present an overview on the critical role that successful leaders play in establishing and maintaining a healthy Process Safety Culture. In addition, it will provide an overview of seven practices that DEKRA has witnessed great process safety leaders use during our 30 years of work. These practices include Vision, Collaboration, Credibility, Communication, Action Orientation, Feedback and Recognition, and Accountability.

All seven practices work together in a way that creates, not only an exemplary Process Safety Culture and an environment where people *want* to work safely, but also a culture that is sustainable.

1. Introduction

On March 23, 2005, the BP Texas City Refinery experienced a catastrophic explosion that resulted in 15 deaths and more than 170 injuries. This landmark event kicked off an investigation by the U.S. Chemical Safety and Hazard Investigation Board (CSB). CSB recommended that BP form an independent panel to conduct a review of the company's safety culture, safety management systems, and corporate safety oversight at their U.S. refineries [1].

The Baker Panel Report, the resulting report, is today recognized as one of the most respected and thorough analysis of the technical, cultural, and management system failures that can occur within an organization [2]. Eight of the 10 recommendations center around Leadership Commitment, Organizational Culture and Oversight topics.

Yet despite the lessons learned from this tragedy, catastrophic events that are tied to hazardous materials continue to occur. At least six of the last 20 investigations completed by the U.S. CSB that occurred within the last decade include key lessons tied to two factors: leadership and culture [3].

2. The Role of Leadership in Establishing a Commitment to Process Safety

DEKRA's research on high performance culture over the past 30 years has found that there are a set of seven leadership practices that lead to a positive safety culture. A culture that embodies a Commitment to Process Safety is critical for organizations that want its employees at all levels to focus on critical work processes that *prevent* catastrophic incidents versus solely focusing on outcomes.

The Authentic Commitment to Process Safety is the cornerstone of process safety excellence. Management commitment has no substitute for embedding this commitment throughout any organization. Organizations generally do not improve without strong leadership and solid organizational commitment.

The entire organization must make the same commitment. A workforce that is convinced, through the demonstrated actions and behaviors of their leaders, that the organization fully supports safety as a core value will tend to do the right things, in the right ways, at the right times, even when no one is looking. Through practice and continual improvement in these seven leadership practices, workers at all levels will find it easier to engage in a culture that embodies a Commitment to Process Safety in their daily routines and practices.

3. Understanding the Role of Safety Culture

According to the UK Health and Safety Executive, safety culture is "the product of the individual and group values, attitudes, competencies, and patterns of behavior that determine the commitment to, and the style and proficiency of, an organization's health and safety programs." [4]

However, a more colloquial way to define safety culture is simply, "how the organization behaves when no one is watching."

Values are what shape beliefs and attitudes. If an organization shows little regard for safety, there will be little regard for it on all levels, from the c-suite all way down to the frontline worker. That's because if the values of the organization affect the individual, the organization plays a significant role in determining individual behaviors.

Logically, that means the organization has great influence on its safety culture. A weak culture, for example, is evident when employees at all levels are inactive in response to common process safety practices. However, this analysis still leaves unanswered questions, such as:

- "Why do operators shortcut the turnover process and why do they feel comfortable in doing so?"
- "Why do supervisors tolerate a practice that jeopardizes the safety of the facility?"

Answering these questions requires an understanding of safety culture. The organization's leadership has the primary responsibility for identifying the need for both cultural change and sustaining a sound safety culture. In the CCPS Risk Based Process Safety Management System, the first pillar focuses on Commitment to Process Safety [5]. Throughout implementation of this pillar, the focus is on Process Safety Culture, Workforce Involvement, and Stakeholder Outreach.

By identifying and addressing the pathologies within the safety culture, we can address the root causes of safety performance problems. Regardless of whether one is seeking to establish a new safety management system, repair an existing underperforming system, or fine-tune a basically sound system to achieve higher performance, it is the actions, or inactions, of an individual working within the system that can ultimately be the limiting performance factor. Creating and sustaining a sound safety culture can be a decisive factor in determining the performance of the individual and the system.

4. What Makes a Great Process Safety Leader?

To be a great process safety leader, you need to care deeply about your people and regularly demonstrate the value of safety through your actions.

When an organization has successfully embedded safety as a value that permeates decisions at every level, leaders continue to push for further improvement. Senior executives push employees to find and address potential exposures throughout the organization. They see it as their personal responsibility to ensure their organizations promote the safety of their employees.

A leader needs to be willing to do the things that are necessary to transform the organization, must take safety personally, and hold a sincere commitment to process safety. Obviously, leaders want high performance and profit. When it comes to process safety, however, a leader has to make decisions based on how it affects their workforce and the community.

DEKRA has identified seven practices used by great process safety leaders during our 30 years of work. Each of these practices play a critical role in establishing and maintaining a health Process Safety Culture in modern organizations.

Vision. Leaders must have the ability to "see" what safety excellence looks like and a capability to articulate it throughout the organization. It must include both occupational and process safety incidents and include a long view past the year's earnings and losses.

Collaboration. Effective leaders work well with employees, promote cooperation and collaboration, actively seek input from people on issues that affect them, and encourage others to implement their decisions to improve safety.

Credibility. Does the leader generate a high level of trust with his or her employees? This requires a willingness to admit mistakes and advocate the safety interests of everyone, from managers to the front line.

Communication. Safety leaders need to be talking about safety every time they speak. Everything they communicate must be within the context of safety. They need to be aware of messages that compete with safe behaviors and eliminate their from their vocabulary.

Action orientation. Is the safety leader ready to tackle safety proactively rather than just react to incidents? Safety leaders need to show urgency even in the absence of incidents to show they're serious about achieving results.

Feedback and recognition. Leaders need honest and accurate feedback on the effect of their behaviors to help them ensure consistency between their passion for people and the message employees receive based on their actions. Recognition should be given to employees who bring exposure to light or who work to instill a culture of safety for themselves and others.

Accountability. An effective leader gives workers a fair appraisal of their safety efforts and results, clearly communicates individual roles in the safety effort, and fosters the sense that every person is responsible for safety throughout the organization.

5. The Importance of 360-degree Feedback

Understanding the seven key factors that drive cultural change is the first step that leaders must take if they are to begin the process of creating high value for doing the things that process safety excellence requires.

This first step, however, is not enough to get the results we need. For maximum effectiveness, leaders must routinely demonstrate these practices. But that can be a challenge. As is often the case, it is almost impossible for leaders to have an accurate view of how well their employees believe they are executing these critical practices.

The second step for leadership is to get feedback from the people around them who watch and follow their cues. Without getting honest feedback that allows leaders to see what they are missing; it becomes nearly impossible for them to improve the execution of these critical practices. A comprehensive 360-degree survey will include a self-assessment, along with surveys of the leader's manager(s), peers, and direct reports.

This data can then be compared against databases related to the best leadership practices and normative results are then generated. The norms are updated regularly as the data set grows and ages. This is to assure that changes in the field and the resulting practices that are necessary to account for performance as these changes occur are built in.

To be helpful, the feedback needs to be honest and accurately define the gaps in performance and must be very specific. Generalities don't work. Throughout the last 30 years, we have found that assessments that provide statistical analysis, along with anonymity to the greatest extent possible, are critical to candor and honesty. (Obviously, if a leader has one manager it will not be possible to provide anonymity for the manager's feedback).

A 360-degree feedback instrument is the most preferred and efficient method for gathering individual feedback for the leader. A second and possibly less efficient method is to have a third party interview the people around the leader. Whichever is used, the leader needs to be able to see how they are doing in each of the practices by those around them.

The advantage of using a validated 360-degree instrument is it gives the leader a benchmark for comparison. It lets them see how they stack up against a large group of leaders in a way that compares their performance relative to other effective leaders.

Figure 1: Individual Leader Percentile Scores

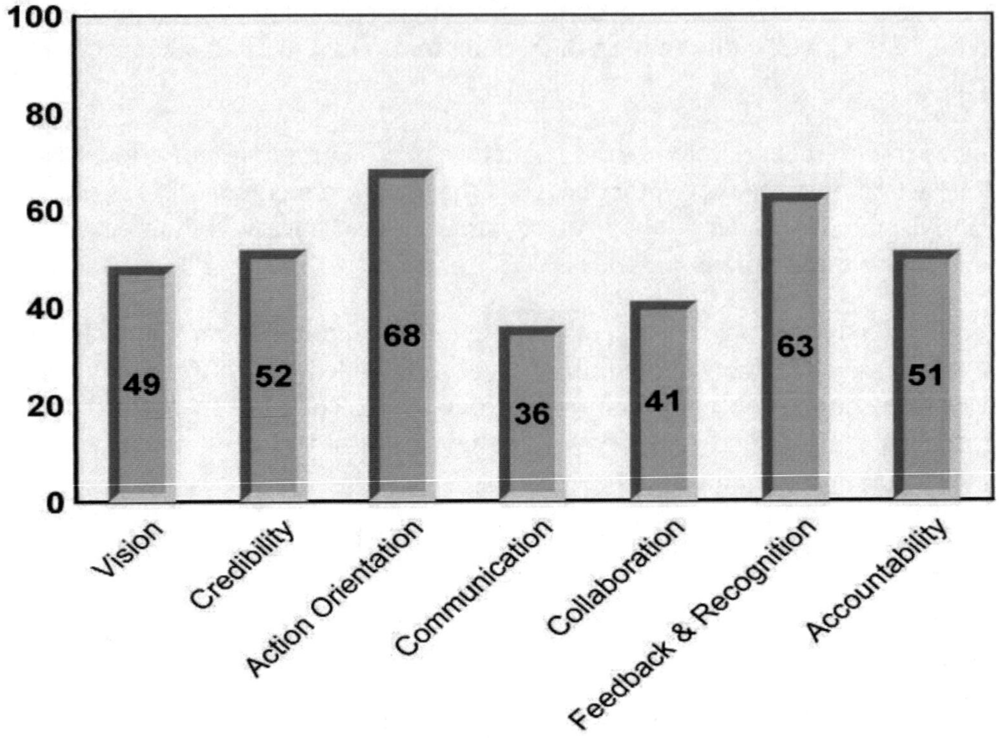

Figure 1 is an example that demonstrates how a senior leader who is overseeing a company struggling with process safety performance compared in the seven key leadership practices against hundreds of other leaders.

This leader's scores by their direct reports are reported in percentiles giving them a clear and unambiguous view of where improvement needs to be made.

The statistical analysis in Figure 1 provides a good starting point that highlights opportunities in each of the leadership dimensions, while highlighting particular opportunities in Communication and Collaboration. To improve the understanding of these results, it is desirable to provide an opportunity for open text comments in the 360-degree survey instrument. While work must be done to remove any identifying characteristics from the comments, direct examples from survey participants can help provide improved insight to the leader.

Figure 2: DEKRA Leadership Survey Results Commentary

List the three most important behaviors this Leader does to improve safety performance.

- Covers all the details, nothing goes unnoticed.
- Highly committed to his responsibilities.
- Assures good planning.
- Experience/intelligence.
- Asks the right questions that assure they have considered all potential issues/problems.
- You can always count on them if you want a difficult job done well and thoroughly.
- Assures follow-thru and meeting commitments.
- Does an outstanding job at recognizing and rewarding people.

Describe the three most important things this Leader could do to improve safety performance.

- Guide upper management without pummeling them.
- Improve team working skills with their colleagues to tap into their enormous experience and expertise.
- Define and articulate specific safety performance goals.
- The old axiom about "It's not what you say, it's how you say it" frequently applies. Their Points are usually excellent, but can at times be delivered in a sarcastic or belittling manner. They would be even more effective if they could keep the content, but change the tone.
- Frame their suggestions in a more positive manner so they are better received by others.
- Be more open-minded.
- Communicate their experience more broadly.
- They are very good at verbal jousting and I think that sometimes this becomes more important than the subject under discussion. Try not to come out on top of every discussion.

Figure 2 represents the specific feedback that was given by the reports that further clarify where there is excellence and what to work on. These details allowed the leader to consciously work on improved collaboration with his subordinates on Process Safety improvement issues and to focus on the tone and approach to his verbal messaging.

This particular leader was surprised by the feedback. After carefully studying it, and reflecting upon the statistical analysis and commentary, he was able to see its validity. The leader was surprised that what they thought was acceptable leadership performance in Communication and Collaboration was actually negatively impacting the group. He revealed that, without this wakeup call, they would likely not been motivated to change.

Ultimately, the changes in leadership approach led to improvements in leading and lagging safety performance in the organization. Moreover, his leadership skills can be resurveyed in the future to identify next steps for continual improvement.

It is the case that leaders and organizations that perform the best and continually strengthen their safety cultures and initiatives use this data and feedback process longitudinally. As time goes and the leader responds to the feedback the focus shifts naturally to the ongoing needs in behavioral change for even more robust leadership capability. This process also strengthens the interdependence and ability to communicate, and the level of trust within workgroups.

6. Case Study: Linking Leadership & Culture Gaps to Catastrophic Incidents

For more than 20 years, the U.S. Chemical Safety Board has been investigating major process safety events and publishing comprehensive reports and supporting learning tools, including case studies and video animation summaries of the events. This source of incident investigation information also provides important insight into leadership and cultural issues that were involved in each event. This information can provide valuable insight for leaders to learn from these accidents to help improve their approach and organizational performance.

Table 1 provides a summary of four investigation reports, highlighting possible avenues of learning for leaders.

Table 1: Leadership Attributes Highlighted during Major Incident Investigations

		Leadership Attribute Opportunities						
CSB Case Study Title	**Leadership & Culture Aspects**	Vision	Collaboration	Credibility	Communications	Action Orientation	Feedback & Recognition	Accountability
(2017) Packaging Corporation of America Hot Work Explosion [6]	PCA did not adequately demonstrate a learning culture with respect to making needed improvements based on lessons learned from past publicized and well understood industry incidents.	x	x			x	x	
(2016) Key Lessons from the ExxonMobil Baton Rouge Refinery Isobutane Release and Fire [7]	Long-standing practices that accepted operators removing malfunctioning plug valve gearboxes despite the lack of detailed procedures and training for safe removal.		x		x	x	x	x

(2013) Williams Olefins Plant Explosion and Fire [8]	Ineffective implementation of the facility's process safety management, as well as weaknesses in written programs themselves, were causal to the incident. Weaknesses in these programs resulted from a culture at the facility that did not foster and support strong process safety performance.	X	X	X	X	X	X	X
(2013) West Fertilizer Company Fire and Explosion [9]	In 2009, the facility's property insurance carrier noted: (1) Weak safety culture, (2) "Written programs are incomplete and outdated", (3) No accident investigation program, and (4) No structured safety program.	X	X	X	X	X	X	X

7. Conclusion

Organizations that deliver outstanding results in avoiding major project safety events, even in environments with high risk factors and complex organizations, consistently have a strong organizational Commitment to Process Safety. This culture is facilitated and nurtured by leaders who demonstrate strong capabilities in the seven key aspects in safety leadership.

All of these elements work together in a way that creates both an exemplary safety culture and an environment where people *want* to work safely, whether it be process safety or occupational safety.

Leaders in world-class process safety organizations must serve as role models in this effort. The leadership of an organization has the primary responsibility for identifying the need for, and fostering, cultural change and for sustaining a sound safety culture once it is established.

Leadership can start this process by making their personal commitment to workers a top priority. This commitment starts with, and can only be accomplished by, assessing and continually improving their individual leadership practices.

8. References

[1] United States Chemical Safety and Hazard Investigation Board, "BP America Refinery Explosion", Retrieved from: https://www.csb.gov/bp-america-refinery-explosion/ , Accessed on February 15, 2021.

[2] United States Chemical Safety and Accident Investigation Board, "Report of the BP Independent Refineries Safety Review Panel (Baker Panel Report)", Retrieved from: https://www.csb.gov/assets/1/20/baker_panel_report1.pdf?13842 , Accessed on February 12, 2021.

[3] United States Chemical Safety and Accident Investigation Board, "Completed Investigations", Retrieved from: https://www.csb.gov/investigations/completed-investigations/?Type=2 , Accessed on February 12, 2021.

[4] United Kingdom Health & Safety Executive, Health & Safety Laboratory, "Safety Culture: A review of the literature (HSL/2002/25), Retrieved from https://www.hse.gov.uk/research/hsl_pdf/2002/hsl02-25.pdf, Accessed on February 1, 2021.

[5] Center for Chemical Process Safety, "Risk Based Process Safety Overview", Retrieved from https://www.aiche.org/ccps/resources/publications/books/guidelines-risk-based-process-safetyccps/documents/overview , Accessed on February 20, 2021.

[6] United States Chemical Safety and Accident Investigation Board, "Packaging Corporate of America Hot Work Explosion", Retrieved from https://www.csb.gov/packaging-corporation-of-america-hot-work-explosion-/ , Accessed on February 1, 2021.

[7] United States Chemical Safety and Accident Investigation Board, "ExxonMobil Refinery Chemical Release and Fire", Retrieved from https://www.csb.gov/exxonmobil-refinery-chemical-release-and-fire/, Accessed on February 3, 2021.

[8] United States Chemical Safety and Accident Investigation Board, "Williams Olefins Plant Explosion and Fire", Retrieved from https://www.csb.gov/williams-olefins-plant-explosion-and-fire-/, Accessed on January 31, 2021.

[9] United States Chemical Safety and Accident Investigation Board, "West Fertilizer Explosion and Fire", Retrieved from https://www.csb.gov/west-fertilizer-explosion-and-fire-/ , Accessed on February 6, 2021

SPRING21
+17TH GCPS
A Joint AIChE and CCPS Meeting

Measuring Vital Signs of Process Safety Culture

Hamza Hameed
Senior Process Safety Engineer
Fatima Fertilizer Company Limited, Pakistan
hamza.hameed@fatima-group.com

Muhammad Ahsan Sarfraz
Manager Health, Safety and Environment
Fatima Fertilizer Company Limited, Pakistan
Ahsan.sarfraz@fatima-group.com

Prepared for Presentation at
American Institute of Chemical Engineers
2020 Spring Meeting and 17th Global Congress on Process Safety
Virtual
April 18 - 22, 2021

Measuring Vital Signs of Process Safety Culture

Hamza Hameed
Senior Process Safety Engineer
Fatima Fertilizer Company Limited, Pakistan
hamza.hameed@fatima-group.com

Muhammad Ahsan Sarfraz
Manager Health, Safety and Environment
Fatima Fertilizer Company Limited, Pakistan
Ahsan.sarfraz@fatima-group.com

Keywords: Process Safety Culture & its core principles, Sense of Vulnerability, Normalization of Deviance, Fostering Mutual Trust, Assessing & advancing Process Safety Culture, Deferring to Expertise, Establishing an Imperative for Safety,

Abstract

In line with Fatima Group's vision and HSE policy, Fatima Fertilizers Company Limited (FFCL) stipulated a Process Safety Management program (PSM) as the major corner-stone. It was purely a self-initiative to implement this model rather following any regulatory requirement. At Fatima Fertilizer Company Limited, Process Safety Performance is tracked using Global Method of Leading and Lagging Indicators. Leading Indicators are a set of data that measures performance across various elements of Process Safety. In 2020, in order to uplift Process Safety performance at FFCL, we invested in activities keeping in view Core Principles of Process Safety Culture. While analyzing Leading Indicators for 2019, FFCL performance across various elements was linked with Core principles of Process Safety; this helped to identify strong and weak areas of our culture. To supplement this analysis in pursuit of strong Process Safety Culture, another study was conducted with title **"What is it in for me, Two Decades of Process Safety Beacons".** All PS beacons till date were critically reviewed and relevant information was extracted. Main objective of this activity was to learn from past mistakes and strengthen Process Safety elements

1 Introduction

Fatima Fertilizer Company Limited (FFCL) is the first and the only green field project which has materialized under the 2001 Fertilizer Policy of the Government of Pakistan. Fatima Fertilizer Limited has successfully implemented ISO 45001:2018, ISO 9001:2015 and ISO 14001:2015 as Integrated Management System (IMS). Moreover, Fatima Fertilizer Limited has implemented DuPont 22 element model of Process Safety Management (PSM). This journey started in 2012 and in span of 5 years, FFCL site achieved a rating of 04 equivalents to Excellence (Maximum rating 05 equivalent to World Class). FFCL site is also IFA Protect & Sustain certified site. Our continuous focus on Health and Process Safety is evident from our Safe Million Man-Hours figure (current number is 59 Million).Last LTI at FFCL was recorded in 2012. The Total Recordable Injury Rate (TRIR) remained zero in 2019 and 0.03 in 2020. FFCL has also achieved the Royal Society for the Prevention of Accidents (RoSPA-UK) health and safety award (Gold) straight for last 04 years. FFCL Site has also the honor of achieving 04 British Safety International Safety Awards in a row since 2014 and achieved distinction in 2019. FFCL laboratory is accredited with ISO/IEC-17025 which is 1st fertilizer lab of Pakistan. Furthermore, FFCL Learning and Development center also known as training center is also certified with ISO-29990 which makes FFCL 1st manufacturing site in the region to have this honor. [1]

This paper will describe FFCL efforts to uplift Process Safety Culture at site. FFCL Process Safety Management journey started in 2012 after a fire incident at an Ammonia plant. Rigorous efforts were made to reduce Process Safety Incidents at the site. Acknowledging the fact that cultural aspect of Process Safety affects all other elements of Process Safety and its improvement is one of the most challenging and time intensive activities, FFCL invested in activities to uplift Process Safety Culture. Moreover, all these activities were aligned with Centre for Chemical Process Safety (CCPS) Process Safety Culture Core Principles.

1.1. Facility Description

FFCL is one of Pakistan's largest fertilizer manufacturing complexes that process natural resources like natural gas, rock phosphate, limestone and water into vital products for farmers and industrial customers. FFCL final products include urea, calcium ammonium nitrate (CAN) and nitro phosphate (NP), while Ammonia and Nitric Acid are the intermediate products. The complex contains integrated utilities section for all 5 production units (Calcium ammonium nitrate, Urea, Nitro phosphate, Nitric Acid and Ammonia) and 1 bagging unit for final product handling and dispatch. The fertilizer complex is a fully integrated production facility occupying 947 acres of land and a total of 1886 employee's including 215 management, 556 staff and 1115 contract employees.

The parent unit of the complex is a 1700 MTPD Ammonia plant which is currently operational using natural gas as the basic raw material. This ammonia plant was relocated from Rozenburg, Netherlands to Pakistan in year 2007. Prior to relocation of the ammonia plant it had gone through several revamps and energy conversation projects resulting in an increase in production capacity from 1350 MTPD to 1650 MTPD. The plant was shut down in 2000 and purchased by Fatima group in year 2006 and relocated to Pakistan in year 2007.

The ammonia and its by-product carbon dioxide are used for the production of 1500 MTPD prilled urea. The urea plant is a Stamicarbon Urea 2000+ technology new plant and was commissioned in 2010. The nitric acid production facility has a capacity of 1500 MTPD (based on 100% conc.), which along with ammonia and limestone, is utilized for the production of 1200 MTPD prilled nitro phosphate and 1400 MTPD granulated calcium ammonium nitrate. The NP plant is designed by Chemical Fertilizer Industry and it was commissioned in June 2011. The Ammonia, CAN and NA plants are relocated plants. The complex is self- sufficient in power generation and also supplies a residential colony. Figure-1 below illustrates the main process units at the FFCL plant. [1]

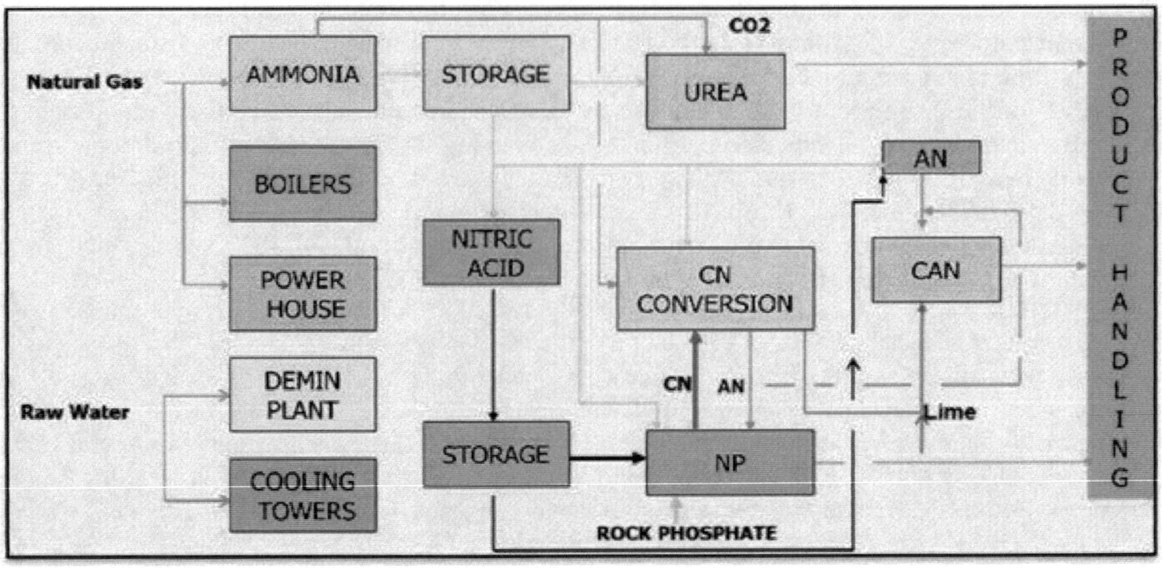

Figure 1. FFCL Flow Diagram- Raw Materials Conversion into Products

1.2. PSM Implementation and Growth History at FFCL Site

FFCL established in 2007 and conventional Safety Management systems were adopted initially which mainly worked around personnel safety performances while process safety was not the prime focus area. No formal process safety management system was launched, and the organization was still young without a rich culture and legacy. Employees from different backgrounds and cultures had joined the organization so overall a very lean safety culture existed. There were no legislative requirements in the region for process safety systems rather legislation are largely focused on labor laws. Consequently the overall site safety performance remained unsatisfactory with major setbacks in the form of severe process safety incidents (major fires & Tier Level-1 Process Safety incidents). Below are a few glimpses.

- FFCL had two fatalities in 2010 and one in 2011.One fatality occurred due to electrocution as there were no regime of electric tools testing and unskilled person tried to rectify the faulty electric asset. The second fatality occurred due to slippage of person in created opening.
- 22 recordable injuries were reported in 2011. Overall number was 70 including both 2010 and 2011. 70% of these injuries were related to contractors.
- 4 Lost Time Injuries, 1 Restricted Work Case, 2 Medical Treatment Cases & 48 First Aid Cases were reported in 2012.
- High number of fires with 29 process and non-process fires occurred in 2012
- Elevated Total Recordable Injury Rate of 0.51 (per 200,000 hours).

Major process safety incidents like an ammonia syngas compressor fire and operational upsets like Catacarb incident occurred due to which the plant remained unavailable for more than 40 days. Following the 2012 series of process safety upsets on the ammonia plant, site management analyzed the situations and identified inadequate process safety management systems as the root cause. Subsequently it was agreed to align exiting systems and practices with Best in class Process safety model (DuPont PSM model) for sustainable business model and to enhance safety, reliability and productivity of site. The whole PSM system was implemented at site in less than four years and the first 3rd party assessment was carried out by DuPont in May 2015 which rated FFCL at 3.5 from scale 1 to 5 which meant that the PSM system was well implemented. The 2nd assessment was done by DuPont in November 2017 and rated FFCL site at Excellence level. [1]

Following the successful implementation and external Third Party Audits of Process Safety Management at FFCL site, focus shifted towards program sustainability. While External Audits are planned at a span of every three years, internal audits (1st Party & 2nd Party) were regularly conducted at site. All these audits pointed towards an alarming fact of **A Stagnant Process Safety Culture.**

After detailed analysis by FFCL Process Safety Engineers, the following causes were at the heart of Process Safety Culture Stagnation:

- Since worker injury rate was at decline at FFCL, workers and management assumed that this was also indicating that the risk of Low Frequency, High Consequence Process Safety was at decline.
- As injuries are easily classified and injury rates calculated through Total Recordable Injury Rate (TRIR), focus sometimes inadvertently shifts towards Personal Safety instead of Process Safety.
- Process Safety Management was implemented as One-Time project. Efforts were not put to embed in FFCL working culture
- FFCL strong Personal Safety record had led its management to complacency. The absence of a major incident had lessened FFCL Sense of Vulnerability. FFCL Statistics had continued to demonstrate that our safety system were better than others working around us.
- FFCL Process Safety Engineers had failed to demonstrate the significance of Process Safety Culture. Although we had developed comprehensive PSM Standard Operating Procedures, training plans and audit regimes, our discussion with FFCL management and employees did not include cultural aspects related to all PSM Elements

2 Focusing Process Safety Culture

In early part of 2019, this grey area of Process Safety was ultimately realized and in second half of 2019, efforts were initiated to uplift Process Safety culture. To ensure we moved towards the right direction, we decided to consult Global Practices. After consulting multiple practices and literature, we narrowed our approach to CCPS Process Safety Culture Core Principles. In the next step, activities aligned with these core principles were agreed upon and were later executed.

2.1 Learn to Assess and Advance Process Safety Culture: [3]

2.1.1 Leading and Lagging Indicators Analysis:

At FFCL, Process Safety Performance is tracked through a set of Leading and Lagging Indicators. This set is summarized and presented to site leadership team in monthly Executive Health, Safety and Environment Committee. FFCL Leading Indicators track site performance across different PSM elements and include followings

- Process Safety Action Items Compliance
  - Incident Investigation Reports and Recommendations' Compliance
  - Quarterly Emergency Drills' Recommendations Compliance
  - Process Hazard Analysis (PHA) and Recommendations' Compliance
  - Process Safety Management Audit Recommendations' Compliance
  - Permit to Work Quality Audit Score
- Mechanical Integrity Leading Indicators
- Safety Instrumented System Integrity
- Management of Change Audits' Score
- Operating and Maintenance Procedure Revision Compliance
- Management Safety Audit and Recommendations' Compliance
- Executive Health, Safety and Environment Committee Recommendations' Compliance
- Un-Safe and Un-Safe conditions' Reporting
- Job Cycle Check Compliance

While lagging Indicator matrix, defined a debit score for followings

- Process and Non-Process Fires
- Process Safety Incidents
- Repeat Incidents
- Environment Incidents
- All categories of Injuries

At start of each year, an annual analysis was conducted by FFCL Process Safety Engineers to highlight strong and weak areas of Process Safety Management and based on this analysis Leading and Lagging Indicators were revised. In 2019, while conducting Leading and Lagging Indicators' Analysis, strong and weak aspects of Process Safety Culture were examined. Figure 2 below summarizes weak areas of Process Safety associated with the element of "Process Hazard Analysis". FFCL, PHA Standard Operating Procedure, categorizes 4 levels of PHA Recommendations based upon associated risk. Level-1 PHA recommendations address highest risk of the concerned area while Level-IV PHA recommendations are associated with the lowest risks. Level 1 & 2 PHA recommendations have different hierarchy of approval and waiver process (approved by highest authority of any department) than Level 3 & 4 (approved by line managers)

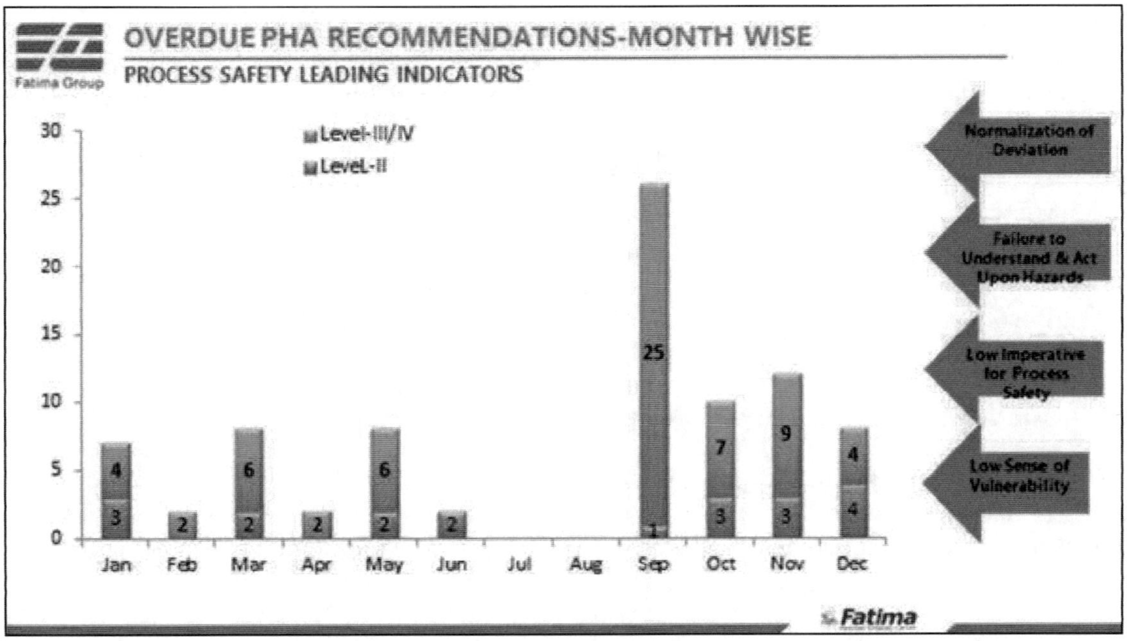

Figure 2 Cultural Aspects Related With Process Hazard Analysis Recommendations

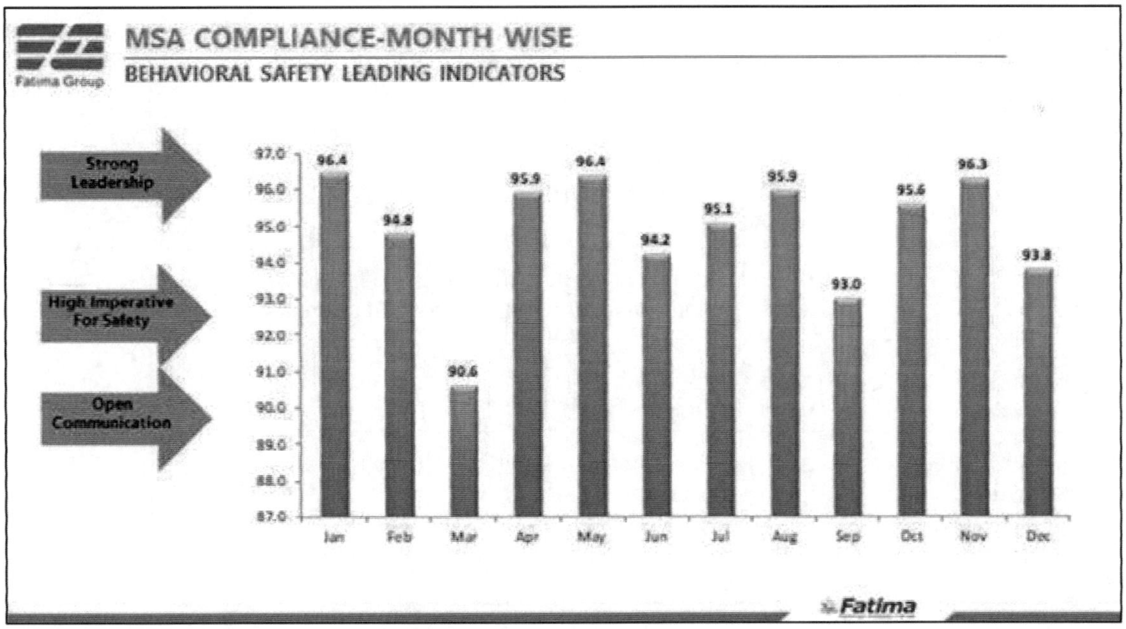

Figure 3 Cultural Aspects Related With Management Safety Audits (MSA) Compliance

This method of Leading and Lagging Analysis unlocked an insight into cultural aspects of Process Safety Activities. Action Plans generated based on this analysis (Some examples are mentioned below) ensured weaker cultural areas were focused on in the coming years.

- Risk Based (High Risk Areas) Management Safety Audit Schedule for Managers, Increased Weightage of Executive Health, Safety and Environment Committee Recommendation ⟶ **Providing Strong Leadership**

- Risk Based Scoring in Leading Indicators for Process Hazard Analysis Recommendations, Increased Score Debit in Leading Indicator for each Temporary Repair ⟶ **Normalization of Deviance**

2.2 Fostering Mutual Trust/ Ensuring Open and Frank Communication: [3]

Keeping in mind the famous saying "What gets measured, Gets Managed", a comprehensive Process Safety Survey was compiled. Sets of close ended questions from all core principles were conceived and it was completed by all FFCL employees either through email or in hard form. Second phase of the survey included evaluating Survey Results and focusing on major concerns, results' communication and sharing frequent feedback with the employees. This whole scheme was targeted to receive true feedback on FFCL Process Safety Culture and to gain the trust of employees. All questions in the survey were divided into 9 core principles of Process Safety. Specimens from each core principle is enlisted in below table

Table 1: Process Safety Culture Survey Specimen

Core Principle	Question	Yes/No
Establish an Imperative for Safety	Is the PSMS a detailed set of management system procedures that represents a "paper only" program that sits on the shelf, or has it actually been implemented and is it being used?	
Providing Strong Leadership	Is process safety an agenda item or is it considered at high level meetings on a regular basis (not just after an incident)?	
Foster Mutual Trust	Does the sharing of information that will reduce safety risks occur without fear of punishment?	
Ensure Open and Frank Communications	Are those bearing negative safety-related news required to "prove it is unsafe?"	
Maintain a Sense of Vulnerability	Do risk analyses routinely eliminate proposed safeguards under the banner of "double jeopardy?"	
Understand and Act Upon Hazards/Risks	Are the appropriate resources applied to the hazard/risk assessment process?	

Empower Individuals to Successfully Fulfill their Safety	Are employees empowered to suggest or initiate improvements?	
Combating the Normalization of Deviance	Is there undue peer pressure or organizational pressure to work extensive amounts of overtime?	
Learn to Assess and Advance the Culture	Has the organization established and nurtured a questioning and learning environment?	

2.3 Providing Strong Leadership [3]

The leadership team's role in strengthening culture is of utmost importance. Leadership team's communication must lay down significance of Process Safety. Acknowledging this fact, a half day workshop on Process Safety was arranged for FFCL Leaders. The initial idea was to include the top layer of FFCL managers but bearing in mind the role of line managers, participation was widened. This workshop was led by Fatima Group Corporate HSE Manager with more than 50 participants. Key topics of workshop included

- Process Safety Leadership Basics
- Characteristics of Leadership and Management in Process Safety Culture
- Consistence of Process Safety Messages
- Case Studies of Past Incidents
- Turnover of Leadership, Succession Planning and Organizational Management of Change

By dedicating the site leadership team's time for a discussion on Process Safety a strong message was sent across the site that "Process Safety will always take the front seat"

2.4 Maintaining a Sense of Vulnerability [3]

A high sense of Vulnerability is an indication of Strong Process Safety Culture. With strong safety statistics and lack of major incident, a sense of complacency was being felt in working patterns and communications. To overcome this problem, a campaign with title **"On This Day in History"** was run at FFCL site. Details of major process safety incidents around the world, FFCL site and other Fatima Group sites were collected and on the day of any particular incident, a memorial flyer was shared with all Fatima Group Sites. These flyers were discussed in Weekly Safety Talks. Total of 58 incidents were shared on their anniversaries with either of below messages:

- IF IT HAPPENED THERE, CAN IT HAPPEN HERE?
- IF IT HAPPENED THEN, CAN IT HAPPEN AGAIN?

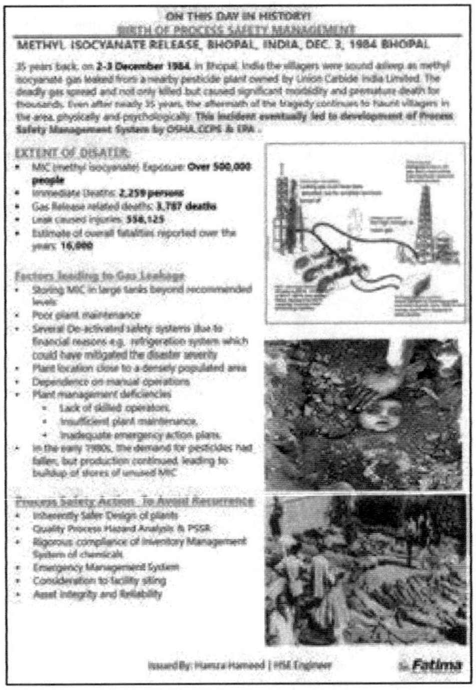

Figure 4 & 5: On This Day in History Flyer on Bhopal Tragedy and FFCL Incident

All "On This Day in History" flyers were discussed in all weekly Safety Meetings (D-Level Talk). This valuable discussion on past incidents provided an excellent opportunity for learning and as well as uplifted Sense of Vulnerability across site. Considering the benefits of this exercise, this practice was later initiated in other Fertilizer Sites of Fatima Group in 2021.

2.5 Defer to Expertise [3]

To ensure all relevant process safety and technical information are communicated to relevant individuals, a detailed study of Process Safety Beacons was conducted in 2020. The assigned individual studied 2 decades (total 226 beacons) of CCPS process safety beacons and developed an action plan for each beacon. The main idea was to learn from mistakes of others and to share Process Safety Beacons with concerned interfaces only. Learning from beacons was divided into 4 categories

- For Discussion in weekly safety talks
- Conducting Field Audits
- Conducting Mock Drill Exercises
- Conducting detailed Trainings

Each beacon covers a Process Safety or technical topic pertinent to a specific interface. For example, Feb-2012 Beacon (Mechanical Integrity-The Bolts Are Too Short) is related to Maintenance teams or Sept-2008 (Plant Security) is related to Security Personnel. Through this study, a matrix was developed that segregated information sharing according to their end users.

Below table includes the number of Process Safety Beacons against each department discussed in their weekly safety talks.

Table 2: Number of Process Safety Beacon Distribution w.r.t End Users

Topics	Number of Beacons to be Shared in Weekly Safety Talks
Relevant to all site	33
Production Related	162
Maintenance Related	31
Electrical and Instrument Related	15
Laboratory Related	06
Plant Security Related	04
Emergency Response Team & Incident Commander Related	11
Material Warehouse Related	21
Technical Services Related	09

2.6 Empowering Individuals to Successfully Fulfill their Process Safety Responsibilities [3]

To ensure all Process safety activities are performed as intended, competency of individuals is one of the key factors. After successful implementation of Process Safety Management, a Training Need Analysis of all employees was conducted w.r.t Process Safety. Training programs for most commonly used Process Safety Activities were developed and external trainers were hired. For instance, Root Cause Analysis and Process Hazard Analysis trainings were delivered by DuPont Specialist trainers. Empowering individuals to fulfill their Process Safety tasks demand high level of expertise. This has been achieved by conducting Process Safety Trainings on cyclic basis. For example FFCL Annual Training Calendar contains below trainings each year

- Process Hazard Analysis- Two day Training
- Incident Investigation and Root Cause Analysis- Two Day Training
- Process Safety Management Workshop- Two Day Training
- Pre-Startup Safety Review- Half Day Training
- FFCL Permit to Work System- One Day Training
- Emergency Response Practical and Theoretical Trainings

Through this cycle of training and refresher sessions, employee's and contractor's process safety knowledge base is strengthened and hence is reflected in strong process safety performances. For example Root Cause Analysis training contains numerous individual and group exercises for finding correct Root Cause, Using Why-Tree technique for Root Cause and generating correct Recommendations at the end. For this particular training, average score for Incident Review increased from **80 to 85%.** This review is conducted by a committee called "Incident Report Review Committee" and is led by a senior manager. This committee randomly selects Detail Investigation Reports and evaluates on a fixed criteria. Some trainings' also include Pre and Post Training tests to quantify improvement in knowledge base of participants.

2.7 Establishing an Imperative for Safety [3]

For a production site, difference of opinion between production and safety groups is a never-ending debate. Prioritizing safety over production depicts a strong safety culture. FFCL is no exception w.r.t this discussion. During initial phases of PSM implementation, numerous events were reported when a process safety activity was by-passed due to production pressures e.g. deliberately skipping a Pre-Start-up Safety Review or delaying a Process Hazard Analysis study etc. All such cultural issues have been gradually addressed by carefully changing some practices that indicated a High Imperative of Process Safety. Few examples include

- A norm was introduced to start all communication meetings for site employees with a SAFETY PAUSE, which can be a short video, a short safety message or a recent safety incident. In 2020, all meetings initiated with a safety pause on Pandemic.
- In 2020, sequence of Plant Morning Meetings was changed with HSE manager starting the session. In these meetings, all on-going major process safety concerns or recent incident are shared with site, giving a strong message across site that Our Work Starts with Safety.
- Similarly Management Review Sessions with CEO are started with Safety Stats that include Current Safe Million Man-Hours Number, Total Recordable Incident Rate and Summary of running year incident.
- Like any other organization, at FFCL incentives are given for achieving Production Targets. Applying same philosophy to safety matters, HSE rewards are distributed for achieving safety related milestones for example gifts distribution across site for achieving each safe million man-hour. Furthermore, in 2019 new categories were added in FFCL Achievers Event for an example Best investigation for an incident conducted by a team.

All these steps have helped to create an aura of Process Safety at FFCL.

2.8 Combatting the Normalization of Deviance [3]

From the start of 2020, FFCL HSE section has put special efforts to highlight areas where site performances have been continuously below par. At FFCL, Process safety performance is monitored through Leading and Lagging Indicators, presented to site top management each month. Process safety engineers continuously look out for abnormal trends across PSM elements, and alarm is raised in monthly Executive HSE Committee to site leaders. By timely raising concerns, normalization of deviances is avoided. For an example in 2020, serious observations relating to Work at Height were being repeatedly reported. A detailed Permit to Work analysis was presented to top management in September, which ultimately led to special focus towards Scaffolding practices. An action plan was developed that included Scaffolding Procedure revision and training of all scaffolding inspectors, supervisors and workers. Below figures are other two examples of concerns being raised in Executive HSE Committee Reviews

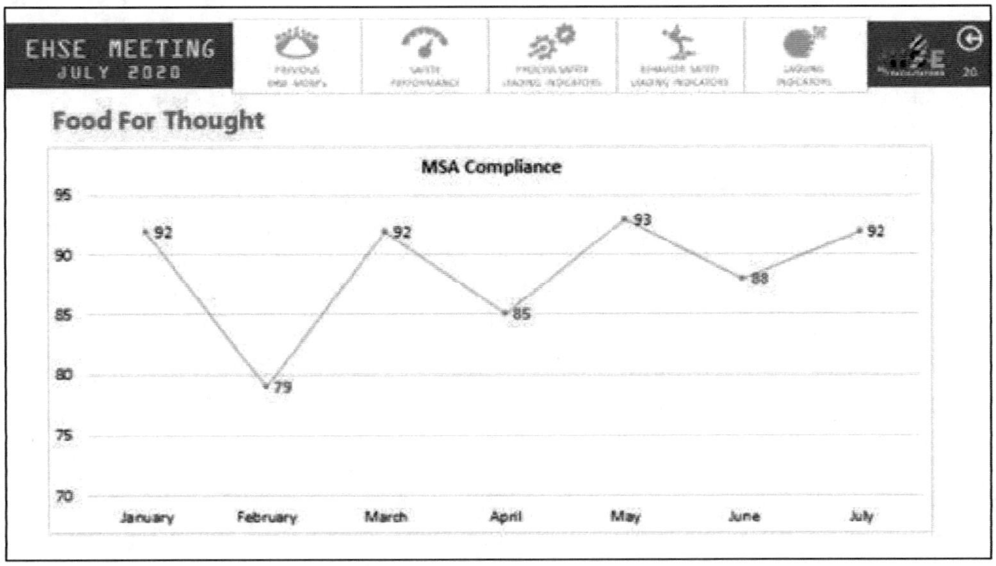

Figure 6: Management Safety Audits Varying Compliance Trend (July-2020)

Figure 6 is extracted from Executive Health, Safety and Environment Committee Review (July-2020). Through this data, a concern was brought in notice of senior management that Compliance of Management Safety Audits is not consistently good. Management Safety Audits are performed by all engineers once per month and historically has been one of key strengths of our Process Safety System. By raising timely concern, a message was communicated across all site (through their managers) that this area needs immediate focus. Improvement was observed immediately in proceeding months

3 Conclusion

Investing in Process Safety Culture has thus far helped to uplift Process Safety Performance at FFCL site. One lesson that FFCL has learnt over years about Process Safety is that Cultural issues are the most difficult tasks to deal with. Over the years different definitions have emerged for Process Safety Culture, most commonly used are "what people do when their boss is not around" or "The combination of group values and behaviors that determine The Way Process Safety is managed"

Cultural issues can go un-noticed and may suddenly result in a significant loss. One of the reasons for these cultural issues going un-noticed is difficulty in measuring an organizations' cultural status. It took FFCL several years to determine cultural status of site by agreeing upon vital signs of Process Safety Culture. For example Leading Indicators' Analysis unlocked strengths and weakness associated with process safety activities for example Process Hazard Analysis,

Management Safety Audits, Pre-Start up Safety reviews etc. Other activities like Process Safety Survey gave qualitative analysis of Process Safety Culture. Some other cultural aspects are harder to predict and afterwards rectify for example Complacency or Lower Sense of Vulnerability. Process safety incidents over the years have taught one important lesson is there is no such thing as "Stagnant Culture" either culture has an upward trend or a downward. To carter this cultural aspect, "On This Day in History" has helped FFCL to reduce complacency at all levels of organization. Since, there are so many variables associated with Process Safety culture, each organization will face different challenges and accordingly they need to work out Vital Signs of Process Safety Culture.
Below are some valuable lessons learnt regarding Process Safety after implementing above mentioned activities

- "Out of the box thinking" in Process Safety is equally important like any other aspect of life.
- A stagnant Process Safety Culture is the one on decline. Employees must be continuously engaged in Process Safety activities to get the best results.
- A strong Process Safety Culture sees off difficult times unharmed for example unplanned extended shutdowns, mergers, acquisitions and even economic crisis

This cultural shift has started to show its positive result and improved results are now shown at various fronts.

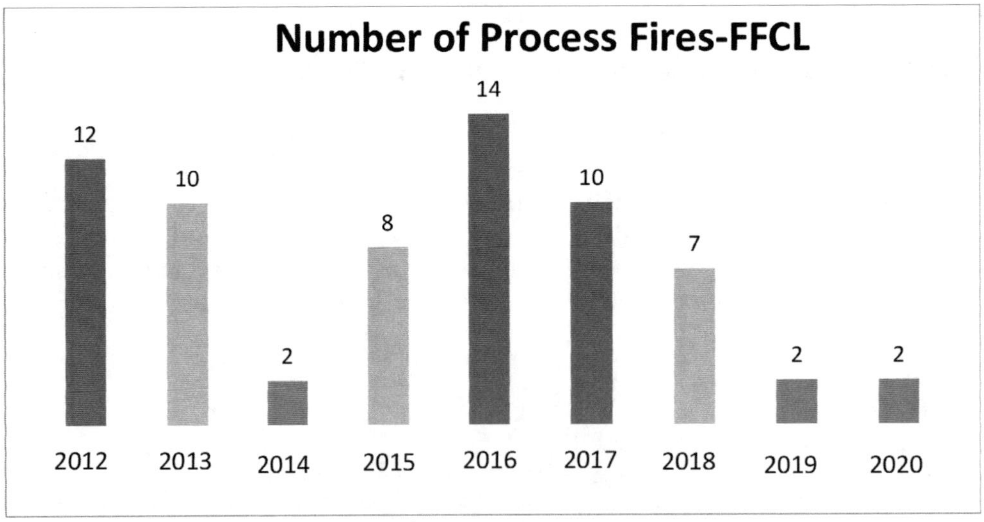

Figure 7: Number of Process Fires- FFCL

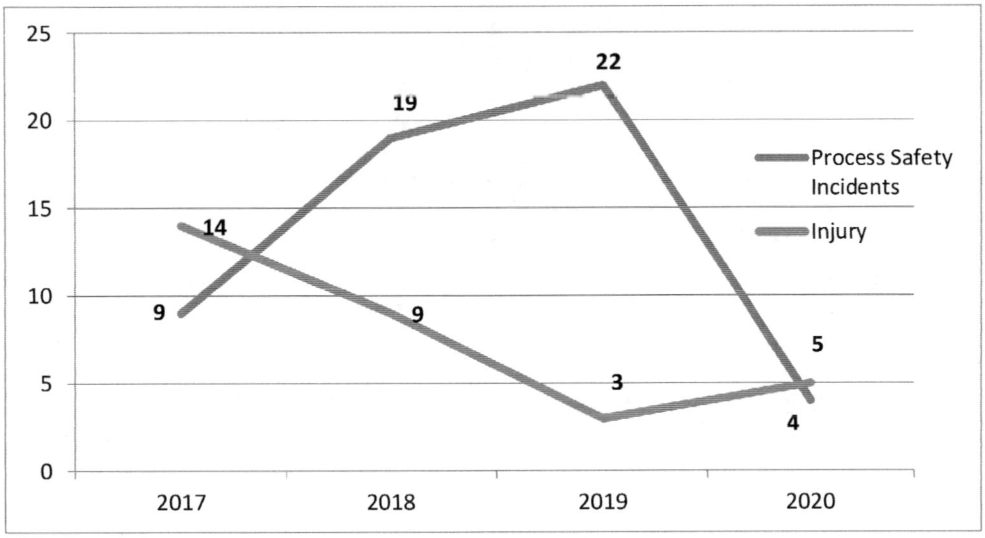

Figure 8: Number of Process Safety Incidents and Injuries- FFCL

4 References

[1] Ahsan Sarfaraz/ Usman Ali Mughal, "50 Million Safe Man Hours without a Zero Loss Time Injury," in 16TH Global Conference on Process Safety, 2020.

[2] Guidelines for Risk Based Process Safety, First Edition. CCPS. ©2007 AIChE. Published by John Wiley & Sons, Inc.

[3] Essential Practices for Creating, Strengthening, and Sustaining Process Safety Culture, First Edition. CCPS. ©2018 AIChE. Published 2018 by John Wiley & Sons, Inc.

SPRING21
+17TH GCPS
A Joint AIChE and CCPS Meeting

In Process Safety - Is 14 audit findings a good vital sign

Larry Lepard, CPSA
Principal Consultant - ERM
9825 Kenwood Road, Suite 100, Cincinnati, OH 45242
Larry.lepard@erm.com

Prepared for Presentation at
American Institute of Chemical Engineers
2020 Spring Meeting and 17th Global Congress on Process Safety
Virtual
April 18 - 22, 2021

In Process Safety - Is 14 audit findings a good vital sign

Larry Lepard, CPSA
Principal Consultant - ERM
9825 Kenwood Road, Suite 100, Cincinnati, OH 45242
Larry.lepard@erm.com

Keywords: Process Safety, Compliance Audit, Metrics, Vital Signs

Abstract

PSM compliance audits are required every three years. What do the audit results tell you about your process safety program? Is the number of audit findings a useful vital sign?

It is critical to know how a company performs, views and uses the results of a process safety compliance audit. This can greatly affect their overall risk management and therefore the protection of their employees and neighbors. There can be a tendency to look at the number of audit findings and not the potential severity.

Some companies focus strictly on the numbers, try to compare results between different operations or incorporate the results into employee performance reviews and/or bonus payouts. These practices may have unintended negative consequences.

The author will share lessons learned, that were made while conducting hundreds of compliance audits, and will discuss how leading organizations manage their audit findings and use the results to improve performance and reduce risks.

1 Introduction

Under the Process Safety Management (PSM) Regulation, 29 CFR 1910.119, every covered facility is required to certify that they have evaluated compliance. This is accomplished by conducting a compliance audit of the provisions of the PSM regulations at least every three years. The main purpose of this exercise is to verify that the procedures and practices developed under the PSM standard are adequate and are being followed.

An audit is a fundamental part of an effective PSM program because its purpose is to verify that "systems to manage process safety" are in place and functioning effectively, and to take corrective action when findings indicate that is warranted [1]. The goal of a PSM program is to prevent process safety related incidents from occurring; or at the least, to minimize the potential frequency and severity of incidents should they occur. Process safety incidents can result in a toxic release, fire or explosion. These incidents can cause deaths, severe injuries, environmental impacts and lost production. Just as process safety incidents are rarely caused by a single event, rather a combination or a series of multiple events and failures, all the elements of a PSM program are designed to work together to prevent incidents. As such, an effective audit must evaluate all the elements to identify where there are gaps. Gaps in this sense can refer to regulatory non-compliance or not implementing what are generally accepted best industry practices. The conclusion of a compliance audit is generally an audit report that identifies findings (gaps) and recommendations for improvement. What a company does with the report and how they manage the recommendations can greatly affect the overall success of the PSM program.

2 PSM Program and Compliance Audits

The OSHA PSM regulation was issued in 1992 consequential to which, the PSM compliance audits have been required since 1995. Unlike some of OSHA's standards, which prescribe precisely what employers must do to comply, PSM is a "performance-based" standard - it outlines key features of existing safety programs for controlling highly hazardous chemicals, and employers have the flexibility to tailor their safety programs to the unique conditions at their facilities.
Specifically, the required process safety elements as per the OSHA PSM standard (29 CFR 1910.119), are Employee Participation, Process Safety Information (PSI), Process Hazard Analysis (PHA), Operating Procedures (OP or SOP), Training, Contractors, Pre-Startup Safety Review (PSSR), Mechanical Integrity (MI), Hot Work Permit (HW), Management of Change (MOC), Incident Investigation, Compliance Audits and Trade Secrets [2]. There are many guidance documents available describing how to develop and implement an effective PSM program. The following are examples: OSHA 3908-03 2017 Process Safety Management for Small Businesses, OSHA 3132 Process Safety Management, CCPS Guidelines for Implementing Process Safety Management, and the IIAR Process Safety Management Guidelines for Ammonia Refrigeration. In general, sites are expected to develop policies and procedures that describe what will be done to comply with each element and provide training to all affected employees and contractors as necessary for implementation. Based on the author's experience and discussion with other PSM professionals, the golden rule for PSM programs is "Say what you do and do what you say". A key piece to a successful audit is having documentation to demonstrate that you have done what you say.

This paper does not focus on PSM program development but rather on the compliance audit element and the role this plays in the overall health of the PSM program. The compliance audit is intended to evaluate both the design and implementation of the PSM program on a periodic basis [3]. It is a measure of whether the program is properly developed, implemented and functioning as intended. A good audit program will identify areas of non-compliance, opportunities for improvement and should help to prioritize corrective actions. All companies have limited resources so there is a need to apply those resources where they will make the biggest impact. A successful audit process will continually improve the overall PSM program. Successful PSM programs have been shown to reduce injuries, regulatory non-compliance, production downtime and the associated costs.

3 What is required in a PSM compliance audit

The regulatory requirements for this element are relatively brief. The regulatory language from 29 CFR 1910.119(o) [2] related to compliance audits is as follows:

1. Employers shall certify that they have evaluated compliance with the provisions of this section at least every three years to verify that the procedures and practices developed under the standard are adequate and are being followed.
2. The compliance audit shall be conducted by at least one person knowledgeable in the process.
3. A report of the findings of the audit shall be developed.
4. The employer shall promptly determine and document an appropriate response to each of the findings of the compliance audit, and document that deficiencies have been corrected.
5. Employers shall retain the two (2) most recent compliance audit reports.

Again, there are many guidance documents available describing how to develop and implement an audit program. The CCPS Guidelines for Auditing Process Safety Management Systems [1] alone is a 900-page document. As such, this paper guides the readers to a successful audit through identification of some key points and sharing of lessons learned. It does not tell the reader everything they need to know to develop and implement an audit program or how to be a successful auditor.

The following are minimal requirements and key points related to each of the required items from above.

Requirement 1 – As the standard requires the employer to "certify" they have evaluated compliance, there needs to be a certification statement signed by the person responsible for PSM implementation at the facility. It is common practice to include a certification statement in the audit report or to attach a certification letter to the report. Audits and certifications must occur at least every three years. A facility may choose to conduct audits more frequently than this. Some companies choose to develop audit frequency requirements based on factors such as the number of repeat findings, high-risk findings &/or other factors.

Requirement 2 – The audit can be conducted by a single person provided that person is knowledgeable in the process. This means the person must have knowledge / experience with the covered process(es) at the facility and knowledge in conducting PSM audits. The selection of effective audit team members is critical to the success of the program. Team members should be chosen for their experience, knowledge, and training and should be familiar with the processes and with auditing techniques, practices and procedures. The size of the team will vary depending on the size and complexity of the process under consideration. The audit should be conducted or led by a person knowledgeable in audit techniques and who is impartial towards the facility or area being audited [3]. It is typical to have members with varied experience on the team, some members having more process knowledge and other members being stronger in regulatory requirements and auditing process. The audit team can consist entirely of personnel from the site or company, third party auditors or a combination of both. In my experience, a combination of internal and third party audit members can produce the best results. Internal auditors generally have a better understanding of the covered process and company requirements where third party auditors may have a better (at least different) understanding of legal requirements, can bring experience / examples from other companies and industries and can offer a fresh set of eyes.

Requirement 3 – Some form of a formal report that identifies the findings must be developed at the conclusion of the audit. An audit report can be as simple as a findings table or can be quite complex. Some companies require the completion of an audit protocol, with a set of questions for each element, which must be answered and included as part of the report. In general, audit reports typically include some form of narrative about the facility, the audit method, audit participants, description of risk ranking (if used), a summary or conclusion of the results and a detailed findings table. It is common practice for the findings table to include the regulatory or company citation, a description of the requirement, a description of the issue / gap observed and recommended corrective actions.

Requirement 4 – A corrective action plan must be developed to document what will be done to address each of the findings and document that deficiencies have been corrected. While it is not specifically stated or required in this section, it is best practice to follow the same requirements as that for PHA recommendations. As such, the corrective action plan should identify the issue, what will be done, the expected completion date and the actual completion date. It is very important to maintain documentation necessary to demonstrate that the stated corrective action was completed. Per Appendix C to §1910.119 -- Compliance Guidelines and Recommendations for Process Safety Management (Non-mandatory) "Corrective action is one of the most important parts of the audit. It includes not only addressing the identified deficiencies, but also planning, follow-up, and documentation. The corrective action process normally begins with a management review of the audit findings. The purpose of this review is to determine what actions are appropriate, and to establish priorities, timetables, resource allocations and requirements and responsibilities. In some cases, corrective action may involve a simple change in procedure or minor maintenance effort to remedy the concern. Management of change procedures need to be used, as appropriate, even for what may seem to be a minor change. Many of the deficiencies can be acted on promptly, while some may require engineering studies or in-depth review of actual procedures and practices. There may be instances where no action is necessary and this is a valid response to an audit finding. All actions taken, including an explanation where no action is taken on a finding, need to be documented as to what was done and why [3]."

Requirement 5 – The two most recent compliance audit reports and documentation necessary to demonstrate the deficiencies have been corrected must be maintained. In practice, some companies choose to keep all audits for the life of the process and others discard prior audit reports and only keep the two most recent.

4 What does a PSM compliance audit tell you and how to use the results

This is where we get to the question in the title of this paper; **"Is 14 audit findings a good vital sign"?** Once a compliance audit has been completed, the focus then shifts to the number of audit findings per element, how to resolve the findings, and the priorities associated with the resolution. Does a fewer number of findings, less than 14, indicate a good vital sign? The answer to that question is maybe. In my opinion, the number of findings alone does not indicate much about the status and health of the PSM program.

Many variables can affect the number of audit findings that have little to do with the health of the program. Audit variables such as the number of auditors, skill / knowledge level of the auditor(s), number of audit days and the site resources available to answer questions; can all affect the number of findings.

The other factor, which can be more telling than the number of findings, is the type and severity of findings. For example, if the audit finding is that the facility has no MI program, this is more serious than having implemented a preventative maintenance (PM) program but failing to inspect one vessel out of ten. This in turn would be more serious than having inspected all ten vessels but having some administrative or recordkeeping errors, if the reports did not include all of the required items (i.e. missing the date or name of who completed the inspection, etc.).

How does one then determine the significance or severity of the finding? There are different approaches to determining the significance or severity of the audit findings. Both quantitative and qualitative methods can be applied to audit findings. In quantitative grading, a protocol or set list of questions is generally used and a value or number is assigned to each question. Normally a weighting is assigned where the questions judged to represent greater risk are worth more points than other questions. With this method, an overall score is normally assigned. In qualitative grading a set criterion for categorizing the findings and their relative importance must be established. A simple example would be a high-medium-low qualitative ranking. Each of these categories could also be assigned qualitative definitions similar to the severity and likelihood definitions used in a PHA.

To apply this to the MI example above; not having implemented a MI program would be scored high, failing to inspect one of ten vessels would be medium and missing required information on a completed inspection record would be low. So in theory having only fourteen findings could represent the worst case scenario of a facility having failed to develop or implement programs to comply with any of the fourteen PSM elements and having one finding for each element or having fully developed and implemented programs but just having some minor administrative or documentation issues.

For this reason, establishing and applying some form of grading scale to the audit findings is key. Scoring the findings will also help in developing the corrective action plan, as higher risk items should be assigned a priority for completion.

There are however potential downsides to both the quantitative and qualitative method. With quantitative scoring having one overall score can lead to comparisons being made between facilities or even at the same facility comparing the score of previous audits to the current one. With qualitative scoring, there is a tendency to try to compare the number of high-risk findings. The total score or number of findings by severity can be useful if you are comparing similar facilities, use a consistent audit team and apply a consistent scoring approach. This can be misleading if management tries to compare the scores of dissimilar facilities, the audit team was not consistent or the scoring criteria changes between audit cycles. If too much emphasis is placed on the score or number of findings by category or unreasonable expectations are assigned this can lead to pressure being applied to the auditor(s) to give more credit than is due in a quantitative system or not assign high risk in a qualitative system.

The author has experienced or observed the following scenarios over the last twenty plus years while working in the process industry and as a consultant. These scenarios have been organized as those causing a negative or a positive impact on the overall PSM program.

Negative: The following are viewed to be negative as they are examples of improper execution of the audit process and/or corrective action of the identified findings. As a result, the companies were not getting the full benefit of the audit process that is intended to identify gaps and take corrective actions that will improve the overall PSM program.

- A company tied the number of high-risk audit findings to the personnel performance metrics for some employees and a percentage of annual bonus for all employees. This led to an extreme amount of pressure being applied on both the audit participants and the auditors, for findings to not be categorized as high-risk. The majority of the discussion in daily and final close out meetings was around the ranking of the findings. Due to the focus on the risk category, participants did not use this time to understand what the findings were telling them about the health of the PSM program. They did not discuss what was contributing to the issues or how they could be corrected. As such, they missed the opportunity to get additional input from the auditors that could help to improve their PSM program.
- In another instance, the time allowed to close action items identified in an audit, was based on the severity of the finding. While it is a good idea to prioritize high-risk items, assigning an arbitrary time to completing them can cause issues. In this case, for all findings assigned as high the corrective action had to be closed within ninety days. In practice, this led to many of the corrective actions being written to address only a part of the problem so that it could be accomplished and closed within the ninety days. As an example, an audit finding identified that based on the relief system study some vessels did not have adequate protection installed. The corrective action assigned was to get a quote to design and install appropriate relief protection. The quote was obtained within the ninety days and the action item was closed. The issue was that three years later it was identified that the capital request to implement the project was not approved and the corrective measures were not implemented. The issue / risk still existed but there was no open action item to track the

actual installation of the required relief protection. The audit finding identified a clear risk or gap in the PSM program (a negative vital sign) but due to the focus on closing action items within the prescribed time, the corrective action was not managed in a way that reduced the risk.

- In an effort to try to learn from past audit findings and focus efforts to areas that showed the greatest gaps across most facilities, a company decided to change their scoring criteria in a quantitative grading system. Audit questions that typically had no findings written against them in prior audits were assigned less points and areas with larger gaps (more findings) were assigned more points. While this can be a good practice, it was poorly communicated to the sites and this led to issues with execution. The sites had developed employee performance goals tied to improving the audit score by a set percentage. Efforts were focused on improving areas where they had lost the most points on the previous audit and/or where they thought they could gain the most points on future audits. They did not know that corporate had changed the scoring criteria. As a result, improvement efforts were not focused on the areas that would have the greatest impact on the score under the new grading system. In practice with the change in grading, the overall score for most sites went down, even though the health of their overall programs had improved or at least stayed the same. Without understanding that the grading and therefore the expectations had changed, management was still trying to compare current audit scores to past results. Due to the poor communication and lack of understanding of the new scoring criteria, the health of the PSM program was incorrectly viewed to have gotten worse. Employees who had worked hard to implement the corrective action plan from the previous audit received negative criticism and performance ratings because of the perceived lack of improvement. This discouraged employee participation and hindered improvements to the PSM program.
- Compliance audits were viewed as a "have to do" activity as opposed to a value-added part of the PSM program. Audits were conducted by third party auditors and contracts awarded primarily based on costs. As the company had multiple facilities in different regions of the country different consulting companies, auditors and audit approaches were used with the quality of the audits varying greatly. As these audits were viewed as a check the box activity, little effort was given to understanding or correcting the issues that were identified. An incident at one facility led to an OSHA inspection with Notice of Violations (NOVs) and proposed penalties being issued. This prompted an OSHA inspection at another facility with the same result and so on. After multiple facilities in different states had been inspected and were facing proposed penalties the company entered into a Consent Decree with OSHA that included required changes to their audit process. In this case, prior to the OSHA penalties, the appropriate resources were not applied to conducting the audits or managing the resulting findings. The company was gaining little to no value from the audits because the audits often did not identify serious gaps which existed in their PSM programs and they did not properly manage the issues that were identified. The audits were not being used effectively to improve the overall health of the PSM program and reduce risks.

Positive: The following are viewed to be positive, as they are examples of proper execution of the audit process and/or corrective action of the identified findings. The companies invested the necessary resources and time to conduct the audits, revise the audit process when necessary, understand what the audit findings were telling them about the health of their programs and took appropriate corrective actions. As a result, the companies were getting the intended benefit of the audit process by identifying gaps and taking corrective actions that improved the overall PSM program thereby reducing risks.

- Past audit findings identified a lack of employee understanding and/or adherence to operating and emergency response procedures. Root cause analysis identified that a formal and consistent training program had not been implemented. Training was largely conducted via on-the-job training but there was no trainer qualification process or set requirements for what had to be communicated before an employee was considered to have completed training. A formal training procedure and plan was developed and implemented which included minimum requirements/qualifications to be a trainer, a train-the-trainer process, training curriculum by job position, written tests to gauge understanding and progress at set stages and a final review and approval process which included a panel interview and demonstration of knowledge before an employee was considered trained. After implementing this program, sites performed much better in this audit area, experienced fewer incidents and stated they observed production improvements related to fewer off-spec batches and less down time.
- Multiple audit cycles had identified issues related to tracking corrective actions to closure. It was identified that recommendations and associated corrective actions were coming from multiple sources or elements but there was no consistent way to track them to closure. The company had an electronic process for managing MOCs and action items were assigned and marked as closed within the system. They however had not established a process to maintain documentation of what had been done to consider the item closed. Therefore, actions like train all affected employees were marked closed but the site could not produce training records. In addition, actions identified in PHAs, incident investigations and prior audit reports were not being tracked in a centralized manner. PHA and audit findings were being tracked in spreadsheets specific to each PHA or prior audit and recommendations from incident investigations were only being tracked in each individual investigation report/folder. The company invested time and resources to implement an electronic tracking system where all recommendations and associated action items had to be entered into the system and assigned an ID number. For an action to be considered closed supporting documentation had to be attached within the system and verified by a second person. Having a centralized system allowed the company to have better visibility and understanding of the open issues at all facilities, prioritize resources between and within facilities to the higher risk items and establish metrics around closure of action items.
- Senior management recognized that their audit program was not delivering the expected results. The number of significant incidents and near misses experienced each year was not improving. Incident investigations identified issues such as lack of operating procedures or employee training as causal factors but these issues were not being identified in the audit process. In fact, the audits generally identified very few issues and rarely identified any as a high-risk. The majority of audits were conducted by internal resources often from the facility and process area being audited. The audit program used a checklist but this was largely based on document review and included limited or no infield verification of implementation. The decision was made to bring in outside resources to help redesign and implement a new audit process. In the new process, a combination of internal (company) and external (consultant) resources made up the audit team. The internal resources came

from other facilities or at least departments / process areas other than the one being audited. The audits continued to include a review of the relevant documentation and process safety information but also included inspection of the physical facilities, interviews with all levels of plant personnel and focused highly on infield observations of actual work practices. These new audits resulted in a significant increase in both the number of findings and those that were identified as high-risk. Despite some initial push back, senior leadership stayed with the process and were actively involved in determining and monitoring completion of corrective actions. Overtime they noted improvements in their overall PSM programs and a reduction in the number of significant incidents.

5 Conclusion

PSM Compliance audits have been required since 1995, despite being required for over 25 years the approaches used to conduct audits, manage the resulting recommendations and the value gained can vary greatly. An effective audit program is an essential element to the overall PSM program as it is the primary method to verify that systems to manage process safety are in place and functioning effectively. A poorly developed or executed audit program may not identify serious gaps that exist in the PSM program and therefore the opportunity to improve is missed. To have a successful audit program there must be support from all levels of the company and appropriate resources need to be assigned and allocated. A consistent audit approach must be developed and implemented that includes at a minimum, an audit method (protocol, experience based or combination), a grading system, auditor qualifications, a report identifying the findings and management of the resulting recommendations. Once a quality audit process is implemented, proper management of the resulting recommendations is arguably the most important aspect. The corrective actions need to address the cause, not just the symptom, to improve the overall health of the PSM program. This will likely require root cause analysis to determine why the systems are not achieving the expected results. The "why" must be understood in order to implement effective corrective actions. Corrective actions need to be prioritized and assigned reasonable expected due dates for completion. Facilities must maintain documentation necessary to demonstrate closure of the actions. The audit identifies the gaps; an effective corrective action plan must be developed and implemented within a timely manner for the benefit on an approved PSM program and reduced risks to employees and the community to be achieved.

As a vital sign or a health indicator of their PSM program, focusing just on the number of findings, establishing the wrong metrics, changing the metrics without proper communication or setting unrealistic deadlines for closing actions can have unintended consequences. Companies need to look beyond just the number of audit findings or score to understand what the audit is telling them about their processes and take appropriate actions to improve the PSM program. Fourteen high severity process related findings is a much worse vital sign than fourteen findings related to administrative or record keeping issues.

6 References

[1] CCPS. *Guidelines for Auditing Process Safety Management Systems, Second Edition.* Center for Chemical Process Safety, American Institute of Chemical Engineers, New York, NY, 2011.

[2] Occupational Safety and Health Administration, Process Safety Management, available at https://www.osha.gov/dsg/psm/index.html#:~:text=Unlike%20some%20of%20OSHA's%20standards,the%20unique%20conditions%20at%20their Accessed on February 9, 2021.

[3] Occupational Safety and Health Administration, Appendix C to §1910.119 -- Compliance Guidelines and Recommendations for Process Safety Management (Non-mandatory), available at https://www.osha.gov/pls/oshaweb/owadisp.show_document?p_id=9763&p_table=STANDARDS Accessed on February 10, 2021.

617884 Making Zero a Reality: Selecting & Measuring the "Right" Zero

Abstract:

Since the inception of the Process Industries (PI), leaders of these organizations have accepted some level of fires, explosions, releases to the environment and other events as the cost of doing business and *inevitable*. This cultural mindset continues to this day and is partially why, in spite of the many advances in process, occupational, and environmental safety made in the PI, it is still experiencing major accidents, fires & environmental releases.

Real safety performance and behaviors cannot be managed, much less produced, by corporate safety programs, many of which can be viewed as "bumper sticker" or "fake" safety, while active and latent hazards go unnoticed and/or unaddressed. The false sense of security that many PI safety programs encourage is evidenced by the continued occurrences of large-scale catastrophes. The dichotomy of having a "safety culture" versus such events is clearly evident, yet the PI continues to focus on and measure the same indices – all of which have little bearing on the cause of these events. *Zero* can be a goal, but it needs to be a realistic zero - a goal where no humans ever make an error is completely unrealistic. For many organizations the mistake isn't establishing Zero as a goal, the mistake is selecting the *wrong* Zero and then compounding the problem by failing to support the organization in order to make Zero a reality.

This paper examines the organizational policies and safety metrics of the PI with that of the single most reliable and safe organization in the world, which has achieved a safety record of ZERO incidents for its entire 65+ year history.

619148 "Why Do I Have Relief System Inadequacies NOW?" Incidents and Industry Events That Have Influenced Changes in Relief System Design

Abstract:

At the management level, many aging facilities are discovering that their long-time relief system elements are no longer deemed adequate. Many are left asking why a relief valve is being found with concerns when they have been running with the valve for years.

The presentation will focus on what has caused relief systems evaluations to change as industry understanding evolved. We will examine Proactive and Reactive changes, Incident-based changes, and changes to long-held assumptions.

Attendees will leave with an understanding of changes to pressure relieving systems and will be better prepared to discuss any inadequacies with site and company management.

620216 Safety Interlock Overview and ChE / EE Collaboration

Abstract:

This presentation will give an overview of the basics of safety interlock design, with more focus on SIL interlocks. This will give an overview of the basic concepts, governing standards, design process, overview of calculations, and other life-cycle considerations. There will also be an example which highlights the need to have collaboration between Chemical and Electrical Engineer discipline expertise.

SPRING21
+17TH GCPS
A Joint AIChE and CCPS Meeting

It Was a Dark and Stormy Night – Investigation of Acrylate Storage Tank Explosions

Kathleen Kas, Dow Chemical Company, Washington Crossing, PA
kkas@dow.com

Delmar "Trey" Morrison, Exponent Inc, Warrenville, IL
tmorrison@exponent.com

Prepared for Presentation at
American Institute of Chemical Engineers
2020 Spring Meeting and 17th Global Congress on Process Safety
Virtual
April 18 - 22, 2021

It Was a Dark and Stormy Night – Investigation of Acrylate Storage Tank Explosions

Kathleen Kas, Dow Chemical Company, Washington Crossing, PA
kkas@dow.com

Delmar "Trey" Morrison, Exponent Inc, Warrenville, IL
tmorrison@exponent.com

Keywords: Lightning, monomer, explosion, ignition

Abstract

It was in the early morning hours of May 16, 2012 when a thunderstorm rolled into the Bristol, Pennsylvania plant where The Dow Chemical Company produced acrylic polymers. As the weather turned stormy, lightning struck in the tank farm area containing inhibited ethyl acrylate, butyl acrylate, and methyl methacrylate. Within seconds, an ethyl acrylate tank exploded and was followed a few minutes later by a butyl acrylate tank explosion. The explosions and ensuing fire damaged the tank farm and led to a lengthy shutdown; fortunately, there were no injuries. This paper details the investigation into the complex sequence of events that led to the two explosions. The investigation team evaluated data gathered from eyewitnesses, security system cameras, weather services, and the process control system. The team also analyzed physical evidence including electrical bonding and grounding systems, deflagration arresters, and various tank components. Although the adjacent ethyl and butyl acrylate tanks were involved in the event, the investigation did find that flame isolation devices functioned as intended and prevented the propagation of flames to surrounding processes via the interconnected venting system. However, there was a puzzle left to unravel. The tank installations were determined to be in compliance with the industry lightning standards such as NFPA 780. If the tanks met industry standards, how did the explosion happen? While the exact mechanism could not be determined with certainty, the ignition of the atmosphere in the ethyl acrylate tank most likely occurred from a spark discharge inside the tank's headspace, likely between internal components at different electrical voltage potentials. This type of mechanism could invalidate the Faraday Cage assumption inherent in current industry guidance and may represent a hazard that is not currently well-recognized in industry. It should also be noted that, while lightning can't be prevented, the outcome of a lightning strike can also be mitigated by the use of safeguards such as inertion, flame arresters, diking, and foam systems.

1 Introduction

On May 16, 2012, at approximately 3:34 am, a lightning strike occurred at the Dow Bristol Plant in Bristol, Pennsylvania (Figure 1). Immediately after a lightning flash was observed, flames were observed on top of the Ethyl Acrylate Monomer (EA) tank about one second before it exploded. Liquid EA spilled into the dike shared with the Butyl Acrylate Monomer (BA) tank, and the BA tank subsequently exploded a few minutes later. The explosions and ensuing fire damaged the tank farm and led to a lengthy shutdown; fortunately, there were no injuries.

This paper details the investigation into the complex sequence of events that led to the two explosions. Key questions that the investigation team needed to answer included:

1. Was lightning the true initiating event?
2. Why did the contents of the EA tank ignite if the tank was grounded?
3. Why did the BA tank explode, since the headspace was not in the flammable region prior to the event?
4. What learnings can we take away from this incident?

This paper summarizes the event, the evidence collection and analyses, the testing of hypotheses, the conclusions from the event, and key learnings.

Figure 1 The dike fire and aftermath of the explosions.

2 Description of the Storage Tank System & Safeguards

The Dow Bristol Plant creates products for the Dow Coating Materials business. Acrylate-based coatings are produced at the facility. The facility receives, stores, and processes acrylate monomers including Ethyl Acrylate ("EA," CAS No. 140-88-5) and Butyl Acrylate ("BA," CAS No. 141-32-2).

2.1 Acrylate Properties

At room temperature, EA monomer is a colorless liquid with a pungent odor and a low odor threshold of approximately 1.2 ppb [1]. BA monomer is also a colorless liquid, with a fruity odor and an odor threshold of about 35 ppb.

The acrylate monomers are inhibited with the methyl ester of hydroquinone (MEHQ) to prevent unintended polymerization during storage. This inhibitor requires the presence of oxygen to retain activity. While systems exist to provide a mixed nitrogen/oxygen gas with a low enough oxygen level concentration to maintain both inhibitor activity and inertion, this type of mixed gas system has the potential to fail to an unsafe state. The limiting oxygen concentration for EA is 9.0 volume percent; thus, the control system to manage the atmospheric concentration would require a high level of reliability. The failure of the atmospheric control system can, and has, resulted in unintended polymerization in industry. In this installation, the risk of a runaway polymerization due to mixed gas failure was determined to be a greater risk than that of headspace ignition, so the tank was blanketed with air to maintain inhibitor efficacy.

The flash point of EA was below the ambient temperature of 19°C at the time of the incident, so the vapor headspace of the tank was flammable. The nearby BA tank's headspace, however, was not in the flammable region. The initial headspace vapor concentrations are shown on a vapor pressure curve for EA and BA in Figure 2.

Table 1. Physical properties of the acrylate monomers.

Property	**Ethyl Acrylate**	**Butyl Acrylate**
Flash Point	8°C	37°C
Lower Explosion Limit (LEL)	1.40%	1.30%
Upper Explosion Limit (UEL)	14.0%	9.90%
Maximum Experimental Safe Gap (MESG)	0.86 mm	0.88 mm
Odor Threshold	1.2 ppb	35 ppb

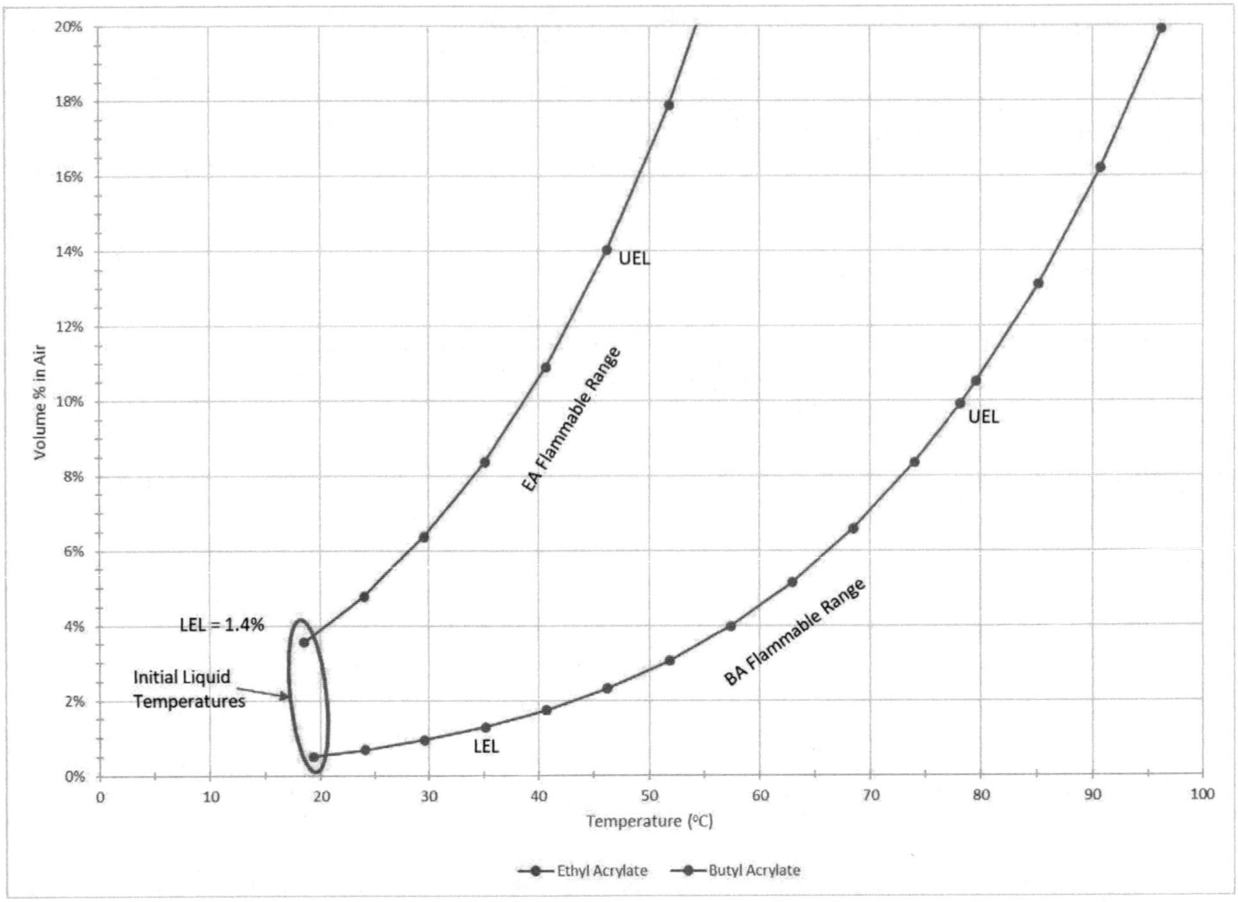

Figure 2 Volume percent of EA and BA vapors in air versus temperature.

2.2 Description of the Tank Farm

EA and BA monomers were stored in Tanks 120 and 119, respectively, in Tank Farm 30A.

Figure 3 is an aerial photograph of the area taken circa 2007. Liquid monomers were pumped into the EA and BA tanks from trucks and rail cars and transferred to the downstream manufacturing processes as needed. Truck and rail unloading stations are located around the perimeter of the tank farm dike walls. Figure 4 is a view of the tanks after the incident.

Figure 3 **Aerial image of EA Tank (south) and BA Tank (north) in Tank Farm circa 2007.**

Figure 4 **View of tank farm after fire suppression efforts. Tanker truck in foreground was unloading MMA at the time of the incident.**

The two atmospheric storage tanks involved in the incident contained inhibited BA liquid and inhibited EA liquid. The EA tank inventory was 673,482 lb and the BA tank held approximately 536,220 lb. At the time of the event, no transfers were occurring in or out of the EA or BA tanks. Truck unloading was occurring to a Methyl Methacrylate (MMA) tank in a separate tank farm area; however, the investigation concluded that MMA did not play a role in the event.

As shown in Figure 5, each storage tank was filled through an overhead dip tube and was equipped with a vapor header connection to the thermal oxidizer, an atmospheric conservation vent, and an atmospheric emergency pressure relief vent. The vapor header connection and atmospheric conservation vent were isolated from the tank using inline deflagration arrestors rated for this service.

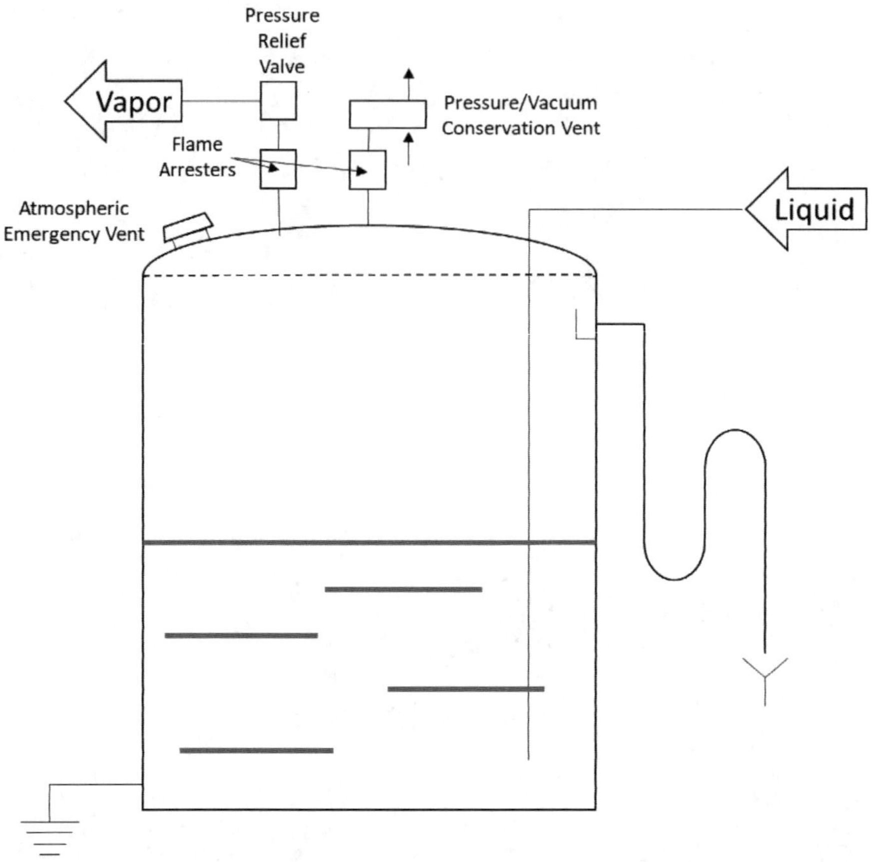

Figure 5 General tank appurtenance configuration and relationship.

2.3 Tank Construction

Tank 120 was in dedicated inhibited Ethyl Acrylate (EA) monomer service, and Tank 119 was in dedicated inhibited Butyl Acrylate (BA) monomer service. The carbon steel tanks were operated as atmospheric storage tanks and were originally constructed in the 1940s. The tanks were 30-feet in diameter by 29-feet tall with dome roofs. The dome roofs were originally constructed from petal sections with riveted overlap joints. At some point in the past, the rivets had been removed and the overlapped joints had been welded. The tanks were both 30 ft in diameter by 29 ft tall carbon steel cylindrical storage tanks with domes roofs. The steel thickness at the roofs and walls was approximately 0.3 inches. The tanks were equipped with a static grounding system and lightning protection consistent with guidelines in NFPA 780.

Table 1. Tank appurtenances.

Vent Device	Liquid Line	Instrumentation
Pressure/vacuum conservation vent to atmosphere	Liquid fill line	Capacitance level probe
Pressure relief valve to vent system (oxidizer)	Liquid withdrawal line	Liquid temperature probe
Pressure relief hatch to atmosphere	Overflow line with loop seal	Liquid temperature gauge
	Fire foam nozzle	

2.4 Tank Farm Grounding Installation

At the time of the event, National Fire Protection Association (NFPA) 780-2011, Standard for the Installation of Lightning Protection Systems, was a recognized industry standard. Section 7.2.2. of NFPA 780 advises: [2]

"Metallic structures that are electrically continuous; tightly sealed to prevent the escape of liquids, vapors, or gases; and of 3/16 in. (4.8mm) thickness or greater to withstand direct strikes... shall be considered to be inherently self-protecting."

American Petroleum Institute (API) Recommended Practice 2003, seventh edition (2008), also provided guidance. Section 5.4.1 stated: [3]

"Metallic tanks, equipment, and structures commonly found in the petroleum industry that are in direct contact with the ground (i.e. no nonconducting membranes) have proved to be sufficiently well grounded to provide for safe propagation to ground of lightning strokes. Supplemental grounding by means of driven ground rode neither decreases nor increases the probability of being struck, nor does it reduce the possibility of ignition of the contents."

The EA and BA tank installations at Bristol were found to be in compliance with these provisions. The vessels were steel tanks exceeding 3/16" thickness that were mounted on concrete pads and attached to grounding rods. Metallic piping and conduit were appropriately bonded to the tanks. There was an insulating membrane installed in the containment area; however, it did not isolate the tank and its associated pad from the connection to earth. Routine maintenance and testing were performed on the grounding system to maintain it in good working order. Therefore, no gap between industry guidance and the tank installations was identified.

2.5 Deflagration Arresters

Deflagration (flame) arresters were installed on the storage tanks to prevent the propagation of deflagrations from an external source into a tank storing combustible or flammable material (Figure 6). Flame arresters typically contain small, high surface area passages through which momentum and heat of a flame front passing through a flammable vapor/air mixture in piping are quenched. Flame arresters are sized according to the chemicals involved. NFPA 497 reports maximum experimental safe gap (MESG) values between the plates of 0.86 mm for ethyl acrylate (inhibited) and 0.88 mm for n-butyl acrylate (inhibited).[4] These gap sizes correspond to Class I Group D materials, or those materials with an MESG greater than 0.8 mm.[5]

Figure 6 Example pressure/vacuum relief device (VPRV) mounted above a deflagration arrester.

3 Investigation

The investigation was initiated once the fire was extinguished and the scene was rendered safe. The focus of the investigation was to gather data to understand the sequence of events, identify the ignition mechanism for the EA tank, and determine the means by which the BA tank also ignited.

3.1 Understanding the Sequence of Events

3.1.1 Eyewitness accounts

Two witnesses were located in the area of the tank farm at the time of the lightning strike and explosion, with positions shown in Figure 7. None of the witnesses saw the location of the lightning strike. One witness (Witness #1) was sitting in a truck by the MMA unloading station completing paperwork. He saw the flash and heard the lightning strike, which was shortly followed by the EA tank explosion. A second witness (Witness #2) was located in a nearby building, sitting at a desk on the east side. He looked up immediately after the strike and noted flames on top of the EA tank, which preceded the explosion.

Figure 7 Witness locations to the lightning strike and ensuing events.

The lightning discharge itself was not captured by surveillance video; however, a flash associated with the lightning strike was recorded by a camera pointing to the northwest.

3.1.2 Evidence collection

Immediately after the emergency response activities were completed, the scene was secured. Photographs were taken of the tank farm and surrounding areas. A chain-of-custody process was established for evidence. Debris from the explosions was located, tagged, photographed, logged, bagged, and retained in a locked area. Security camera footage from all site cameras was secured. The Distributed Control System (DCS) data was retrieved and archived so that it would not be overwritten.

3.1.3 Verification of lightning strike

At the time identified in security video and supported by witness testimony, a lightning strike to ground was recorded by two lightning verification services: StrikeNet and TOA. Both organizations recorded a strike to ground within the Bristol site with a magnitude between -36 kA and -38kA. Although the sensors used by the two organizations placed the lightning strike within the Bristol site at slightly different locations, the tank farm fell within the confidence zones for both sets of data.

The equipment, debris, and surrounding areas were inspected after the event to determine if the locations of one or more lightning attachment points could be located; however, no burn marks or other indications of a lightning strike were evident on the steel piping or equipment in the tank farm or connected to equipment in the tank farm. Attempts to detect erratic magnetic fields on steel equipment were also inconclusive. However, based on the eyewitness testimony, security camera footage, and lightning strike data, it was concluded that a lightning strike had occurred and initiated the event.

3.2 Evaluation of Physical Evidence

3.2.1 Review of tank damage

About a second after the lightning strike, the EA tank exploded (Figure 8). The explosion displaced the dome roof, separated piping systems, and damaged the lower tank shell. The explosion yielded an immediate fireball followed by a tank fire and pool fire in the surrounding dike area from burning EA monomer liquid. A post-incident photograph of the EA tank is shown in Figure 9. Flames were observed at dome level immediately after the lightning strike by an employee in a building about 100 feet southwest of the tank. The roof of the EA tank fractured into multiple pieces, and a portion was projected outward and hit a nearby trailer. A large section of the roof, including the deflagration arrester and vacuum/pressure relief assembly, fell into the tank. The northwestern and western portions of the upper-most segment of the sidewall of the tank, as well as the external ladder of the tank, were bent inward. The flat tank bottom bowed due to the internal pressure prior to rupture of the tank roof.

Figure 8 **First frame of surveillance video showing the EA tank explosion. This frame is time-stamped 1 second after the frame capturing the lightning strike.**

Figure 9 **Ethyl acrylate tank after the incident. The upper portion of the tank was curled into the tank from when the roof fell back into the tank after the explosion. Oxidation and soot deposition are evident on the outside of the tank wall.**

Approximately two minutes after the EA tank exploded, the BA tank also ruptured (Figure 10). The roof was found on a nearby loading rack with apparent soot deposition but with paint intact. The dome was curved inward and several of the radial seams of the roof were separated. A photograph of the dome appears in Figure 11.

Figure 10 BA tank from the west.

Figure 11 BA tank dome, which landed on a nearby loading rack.

3.2.2 Deflagration Arresters

The EA tank conservation vent device was found to be damaged but still attached to the tank's dome. The overpressure relief component of the conservation vent was separated from the rest of the assembly. The pressure relief valve pallet had been displaced upward, and the weather hood studs had been pulled out of the assembly flange. There was evidence that the guide cap impacted on the top weight, and the guide cap rivet was pulled out. The fluorinated ethylene propylene (FEP) polymer seal on the pallet appeared heat-affected.

As shown in Figure 12, the EA tank's conservation vent flame arrester internal cartridge was damaged mechanically – the plates were displaced. Examination of the plates in the cartridge revealed that the tank headspace side of the cartridge displayed a circular pattern of erosion indicative of an unknown duration exposure to flames venting from the tank into the flame arrester. The atmospheric side of the cartridge did not exhibit a similar heat-affected damage pattern. The presence of erosion on the tank side of the cartridge but not on the opposing side indicated that flame travel through the arrester was directed from the inside of the tank to the outside. This confirmed that the flame arrester operated as intended and did not allow flames to vent from the tank. This also supports a conclusion that the flame arrester cartridge would have prevented an external fire from passing into the tank headspace.

Figure 12 EA tank conservation vent deflagration arrester cartridge. Tank side on the left and the outside is on the right. Flame impingement pattern highlighted on inside of cartridge.

As shown in Figure 13, the BA tank conservation vent flame arrester showed similar flame impingement patterns from the inside of the tank as the EA tank.

Figure 13 BA conservation vent deflagration arrester cartridge. Tank side on left and outside on right. Flame impingement pattern highlighted on inside of cartridge.

The BA tank's vent line inline deflagration arrester was ejected from the tank and later found inside the tank after being subjected to the pool fire inside the tank. The deflagration arrester was too heavily damaged to determine if it was attacked from an internal or an external fire. However, the EA tank vent line flame arrester was intact enough to conclude that the EA tank fire was unlikely to communicate with the BA tank through the vent header. The EA tank vent line was equipped with a 3-inch inline flame arrester, shown in Figure 14. The flame arrester was located between the tank and the vent header, which went to the thermal oxidizer. The flame arrester assembly was fragmented and pieces were projected in multiple directions away from the tank. The impact caused collapse and deformation of the corrugated aluminum plates in the internal cartridge based upon the lack of fire/explosion damage to the edges of the plates. The plate array also displayed a circular pattern of erosion indicative of an unknown duration of exposure to flames venting from the tank into the flame arrester. The atmospheric side of the cartridge did not exhibit a similar heat-affected damage pattern. The plate surfaces were also free of heat-affected damage patterns. The presence of erosion on the tank side of the cartridge but not on the opposing side indicates that flame travel through the arrester was directed from the inside of the tank to the outside. As shown in Figure **15**, the clean plate surfaces indicate that the flame penetrated a minimal depth into the cartridge. This indicates that the flame arrester operated as intended and did not allow flames to vent from the tank. This also indicates that an external fire, for example due to the ignition of a release of flammable vapor through the shared vent line, did not pass into the tank through the arrester.

Figure 14 EA vent line inline deflagration arrester cartridge. Tank side on left and outside on right. Flame impingement pattern highlighted on inside of cartridge.

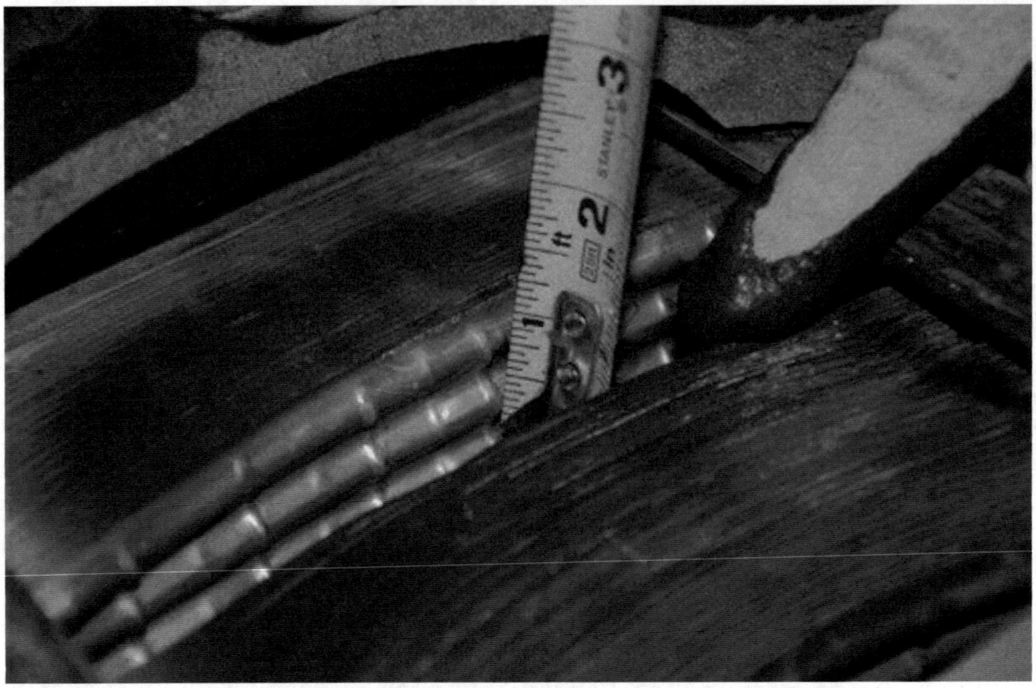

Figure 15 Image of clean plate surface inside cartridge indicating that flame penetration depth into the cartridge was minimal.

3.2.3 Location of EA tank headspace ignition

After the lightning strike, there was a time delay before the EA tank exploded. Based on the witness statements, security camera footage, and recorded process data, the following sequence of events was identified for the explosion. Table 3 provides the sequence of events.

Table 3. Sequence of events.

3:47:08	Lightning strike
	Flames visible on roof of EA tank
3:47:09	Explosion of EA tank
	EA tank fire and pool fire in dike
	BA tank explosion

The duration of the delay was approximately 1 to 1.2 seconds. During this interval, Witness #2 reported seeing flames on top of the EA tank but could not be more specific about the nature of those flames. Possible explanations include ignition of a pre-existing vapor release from one of the roof appurtenances or a crack in the roof, burning of a combustible solid material such as a plastic component, or venting of flames from the emergency vent. There are many parameters that may have affected this timing, including the flame speed in EA vapor/air mixtures and the effects of pressure relief devices. The approximate method presented in the SFPE Handbook was used to estimate the pressure rise with time inside the EA tank head space by assuming propane combustion properties for the EA vapors (Equation 1).[6] Given these calculations, the estimated burst pressure for the vessel of approximately 2 psig would be reached in about 1 second, as shown in the graph in Figure 16. Thus, the 1 to 1.2 second delay between the lightning strike (ignition) and explosion is reasonable and consistent with the delay expected for an internal deflagration to cause the tank to rupture.

Equation 1:

$$\frac{P - P_0}{P_0} = \left(\frac{1}{\gamma} + \frac{P_0}{P_m - P_0}\right)^2 \left(\frac{P_m - P_0}{P_0}\right)^3 \left(\frac{\chi S_u t}{a}\right)^3 \ll 1 \quad (7)$$

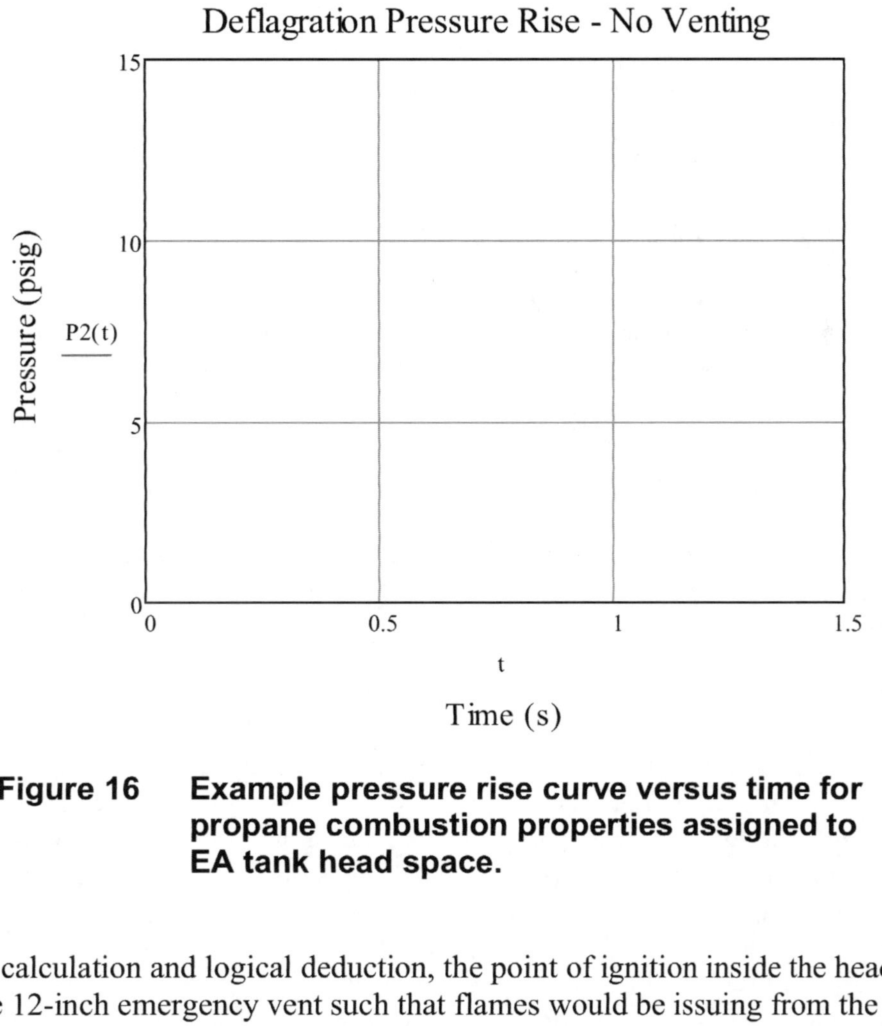

Figure 16 Example pressure rise curve versus time for propane combustion properties assigned to EA tank head space.

Based on this calculation and logical deduction, the point of ignition inside the headspace was likely near the 12-inch emergency vent such that flames would be issuing from the vent for a period of time prior to tank rupture pressure being reached. The emergency vent would be anticipated to open quickly after an internal pressure rise begins, which would allow flames to vent. If the internal deflagration had originated away from this area, the witness would have only observed flames close to or coincident with the explosion.

3.3 Testing Hypotheses

For a fire to occur, it is necessary to have a fuel, an oxidant, and an ignition source. In this case, the flashpoint of EA was 8°C – well below the 19°C temperature of the liquid. The tank contents were stored under an air atmosphere. Since fuel and oxygen were both present in the EA tank, the investigation focused on determining potential ignition mechanisms for the EA tank.

While the headspace of the EA tank was flammable, the flash point of BA is 37°C – above the ambient temperature at the time of the event. As a result, hypotheses were developed to determine not only the ignition source for the BA tank but also how the tank temperature increased to the point where the contents became flammable.

3.4 Hypotheses: Ignition of the EA Tank

Three potential sources of the EA tank ignition were evaluated:

1. Ignition source entered the EA tank via the vent system
2. Ignition source entered the EA tank while it was breathing/venting to atmosphere
3. Ignition source internal to the tank.

3.4.1 EA Hypothesis 1: Ignition source entered the tank via the vent system

The vent header could possibly have contained a flammable mixture due to the MMA truck unloading activities that would displace the MMA tank flammable headspace into the vent header. However, ignition in the vent header system outside of the tanks was eliminated as a potential cause based upon several observations. The fire pattern present on the EA tank vent header deflagration arrester was located on the inside (i.e., tank side) of the deflagration arrester, indicating that flames were venting into the arrester cartridge for some period of time prior to the explosion. Additional evidence that discounted external ignition in the vent system included the lack of soot in examined piping, the lack of overpressure damage in piping and connected tank components, and the lack of indication of failure of the PRV isolating the tank from the vent header.

3.4.2 EA Hypothesis 2: External ignition source while EA tank was breathing/venting

Although the EA tank was not directly associated with the filling activities ongoing at the time of the incident, the possibility for flammable EA vapors to leak from the tank was evaluated as a potential path to internal ignition. Since the EA tank was anticipated to act as a Faraday-like cage and prevent penetration of an electrical discharge, the possibility of an external discharge igniting a vapor leak was considered. Four primary leak locations were identified and evaluated:

1. Release from a crack in the shell or dome roof
2. Release from an incompletely filled overflow line loop seal

3. Release from EA conservation vent
4. Release from EA 12-inch emergency vent

EA vapors have a very low detection threshold (0.5 ppb), and from the investigation team's experience, a leak in the tank would have been detected during normal operator inspections if it was near ground level. A small leak at roof level could not be as easily detected. A recent tank inspection did not detect cracks in the tank shell or vapor leaks in the tank. If such a small crack existed, it would be unlikely to be larger than the minimum experimental safe gap (MESG) of 0.86 mm; thus, it was unlikely that ignition could propagate back into the tank through such a crack. Therefore, a leak from a crack in the shell or dome roof was refuted as a potential ignition pathway.

The loop seal fluid had been replaced in 2010, and a leak past the seal fluid would be anticipated to be detected by field operators since the opening is near ground level adjacent to the tank. Thus, a leak from the loop seal was refuted as a potential source for vapors.

An ignition pathway through the EA conservation vent was also considered. However, the deflagration arrester did not exhibit signs of impingement by an external flame. Additionally, the deflagration arrester appeared to extinguish the internal flames as intended before they could pass through the cartridge. The deflagration arrester cartridge was packed with parallel plates and suitable for suppressing flames in either flow direction. Thus, an external ignition at the VPRV that flashed back into the tank to ignite the tank headspace vapors was refuted as a possible cause.

The larger 12-inch diameter emergency vent was not equipped with a deflagration arrester. The pallet assembly had a rigid FEP film diaphragm seal, which was an electrical insulator. The device was destroyed in the incident, but most major components were recovered. From the recovered components, no evidence was found of electrical bonding (e.g., bonding strap) between the pallet and the frame or tank. The listed set point for the emergency vent was the same (1.5 psig) as that for the conservation vent; thus, the tank could have breathed out through either vent. An ignition at this emergency pressure vent assembly could not be refuted during the investigation.

3.4.3 EA Hypothesis 3: Ignition source generated from an internal spark

As indicated in NFPA 780-2011 and API 2003 (2008), metallic tanks that are electrically continuous, sealed, and in direct contact with the ground are considered to be protected from lightning. This is based on the assumption that the steel tank system acts like a Faraday cage. A Faraday cage is an electrically conductive enclosure that protects the interior of the enclosure from the effects of external electrical fields. Electrical charges are distributed and flow to ground, cancelling the field's effect on the interior.

However, the protection offered by the Faraday cage can be negated by the presence of an unbonded internal component. As API 2003 (2008) pointed out:

"Metal tanks with fixed metal roofs and horizontal metal tanks that are maintained in good condition are generally protected from damage by direct-stroke lightning and ignition of its contents, if all metal components are in electrical contact (i.e. bonded)."

It was therefore possible that direct internal ignition of the EA vapors inside the tank could have occurred due to a spark discharge across a gap between two internal tank components that were not electrically bonded with each other. A detailed analysis of the tank instrumentation, equipment, and appurtenances yielded a list of several potential ignition locations. Piping penetrations into the headspace were either refuted as non-credible or assigned a low likelihood given the nature of mechanical contact between piping and the tank shell.

Conservation and emergency relief vents

The conservation vent and the emergency relief vent both utilized weighted floating pallets with FEP diaphragm seals. The pallets were not electrically bonded to the device bodies or the tank shell. Given the discontinuous electrical nature, it is possible that a spark discharge occurred inside either assembly. However, the conservation vent was separated from the tank headspace by the deflagration arrester. Thus, the conservation vent was refuted as a potential ignition source.

The pallet manway was assumed to behave as a capacitor in the event of a lightning strike, and calculations were performed to estimate the energy of a spark that could have occurred between the pallet and the tank. The calculations indicated that the energy generated would only reach about half the minimum ignition energy (MIE) of ethyl acrylate. However, due to uncertainty in the estimation, the emergency relief vent could not be refuted based on the available information.

High level capacitance probe

The tank was also equipped with a high-level capacitance probe. It was determined that the electrical ground lead inside the sensor housing was not connected to an earth ground, which may have affected the device's ability to dissipate electrical charge. It was noted, however, that the circuit breaker remained active and did not trip during the event. An examination of the level switch showed that the fuse was intact and there were no signs of overvoltage/overcurrent. However, it appeared that the diodes between probe and shield inputs may have been shorted, which could have occurred if the arc were limited to a few microseconds of exposure. Although evidence was not found to indicate that a spark actually occurred between the high-level probe and the tank, the potential for a spark discharge from the probe to a surface inside the tank could not be entirely refuted.

Surface charge

The team also considered the possibility of a liquid surface charge creating a spark discharge. However, this was deemed unlikely based on the conductivity of EA.

3.4.4 Plausible hypotheses: EA tank ignition

Based on this evaluation, the most likely ignition pathway was an electrical discharge from one of two electrically isolated equipment components internal to the EA tank – the high-level probe or the pallet manway.

3.5 *Hypotheses: Ignition of the BA Tank*

Although the headspace in the EA tank was flammable, the BA liquid was below its flashpoint and was not in a flammable region. Thus, only one side of the fire triangle was present. For ignition to have occurred within the BA tank approximately two minutes after the lightning strike, a source of ignition energy must have made its way into the tank head space, and the tank head space must have had a flammable concentration of vapors. Hypotheses were therefore generated to explain both the generation of an ignitable mixture of BA in air and the source of ignition.

3.5.1 Creating a flammable atmosphere in the BA tank

The vapor space of the BA tank was not flammable under normal operating conditions. The measured temperature of 19.4°C prior to the incident was well below the flash point of 37°C reported in the Safety Data Sheet. However, there must have been a flammable composition in the vapor space of the BA tank at the time when the tank exploded approximately two minutes after the EA tank explosion. The BA tank was in close proximity to the burning EA tank, and the two tanks shared a common dike area. Heat output from the EA fire in the dike, and possibly heat from the EA tank itself, could have raised the temperature of the BA tank, enriching the vapor space in BA vapors. The temperature probe in the BA tank was located in the lower part of the tank and ceased its output to the control system a little more than a minute after the EA tank erupted. Prior to its failure, the thermocouple had registered no appreciable change in temperature. However, given the size of the tank and proximity to the fire, it is very likely that a portion of the tank contents was raised significantly in temperature from the radiant heat of the EA tank fire and potentially direct flame impingement, possibly creating boiling at the shell interface.

One potential outcome of boiling inside a tank due to pool fire exposure is a Boiling Liquid Expanding Vapor Explosion (BLEVE). This type of explosion is possible if the tank is a rated pressure vessel and pressure relief systems are unable to maintain a safe pressure during fire exposure. Since the BA tank was an atmospheric storage tank and the shell did not display bulges, a BLEVE was refuted as a potential cause of explosion.

Three hypothetical paths to ignition of the BA tank vapor space were identified: ignition in the vent pipe which entered the tank, ignition inside of the tank, and ignition outside of the tank which propagated inside.

3.5.2 BA Hypothesis 1: Communication through the vent header

At the time of explosion of the BA tank, the vent header system had been compromised at the EA tank. Despite this damage, ignition in the vent header system outside of the BA tank was eliminated as a potential cause based upon observations of the deflagration arresters. Although the vent line deflagration arrester was heavily damaged and conclusions could not be drawn, the counterpart deflagration arrester beneath the conservation vent displayed internal fire patterns, not external patterns. This model of flame arrester appeared to successfully extinguish flames in EA and BA vapor streams based upon analysis of the surviving components.

3.5.3 BA Hypothesis 2: Ignition through the conservation vent

Because of the fire damage to the conservation vent flame arrester on the BA tank, no direct evidence of functionality could be determined. However, there were no apparent gaps in the plates in the cartridge. Given its rating for this service, it is unlikely that BA vapors venting through the conservation vent could flash back through the device into the tank headspace.

3.5.4 BA Hypothesis 3: Hot surface ignition by the exterior pool fire

Ignition inside of the tank could have occurred due to hot surface ignition within the vapor space. The low BA auto-ignition temperature of 292°C could possibly have been achieved on the inner tank shell surface from radiant heat or direct flame impingement at or near the liquid level where BA vapors would be anticipated to be within the flammable range. The BA tank shell was heavily heat-affected by the fire, and all paint had been heavily damaged. The tank roof was still painted, but the side facing the EA tank was soot covered indicating fire exposure. Thus, this pathway of hot surface ignition inside the tank head space could not be eliminated as a hypothetical cause.

3.5.5 BA Hypothesis 4: Direct ignition by the surrounding fire

Given exposure of the tank to the surrounding fire and the evidence of an internal deflagration, it is apparent that the liquid inside the BA tank was generating vapors. The BA tank was equipped with both a conservation vent and an emergency pressure vent. Since the conservation vent and the 12-inch emergency pressure vent pallet manway were set at the same relief pressure; it is likely that both vents were relieving BA vapors during the incident. Under this condition, the nearby EA tank and EA liquid pool fires could have directly ignited the vented vapors if flames contacted the vapors. The resulting ignition at the pallet manway would have allowed a flash back into the tank head space. This potential ignition pathway was a likely cause for the BA tank explosion.

3.5.6 Plausible hypotheses: Ignition of the BA tank

Although autoignition of the BA as a result of direct flame impingement could not be ruled out, the more likely ignition pathway was ignition of BA vapors venting from the emergency pressure vent pallet manway and flashing back into the tank.

4 Analysis and Conclusions

While the location of the lighting strike could not be precisely determined, the investigation concluded that the lightning strike caused ignition of the flammable head space vapors in the EA tank. The resulting internal deflagration caused the tank's internal pressure to quickly rise to a level causing the tank dome roof to rupture. The subsequent EA liquid pool fire in the containment dike and in the ruptured EA tank caused the BA tank, which did not normally contain a flammable vapor head space, to raise the head space concentration into the flammable range such that internal deflagration could occur.

The most likely cause of ignition of the EA vapors was determined to be an internal spark discharge from a non-electrically bonded component. The two most likely candidate components were the high-level capacitance probe and the 12-inch diameter emergency pressure vent pallet manway. The most likely cause of ignition of the BA vapors was determined to be either hot surface ignition by the internal shell above the liquid level or direct ignition of vapors vented from the 12-inch diameter emergency pressure vent pallet manway by the surrounding fire.

Following the event, Dow created a corporate learning experience report that was shared throughout the organization. It described the event, detailed the root and contributing causes, and discussed key learnings. A number of follow-up actions were also created. These included:

- Review operations policies and procedures concerning suspending manual loading and unloading of flammable materials when there is nearby lightning. Lightning guidance from the National Weather Service [7] recommends stopping outdoor activities from the time when lightning is first seen/heard until 30 minutes after the last clap of thunder.
- Review flame arrester maintenance procedures to ensure that they include plate spacing observations and proper methods for cleaning. Although the flame arresters were found to have worked as designed in this event, it was a good opportunity for the organization to reaffirm that all flames arresters are being properly serviced.
- Review maintenance procedures for inspection and testing of tank grounds and lightning apparatus to confirm that all are being maintained and work as designed.
- For storage tanks that can have a flammable headspace, complete an assessment of the tank, which included:
    - Ensuring that hatches are closed when not in use
    - Ensuring that roofs are in good condition, with no non-conductive patches
    - Evaluating the installation for the appropriate use of vacuum vent valves, flame arresters, gas padding, and other means to prevent flame propagation into the tank.
    - Assessing the equipment for tank penetrations that are not bonded to the tank shell or are not electrically continuous throughout their construction. These could include dip pipes, float level indicators, etc.

Action plans were developed based on the opportunities that were identified from this review.

5 Lessons Learned

Key takeaway messages from this investigation include:

- Lighting strikes are powerful natural events that can't be prevented. It has been estimated that lightning strikes would be expected to increase by about 12 percent per degree Celsius of warming [8]. It is critical to design our equipment and facilities to withstand extreme weather events like lightning.

- Following industry standards on lighting will not prevent all ignition events.

  As stated in API Recommended Practice 2003 (2008), Protection Against Ignitions Arising Out of Static, Lightning, and Stray Currents Section 5.1, it should be noted that: *"Even when all known precautions are employed, prevention or safe dissipation of direct-stroke lightning cannot be absolutely assured. In the case of indirect lightning currents, incendive sparks may occur in some segments of a system that use the best-known precautionary methods and devices."*

- Current standards may not protect against the potential hazard of unbonded components.

  The Faraday cage assumption inherent in the assumption of grounding in the event of a lightning strike assumes that there are no unbonded components. Should that assumption be invalidated, a system that otherwise complies with industry lightning guidance and best practices may not actually be protective. An evaluation of uninerted tanks in flammable service is recommended to identify and remediate any unbonded components.

- To mitigate the risk of lightning, it is prudent to employ good industry practices against ignition, such as flame arresters, closed operation and venting, and grounding. Also, when installing new tanks, consider the proximity to existing tanks. The BA tank likely would not have ignited had it not been in close proximity to the EA tank and had not shared a common dike.

- Consider inerting flammable tanks in situations where the risk of ignition outweighs the risk of failure of the inerting or mixed gas system.

6 References

[1] ACRYLATE ESTERS A SUMMARY OF SAFETY AND HANDLING 3RD EDITION Acrylate Esters, A Summary of Safety and Handling, 3rd edition, 2002. Available at ICSHAM Acrylic Esters Safe Handling Guide (dow.com).

[2] National Fire Protection Association (NFPA) 780, Standard for the Installation of Lightning Protection Systems, 2020 edition.

[3] American Petroleum Institute (API), Recommended Practice 2003, seventh edition (2008).

[4] NFPA 497, Recommended Practices for the Classification of Flammable Liquids, Gases, or Vapors and of Hazardous (Classified) Locations for Electrical Installations in Chemical Process Areas, 2012 edition.

[5] Grossel, S.S., *Deflagration and Detonation Flame Arrestors*, CCPS (2002).

[6] Society of Fire Protection Engineers (SFPE), Handbook of Fire Protection Engineering, 3rd Edition, section “Closed Vessel Deflagrations,” pages 3-406 to 3-410 (2002).

[7] National Weather Service, https://www.weather.gov/safety/lightning-safety-overview, accessed March 14, 2021.

[8] Scientific American, https://www.scientificamerican.com/article/lightning-may-increase-with-global-warming/, accessed March 14, 2021.

SPRING21 +17TH GCPS
A Joint AIChE and CCPS Meeting

Which eye are you looking through? The impact of cognitive bias on process safety

Trish Kerin
IChemE Safety Centre
Melbourne, Australia
tkerin@icheme.org

Prepared for Presentation at
American Institute of Chemical Engineers
2021 Spring Meeting and 17th Global Congress on Process Safety
Virtual
April 18 - 22, 2021

Which eye are you looking through? The impact of cognitive bias on process safety

Trish Kerin
IChemE Safety Centre
Melbourne, Australia
tkerin@icheme.org

Keywords: bias, culture, leadership, lessons learned

Abstract

Bias surrounds us every day. Everything we look at has a lens of our experience coloring our perception, and everyone has their own lens. This impacts how we share and receive information. So why is this important for process safety? Cognitive biases cover a range of aspects, each of which can have an impact on process safety, from what we remember, to coping with too much information, to a lack of context and meaning or a need to act with great haste. There are many identified specific cognitive biases, not every one of them impacts on process safety, but many can and have. This paper will review several process safety type incidents that had cognitive bias as a factor and discuss ideas on how to manage the bias, so we do not fall into its trap.

1 Cognitive bias

Cognitive bias is defined as a systematic deviation from norms or rational judgement, a way of creating a subjective based reality based on an individual's perceptions [1]. While on the surface this sounds like a pitfall in human decision making, our cognitive biases allow us to make quick decisions. This can be advantageous when time is of the essence, provided the bias does not lead us in the wrong direction. Making decisions based on biases, or heuristics to estimate outcomes is a faster way to make decision than rational decision making, and it also fits in with the idea that we have two mind systems, system 1 and system 2. "System 1 operates automatically and quickly, with little or no effort and no sense of voluntary control. System 2 allocates attention to the effortful mental activities that demand it, including complex computations." [2, p. 20]. The problem with system 1 is that it is sometimes wrong, and unless we use system 2 to override the wrong answer, we can end up with the wrong decision. Take for example Figure 1. Quickly glance at it an decide which figure is the largest. Your system 1 would have said the one on the right.

Figure 1 – Which figure is larger? [2, p. 100]

If you were to engage your system 2, and take a ruler and measure the figures you would see they are all the same size. System 1 was misled by a perspective perception. The challenge for managing bias is to engage system 2 prior to making critical decisions.

Another concept to describe this System 1 thinking is called "thin slicing". This "refers to the ability of our unconscious to find patterns in situations and behavior based on very narrow slices of experience." [3, p. 23] It can be summarized as that initial gut decision we may feel in an instant.

1.1 Most common biases

There are 188 identified different cognitive biases, though they can be grouped into the 4 categories of:

- What should we remember?
- We need to act fast
- Too much information
- Not enough meaning [4]

Common biases that can have an impact on safety are:

- Anchoring
- Confirmation bias
- Framing
- Hindsight bias
- Status quo bias
- Sunk-cost bias [5]

In addition to these six, Halo bias can lead to giving more weight to information from certain people than others.

Table 1 shows the linkages between the groups, sub-groups and biases as categorized by Benson (2017). Note none of the biases discussed in this paper fit into the "What should we remember?" group.

Table 1. Groups, sub-groups and biases [4]

Group	Sub-group	Bias
We need to act fast	To avoid mistakes we aim to preserve autonomy and group status to avoid irreversible decisions	Status quo bias
	To get things done, we tend to complete things we've invested time & energy in	Sunk cost fallacy
Too much information	We notice when something has changed	Framing
		Anchoring
	We are drawn to details that confirm our own existing beliefs	Confirmation bias
Not enough meaning	We project our current mindset and assumptions onto the past and future	Hindsight bias
	We imagine things and people we're familiar with or fond of as better	Halo effect

2 How biases impact decision making

System 1 is the part of our mind that makes decisions quickly, and without active thought. It is doing this based on our past experiences and various cognitive biases that exist for each of us. This means that our biases have an effect on our decision making when our system 1 is in control.

2.1 Status quo bias

This is a tendency to want things to remain the same as they have been, an avoidance of change [5]. An everyday example here may be remaining with your current cellular phone provider even though a competitor may offer better rates or service. This type of bias can become an issue if the current way is not working or has unmanaged hazards in it.

2.2 Sunk cost fallacy

This is a tendency to want to continue with a course of action, even when evidence suggest otherwise, in an attempt to recoup losses [5]. It can occur when large sums of money or effort have been invested in something and it has not shown the expected rewards. There can be a tendency to invest more, even though evidence suggest that it will not be successful or indeed hazardous to continue. An everyday example here could be to 'double or nothing' in a game of chance. The odds of winning have not got better, but an attempt is made to win back previous losses by betting higher amounts. This type of bias can become an issue where it is potentially not safe to continue.

2.3 Framing

How information is presented can change how the information is interpreted [5]. An everyday example here may be that a yoghurt brand markets it product as 90% fat free, the corollary to this is that the product is 10% fat. If you are conscious about the amount of fat in your food you would more likely choose the 90% fat free rather than the 10% fat product. This type of bias can become an issue where data is presented in a misleading way. It may also have an opposite effect when some data is presented in a way to prime the mind so that it is more likely to make a favorable decision when caught up in the "thin slicing" stage of thinking [3].

2.4 Anchoring

This is based on placing more emphasis on the first information received, even if subsequent information shows it to be false [5]. An everyday example here might be when shopping and seeing a marked down price – 'was $200 now $150'. Seeing the first price of $200 may make you think that $150 is a good deal, because your expectation is anchored at $200. This type of bias can become an issue where the first piece of data is incorrect, and should be discounted from the decision making process.

2.5 Confirmation bias

This relates to placing more emphasis on data that supports your initial conclusion [5]. It can lead to the discounting of contrary data or searching for data that confirms your thoughts. An everyday example here is related to social media algorithms. We are presented with streams that typically confirm our own beliefs and we do not see counter information, which can lead us to believe that our thoughts are the same as the broader community. This type of bias can lead us to make the wrong decision because we do not accept data that disagrees with our initial thoughts.

2.6 *Hindsight bias*

This bias "is defined as the belief that an event is more predictable after it becomes known than it was before it became known" [6, p. 411]. Hindsight makes it difficult to understand why a person took the action they did when it led to an incident occurring. We are left asking questions like 'why did they do that; they must have known what would happen?' An everyday example here is preparing to go away for a camping trip, and on arrival realizing that an important piece of equipment was left behind, only to exclaim 'I knew I would forget something!' If this was actually known ahead of time it could have been prevented. This type of bias can lead us to believe that an incident could not happen to us, because would not make that kind of mistake that led to an incident.

2.7 *Halo effect*

This is a tendency to like everything about a particular person, even aspects of them you have not observed [2]. It can lead to following the advice of a person because you do not consider that they may be wrong. The corollary of this is the horns effect, where you do not believe the advice of a person because you do not believe they could be right. An everyday example of this may be thinking that your favorite actor must be a very nice person based on their attractive physical appearance, even though you have never met them. This type of bias can lead us to make an incorrect decision based on the opinion of others.

3 Biases in process safety incidents

There are several well know incidents where cognitive biases contributed to the event occurring. An incident has been selected for each bias, with the exception of hindsight bias, as the typical concern with hindsight bias is that it can inhibit learning from incidents after they have occurred and while this can result in a repeat incident, that has not been explored here.

The summary of incidents and their associate biases are contained in Table 2. This paper does not go into detail about the incidents, only mentioning the bias impacts. For more information on each incident refer to the references.

Table 2. Incident examples and their associated bias.

Bias	Incident	Example
Status quo bias	Columbia	Over the years that the space shuttle program had been running, NASA has normalized foam strikes as a part of launch. This because the status quo, and it was determined that they were not a safety of flight issue, and therefore there

		was no need to change the status quo [7].
Sunk cost fallacy	Bhopal	The Bhopal facility had a history of financial losses and equipment failures resulting in leaks. The facility had been shut down for a period of time and a decision was made to restart it to consume the feedstock, rather than dispose of it. During the final operation the toxic cloud was released. The decision to do the final produce run is an example of the sunk cost fallacy, continuing to run the facility, when it should have been shut down [8].
Framing	Challenger Space Shuttle	Temperature criteria had not been established for the solid rocket booster o-rings, but there was concern that the o-rings were brittle at low temperature. The initial framing suggested that the temperature effect was inconclusive, leading to concerns about the potential blow by of the o-rings being discarded [9].
Anchoring	Texas City Refinery	During the start up of the ISOM unit, the Raffinate Splitter Tower was filled with heated raffinate. The level indicator in the tower only covered the bottom section of the tower, and was also providing faulty readings. The operators became anchored to the level indicator results, not realizing the tower was in fact over filling [10].
Confirmation bias	Macondo	During the abandonment activities a cement barrier was installed. At the time it was believed that the installation went well and therefore the barrier would work. This led to test results being misinterpreted when assessing the barrier integrity. The belief that the barrier installation had been successful led to confirmation bias discounting the failed test results as incorrect [11].

Halo effect	Dreamworld	The Halo effect can exist when a junior or inexperienced worker is working at the direction of a more experienced person or being trained. As the junior person does not have the knowledge themselves, they will often do as they are told, deferring to the experience – the halo. The junior operator was told during her training she did not need to operate the emergency stop as that was the senior operator's role. When the raft became stuck on the rails no one hit the emergency stop, allowing the following raft to collide with it. Four people were killed when the second raft flipped as a result of the collision [12].

3.2 Hindsight bias

Hindsight bias often inhibits our ability to learn from an incident, because "the very outcome knowledge which gives us the feeling that we understand what the past was all about may prevent us from learning anything from it." [13, p. 311] In summary, we do not believe we would make the same mistakes. This phenomenon has led the IChemE Safety Centre to develop interactive case studies, where an attempt is made to remove the hindsight related to an incident, by telling the story without disclosing the outcome. Anecdotally these case studies appear to create an incident experience for the participant, without the consequences, which may assist with the learning being retained [14].

4 Ideas to avoid bias

The term often used for avoiding biases is called "debiasing". When considering the idea of system 1 and system 2 in our minds, to effectively debias we need to learn to engage system 2 to verify, correct and override where necessary our system 1 response. System 2 can be engaged by being asked questions that require thought, as this engages it to start to process. It has been found in studies that frowning while undertaking a task can increase the vigilance of system 2. This is because frowning usually occurs with cognitive strain, and this strain can be induced by frowning [2]. There are several ideas that can be tried to counter biases shown in Table 3.

Table 3. Ideas to counter bias

Bias	**Ideas to counter it**
Status quo bias	Appoint a person in the discussion to have the role of challenging the ideas, ensuring they have psychological safety to speak up without fear. This is akin to the "black hat" idea [15]. In this way there is someone free to ask questions about why the status quo exists as it is. The exploration of the status quo may highlight aspects that need changing.
Sunk cost fallacy	Engage someone from outside the issue, a coworker who is not tied to the outcome, to challenge what the next steps should be. They can help establish the pros and cons for each action, leading to a risk based decision on progression or not.
Framing	Reframe the data and discuss both aspects, identifying the counterfactual and assess both sides. From a priming aspect, it is possible to place subconscious thought in the mind prior to making decision. [3] An example might be that prior to undertaking a high risk task, the hazards and examples when things have previously gone wrong are specifically discussed. This act may prime the mind to look for and identify similar patterns.
Anchoring	When a decision needs to be reached, instead of a free discussion, have each person write down their initial ideas first. This way they have their thoughts that are not influenced by others during the discussion.
Confirmation bias	Engage someone from outside the issue, a coworker who is not tied to the outcome, to challenge what the data is telling you – do they reach the same conclusion without the background confirmation?
Halo effect	Attempting to slow down the thought process can assist with the time to look for more data to draw a conclusion from. If you are aware of the halo effect, you can try to think of different impressions of the person. This may be done by asking questions to try and establish a better understanding of them.
Hindsight bias	Attempt to understand the context in which decisions are made, not just the decision. This is useful during incident investigation, understand why an action was taken rather then what the action was. When we understand the context of

	decisions, we have the potential to see how it occurred instead of dismissing and thinking we would not have done it.

5 Conclusion

All actions and decisions we make are subject to biases, and we are often not aware of them. While most of the time this does not have an impact on the safety at a facility, sometimes the biases can lead to decisions which cause incidents. For this reason, it is important to understand the biases that may be present, as well as how our minds work with system 1 and system 2 and develop strategies to check if our system 1 decisions are likely to result in an incident.

6 References

[1] M. Haselton, N. D and A. PW, "The evolution of cognitive bias," in *The Handbook of Evolutionary Psychology*, Hoboken NJ, Wiley, 2005, pp. 724-746.

[2] D. Kahneman, Thinking, Fast and Slow, New York: Farrar, Straus and Giroux, LLC, 2011.

[3] M. Gladwell, Blink, Boston: Little, Brown and Company, 2005.

[4] B. Benson, "Cognitive Bias Codex," September 2017. [Online]. Available: https://www.visualcapitalist.com/wp-content/uploads/2017/09/cognitive-bias-infographic.html . [Accessed 19 July 2020].

[5] W. Hersing, "Managing cognitive bias in safety decision making: Application of emotional intelligence competencies," *Journal of Space Safety Engineering,* vol. 4, pp. 124-128, 2017.

[6] N. Roese and K. Vosh, "Hindsight bias," *Perspectives on Psychological Science,* pp. 411-426, 2012.

[7] Columbia Accident Investigation Board, "Columbia Accident Investigation Board," Columbia Accident Investigation Board, Washington DC, 2003.

[8] K. Bloch, Rethinking Bhopal, Amsterdam: Elsevier, 2016.

[9] Roger's Commission Report, "Report of the Presidential Commission on the Space Shuttle Challenger Accident," Presidential Commission on the Space Shuttle Challenger Accident, Washington DC, 1986.

[10] CSB, "Investigation Report Refinery Explosion and Fire," CSB, WashingtonDC, 2007.

[11] CSB, "Investigation Report Executive Summary Drilling Rig Explosion and Fire at the Macondo Well," CSB, Washington DC, 2016.

[12] Coroners Court of Queensland, "Inquest into the deaths of Kate Goodchild, Luke Dorsett, Cindy Low and Roozbeh Araghi at Dreamworld, October 2016," Queensland Courts, Southport, 2020.

[13] B. Fischhoff, "Hindsight ≠ foresight: the effect of outcome knowledge on judgement under uncertainity," *Quality and Safety in Health Care,* vol. 12, pp. 304-312, 2003.

[14] T. Kerin, "Accounting for hindsight bias: improving learning through interactive case studies," *Loss Prevention Bulletin,* no. 264, pp. 17-20, 2018.

[15] E. de Bono, Six thinking hats, Boston: Little, Brown and Company, 1985.

Unusual Hydrogen Explosions Due to Unanticipated Metal-Water Reactions

Robert Zalosh
Firexplo LLC
Wellesley, MA 02481
bzalosh@firexplo.com

Prepared for Presentation at
American Institute of Chemical Engineers
2020 Spring Meeting and 17th Global Congress on Process Safety
Virtual
April 18 - 22, 2021

Unusual Hydrogen Explosions Due to Unanticipated Metal-Water Reactions

Robert Zalosh
Firexplo LLC
Wellesley, MA 02481
bzalosh@firexplo.com

Keywords: hydrogen explosions, electrochemical reactions, metal-water reactions.

Abstract

Inadvertent generation of hydrogen from ambient temperature metal-water chemical and electrochemical reactions have been responsible for a surprising number of hydrogen explosions. This paper describes three sets of incidents involving different metals in applications that are normally considered free of explosion hazards. The first set of incidents involves wet alkaline incinerator ash with an aluminum content of at least two percent. The second set of incidents involves copper conductor electrochemical reactions in subsea electrical cable. The third set involves electrochemical corrosion reactions in galvanized dry sprinkler pipe. In all three applications, test data are discussed to elucidate the specific reactions and potential preventive and mitigation measures.

1 Introduction

Inadvertent generation of hydrogen from reactive powdered metal-water reactions and molten metal-water reactions is a well-known hazard in metal production and processing facilities. Prevention and protection measures for this hazard are described, among other places, in the NFPA 484 Standard on Combustible Metals [1]. However, inadvertent hydrogen generation from ambient temperature metal-water chemical and electrochemical reactions can also occur in many unanticipated settings and scenarios, and such scenarios have been responsible for a surprising number of hydrogen explosions.

Each of the following three sections of this paper provides details of one incident and then summarizes several other incidents for a given metal and application. It then describes the specific reactions likely responsible for the inadvertent hydrogen generation, including pertinent test data. The first set of incidents involve hydrogen generated from reactions of water with Municipal Solid Waste incinerator ash containing aluminum. The second type of metal-water electrochemical reaction involves copper cable in a seawater atmosphere. The third set of incidents entails hydrogen generation from galvanized steel corrosion reactions in dry pipe sprinkler systems.

2 Water-Incinerator Ash Reaction Incidents

2.1 Nortrader Cargo Vessel Explosion

A prime example of this type of incident is the January 2017 explosion on the dry cargo ship *Nortrader* that was the subject of a comprehensive investigation and report by the British Marine Accident Investigation Branch (MAIB) [2]. The *Nortrader* cargo of unprocessed incinerator bottom ash was generated in a nearby Plymouth, England power plant burning Municipal Solid Waste (MSW). Prior to the January 2017 incident, there had been 34 uneventful shipments of MSW bottom ash on 26 vessels similar in size to the 88 m length, nominal 1900 metric ton capacity *Nortrader*.

On January 12, 2017, over a period of about eight hours, 2,333 metric tons of wet MSW bottom ash were loaded into the *Nortrader's* 3409 m^3 cargo hold during steady heavy rain. After loading, the vessel left the dock and anchored overnight in Plymouth Sound while it waited for the weather to improve. During this time, the cargo hold ventilation fans and inlets were closed because of the heavy rain.

A large propagating explosion occurred the next day while the *Nortrader* was still at anchor. The ignition source for the explosion was the actuation of a stop button on an electrical panel situated in a ship compartment immediately forward of the large cargo hold. The ship's chief engineer, who was standing in front of the electrical panel suffered first degree burns to his face and second degree burns to his body, both hands, and lower extremities. The explosion then propagated into the cargo hold and lifted and bent all nine of the cargo hold steel latch covers, which had been held in place by latch cleats. The displaced hatch covers over the cargo hold the are visible in Figure 1.

Figure 1. *Nortrader* damaged cargo hold latch covers after explosions.

After the incident, the MSW from the Nortrader cargo hold was unloaded for inspection and testing. Employees of a consulting company retained by one of the ship's insurers drilled holes into the unloaded pile of MSW bottom ash and tested for the presence of flammable gas. They detected 1.4% hydrogen and 0.2 % carbon monoxide in the pile void spaces.

The MAIB hired another consulting company to acquire samples of the unloaded bottom ash for two types of water reactivity tests. One test series was UN Test N.5, the test method for substances which in contact with water emit flammable gases. It is described in section 33.4.1.4 (Annex E) of the UN Manual of Tests and Criteria [3]. The second test series involved an ad-hoc test with a much higher water-to-ash ratio (6:1) than in UN Test N.5 (1.5:1). Sample preparation for both tests entailed grinding the sample particulate to less than 250 micrometers for UN Test N.5, and to less than 1 mm for the ad-hoc reactivity test.

Several of the ash subsample water reactivity tests resulted in measured hydrogen production rates between 0.41 l/kg-hr and an average of 0.77 l/kg-hr. These rates are slightly less than the UN Test N.5 threshold of greater than 1 l/kg-hr of flammable gas to classify a material as a Division 4.3, substance which emits flammable gas when in contact with water [3]. Nevertheless, these hydrogen production rates indicate a potential for hydrogen accumulation in cargo holds and other enclosures with negligible ventilation and large quantities of bottom ash. Perhaps the UN Manual test criteria should be re-evaluated to provide a separate, less restrictive, category for this type of material that generates lower rates of flammable gas when in contact with large quantities of water.

The unventilated *Nortrader* cargo hold containing water-soaked bottom ash allowed the slowly generated hydrogen to accumulate over a period of about 19 hours while it was anchored in Plymouth Sound. An unsealed cargo hold light fixture access cover provided a path from the cargo hold to the adjacent ship compartment, which was also unventilated during this period. According to the MAIB report, the electrical panel in this compartment had an unsecured door, which allowed hydrogen to enter the panel interior. When an electrical circuit within the panel was energized and an arc or spark developed across a pair of contacts, the hydrogen-air mixture within the panel was ignited. The resulting explosion propagated from the panel into the ship compartment and then, via open light fixture cover, into the large cargo hold. An analysis by a MAIB contractor concluded that the hydrogen concentration in the cargo hold at the time of the explosion was in the range 6% to 9%, corresponding to a hydrogen volume between 114 m^3 and 171 m^3. The two explosions felt and heard by the *Nortrader* crew were the explosions in the electrical panel compartment and in the cargo hold itself.

2.2 MSW incinerator facility explosions

There have been several reported hydrogen explosions at MSW incinerator facilities. One such incident in Japan in 1995 is well documented [4,5,6]. A large amount of aluminum in the waste stream (9.6% Al) produced a clinker, i.e., a hard chunk of solidified molten metal and bottom ash, that was blocking flow in an ash discharge chute leading from the bottom of the incinerator to an ash hopper. Employees attempted to remove the clinker by striking it with a chisel and intermittently pouring water on it. This caused an explosion with ejected hot gas and ash burning three of the workers, one of whom subsequently died from the burn injuries [4,6]. The explosion was attributed to hydrogen formed from the aluminum-water reaction, with the subsequent hydrogen-air mixture in the chute being ignited by the chisel impacting the still hot clinker and ash.

A similar explosion occurred in a Japanese MSW incinerator in 1983 [6]. This explosion occurred in a dust bunker when an iron bar was used to crush some blocks of ash. The worker suffered severe injuries when he was thrown from the explosion site. The ash moisture content was about 10%, and there were significant amounts of aluminum in the ash.

Larson [7] briefly described a 2003 explosion in the basement ash storage bunker at a Swedish MSW incinerator due to hydrogen generation. Fortunately, there were no injuries, but the explosion damaged the bunker walls. Other incinerator facility explosions in Sweden have occurred during conveyor transport of warm wet bottom ash [8].

2.3 Other ash-water reaction explosions

Arm and Lindeburg [8] mention the occurrence of hydrogen explosions due to ash-water reactions in some Swedish ash storage facilities such as rock caverns and former oil wells. They also mention some explosion incidents associated with wetted fly ash in landfills.

There have been some flash fire and explosion incidents involving hydrogen generated from the ash aggregate used to make lightweight/foamed concrete. One such incident in 2009 in the UK occurred in a former dry well that had recently been filled with foamed concrete [9]. A grinder being used to cut reinforcing rods above the concrete ignited the hydrogen-air mixture that had accumulated there. This and other UK incidents involving foamed concrete and bricks caused the UK Highways Agency to ban the use of foamed concrete made with incinerator bottom ash.

2.4 Ash-water reactions producing hydrogen.

Although a variety of metals can react with water to produce hydrogen, most of the incinerator ash generated hydrogen has been found to be due to aluminum-water reactions. The primary reaction responsible for hydrogen generation is believed to be [10]:

$$2Al + 6H_2O \rightarrow 2Al(OH)_3 + 3H_2 \quad (1)$$

This reaction is promoted in an alkaline environment, with the major alkali source in MSW incinerator ash being calcium hydroxide. According to Nithiya et [10], the calcium hydroxide is consumed and regenerated in the following reactions, which combine to produce reaction (1).

$$2Al + 6H_2O + 2Ca(OH)_2 \rightarrow 2CaAl(OH)_5 + 3H_2 \quad (2)$$

$$2CaAl(OH)_5 \rightarrow 2Ca(OH)_2 + 2Al(OH)_3 \quad (3)$$

The presence of these compounds in ash-water reaction products has been confirmed by Nithiya et al [10], and the promoting effect of pH values > 10 in ash has been demonstrated in test data reviewed by Larsen [7] and others. Several researchers have reported significantly greater hydrogen generation rates when the ash is mixed with alkaline solutions compared to mixing with distilled water. Nithiya et al [10] cite research showing that the alkaline ash conditions cause the oxide layer on the aluminum surface to be perforated and dissolved, allowing direct contact and reaction of aluminum and water.

As implied by reactions (1) and (2), hydrogen generation rates from ash-water reactions increase with increasing aluminum content and increasing pH of the ash. The reactions have also been shown to be faster for freshly generated warm ash, and to decay after several days of ash exposure to ambient air (allowing greater oxide layer formation). These effects produce a wide variation in reported hydrogen generation yields obtained in tests with MSW ash samples.

Besides hydrogen generation rate data, reported test data include total hydrogen generated per unit mass of ash. The hydrogen yield data show the yield increasing with increasing ash aluminum content as well as other factors mentioned above. Yield data have been compiled from a variety of sources including data for bottom ash from Japanese incinerators [10], fly ash from Japanese incinerators [6], and both bottom ash and fly ash from Swedish incinerators [8]. The hydrogen generation data cumulative distribution for ambient temperature reactions obtained by mixing with distilled water are shown in Figure 2. The median hydrogen yield is 2.4 liter-H_2 per kg-ash for bottom ash and 10 l/kg-ash for fly ash. The highest measured hydrogen yields are 24 l/kg-ash for bottom ash and 37 l/kg-ash for fly ash.

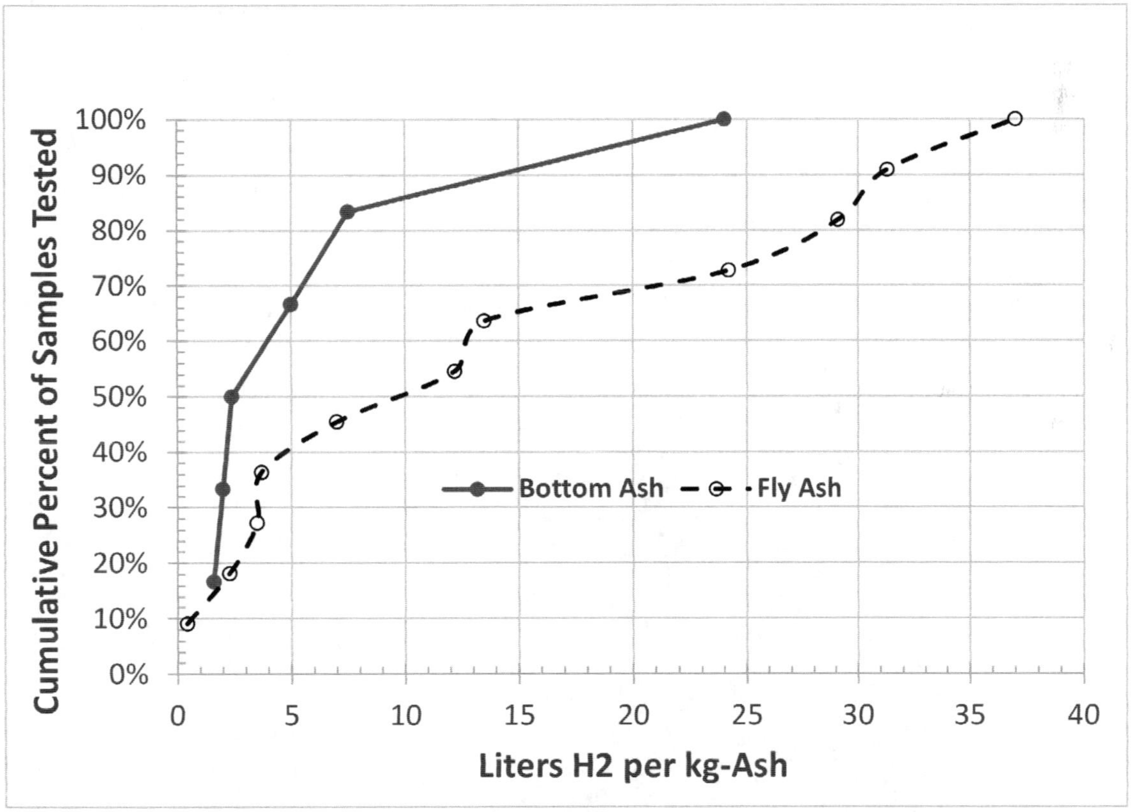

Figure 2. Hydrogen generation data cumulative distribution for bottom ash and fly ash reactions with water.

The data shown in Figure 2 can potentially be used to provide a risk-based estimate of the hydrogen accumulation hazard potential in MSW incinerator ash bunkers and storage and transportation facilities. Daily generation data reported in the cited references show that most of the hydrogen is generated in the first few days after ash formation and contact with copious quantities of water. (Based on the test methods cited above, copious water implies having a water/solid ratio of about two or greater at some point during handling and storage.) Therefore, this data can also be used as a basis for estimating MSW ash storage facility ventilation requirements to prevent flammable mixture formation. These requirements are not needed for ash obtained from other fuels providing those fuels do not have aluminum contents of about 2 percent or higher.

3 Electric Cable – Seawater Reaction Hydrogen Incidents

3.1 Offshore Junction Box Explosions

Offshore oil and gas exploration and production platforms have many electrical junction boxes exposed to sea air and aerosols that produce a saline environment within the box. Several of these boxes have experienced explosions that damage the box. Incident investigations have determined that hydrogen generation within the box produced a flammable atmosphere that was probably ignited by an arc, spark, or electrostatic discharge.

One junction box explosion incident [11] occurred in a box installed above the boat landing level of an unmanned platform in a tropical environment. The box had an IEC 60079 [12] designation "Ex e", meaning that it is intended for normally non-sparking electrical equipment with a rated voltage no greater than 11 kV, to be installed in a potentially flammable gas atmosphere. This particular box contained a 6.6 kV submarine power cable, and had breather ventilation.

During a maintenance crew visit about a year after the cable was first energized, the junction box hinged door was observed to be bulged outward and blown open, as can be seen in Figure 3. There was no interior damage or char deposits and the electrical connections remained intact. The fact that the box did not withstand the explosion is not surprising since it was not constructed or rated as an "Ex d" box.

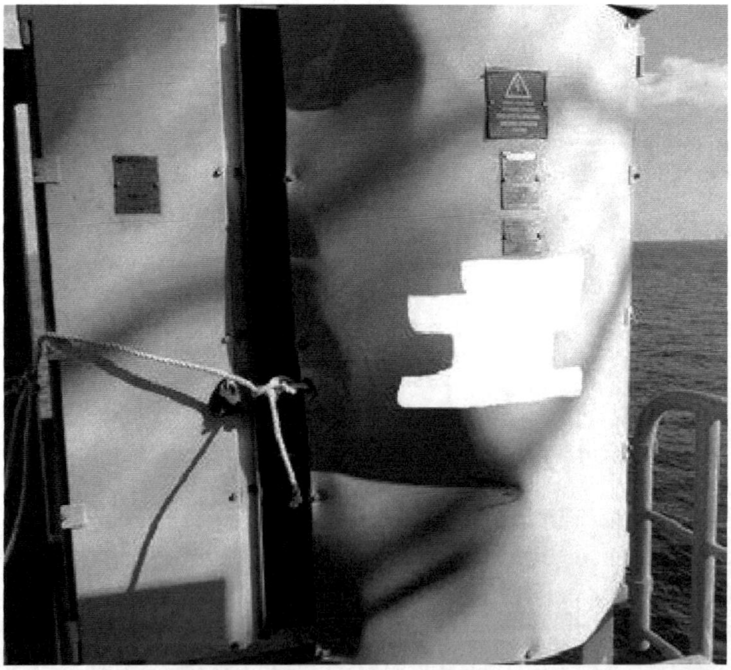

Figure 3. Damaged junction box door on offshore platform [11].

There have been at least two similar reported junction box explosion incidents at other platforms and locations. After one of those incidents, flammable gas was detected in the box [11].

Hydrogen generation from the copper conductor within seawater cables is a recognized phenomenon in subsea electrical cables with the generation and associated copper loss being more pronounced when there is cable insulation damage. This is illustrated in Figure 4, which shows the hydrogen bubbles originating from damaged cables in simulated seawater [13]. Hydrogen is also formed from seawater reactions with the copper and steel components of optical fiber cable [14]. Some seawater cable manufacturers provide hydrogen absorbing materials in their cables to prevent the hydrogen from being released from the cables.

Figure 4. Hydrogen bubbles from a damaged cable in simulated seawater [13]

3.2 Buried Cable Seawater Reaction Hydrogen Incident

An unusual electric cable-seawater reaction hydrogen explosion occurred at a Rhode Island beach in July 2015 [15]. The hydrogen was generated by the seawater corrosion of an abandoned electrical cable that had been used until 2007 to power Coast Guard navigational lights on rock jetties near the beach. People at the beach were not aware of the cable which was buried in the sand. When a hydrogen-air mixture was somehow ignited, the explosion blew a woman out of her beach chair. She suffered broken ribs and bruises upon landing on a jetty about 10 feet away.

Figure 5 shows a portion of the cable after the sand that had obscured it was removed. University of Rhode Island geochemists tested sand samples from the area and measured hydrogen concentrations as high as 5%, i.e., slightly above the Lower Flammable Limit. They also observed that the corroded cable did not have a green patina, indicating that the corrosion was due to oxide, rather than chloride formation.

Figure 5. Corroded electric cable under beach sand [15]

3.3 Seawater Induced Copper Hydrogen Evolution Reactions

Copper exposure to a seawater alkaline environment promotes hydrogen evolution via two different electrochemical reactions. First, copper serves as a catalyst for the cathodic reaction

$$2H_2O + 2e^- \rightarrow H_2 + 2[OH]^- \ (4)$$

Sherif and Park [17] maintain that this reaction is more likely to occur in de-aerated solutions, whereas aerated NaCl solutions are more likely to produce the water and oxygen reduction reaction

$$O_2 + 2H_2O + 4e^- \rightarrow 4[OH]^- \ \ (5)$$

However, Martins et al, [18] have shown that reaction (4) can produce copious amounts of hydrogen when the polycrystalline copper surface allows the formation of a copper-hydroxide activated complex. They demonstrate that copper surface treatment via polishing and annealing allows this activated complex to form and render the copper surface more hydrophilic.

The second seawater induced hydrogen evolution pathway is the following direct copper reaction with water [17]

$$2Cu + H_2O \rightarrow Cu_2O + 2H^+ + 2e^- \rightarrow Cu_2O + H_2 \ \ (6)$$

Reaction (6) is suppressed in the absence of seawater because of the compact film of the cuprous oxide reaction product. Upon seawater exposure, chloride ions cause the formation of a cuprous chloride ion that partially dissolves the cuprous oxide film and forms the more porous Cu_2O and layer of atacamite ($Cu_2(OH)_3$- Cl), which readily admits more chlorine ions to reach the copper underneath it [17, 19]. The dissolution/porosity of the oxide film allows the oxidation reaction to continue and also results in further copper dissolution. Thus, copper corrosion can be often seen in the form of the green patina due to the cuprous chloride, and sometimes seen merely as a loss of copper without any discoloration. The latter was reported to be the case in the buried cable beach incident summarized above [16].

Reported test data on copper seawater corrosion rates can be used in conjunction with Reaction (6) to quantify the corresponding expected hydrogen generation rates. Corrosion rate data used for this purpose was obtained from copper exposure laboratory tests with natural and artificial and aerated seawater [19, 20], and from field tests with copper samples in seawater and sea aerosols in North Atlantic and tropical environments [21, 22]. The calculated hydrogen generation rates for exposures of several months varied from about 2 cc-H_2-per-hr-m^2-Cu to almost 14 cc-H_2 per hr-m^2. The cumulative distribution of the calculated results is shown in Figure 6. These results can be used to determine ventilation rates needed to prevent flammable hydrogen-air mixtures in electrical junction boxes or other small electrical enclosures in a seawater or sea atmosphere environment.

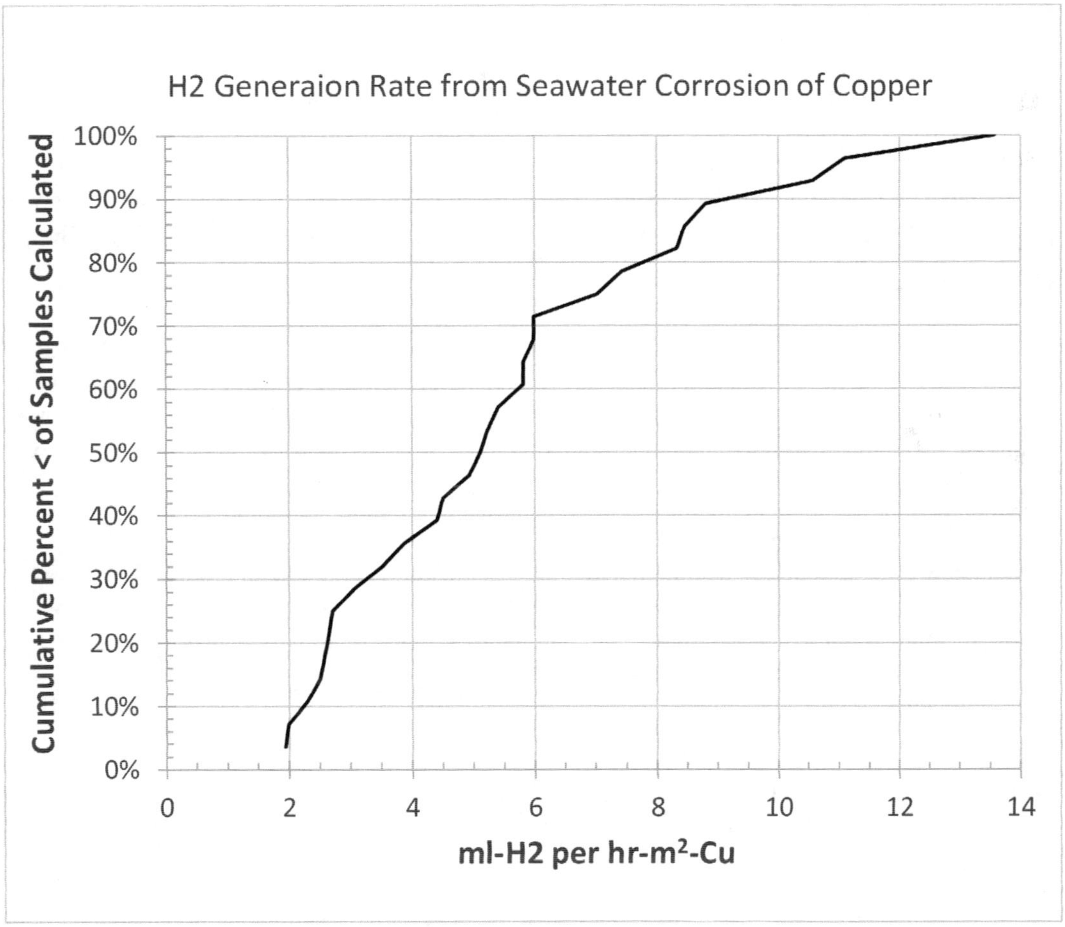

Figure 6. Calculated hydrogen generation rates from seawater exposure of copper.

4 Galvanized Sprinkler Piping Hydrogen Incidents

4.1 Helsinki Theater Sprinkler Room Explosion

On June 29, 2020, a sprinkler contractor was conducting routine maintenance on a dry pipe sprinkler system in the basement of a Helsinki theater. Shortly after the contractor opened the pipe, an explosion occurred that caused burn injuries to the contractor, who required hospitalization. The explosion caused minor damage to the stairwell adjacent to the sprinkler room [23].

An investigation by the Finnish Safety and Chemical Agency (Tukes) determined that the hydrogen was generated from a corrosive reaction involving zinc, water, and air within the galvanized steel pipe [24]. It was the first incident of its kind in Finland, but Tukes was sufficiently concerned about possible future incidents to issue a ban on the use of zinc-coated pipes in new sprinkler systems. Tukes also issued new requirements for maintenance involving opening of sprinkler piping; ventilation and a hydrogen detector are now required.

4.2 Other Sprinkler Pipe Hydrogen Fire and Explosion Incidents

Although the Helsinki theater explosion was the first incident in Finland involving hydrogen generated in galvanized sprinkler piping, there have been several such incidents in other Scandinavian countries, where galvanized dry pipe sprinkler systems are common. The following is a list of some European incidents.

- A 2014 incident occurred at a Danish industrial company when a sprinkler contractor was burned by a hydrogen flame vented from sprinkler piping while it was being drained [22].
- A second 2014 incident in Denmark; this one involved a hydrogen release from a pressurized water tank that had recently been drained. A sprinkler contractor suffered burn injuries from the resulting room explosion which may have been ignited by sparks from the contractors' tools [25].
- There was an explosion in a sprinkler fire pumphouse in France in October 2016 after a sprinkler contractor had drained the sprinkler piping in preparation for doing work on the system. The explosion occurred about 15 minutes after the drain was opened and the contractor left the pumphouse. Fortunately, there were no injuries, but the pumphouse was destroyed [26].

There have also been similar reported incidents in Japan. Yamate et al. [24] mention ignitions occurring when wet galvanized pipe have been cut or subjected to grinding during sprinkler system renovations.

4.3 Hydrogen generation reactions in sprinkler piping

The electrochemical reactions responsible for hydrogen generation in galvanized sprinkler pipe are analogous to the reactions described above for copper conductor exposure to seawater. The reactions described by Yamate et al. [27] are:

$$2H_2O \rightarrow 2H^+ + 2[OH]^- \quad (7)$$

$$2H^{+} + 2e^{-} \rightarrow 2H \rightarrow H_2 \quad (8)$$

$$Zn \rightarrow Zn^{2+} + 2e^{-} \quad (9)$$

$$Zn^{2+} + 2[OH]^{-} \rightarrow Zn(OH)_2 \quad (10)$$

Reactions (9) and (10) produce the zinc corrosion that results in formation of the white zinc hydroxide layer seen below the indicated water line in Figure 7. Su and Fuller [28] did not observe this corrosion when the air above the water was replaced by compressed nitrogen. This suggests that oxygen is required for accelerated corrosion in galvanized pipe, and that oxygen-water electrochemical reaction (5) is a key aspect of both corrosion and hydrogen production in galvanized dry pipe sprinkler systems in which there is residual water together with air in the piping. This often occurs when the piping is not completely drained following a dry pipe valve trip test or inadvertent actuation.

Figure 7 Zinc corrosion produced in 6 weeks in half-filled pipe [25]

Yamate et al. [24] have shown that rapid hydrogen generation can also occur from galvanic corrosion associated with water filled galvanized sprinkler pipe in contact with another metal component such as a stainless steel flexible pipe section. Gas samples from the galvanic coupling pipe revealed a hydrogen concentration of 20% and significant concentrations of zinc in the water. They also obtained visual evidence of hydrogen bubbles leading to a flammable atmosphere when the piping was opened, as shown in Figure 8.

Figure 8 Hydrogen bubbles and flames in opened galvanic coupled piping [24]

Although the hydrogen generation incidents and accelerated corrosion observations have led to some bans on galvanized sprinkler piping, FM Global [28] notes that corrosion products in ordinary black carbon steel dry piping have been responsible for the vast majority of their fire losses due to obstructed sprinkler pipes. Therefore, FM Global continues to recommend use of galvanized steel pipe for dry and pre-action sprinkler systems. In view of the exposure test results noted above, FM Global recommends use of compressed nitrogen, rather than just air, in these dry pipe systems. This is different from the recommendations and requirements of many other organizations, as summarized in a recent Sprinkler Age article [29].

5 Conclusions

Metal-water chemical and electrochemical reactions have been responsible for hydrogen generation and subsequent ignition of hydrogen-air mixtures in many explosion incidents. Many incinerator ash-water incidents have occurred when the ash had a substantial aluminum and alkaline content. A compilation of test data has led to a statistical distribution of the expected hydrogen generation per unit mass of ash, with the median rate being greater for fly ash than for bottom ash.

Copper-seawater electrochemical reactions in offshore facility electrical junction boxes have led to hydrogen accumulations and subsequent ignitions associated with electrical connector arcs and sparks. Copper corrosion test data together with reported electrochemical reactions have allowed calculations of hydrogen generation rates per unit of copper cable surface area. Special cable features, such as hydrogen absorption additives, and flameproof junction box construction can help prevent and mitigate the consequences of these incidents.

Electrochemical corrosion reactions in galvanized dry pipe sprinkler systems containing residual water have led to many reported hydrogen fire and explosion incidents, particularly in Scandinavia, where there are many of these sprinkler systems. Testing has shown that the accelerated corrosion in galvanized dry pipe can be greatly reduced by replacing air in the pipe with compressed nitrogen, and by avoiding the galvanic corrosion resulting from contact between dissimilar metal components in sprinkler piping.

6 References.

[1] NFPA 484 Standard on Combustible Metals, National Fire Protection Association, 2019.

[2] Marine Accident Investigation Branch, Serious Marine Casualty Report Number 26/2017, December 2017.

[3] UN, Recommendations on the transport of dangerous goods, Manual of tests and criteria, United Nations, 5th Edition, 2009.

[4] https://h2tools.org/lessons/hydrogen-gas-explosion-municipal-refuse-incineration-facility. Accessed on 2/8/2021.

[5] http://www.shippai.org/fkd/en/cfen/CC1200025.html. Accessed 2/8/2021.

[6] S. Mizutani,, S. Sakai,, and H. Takatsuki, Investigation of hydrogen generation from municipal solid waste incineration fly ash, J Mater Cycles Waste Manag (2000) 2:16–23.

[7] R. Larsen, Energy recovery of metallic aluminium in MSWI bottom ash, Umea University M.S. Thesis, 2014.

[8] M. Arm, and J. Lindeberg, Gas generation in incinerator ash, 2015, Available at https://www.researchgate.net/publication/266467653. Accessed 2/8/2021.

[9] UK Highways Agency Interim Advice Notice 127/09, Appendix B: The Use of Foamed Concrete, October 2009.

[10] A. Nithiya, A.Saffarzadeh, and T. Shimaoka, Hydrogen gas generation from metal aluminum-water interaction in municipal solid waste incineration (MSWI) bottom ash, Waste Management 73 (2018) 342–350.

[11] Rupturing of 6.6kV Power Junction Box installed at an offshore oil production Well Head Tower, https://www.eng-tips.com/viewthread.dm?qid=434654. Accessed 2/12/2021.

[12] IEC 60079-7, Explosive atmospheres - Part 7: Equipment protection by increased safety "e", IEC, 2015.

[13] N. Douglas, Copper Loss from Electrical Conductors in a Subsea Environment as a Result of Low Insulation Resistance, Oilfield Technology, January 2019. Available at https://www.researchgate.net/publication/331951501.

[14] ITU-T Recommendation L.27, Method for estimating the concentration of hydrogen in optical fibre cables, International Telecommunication Union, 1997.

[15] State: Hydrogen from corrosion of copper cable caused Salty Brine beach blast, Providence Journal, July 24, 2015.

[16] Case Closed: Hydrogen Gas Combustion Caused Salty Brine Beach Blast, National Public Radio, July 24, 2015. Accessed 2/12/2021.

[17] E-S. Sherif, and S-M. Park, Inhibition of Copper Corrosion in 3.0% NaCl Solution by N-Phenyl-1,4-phenylenediamine, Journal of The Electrochemical Society, 152 10 B428-B433 (2005).

[18] P. Martins, et al, Hydrogen evolution reaction on copper: Promoting water dissociation by tuning the surface oxophilicity, Electrochemistry Communications 100 (2019) 30–33.

[19] E-S Sherif, R. Erasmus, and J. Comins, Corrosion of copper in aerated synthetic sea water solutions and its inhibition by 3-amino-1,2,4-triazole, Journal of Colloid and Interphase Science, 309 (2007) 470–477.

[20] F. Mansfield, et al., The Corrosion Behavior of Copper Alloys, Stainless Steels, and Titanium in Seawater, Corrosion Science, Vol. 36, No. 12, pp. 2063-2095, 1994.

[21] A. Drach, et al., Field studies of corrosion behaviour of copper alloys in natural seawater, Corrosion Science 76 (2013) 453–464.

[22] L. Nunez, et al, Corrosion of copper in seawater and its aerosols in a tropical island, Corrosion Science 47 (2005) 461–484.

[23] Explosion in Helsinki City Theater, https://www.foreigner.fi/content/print/explosion-in-helsinki-theater-at-least-one-person-in-hospital/202006291506576691. Accessed 11/17/2020.

[24] Hydrogen explosion caused the Helsinki City Theatre incident, https://tukes.fi/en/-/hydrogen-explosion-caused-the-helsinki-city-theatre-incident. Accessed 2/14/2021.

[25] Explosions in sprinkler system probably caused by hydrogen gas, https://dpa-e.eu/explosions-in-sprinkler-system-probably-caused-by-hydrogen-gas/ Accessed 2/14/2021.

[26] Internally Galvanized Steel Pipe Blamed for Explosions, https://www.complianceonegroup.com/ lifesafety/ internally-galvanized-steel-pipe-blamed-for-explosions. Accessed 2/14/2021.

[27] T. Yamate, et al., Experimental Study on Hydrogen Generation Phenomenon in Fire Piping System in Buildings, Corrosion Engineering, 59, 436-439, 2010.

[28] P. Su, and D. Fuller, Corrosion and Corrosion Mitigation in Fire Protection Systems, FM Global Report, July 2014.

[29] Galvanized Steel Sprinkler Piping, Sprinkler Age, November 2018, Available at https:// sprinkle rage.com/ galvanized-steel-sprinkler-piping/ Accessed 2/15/2021.

619874 Port of Beirut Lebanon Explosion – Analysis and Impacts

Abstract:

On 4 August 2020, 2,750 tons of ammonium nitrate stored at the Port of Beirut (POB), Lebanon, exploded, reportedly resulting in over 190 deaths, several thousand injuries, $10-15 billion USD in property damage, and leaving an estimated 300,000 people homeless. The cargo of ammonium nitrate had been stored in Warehouse 12 at the POB, after having been confiscated by the Lebanese authorities from an abandoned ship, the MV *Rhosus,* in September, 2013. It is assumed that the ammonium nitrate was stored in the warehouse without proper security and safety measures, and without consideration for the need for segregation from other incompatible materials. From recorded videos of the blast, the explosion was preceded by a fire in the warehouse, but the exact cause of the detonation is still under investigation. This was one of the largest non-nuclear explosions in modern history and the lessons learned have widespread implications for facility siting, the handling of explosives and other hazardous materials, the management of chemicals in transportation, emergency management and recovery, and building resilience for large-scale emergencies.

In response to this event, AcuTech participated as part of an assessment team, and was tasked with examining current and previous practices for the handling of chemicals and hazardous materials at the POB, taking an all-hazards approach to examine safety, security, and emergency management concerns in a unified manner.

This paper examines the multifaceted impacts of the blast, the most critical factors that may have contributed to the explosion, and best practices and lessons learned that are applicable not only to ports and facilities handling ammonium nitrate, but all facilities handling hazardous materials operating in challenging urban environments, and which maintain complex interdependencies with critical infrastructure.

AUTHOR INDEX

AIChE
120 Wall Street, FL 23
New York, NY 10005-4020

ISBN 978-1-7138-5727-3

CURRAN ASSOCIATES INC.
proceedings.com

Next-Gen Manufacturing 2019

Topical Conference at the 2019 AIChE Annual Meeting

Orlando, Florida, USA
10 - 15 November 2019

AIChE
The Global Home of Chemical Engineers